Complex Networks
Principles, Methods and Applications

Networks constitute the backbone of complex systems, from the human brain to computer communications, transport infrastructures to online social systems, metabolic reactions to financial markets. Characterising their structure improves our understanding of the physical, biological, economic and social phenomena that shape our world.

Rigorous and thorough, this textbook presents a detailed overview of the new theory and methods of network science. Covering algorithms for graph exploration, node ranking and network generation, among the others, the book allows students to experiment with network models and real-world data sets, providing them with a deep understanding of the basics of network theory and its practical applications. Systems of growing complexity are examined in detail, challenging students to increase their level of skill. An engaging presentation of the important principles of network science makes this the perfect reference for researchers and undergraduate and graduate students in physics, mathematics, engineering, biology, neuroscience and social sciences.

Vito Latora is Professor of Applied Mathematics and Chair of Complex Systems at Queen Mary University of London. Noted for his research in statistical physics and in complex networks, his current interests include time-varying and multiplex networks, and their applications to socio-economic systems and to the human brain.

Vincenzo Nicosia is Lecturer in Networks and Data Analysis at the School of Mathematical Sciences at Queen Mary University of London. His research spans several aspects of network structure and dynamics, and his recent interests include multi-layer networks and their applications to big data modelling.

Giovanni Russo is Professor of Numerical Analysis in the Department of Mathematics and Computer Science at the University of Catania, Italy, focusing on numerical methods for partial differential equations, with particular application to hyperbolic and kinetic problems.

Complex Networks

Principles, Methods and Applications

VITO LATORA

Queen Mary University of London

VINCENZO NICOSIA

Queen Mary University of London

GIOVANNI RUSSO

University of Catania, Italy

CAMBRIDGE
UNIVERSITY PRESS

CAMBRIDGE
UNIVERSITY PRESS

University Printing House, Cambridge CB2 8BS, United Kingdom

One Liberty Plaza, 20th Floor, New York, NY 10006, USA

477 Williamstown Road, Port Melbourne, VIC 3207, Australia

4843/24, 2nd Floor, Ansari Road, Daryaganj, Delhi – 110002, India

79 Anson Road, #06–04/06, Singapore 079906

Cambridge University Press is part of the University of Cambridge.

It furthers the University's mission by disseminating knowledge in the pursuit of education, learning, and research at the highest international levels of excellence.

www.cambridge.org
Information on this title: www.cambridge.org/9781107103184
DOI: 10.1017/9781316216002

First published 2017

Printed in the United Kingdom by TJ International Ltd. Padstow Cornwall

A catalogue record for this publication is available from the British Library.

Library of Congress Cataloging-in-Publication Data

Names: Latora, Vito, author. | Nicosia, Vincenzo, author. | Russo, Giovanni, author.
Title: Complex networks : principles, methods and applications / Vito Latora, Queen Mary University of London, Vincenzo Nicosia, Queen Mary University of London, Giovanni Russo, Università degli Studi di Catania, Italy.
Description: Cambridge, United Kingdom ; New York, NY : Cambridge University Press, 2017. | Includes bibliographical references and index.
Identifiers: LCCN 2017026029 | ISBN 9781107103184 (hardback)
Subjects: LCSH: Network analysis (Planning)
Classification: LCC T57.85 .L36 2017 | DDC 003/.72–dc23
LC record available at https://lccn.loc.gov/2017026029

ISBN 978-1-107-10318-4 Hardback

Additional resources for this publication at www.cambridge.org/9781107103184.

To Giusi, Francesca and Alessandra

Contents

Preface

Social systems, the human brain, the Internet and the World Wide Web are all examples of complex networks, i.e. systems composed of a large number of units interconnected through highly non-trivial patterns of interactions. This book is an introduction to the beautiful and multidisciplinary world of complex networks. The readers of the book will be exposed to the fundamental principles, methods and applications of a novel discipline: *network science*. They will learn how to characterise the architecture of a network and model its growth, and will uncover the principles common to networks from different fields.

The book covers a large variety of topics including elements of graph theory, social networks and centrality measures, random graphs, small-world and scale-free networks, models of growing graphs and degree–degree correlations, as well as more advanced topics such as motif analysis, community structure and weighted networks. Each chapter presents its main ideas together with the related mathematical definitions, models and algorithms, and makes extensive use of network data sets to explore these ideas.

The book contains several practical applications that range from determining the role of an individual in a social network or the importance of a player in a football team, to identifying the sub-areas of a nervous systems or understanding correlations between stocks in a financial market.

Thanks to its colloquial style, the extensive use of examples and the accompanying software tools and network data sets, this book is the ideal university-level textbook for a first module on complex networks. It can also be used as a comprehensive reference for researchers in mathematics, physics, engineering, biology and social sciences, or as a historical introduction to the main findings of one of the most active interdisciplinary research fields of the moment.

This book is fundamentally on the structure of complex networks, and we hope it will be followed soon by a second book on the different types of dynamical processes that can take place over a complex network.

Vito Latora
Vincenzo Nicosia
Giovanni Russo

Introduction

The Backbone of a Complex System

Imagine you are invited to a party; you observe what happens in the room when the other guests arrive. They start to talk in small groups, usually of two people, then the groups grow in size, they split, merge again, change shape. Some of the people move from one group to another. Some of them know each other already, while others are introduced by mutual friends at the party. Suppose you are also able to track all of the guests and their movements in space; their head and body gestures, the content of their discussions. Each person is different from the others. Some are more lively and act as the centre of the social gathering: they tell good stories, attract the attention of the others and lead the group conversation. Other individuals are more shy: they stay in smaller groups and prefer to listen to the others. It is also interesting to notice how different genders and ages vary between groups. For instance, there may be groups which are mostly male, others which are mostly female, and groups with a similar proportion of both men and women. The topic of each discussion might even depend on the group composition. Then, when food and beverages arrive, the people move towards the main table. They organise into more or less regular queues, so that the shape of the newly formed groups is different. The individuals rearrange again into new groups sitting at the various tables. Old friends, but also those who have just met at the party, will tend to sit at the same tables. Then, discussions will start again during the dinner, on the same topics as before, or on some new topics. After dinner, when the music begins, we again observe a change in the shape and size of the groups, with the formation of couples and the emergence of collective motion as everybody starts to dance.

The *social system* we have just considered is a typical example of what is known today as a *complex system* [16, 44]. The study of complex systems is a new science, and so a commonly accepted formal definition of a complex system is still missing. We can roughly say that a complex system is a system made by a large number of single units (individuals, components or agents) interacting in such a way that the behaviour of the system is not a simple combination of the behaviours of the single units. In particular, some collective behaviours emerge without the need for any central control. This is exactly what we have observed by monitoring the evolution of our party with the formation of social groups, and the emergence of discussions on some particular topics. This kind of behaviour is what we find in human societies at various levels, where the interactions of many individuals give rise to the emergence of civilisation, urban forms, cultures and economies. Analogously, animal societies such as, for instance, ant colonies, accomplish a variety of different tasks, from nest maintenance to the organisation of food search, without the need for any central control.

Let us consider another example of a complex system, certainly the most representative and beautiful one: the human brain. With around 10^2 billion neurons, each connected by synapses to several thousand other neurons, this is the most complicated organ in our body. Neurons are cells which process and transmit information through electrochemical signals. Although neurons are of different types and shapes, the "integrate-and-fire" mechanism at the core of their dynamics is relatively simple. Each neuron receives synaptic signals, which can be either excitatory or inhibitory, from other neurons. These signals are then integrated and, provided the combined excitation received is larger than a certain threshold, the neuron fires. This firing generates an electric signal, called an action potential, which propagates through synapses to other neurons. Notwithstanding the extreme simplicity of the interactions, the brain self-organises collective behaviours which are difficult to predict from our knowledge of the dynamics of its individual elements. From an avalanche of simple integrate-and-fire interactions, the neurons of the brain are capable of organising a large variety of wonderful emerging behaviours. For instance, sensory neurons coordinate the response of the body to touch, light, sounds and other external stimuli. Motor neurons are in charge of the body's movement by controlling the contraction or relaxation of the muscles. Neurons of the prefrontal cortex are responsible for reasoning and abstract thinking, while neurons of the limbic system are involved in processing social and emotional information.

Over the years, the main focus of scientific research has been on the characteristics of the individual components of a complex system and to understand the details of their interactions. We can now say that we have learnt a lot about the different types of nerve cells and the ways they communicate with each other through electrochemical signals. Analogously, we know how the individuals of a social group communicate through both spoken and body language, and the basic rules through which they learn from one another and form or match their opinions. We also understand the basic mechanisms of interactions in social animals; we know that, for example, ants produce chemicals, known as pheromones, through which they communicate, organise their work and mark the location of food. However, there is another very important, and in no way trivial, aspect of complex systems which has been explored less. This has to do with the structure of the interactions among the units of a complex system: which unit is connected to which others. For instance, if we look at the connections between the neurons in the brain and construct a similar network whose nodes are neurons and the links are the synapses which connect them, we find that such a network has some special mathematical properties which are fundamental for the functioning of the brain. For instance, it is always possible to move from one node to any other in a small number of steps, and, particularly if the two nodes belong to the same brain area, there are many alternative paths between them. Analogously, if we take snapshots of who is talking to whom at our hypothetical party, we immediately see that the architecture of the obtained networks, whose nodes represent individuals and links stand for interactions, plays a crucial role in both the propagation of information and the emergence of collective behaviours. Some sub-structures of a network propagate information faster than others; this means that nodes occupying strategic positions will have better access to the resources of the system. In practice, what also matters in a complex system, and it matters a lot, is the *backbone* of the system, or, in other words, the architecture of the network of interactions. It is precisely on these *complex networks*, i.e. on the networks of the various complex systems that populate our world, that we will be focusing in this book.

Complex Networks Are All Around Us

Networks permeate all aspects of our life and constitute the backbone of our modern world. To understand this, think for a moment about what you might do in a typical day. When you get up early in the morning and turn on the light in your bedroom, you are connected to the *electrical power grid*, a network whose nodes are either power stations or users, while links are copper cables which transport electric current. Then you meet the people of your family. They are part of your *social network* whose nodes are people and links stand for kinship, friendship or acquaintance. When you take a shower and cook your breakfast you are respectively using a *water distribution network*, whose nodes are water stations, reservoirs, pumping stations and homes, and links are pipes, and a *gas distribution network*. If you go to work by car you are moving in the *street network* of your city, whose nodes are intersections and links are streets. If you take the underground then you make use of a *transportation network*, whose nodes are the stations and links are route segments.

When you arrive at your office you turn on your laptop, whose internal circuits form a complicated microscopic *network of logic gates*, and connect it to the *Internet*, a worldwide network of computers and routers linked by physical or logical connections. Then you check your emails, which belong to an *email communication network*, whose nodes are people and links indicate email exchanges among them. When you meet a colleague, you and your colleague form part of a *collaboration network*, in which an edge exists between two persons if they have collaborated on the same project or coauthored a paper. Your colleagues tell you that your last paper has got its first hundred citations. Have you ever thought of the fact that your papers belong to a *citation network*, where the nodes represent papers, and links are citations?

At lunchtime you read the news on the website of your preferred newspaper: in doing this you access the *World Wide Web*, a huge global information network whose nodes are webpages and edges are clickable hyperlinks between pages. You will almost surely then check your *Facebook* account, a typical example of an *online social network*, then maybe have a look at the daily trending topics on *Twitter*, an information network whose nodes are people and links are the "following" relations.

Your working day proceeds quietly, as usual. Around 4:00pm you receive a phone call from your friend John, and you immediately think about the *phone call network*, where two individuals are connected by a link if they have exchanged a phone call. John invites you and your family for a weekend at his cottage near the lake. Lakes are home to a variety of fishes, insects and animals which are part of a *food web network*, whose links indicate predation among different species. And while John tells you about the beauty of his cottage, an image of a mountain lake gradually forms in your mind, and you can see a white waterfall cascading down a cliff, and a stream flowing quietly through a green valley. There is no need to say that "lake", "waterfall", "white", "stream", "cliff", "valley" and "green" form a *network of words associations*, in which a link exists between two words if these words are often associated with each other in our minds. Before leaving the office, you book a flight to go to Prague for a conference. Obviously, also the *air transportation system* is a network, whose nodes are airports and links are airline routes.

When you drive back home you feel a bit tired and you think of the various networks in our body, from the *network of blood vessels* which transports blood to our organs to the intricate set of relationships among genes and proteins which allow the perfect functioning of the cells of our body. Examples of these genetic networks are the *transcription regulation networks* in which the nodes are genes and links represent transcription regulation of a gene by the transcription factor produced by another gene, *protein interaction networks* whose nodes are protein and there is a link between two proteins if they bind together to perform complex cellular functions, and *metabolic networks* where nodes are chemicals, and links represent chemical reactions.

During dinner you hear on the news that the total export for your country has decreased by 2.3% this year; the system of *commercial relationships* among countries can be seen as a network, in which links indicate import/export activities. Then you watch a movie on your sofa: you can construct an *actor collaboration network* where nodes represent movie actors and links are formed if two actors have appeared in the same movie. Exhausted, you go to bed and fall asleep while images of networks of all kinds still twist and dance in your mind, which is, after all, the marvellous combination of the activity of billions of neurons and trillions of synapses in your *brain network*. Yet another network.

Why Study Complex Networks?

In the late 1990s two research papers radically changed our view on complex systems, moving the attention of the scientific community to the study of the architecture of a complex system and creating an entire new research field known today as *network science*. The first paper, authored by Duncan Watts and Steven Strogatz, was published in the journal *Nature* in 1998 and was about *small-world networks* [311]. The second one, on *scale-free networks*, appeared one year later in *Science* and was authored by Albert-László Barabási and Réka Albert [19]. The two papers provided clear indications, from different angles, that:

- the networks of real-world complex systems have non-trivial structures and are very different from lattices or random graphs, which were instead the standard networks commonly used in all the current models of a complex system.
- some structural properties are universal, i.e. are common to networks as diverse as those of biological, social and man-made systems.
- the structure of the network plays a major role in the dynamics of a complex system and characterises both the emergence and the properties of its collective behaviours.

Table 1 A list of the real-world complex networks that will be studied in this book. For each network, we report the chapter of the book where the corresponding data set will be introduced and analysed.

Complex networks	Nodes	Links	Chapter
Elisa's kindergarten	Children	Friendships	1
Actor collaboration networks	Movie actors	Co-acting in a film	2
Co-authorship networks	Scientists	Co-authoring a paper	3
Citation networks	Scientific papers	Citations	6
Zachary's karate club	Club members	Friendships	9
C. elegans neural network	Neurons	Synapses	4
Transcription regulation networks	Genes	Transcription regulation	8
World Wide Web	Web pages	Hyperlinks	5
Internet	Routers	Optical fibre cables	7
Urban street networks	Street crossings	Streets	8
Air transport network	Airports	Flights	10
Financial markets	Stocks	Time correlations	10

Both works were motivated by the empirical analysis of real-world systems. Four networks were introduced and studied in these two papers. Namely, the neural system of a few-millimetres-long worm known as the *C. elegans*, a social network describing how actors collaborate in movies, and two man-made networks: the US electrical power grid and a sample of the World Wide Web. During the last decade, new technologies and increasing computing power have made new data available and stimulated the exploration of several other complex networks from the real world. A long series of papers has followed, with the analysis of new and ever larger networks, and the introduction of novel measures and models to characterise and reproduce the structure of these real-world systems. Table 1 shows only a small sample of the networks that have appeared in the literature, namely those that will be explicitly studied in this book, together with the chapter where they will be considered. Notice that the table includes different types of networks. Namely, five networks representing three different types of social interactions (namely friendships, collaborations and citations), two biological systems (respectively a neural and a gene network) and five man-made networks (from transportation and communication systems to a network of correlations among financial stocks).

The ubiquitousness of networks in nature, technology and society has been the principal motivation behind the systematic quantitative study of their structure, their formation and their evolution. And this is also the main reason why a student of any scientific discipline should be interested in complex networks. In fact, if we want to master the interconnected world we live in, we need to understand the structure of the networks around us. We have to learn the basic principles governing the architecture of networks from different fields, and study how to model their growth.

It is also important to mention the high interdisciplinarity of network science. Today, research on complex networks involves scientists with expertise in areas such as mathematics, physics, computer science, biology, neuroscience and social science, often working

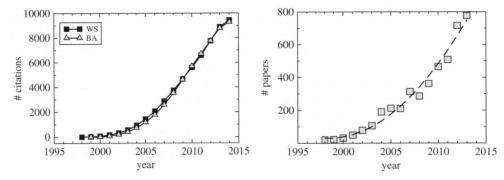

Fig. 1 Left panel: number of citations received over the years by the 1998 Watts and Strogatz (WS) article on small-world networks and by the 1999 Barabási and Albert (BA) article on scale-free networks. Right panel: number of papers on complex networks that appeared each year in the public preprint archive arXiv.org.

side by side. Because of its interdisciplinary nature, the generality of the results obtained, and the wide variety of possible applications, network science is considered today a necessary ingredient in the background of any modern scientist.

Finally, it is not difficult to understand that complex networks have become one of the hottest research fields in science. This is confirmed by the attention and the huge number of citations received by Watts and Strogatz, and by Barabási and Albert, in the papers mentioned above. The temporal profiles reported in the left panel of Figure 1 show the exponential increase in the number of citations of these two papers since their publication. The two papers have today about 10,000 citations each and, as already mentioned, have opened a new research field stimulating interest for complex networks in the scientific community and triggering an avalanche of scientific publications on related topics. The right panel of Figure 1 reports the number of papers published each year after 1998 on the well-known public preprint archive arXiv.org with the term "complex networks" in their title or abstract. Notice that this number has gone up by a factor of 10 in the last ten years, with almost a thousand papers on the topic published in the archive in the year 2013. The explosion of interest in complex networks is not limited to the scientific community, but has become a cultural phenomenon with the publications of various popular science books on the subject.

Overview of the Book

This book is mainly intended as a textbook for an introductory course on complex networks for students in physics, mathematics, engineering and computer science, and for the more mathematically oriented students in biology and social sciences. The main purpose of the book is to expose the readers to the fundamental ideas of network science, and to provide them with the basic tools necessary to start exploring the world of complex networks. We also hope that the book will be able to transmit to the reader our passion for this stimulating new interdisciplinary subject.

The standard tools to study complex networks are a mixture of mathematical and computational methods. They require some basic knowledge of graph theory, probability, differential equations, data structures and algorithms, which will be introduced in this book from scratch and in a friendly way. Also, network theory has found many interesting applications in several different fields, including social sciences, biology, neuroscience and technology. In the book we have therefore included a large variety of examples to emphasise the power of network science. This book is essentially on the structure of complex networks, since we have decided that the detailed treatment of the different types of dynamical processes that can take place over a complex network should be left to another book, which will follow this one.

The book is organised into ten chapters. The first six chapters (Chapters 1–6) form the core of the book. They introduce the main concepts of network science and the basic measures and models used to characterise and reproduce the structure of various complex networks. The remaining four chapters (Chapters 7–10) cover more advanced topics that could be skipped by a lecturer who wants to teach a short course based on the book.

In Chapter 1 we introduce some basic definitions from graph theory, setting up the language we will need for the remainder of the book. The aim of the chapter is to show that complex network theory is deeply grounded in a much older mathematical discipline, namely *graph theory*.

In Chapter 2 we focus on the concept of centrality, along with some of the related measures originally introduced in the context of *social network analysis*, which are today used extensively in the identification of the key components of any complex system, not only of social networks. We will see some of the measures at work, using them to quantify the centrality of movie actors in the *actor collaboration network*.

Chapter 3 is where we first discuss network models. In this chapter we introduce the classical *random graph models* proposed by Erdős and Rényi (ER) in the late 1950s, in which the edges are randomly distributed among the nodes with a uniform probability. This allows us to analytically derive some important properties such as, for instance, the number and order of *graph components* in a random graph, and to use ER models as term of comparison to investigate *scientific collaboration networks*. We will also show that the *average distance* between two nodes in ER random graphs increases only logarithmically with the number of nodes.

In Chapter 4 we see that in real-world systems, such as the *neural network of the* C. elegans or the movie actor collaboration network, the neighbours of a randomly chosen node are directly linked to each other much more frequently than would occur in a purely random network, giving rise to the presence of many triangles. In order to quantify this, we introduce the so-called *clustering coefficient*. We then discuss the Watts and Strogatz (WS) *small-world model* to construct networks with both a small average distance between nodes and a high clustering coefficient.

In Chapter 5 the focus is on how the degree k is distributed among the nodes of a network. We start by considering the graph of the *World Wide Web* and by showing that it is a *scale-free network*, i.e. it has a power–law *degree distribution* $p_k \sim k^{-\gamma}$ with an exponent $\gamma \in [2, 3]$. This is a property shared by many other networks, while neither ER random graphs nor the WS model can reproduce such a feature. Hence, we introduce the so-called *configuration model* which generalises ER random graph models to incorporate arbitrary degree distributions.

In Chapter 6 we show that real networks are not static, but grow over time with the addition of new nodes and links. We illustrate this by studying the basic mechanisms of growth in *citation networks*. We then consider whether it is possible to produce scale-free degree distributions by modelling the dynamical evolution of the network. For this purpose we introduce the *Barabási–Albert model*, in which newly arriving nodes select and link existing nodes with a probability linearly proportional to their degree. We also consider some extensions and modifications of this model.

In the last four chapters we cover more advanced topics on the structure of complex networks.

Chapter 7 is about networks with *degree–degree correlations*, i.e. networks such that the probability that an edge departing from a node of degree k arrives at a node of degree k' is a function both of k' and of k. Degree–degree correlations are indeed present in real-world networks, such as the *Internet*, and can be either positive (assortative) or negative (disassortative). In the first case, networks with small degree preferentially link to other low-degree nodes, while in the second case they link preferentially to high-degree ones. In this chapter we will learn how to take degree–degree correlations into account, and how to model correlated networks.

In Chapter 8 we deal with the *cycles* and other small subgraphs known as *motifs* which occur in most networks more frequently than they would in random graphs. We consider two applications: firstly we count the number of short cycles in *urban street networks* of different cities from all over the world; secondly we will perform a motif analysis of the *transcription network of the bacterium* E. coli.

Chapter 9 is about network mesoscale structures known as *community structures*. Communities are groups of nodes that are more tightly connected to each other than to other nodes. In this chapter we will discuss various methods to find meaningful divisions of the nodes of a network into communities. As a benchmark we will use a real network, the *Zachary's karate club*, where communities are known a priori, and also models to construct networks with a tunable presence of communities.

In Chapter 10 we deal with *weighted networks*, where each link carries a numerical value quantifying the intensity of the connection. We will introduce the basic measures used to characterise and classify weighted networks, and we will discuss some of the models of weighted networks that reproduce empirically observed topology–weight correlations. We will study in detail two weighted networks, namely the *US air transport network* and a *network of financial stocks*.

Finally, the book's *Appendix* contains a detailed description of all the main graph algorithms discussed in the various chapters of the book, from those to find shortest paths, components or community structures in a graph, to those to generate random graphs or scale-free networks. All the algorithms are presented in a C-like pseudocode format which allows us to understand their basic structure without the unnecessary complication of a programming language.

The *organisation* of this textbook is another reason why it is different from all the other existing books on networks. We have in fact avoided the widely adopted separation of the material in theory and applications, or the division of the book into separate chapters respectively dealing with empirical studies of real-world networks, network measures, models, processes and computer algorithms. Each chapter in our book discusses, at the same time, real-world networks, measures, models and algorithms while, as said before, we have left the study of processes on networks to an entire book, which will follow this one. Each chapter of this book presents a new idea or network property: it introduces a network data set, proposes a set of mathematical quantities to investigate such a network, describes a series of network models to reproduce the observed properties, and also points to the related algorithms. In this way, the presentation follows the same path of the current research in the field, and we hope that it will result in a more logical and more entertaining text. Although the main focus of this book is on the mathematical modelling of complex networks, we also wanted the reader to have direct access to both the most famous *data sets of real-world networks* and to the *numerical algorithms* to compute network properties and to construct networks. For this reason, the data sets of all the real-world networks listed in Table 1 are introduced and illustrated in special DATA SET Boxes, usually one for each chapter of the book, and can be downloaded from the book's webpage at www. complex-networks.net. On the same webpage the reader can also find an implementation in the C language of the graph algorithms illustrated in the Appendix (in C-like pseudocode format). We are sure that the student will enjoy experimenting directly on real-world networks, and will benefit from the possibility of reproducing all of the numerical results presented throughout the book.

The style of the book is informal and the ideas are illustrated with examples and applications drawn from the recent research literature and from different disciplines. Of course, the problem with such examples is that no-one can simultaneously be an expert in social sciences, biology and computer science, so in each of these cases we will set up the relative background from scratch. We hope that it will be instructive, and also fun, to see the connections between different fields. Finally, all the mathematics is thoroughly explained, and we have decided never to hide the details, difficulties and sometimes also the incoherences of a science still in its infancy.

Acknowledgements

Writing this book has been a long process which started almost ten years ago. The book has grown from the notes of various university courses, first taught at the Physics Department of the University of Catania and at the Scuola Superiore di Catania in Italy, and more recently to the students of the Masters in "Network Science" at Queen Mary University of London.

The book would not have been the same without the interactions with the students we have met at the different stages of the writing process, and their scientific curiosity. Special thanks go to Alessio Cardillo, Roberta Sinatra, Salvatore Scellato and the other students and alumni of Scuola Superiore, Salvatore Assenza, Leonardo Bellocchi, Filippo Caruso, Paolo Crucitti, Manlio De Domenico, Beniamino Guerra, Ivano Lodato, Sandro Meloni,

Andrea Santoro and Federico Spada, and to the students of the Masters in "Network Science".

We acknowledge the great support of the members of the Laboratory of Complex Systems at Scuola Superiore di Catania, Giuseppe Angilella, Vincenza Barresi, Arturo Buscarino, Daniele Condorelli, Luigi Fortuna, Mattia Frasca, Jesús Gómez-Gardeñes and Giovanni Piccitto; of our colleagues in the Complex Systems and Networks research group at the School of Mathematical Sciences of Queen Mary University of London, David Arrowsmith, Oscar Bandtlow, Christian Beck, Ginestra Bianconi, Leon Danon, Lucas Lacasa, Rosemary Harris, Wolfram Just; and of the PhD students Federico Battiston, Moreno Bonaventura, Massimo Cavallaro, Valerio Ciotti, Iacopo Iacovacci, Iacopo Iacopini, Daniele Petrone and Oliver Williams.

We are greatly indebted to our colleagues Elsa Arcaute, Alex Arenas, Domenico Asprone, Tomaso Aste, Fabio Babiloni, Franco Bagnoli, Andrea Baronchelli, Marc Barthélemy, Mike Batty, Armando Bazzani, Stefano Boccaletti, Marián Boguñá, Ed Bullmore, Guido Caldarelli, Domenico Cantone, Gastone Castellani, Mario Chavez, Vittoria Colizza, Regino Criado, Fabrizio De Vico Fallani, Marina Diakonova, Albert Díaz-Guilera, Tiziana Di Matteo, Ernesto Estrada, Tim Evans, Alfredo Ferro, Alessandro Fiasconaro, Alessandro Flammini, Santo Fortunato, Andrea Giansanti, Georg von Graevenitz, Paolo Grigolini, Peter Grindrod, Des Higham, Giulia Iori, Henrik Jensen, Renaud Lambiotte, Pietro Lió, Vittorio Loreto, Paolo de Los Rios, Fabrizio Lillo, Carmelo Maccarone, Athen Ma, Sabato Manfredi, Massimo Marchiori, Cecilia Mascolo, Rosario Mantegna, Andrea Migliano, Raúl Mondragón, Yamir Moreno, Mirco Musolesi, Giuseppe Nicosia, Pietro Panzarasa, Nicola Perra, Alessandro Pluchino, Giuseppe Politi, Sergio Porta, Mason Porter, Giovanni Petri, Gaetano Quattrocchi, Daniele Quercia, Filippo Radicchi, Andrea Rapisarda, Daniel Remondini, Alberto Robledo, Miguel Romance, Vittorio Rosato, Martin Rosvall, Maxi San Miguel, Corrado Santoro, M. Ángeles Serrano, Simone Severini, Emanuele Strano, Michael Szell, Bosiljka Tadić, Constantino Tsallis, Stefan Thurner, Hugo Touchette, Petra Vértes, Lucio Vinicius for the many stimulating discussions and for their useful comments. We thank in particular Olle Persson, Luciano Da Fontoura Costa, Vittoria Colizza, and Rosario Mantegna for having provided us with their network data sets.

We acknowledge the European Commission project LASAGNE (multi-LAyer SpAtiotemporal Generalized NEtworks), Grant 318132 (STREP), the EPSRC project GALE, Grant EP/K020633/1, and INFN FB11/TO61, which have supported and made possible our work at the various stages of this project.

Finally, we thank our families for their never-ending support and encouragement.

Life is all mind, heart and relations

Salvatore Latora
Philosopher

1 Graphs and Graph Theory

Graphs are the mathematical objects used to represent networks, and *graph theory* is the branch of mathematics that deals with the study of graphs. Graph theory has a long history. The notion of the graph was introduced for the first time in 1763 by Euler, to settle a famous unsolved problem of his time: the so-called Königsberg bridge problem. It is no coincidence that the first paper on graph theory arose from the need to solve a problem from the real world. Also subsequent work in graph theory by Kirchhoff and Cayley had its root in the physical world. For instance, Kirchhoff's investigations into electric circuits led to his development of a set of basic concepts and theorems concerning trees in graphs. Nowadays, graph theory is a well-established discipline which is commonly used in areas as diverse as computer science, sociology and biology. To give some examples, graph theory helps us to schedule airplane routing and has solved problems such as finding the maximum flow per unit time from a source to a sink in a network of pipes, or colouring the regions of a map using the minimum number of different colours so that no neighbouring regions are coloured the same way. In this chapter we introduce the basic definitions, setting up the language we will need in the rest of the book. We also present the first data set of a real network in this book, namely *Elisa's kindergarten network*. The two final sections are devoted to, respectively, the proof of the Euler theorem and the description of a graph as an array of numbers.

1.1 What Is a Graph?

The natural framework for the exact mathematical treatment of a complex network is a branch of *discrete mathematics* known as *graph theory* [48, 47, 313, 150, 272, 144]. Discrete mathematics, also called finite mathematics, is the study of mathematical structures that are fundamentally *discrete*, i.e. made up of distinct parts, not supporting or requiring the notion of continuity. Most of the objects studied in discrete mathematics are countable sets, such as integers and *finite graphs*. Discrete mathematics has become popular in recent decades because of its applications to computer science. In fact, concepts and notations from discrete mathematics are often useful to study or describe objects or problems in computer algorithms and programming languages. The concept of the graph is better introduced by the two following examples.

Example 1.1   *(Friends at a party)*   Seven people have been invited to a party. Their names are Adam, Betty, Cindy, David, Elizabeth, Fred and George. Before meeting at the party, Adam knew Betty, David and Fred; Cindy knew Betty, David, Elizabeth and George; David knew Betty (and, of course, Adam and Cindy); Fred knew Betty (and, of course, Adam).

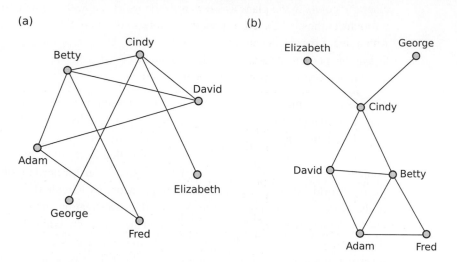

The network of acquaintances can be easily represented by identifying a person by a point, and a relation as a link between two points: if two points are connected by a link, this means that they knew each other before the party. A pictorial representation of the acquaintance relationships among the seven persons is illustrated in panel (a) of the figure. Note the symmetry of the link between two persons, which reflects that if person "A" knows person "B", then person "B" knows person "A". Also note that the only thing which is relevant in the diagram is whether two persons are connected or not. The same acquaintance network can be represented, for example, as in panel (b). Note that in this representation the more "relevant" role of Betty and Cindy over, for example, George or Fred, is more immediate.

Example 1.2   *(The map of Europe)*   The map in the figure shows 23 of Europe's approximately 50 countries. Each country is shown with a different shade of grey, so that from

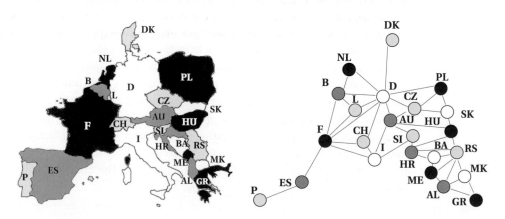

the image we can easily distinguish the borders between any two nations. Let us suppose now that we are interested not in the precise shape and geographical position of each country, but simply in which nations have common borders. We can thus transform the map into a much simpler representation that preserves entirely that information. In order to do so we need, with a little bit of abstraction, to transform each nation into a point. We can then place the points in the plane as we want, although it can be convenient to maintain similar positions to those of the corresponding nations in the map. Finally, we connect two points with a line if there is a common boundary between the corresponding two nations. Notice that in this particular case, due to the placement of the points in the plane, it is possible to draw all the connections with no line intersections.

The mathematical entity used to represent the existence or the absence of links among various objects is called the *graph*. A graph is defined by giving a set of elements, the graph nodes, and a set of links that join some (or all) pairs of nodes. In Example 1.1 we are using a graph to represent a network of social acquaintances. The people invited at a party are the nodes of the graph, while the existence of acquaintances between two persons defines the links in the graph. In Example 1.1 the nodes of the graph are the countries of the European Union, while a link between two countries indicates that there is a common boundary between them. A *graph* is defined in mathematical terms in the following way:

> **Definition 1.1 (Undirected graph)** *A graph, more specifically an undirected graph, $G \equiv (\mathcal{N}, \mathcal{L})$, consists of two sets, $\mathcal{N} \neq \emptyset$ and $\mathcal{L}$. The elements of $\mathcal{N} \equiv \{n_1, n_2, \ldots, n_N\}$ are distinct and are called the nodes (or vertices, or points) of the graph G. The elements of $\mathcal{L} \equiv \{l_1, l_2, \ldots, l_K\}$ are distinct unordered pairs of distinct elements of $\mathcal{N}$, and are called links (or edges, or lines).*

The number of vertices $N \equiv N[G] = |\mathcal{N}|$, where the symbol $|\cdot|$ denotes the cardinality of a set, is usually referred as the *order* of G, while the number of edges $K \equiv K[G] = |\mathcal{L}|$ is the *size* of G.[1] A node is usually referred to by a label that identifies it. The label is often an integer index from 1 to N, representing the order of the node in the set $\mathcal{N}$. We shall use this labelling throughout the book, unless otherwise stated. In an undirected graph, each of the links is defined by a pair of nodes, i and j, and is denoted as (i, j) or (j, i). In some cases we also denote the link as l_{ij} or l_{ji}. The link is said to be *incident* in nodes i and j, or to join the two nodes; the two nodes i and j are referred to as the *end-nodes* of link (i, j). Two nodes joined by a link are referred to as *adjacent* or *neighbouring*.

As shown in Example 1.1, the usual way to picture a graph is by drawing a dot or a small circle for each node, and joining two dots by a line if the two corresponding nodes are connected by an edge. How these dots and lines are drawn in the page is in principle irrelevant, as is the length of the lines. The only thing that matters in a graph is which pairs of nodes form a link and which ones do not. However, the choice of a clear drawing can be

---

[1] Sometimes, especially in the physical literature, the word *size* is associated with the number of nodes, rather than with the number of links. We prefer to consider K as the size of the graph. However, in many cases of interest, the number of links K is proportional to the number of nodes N, and therefore the concept of size of a graph can equally well be represented by the number of its nodes N or by the number of its edges K.

very important in making the properties of the graph easy to read. Of course, the quality and usefulness of a particular way to draw a graph depends on the type of graph and on the purpose for which the drawing is generated and, although there is no general prescription, there are various standard drawing setups and different algorithms for drawing graphs that can be used and compared. Some of them are illustrated in Box 1.1.

Figure 1.1 shows four examples of small undirected graphs. Graph G_1 is made of $N = 5$ nodes and $K = 4$ edges. Notice that any pair of nodes of this graph can be connected in only one way. As we shall see later in detail, such a graph is called a *tree*. Graphs G_2 has $N = K = 4$. By starting from one node, say node 1, one can go to all the other nodes 2, 3, 4, and back again to 1, by visiting each node and each link just once, except of course node 1, which is visited twice, being both the starting and ending node. As we shall see, we say that the graph G_2 contains a *cycle*. The same can be said about graph G_3. Graph G_3 contains an isolated node and three nodes connected by three links. We say that graphs G_1 and G_2 are connected, in the sense that any node can be reached, starting from any other node, by "walking" on the graph, while graph G_3 is not.

Notice that, in the definition of graph given above, we deliberately avoided *loops*, i.e. links from a node to itself, and *multiple edges*, i.e. pairs of nodes connected by more than one link. Graphs with either of these elements are called *multigraphs* [48, 47, 308]. An example of multigraph is G_4 in Figure 1.1. In such a multigraph, node 1 is connected to itself by a loop, and it is connected to node 3 by two links. In this book, we will deal with graphs rather than multigraphs, unless otherwise stated.

For a graph G of order N, the number of edges K is at least 0, in which case the graph is formed by N isolated nodes, and at most $N(N - 1)/2$, when all the nodes are pairwise adjacent. The ratio between the actual number of edges K and its maximum possible number $N(N - 1)/2$ is known as the *density* of G. A graph with N nodes and no edges has zero

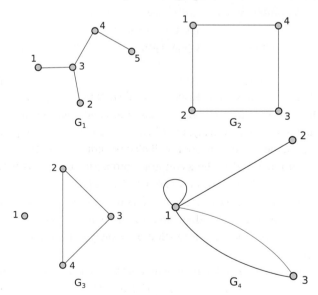

Fig. 1.1   Some examples of undirected graphs, namely a tree, G_1; two graphs containing cycles, G_2 and G_3; and an undirected multigraph, G_4.

Graph Drawing

A good drawing can be very helpful to highlight the properties of a graph. In one standard setup, the so called *circular layout*, the nodes are placed on a circle and the edges are drawn across the circle. In another set-up, known as the *spring model*, the nodes and links are positioned in the plane by assuming the graph is a physical system of unit masses (the nodes) connected by springs (the links). An example is shown in the figure below, where the same graph is drawn using a circular layout (left) and a spring-based layout (right) based on the *Kamada–Kawai algorithm* [173].

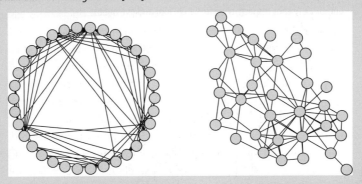

By nature, springs attract their endpoints when stretched and repel their endpoints when compressed. In this way, adjacent nodes on the graph are moved closer in space and, by looking for the equilibrium conditions, we get a layout where edges are short lines, and edge crossings with other nodes and edges are minimised. There are many software packages specifically focused on graph visualisation, including *Pajek* (`http://mrvar.fdv.uni-lj.si/pajek/`), *Gephi* (`https://gephi.org/`) and *GraphViz* (`http://www.graphviz.org/`). Moreover, most of the libraries for network analysis, including *NetworkX* (`https://networkx.github.io/`), *iGraph* (`http://igraph.org/`) and *SNAP* (Stanford Network Analysis Platform, `http://snap.stanford.edu/`), support different algorithms for network visualisation.

density and is said to be *empty*, while a graph with $K = N(N-1)/2$ edges, denoted as $\mathbb{K}_N$, has density equal to 1 and is said to be *complete*. The complete graphs with $N = 3$, $N = 4$ and $N = 5$ respectively, are illustrated in Figure 1.2. In particular, $\mathbb{K}_3$ is called a *triangle*, and in the rest of this book will also be indicated by the symbol $\triangle$. As we shall see, we are often interested in the asymptotic properties of graphs when the order N becomes larger and larger. The maximum number of edges in a graph scales as N^2. If the actual number of edges in a sequence of graphs of increasing number of nodes scales as N^2, then the graphs of the sequence are called *dense*. It is often the case that the number of edges in a graph of a given sequence scales much more slowly than N^2. In this case we say that the graphs are *sparse*.

We will now focus on how to compare graphs with the same order and size. Two graphs $G_1 = (\mathcal{N}_1, \mathcal{L}_1)$ and $G_2 = (\mathcal{N}_2, \mathcal{L}_2)$ are *the same* graph if $\mathcal{N}_1 = \mathcal{N}_2$ and $\mathcal{L}_1 = \mathcal{L}_2$; that is, if both their node sets and their edge sets (i.e. the sets of unordered pairs of nodes defining $\mathcal{L}$) are the same. In this case, we write $G_1 = G_2$. For example, graphs (a) and (b) in Figure 1.3

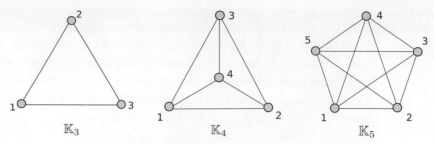

Fig. 1.2   Complete graphs respectively with three, four and five nodes.

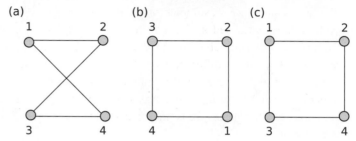

Fig. 1.3   Isomorphism of graphs. Graphs (a) and (b) are the same graph, since their edges are the same. Graphs (b) and (c) are isomorphic, since there is a bijection between the nodes that preserves the edge set.

are the same. Note that the position of the nodes in the picture has no relevance, nor does the shape or length of the edges. Two graphs that are not the same can nevertheless be *isomorphic*.

Definition 1.2 (Isomorphism)   *Two graphs, $G_1 = (\mathcal{N}_1, \mathcal{L}_1)$ and $G_2 = (\mathcal{N}_2, \mathcal{L}_2)$, of the same order and size, are said to be* isomorphic *if there exists a bijection $\phi : \mathcal{N}_1 \rightarrow \mathcal{N}_2$, such that $(u, v) \in \mathcal{L}_1$ iff $(\phi(u), \phi(v)) \in \mathcal{L}_2$. The bijection ϕ is called an* isomorphism.

In other words, G_1 and G_2 are isomorphic if and only if a one-to-one correspondence between the two vertex sets $\mathcal{N}_1, \mathcal{N}_2$, which preserves adjacency, can be found. In this case we write $G_1 \simeq G_2$. Isomorphism is an equivalence relation, in the sense that it is reflexive, symmetric and transitive. This means that, given any three graphs G_1, G_2, G_3, we have $G_1 \simeq G_1$, $G_1 \simeq G_2 \Rightarrow G_2 \simeq G_2$, and finally $G_1 \simeq G_2$ and $G_2 \simeq G_3 \Rightarrow G_1 \simeq G_3$. For example, graph (c) in Figure 1.3 is not the same as graphs (a) and (b), but it is isomorphic to (a) and (b). In fact, the bijection $\phi(1) = 1$, $\phi(2) = 2$, $\phi(3) = 4$, and $\phi(4) = 3$ between the set of nodes of graph (c) and that of graph (a) satisfies the property required in Definition 1.2. It is easy to show that, once the nodes of two graphs of the same order are labelled by integers from 1 to N, a bijection $\phi : \mathcal{N}_1 \rightarrow \mathcal{N}_2$ can be always represented as a permutation of the node labels. For instance, the bijection just considered corresponds to the permutation of node 3 and node 4.

In all the graphs we have seen so far, a label is attached to each node, and identifies it. Such graphs are called *labelled graphs*. Sometimes, one is interested in the relation between nodes and their connections irrespective of the name of the nodes. In

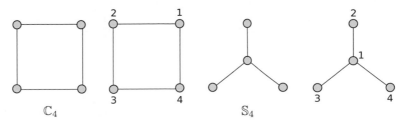

Fig. 1.4 Two unlabelled graphs, namely the cycle $\mathbb{C}_4$ and the star graph $\mathbb{S}_4$, and one possible labelling of such two graphs.

this case, no label is attached to the nodes, and the graph itself is said to be *unlabelled*. Figure 1.4 shows two examples of unlabelled graphs with $N = 4$ nodes, namely the cycle, usually indicated as $\mathbb{C}_4$, and the star graph with a central node and three links, $\mathbb{S}_4$, and two possible labellings of their nodes. Since the same unlabelled graph can be represented in several different ways, how can we state that all these representations correspond to the same graph? By definition, two unlabelled graphs are *the same* if it is possible to label them in such a way that they are the same labelled graph. In particular, if two labelled graphs are isomorphic, then the corresponding unlabelled graphs are the same.

It is easy to establish whether two labelled graphs with the same number of nodes and edges are the same, since it is sufficient to compare the ordered pairs that define their edges. However, it is difficult to check whether two unlabelled graphs are isomorphic, because there are $N!$ possible ways to label the N nodes of a graph. In graph theory this is known as the *isomorphism problem*, and to date, there are no known algorithms to check if two generic graphs are isomorphic in polynomial time.

Another definition which has to do with the permutation of the nodes of a graph, and is useful to characterise its symmetry, is that of graph *automorphism*.

Definition 1.3 (Automorphism) *Given a graph $G = (\mathcal{N}, \mathcal{L})$, an automorphism of G is a permutation $\phi : \mathcal{N} \to \mathcal{N}$ of the vertices of G so that if $(u, v) \in \mathcal{L}$ then $(\phi(u), \phi(v)) \in \mathcal{L}$. The number of different automorphisms of G is denoted as a_G.*

In other words, an automorphism is an isomorphism of a graph on itself. Consider the first labelled graph in Figure 1.4. The simplest automorphism is the one that keeps the node labels unchanged and produces the same labelled graph, shown as the first graph in Figure 1.5. Another example of automorphism is given by $\phi(1) = 4$, $\phi(2) = 1$, $\phi(3) = 2$, $\phi(4) = 3$. Note that this automorphism can be compactly represented by the permutation $(1, 2, 3, 4) \to (4, 1, 2, 3)$. The action of such automorphism would produce the second graph shown in Figure 1.5. There are eight distinct permutations of the labels $(1, 2, 3, 4)$ which change the first graph into an isomorphic one. The graph $\mathbb{C}_4$ has therefore $a_{\mathbb{C}_4} = 8$. The figure shows all possible automorphisms. Note that the permutation $(1, 2, 3, 4) \to (1, 3, 2, 4)$ is not an automorphism of the graph, because while $(1, 2) \in \mathcal{L}$, $(\phi(1), \phi(2)) = (1, 3) \notin \mathcal{L}$. Analogously, it is easy to prove that the number of different automorphisms of a triangle $\mathbb{C}_3 = \mathbb{K}_3$ is $a_\triangle = 6$, and more in general, for a cycle of N nodes, $\mathbb{C}_N$, we have $a_{\mathbb{C}_N} = 2N$.

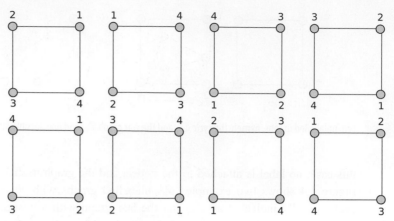

Fig. 1.5   All possible automorphisms of graph $\mathbb{C}_4$ in Figure 1.4.

Example 1.3   Consider the star graph $\mathbb{S}_4$ with a central node and three links shown in Figure 1.4. There are six automorphisms, corresponding to the following transformations: identity, rotation by 120° counterclockwise, rotation by 120° clockwise and three specular reflections, respectively around edge $(1, 2)$, $(1, 3)$, $(1, 4)$. There are no more automorphisms, because in all permutations, node 1 has to remain fixed. Therefore, the number a_G of possible automorphisms is given by the number of permutations of the three remaining labels, that is, $3! = 6$.

Finally, we consider some basic operations to produce new graphs from old ones, for instance, by merging together two graphs or by considering only a portion of a given graph. Let us start by introducing the definition of the *union* of two graphs. Let $G_1 = (\mathcal{N}_1, \mathcal{L}_1)$ and $G_2 = (\mathcal{N}_2, \mathcal{L}_2)$ be two graphs. We define graph $G = (\mathcal{N}, \mathcal{L})$, where $\mathcal{N} = \mathcal{N}_1 \cup \mathcal{N}_2$ and $\mathcal{L} = \mathcal{L}_1 \cup \mathcal{L}_2$, as the *union* of G_1 and G_2, and we denote it as $G = G_1 + G_2$. A concept that will be very useful in the following is that of *subgraph* of a given graph.

Definition 1.4 (Subgraph)   *A subgraph of $G = (\mathcal{N}, \mathcal{L})$ is a graph $G' = (\mathcal{N}', \mathcal{L}')$ such that $\mathcal{N}' \subseteq \mathcal{N}$ and $\mathcal{L}' \subseteq \mathcal{L}$. If G' contains all links of G that join two nodes in $\mathcal{N}'$, then G' is said to be the* subgraph induced *or* generated by $\mathcal{N}'$, *and is denoted as $G' = G[\mathcal{N}']$.*

Figure 1.6 shows some examples of subgraphs. A subgraph is said to be *maximal* with respect to a given property if it cannot be extended without losing that property. For example, the subgraph induced by nodes $2, 3, 4, 6$ in Figure 1.6 is the maximal complete subgraph of order four of graph G. Of particular relevance for some of the definitions given in the following is the *subgraph of the neighbours* of a given node i, denoted as G_i. G_i is defined as the subgraph induced by $\mathcal{N}_i$, the set of nodes adjacent to i, i.e. $G_i = G[\mathcal{N}_i]$. In Figure 1.6, graph (c) represents the graph G_6, induced by the neighbours of node 6.

Let $G = (\mathcal{N}, \mathcal{L})$, and let $s \in \mathcal{L}$. If we remove edge s from G we shall denote the new graph as $G' = (\mathcal{N}, \mathcal{L} - s)$, or simply $G' = G - s$. Analogously, let $\mathcal{L}' \subseteq \mathcal{L}$. We denote as

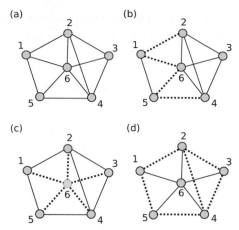

Fig. 1.6 A graph G with $N = 6$ nodes (a), and three subgraphs of G, namely an unconnected subgraph obtained by eliminating four of the edges of G (b), the subgraph generated by the set $\mathcal{N}_6 = \{1, 2, 3, 4, 5\}$ (c), and a spanning tree (d) (one of the connected subgraphs which contain all the nodes of the original graph and have the smallest number of links, i.e. $K = 5$).

$G' = (\mathcal{N}, \mathcal{L} - \mathcal{L}')$, or simply $G' = G - \mathcal{L}'$, the new graph obtained from G by removing all edges $\mathcal{L}'$.

1.2 Directed, Weighted and Bipartite Graphs

Sometimes, the precise order of the two nodes connected by a link is important, as in the case of the following example of the shuttles running between the terminals of an airport.

Example 1.4 *(Airport shuttle)* A large airport has six terminals, denoted by the letters A, B, C, D, E and F. The terminals are connected by a shuttle, which runs in a circular path, $A \rightarrow B \rightarrow C \rightarrow D \rightarrow E \rightarrow F \rightarrow A$, as shown in the figure. Since A and D are the main terminals, there are other shuttles that connect directly A with D, and vice versa. The network of connections among airport terminals can be properly described by a graph

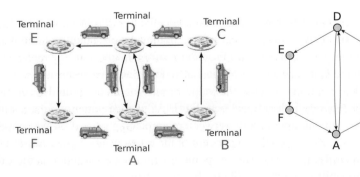

where the $N = 6$ nodes represent the terminals, while the links indicate the presence of a shuttle connecting one terminal to another. Notice, however, that in this case it is necessary to associate a direction with each link. A directed link is usually called an arc. The graph shown in the right-hand side of the figure has indeed $K = 8$ arcs. Notice that there can be two arcs between the same pair of nodes. For instance, arc (A, D) is different from arc (D, A).

We lose important information if we represent the system in the example as a graph according to Definition 1.1. We need therefore to extend the mathematical concept of graph, to make it better suited to describe real situations. We introduce the following definition of the *directed graph*.

> **Definition 1.5 (Directed graph)**   *A directed graph $G \equiv (\mathcal{N}, \mathcal{L})$ consists of two sets, $\mathcal{N} \neq \emptyset$ and $\mathcal{L}$. The elements of $\mathcal{N} \equiv \{n_1, n_2, \ldots, n_N\}$ are the nodes of the graph G. The elements of $\mathcal{L} \equiv \{l_1, l_2, \ldots, l_K\}$ are distinct ordered pairs of distinct elements of $\mathcal{N}$, and are called directed links, or arcs.*

In a directed graph, an arc between node i and node j is denoted by the ordered pair (i, j), and we say that the link is *ingoing* in j and *outgoing* from i. Such an arc may still be denoted as l_{ij}. However, at variance with undirected graphs, this time the order of the two nodes is important. Namely, $l_{ij} \equiv (i, j)$ stands for an arc from i to j, and $l_{ij} \neq l_{ji}$, or in other terms the arc (i, j) is different from the arc (j, i).

As another example of a directed network we introduce here the first data set of this book, namely DATA SET 1. As with all the other data sets that will be provided and studied in this book, this refers to the network of a real system. In this case, the network describes friendships between children at the kindergarten of Elisa, the daughter of the first author of this book. The choice of this system as an example of a directed network is not accidental. Friendship networks of children are, in fact, among social systems, cases in which the directionality of a link can be extremely important. In the case under study, friendships have been determined by interviewing the children. As an outcome of the interview, friendship relations are directed, since it often happens that child A indicates B as his/her friend, without B saying that A is his/her friend. The basic properties of *Elisa's kindergarten network* are illustrated in the DATA SET Box 1.2, and the network can be downloaded from the book's webpage. Of course, one of the first things that catches our eye in the directed graph shown in Box 1.2 is that many of the relations are not reciprocated. This property can be quantified mathematically. A traditional measure of *graph reciprocity* is the ratio r between the number of arcs in the network pointing in both directions and the total number of arcs [308] (see Problem 1.2 for a mathematical expression of r, and the work by Diego Garlaschelli and Maria Loffredo for alternative measures of the reciprocity [128]). The reciprocity r takes the value $r = 0$ for a purely unidirectional graph, while $r = 1$ for a purely bidirectional one. For Elisa's kindergarten we get a value $r = 34/57 \approx 0.6$, since the number of arcs between reciprocating pairs is 34 while we have 57 arcs in total. This means that only 60 per cent of the relations are reciprocated in this network, or, more precisely, if there is an arc pointing from node i to node j, then there is a 60 per cent probability that there will also be an arc from j to i.

Box 1.2   DATA SET 1: Elisa's Kindergarten Network

Elisa's kindergarten network describes $N = 16$ children between three and five years old, and their declared friendship relations. The network given in this data set is a directed graph with $K = 57$ arcs and is shown in the figure. The nine girls are represented as circles, while the seven boys are squares. Bidirectional relations are indicated as full-line double arrows, while purely unidirectional ones as dashed-line arrows. Notice that only a certain percentage of the relations are reciprocated.

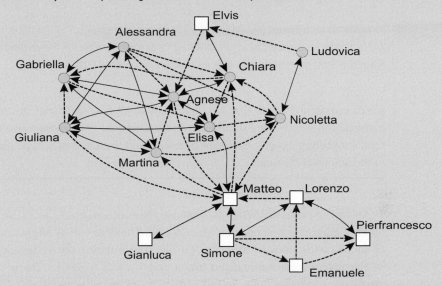

It is interesting to notice that, with the exception of Elvis, the youngest boy in the class, there is almost a split between two groups, the boys and the girls. You certainly would not observe this in a network of friendship in a high school. In the kindergarten network, Matteo is the child connecting the two communities.

Summing up, the most basic definition is that of undirected graph, which describes systems in which the links have no directionality. In the case, instead, in which the directionality of the connections is important, the directed graph definition is more appropriate. Examples of an undirected graph and of a directed graph, with $N = 7$ nodes, and $K = 8$ links and $K = 11$ arcs respectively, are shown in Figure 1.7 (a) and (b). The directed graph in panel (b) does not contain loops, nor multiple arcs, since these elements are not allowed by the standard definition of directed graph given above. Directed graphs with either of these elements are called *directed multigraphs* [48, 47, 308].

Also, we often need to deal with networks displaying a large heterogeneity in the relevance of the connections. Typical examples are social systems where it is possible to measure the strength of the interactions between individuals, or cases such as the one discussed in the following example.

Example 1.5   Suppose we have to construct a network of roads to connect N towns, so that it is possible to go from each town to any other. A natural question is: what is the

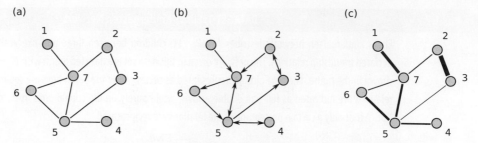

Fig. 1.7 An undirected (a), a directed (b), and a weighted undirected (c) graph with $N = 7$ nodes. In the directed graph, adjacent nodes are connected by arrows, indicating the direction of each arc. In the weighted graph, the links with different weights are represented by lines with thickness proportional to the weight.

set of connecting roads that has minimum cost? It is clear that in determining the best construction strategy one should take into account the construction cost of the hypothetical road connecting directly each pair of towns, and that the cost will be roughly proportional to the length of the road.

All such systems are better described in terms of *weighted graphs*, i.e. graphs in which a numerical value is associated with each link. The edge values might represent the strength of social connections or the cost of a link. For instance, the systems of towns and roads in Example 1.5 can be mapped into a graph whose nodes are the towns, and the edges are roads connecting them. In this particular example, the nodes are assigned a location in space and it is natural to assume that the weight of an edge is proportional to the length of the corresponding road. We will come back to similar examples when we discuss *spatial graphs* in Section 8.3. Weighted graphs are usually drawn as in Figure 1.7 (c), with the links with different weights being represented by lines with thickness proportional to the weight. We will present a detailed study of weighted graphs in Chapter 10. We only observe here that a multigraph can be represented by a weighted graph with integer weights.

Finally, a *bipartite graph* is a graph whose nodes can be divided into two disjoint sets, such that every edge connects a vertex in one set to a vertex in the other set, while there are no links connecting two nodes in the same set.

Definition 1.6 (Bipartite graph) *A bipartite graph, $G \equiv (\mathcal{N}, \mathcal{V}, \mathcal{L})$, consists of three sets, $\mathcal{N} \neq \emptyset$, $\mathcal{V} \neq \emptyset$ and $\mathcal{L}$. The elements of $\mathcal{N} \equiv \{n_1, n_2, \ldots, n_N\}$ and $\mathcal{V} \equiv \{v_1, v_2, \ldots, v_V\}$ are distinct and are called the* nodes *of the bipartite graph. The elements of $\mathcal{L} \equiv \{l_1, l_2, \ldots, l_K\}$ are distinct unordered pairs of elements, one from $\mathcal{N}$ and one from $\mathcal{V}$, and are called* links *or* edges.

Many real systems are naturally bipartite. For instance, typical bipartite networks are systems of users purchasing items such as books, or watching movies. An example is shown in Figure 1.8, where we have denoted the user-set as $U = \{u_1, u_2, \cdots, u_N\}$ and the object-set as $O = \{o_1, o_2, \cdots, o_V\}$. In such a case we have indeed only links between users and items, where a link indicates that the user has chosen that item. Notice that,

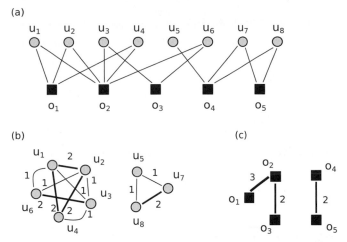

Fig. 1.8 Illustration of a bipartite network of $N = 8$ users and $V = 5$ objects (a), as well as its user-projection (b) and object-projection (c). The link weights in (b) and (c) are set as the numbers of common objects and users, respectively.

Box 1.3   **Recommendation Systems**

Consider a system of users buying books or selecting other items, similar to the one shown Figure 1.8. A reasonable assumption is that the users buy or select objects they like. Based on this, it is possible to construct *recommendation systems*, i.e. to predict the user's opinion on those objects not yet collected, and eventually to recommend some of them. The simplest recommendation system, known as *global ranking method* (GRM), sorts all the objects in descending order of degree and recommends those with the highest degrees. Such a recommendation is based on the assumption that the most-selected items are the most interesting for the average user. Despite the lack of personalisation, the GRM is widely used since it is simple to evaluate even for large networks. For example, the well-known *Amazon List of Top Sellers* and *Yahoo Top 100 MTVs*, as well as the list of most downloaded articles in many scientific journals, can all be considered as results of GRM. A more refined recommendation algorithm, known as *collaborative filtering* (CF), is based on similarities between users and is discussed in Example 1.13 in Section 1.6, and in Problem 1.6(c).

---

starting from a bipartite network, we can derive at least two other graphs. The first graph is a projection of the bipartite graph on the first set of nodes: the nodes are the users and two users are linked if they have at least one object in common. We can also assign a weight to the link equal to the number of objects in common; see panel (b) in the figure. In such a way, the weight can be interpreted as a similarity between the two users. Analogously, we can construct a graph of similarities between different objects by projecting the bipartite graph on the set of objects; see panel (c) in the figure.

1.3 Basic Definitions

The simplest way to characterise and eventually distinguishing the nodes of a graph is to count the number of their links, i.e. to evaluate their so-called *node degree*.

Definition 1.7 (Node degree)   *The* degree k_i *of a node i is the number of edges incident in the node. If the graph is directed, the degree of the node has two components: the number of outgoing links k_i^{out}, referred to as the* out-degree *of the node, and the number of ingoing links k_i^{in}, referred to as the* in-degree *of node i. The total degree of the node is then defined as $k_i = k_i^{out} + k_i^{in}$.*

In an undirected graph the list of the node degrees $\{k_1, k_2, \ldots, k_N\}$ is called the *degree sequence*. The average degree $\langle k \rangle$ of a graph is defined as $\langle k \rangle = N^{-1} \sum_{i=1}^{N} k_i$, and is equal to $\langle k \rangle = 2K/N$. If the graph is directed, the degree of the node has two components: the average in- and out-degrees are respectively defined as $\langle k^{out} \rangle = N^{-1} \sum_{i=1}^{N} k_i^{out}$ and $\langle k^{in} \rangle = N^{-1} \sum_{i=1}^{N} k_i^{in}$, and are equal.

Example 1.6   *(Node degrees in Elisa's kindergarten)*   Matteo and Agnese are the two nodes with the largest in-degree ($k^{in} = 7$) in the kindergarten friendship network introduced in Box 1.2. They both have out-degrees $k^{out} = 5$. Gianluca has the smallest in and out degree, $k^{out} = k^{in} = 1$. The graph average degree is $\langle k^{out} \rangle = \langle k^{in} \rangle = 3.6$

Another central concept in graph theory is that of the reachability of two different nodes of a graph. In fact, two nodes that are not adjacent may nevertheless be reachable from one to the other. Following is a list of the different ways we can explore a graph to visit its nodes and links.

Definition 1.8 (Walks, trails, paths and geodesics)   *A* walk $W(x, y)$ *from node x to node y is an alternating sequence of nodes and edges (or arcs) $W = (x \equiv n_0, e_1, n_1, e_2, \ldots, e_l, n_l \equiv y)$ that begins with x and ends with y, such that $e_i = (n_{i-1}, n_i)$ for $i = 1, 2, \ldots, l$. Usually a walk is indicated by giving only the sequence of traversed nodes: $W = (x \equiv n_0, n_1, .., n_l \equiv y)$. The length of the walk, $l = \ell(W)$, is defined as the number of edges (arcs) in the sequence. A* trail *is a walk in which no edge (arc) is repeated. A* path *is a walk in which no node is visited more than once. A* shortest path *(or geodesic) from node x to node y is a walk of minimal length from x to y, and in the following will be denoted as $\mathbb{P}(x, y)$.*

Basically, the definitions given above are valid both for undirected and for directed graphs, with the only difference that, in an undirected graph, if a sequence of nodes is a walk, a trail or a path, then also the inverse sequence of nodes is respectively a walk, a trail or a path, since the links have no direction. Conversely, in a directed graph there might be a directed path from x to y, but no directed path from y to x.

Based on the above definitions of shortest paths, we can introduce the concept of *distance* in a graph.

Definition 1.9 (Graph distances)   *In an undirected graph the* distance *between two nodes x and y is equal to the length of a shortest path $\mathbb{P}(x, y)$ connecting x and y. In a directed graph the distance from x to y is equal to the length of a shortest path $\mathbb{P}(x, y)$ from x to y.*

Notice that the definition of shortest paths is of crucial importance. In fact, the very same concept of distance between two nodes in a graph is based on the length of the shortest paths between the two nodes.

Example 1.7   Let us consider the graph shown in Figure 1.6(a). The sequence of nodes $(5, 6, 4, 2, 4, 5)$ is a walk of length 5 from node 5 back to node 5. This sequence is a walk, but not a trail, since the edge $(2, 4)$ is traversed twice. An example of a trail on the same graph is instead $(5, 6, 4, 5, 1, 2, 4)$. This is not a path, though, since node 5 is repeated. The sequence $(5, 4, 3, 2)$ is a path of length 3 from node 5 to node 2. However, this is not a shortest path. In fact, we can go from node 5 to node 2 in two steps in three different ways: $(5, 1, 2)$, $(5, 6, 2)$, $(5, 4, 2)$. These are the three shortest paths from 5 to 2.

Definition 1.10 (Circuits and cycles)   *A circuit is a closed trail, i.e. a trail whose end vertices coincide. A cycle is a closed walk, of at least three edges (or arcs) $W = (n_0, n_1, .., n_l)$, $l \geq 3$, with $n_0 = n_l$ and n_i, $0 < i < l$, distinct from each other and from n_0. An undirected cycle of length k is usually said a k-cycle and is denoted as $\mathbb{C}_k$. $\mathbb{C}_3$ is a triangle ($\mathbb{C}_3 = \mathbb{K}_3$), $\mathbb{C}_4$ is called a quadrilater, $\mathbb{C}_5$ a pentagon, and so on.*

Example 1.8   An example of circuit on graph 1.6(a) is $W = (5, 4, 6, 1, 2, 6, 5)$. This example is not a path on the graph, because some intermediate vertex is repeated. An example of cycle on graph 1.6(a) is $(1, 2, 3, 4, 5, 6, 1)$. Roughly speaking a cycle is a path whose end vertices coincide.

We are now ready to introduce the concept of connectedness, first for pairs of nodes, and then for graphs. This will allow us to define what is a component of a graph, and to divide a graph into components. We need here to distinguish between undirected and directed graphs, since the directed case needs more attention than the undirected one.

Definition 1.11 (Connectedness and components in undirected graphs)   *Two nodes i and j of an undirected graph G are said to be* connected *if there exists a path between i and j. G is said to be* connected *if all pairs of nodes are connected; otherwise it is said to be* unconnected *or* disconnected. *A* component *of G associated with node i is the maximal connected induced subgraph containing i, i.e. it is the subgraph which is induced by all nodes which are connected to node i.*

Of course, the first thing we will be interested in looking at, in a graph describing a real network or produced by a model, is the number of components of the graph and their sizes. In particular, when we consider in Chapter 3 families of graphs with increasing order N, a natural question to ask will be how the order of the components grows with the order of the graph. We will therefore find it useful there to introduce the definition of the *giant component*, namely a component whose number of nodes is of the same order as N.

Box 1.4	Path-Finding Behaviours in Animals

Finding the shortest route is extremely important also for animals moving regularly between different points. How can animals, with only limited local information, achieve this? Ants, for instance, find the shortest path between their nest and their food source by communicating with each other via their pheromone, a chemical substance that attracts other ants. Initially, ants explore all the possible paths to the food source. Ants taking shorter paths will take a shorter time to arrive at the food. This causes the quantity of pheromone on the shorter paths to grow faster than on the longer ones, and therefore the probability with which any single ant chooses the path to follow is quickly biased towards the shorter ones. The final result is that, due to the social cooperative behaviour of the individuals, very quickly all ants will choose the shortest path [141].

Even more striking is the fact that unicellular organisms can also exhibit similar path-finding behaviours. A well-studied case is the plasmodium of a slime mould, the *Physarum polycephalum*, a large amoeba-like cell. The body of the plasmodium contains a network of tubes, which enables nutrients and chemical signals to circulate through the organism. When food sources are presented to a starved plasmodium that has spread over the entire surface of an agar plate, parts of the organism concentrate over the food sources and are connected by only a few tubes. It has been shown in a series of experiments that the path connecting these parts of the plasmodium is the shortest possible, even in a maze [224]. Check Ref. [296] if you want to see path-finding algorithms inspired by the remarkable process of cellular computation exhibited by the *P. polycephalum*.

In a directed graph, the situation is more complex than in an undirected graph. In fact, as observed before, a directed path may exist through the network from vertex i to vertex j, but that does not guarantee that one exists from j to i. Consequently, we have various definitions of connectedness between two nodes, and we can define *weakly* and *strongly connected components* as below.

Definition 1.12 (Connectedness and components in directed graphs) *Two nodes i and j of a directed graph G are said to be* strongly connected *if there exists a path from i to j and a path from j to i. A directed graph G is said to be* strongly connected *if all pairs of nodes (i, j) are strongly connected. A* strongly connected component *of G associated with node i is the maximal strongly connected induced subgraph containing node i, i.e. it is the subgraph which is induced by all nodes which are strongly connected to node i.*

The undirected graph G^u obtained by removing all directions in the arcs of G is called the underlying undirected graph *of G. A directed graph G is said to be* weakly connected *if the underlying undirected graph G^u is connected. A* weakly connected component *of G is a component of its underlying undirected graph G^u.*

Example 1.9 Most graphs shown in the previous figures are connected. Examples of disconnected graphs are graph G3 in Figure 1.1 and graph (b) in Figure 1.6. Graph G3 in Figure 1.1 has two components, one given by node 1 and the other given by the subgraph induced by nodes $\{2, 3, 4\}$. Graph (b) in Figure 1.6 has also two components, one given by the subgraph generated by nodes $\{1, 5\}$ and the other generated by nodes $\{2, 3, 4, 6\}$. The directed graph in Example 1.4 is strongly connected, as it should be, since in an airport one wants to join any pair of terminals in both directions.

We will come back to component analysis and to the study of number and size of components in real-world networks in the next chapters.

1.4 Trees

Trees are a particular kind of graph that appear very commonly both in the analysis of other graphs and in various applications. Trees are important because, among all connected graphs with N nodes, are those with the smallest possible number of links. Usually a *tree* is defined as a connected graph containing no cycles. We then say that a tree is a connected *acyclic* graph. The simplest possible non-trivial tree is a graph with two nodes and two links, known as a *triad*, and usually indicated in this book with the symbol $\wedge$. Triads and triangles play an important role in complex networks, and we will come back to them in Section 4.3.

Together with the concept of the tree, we can also introduce that of the *forest*, that is, a graph whose connected components are all trees. Various examples of trees are shown in Figure 1.9. The first graph is a tree with $N = 17$ nodes. The second graph is one of the possible spanning trees of a graph with $N = 7$ nodes and $K = 12$ links. A *spanning tree* of a graph G is a tree that contains all the nodes of G, i.e. a connected subgraph which contains all the nodes of the original graph and has the smallest number of links. Finally, the third graph is a sketch of a *Cayley tree*, an infinite tree in which each node is connected to z neighbours, where z is called the coordination number. Namely, we plot the first three iterations to construct a Cayley tree with $z = 3$, starting with an origin node placed in the centre of the figure. Even if the concept of the tree graph is not familiar to you, you are bound to be familiar with many examples.

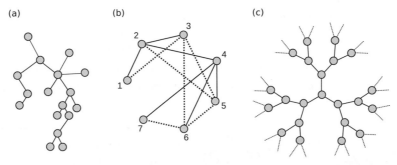

Fig. 1.9   Three examples of trees. A tree with $N = 17$ nodes (a). A spanning tree (solid lines) of a graph with $N = 7$ nodes and 12 links (solid and dashed lines) (b). Three levels of a Cayley tree with $z = 3$ (c).

Example 1.10   *(Trees in the real world)*   Any printed material and textbook that is divided into sections and subsections is organised as a tree. Large companies are organised as trees, with a president at the top, a vice-president for each division, and so on. Mailing addresses, too, are trees. To send a mail we send it first to the correct country, then to the correct state, and similarly to the correct city, street and house number. There are abundant examples of trees on computers as well. File systems are the canonical example. Each drive is the root of an independent tree of files and directories (folders). File systems also provide a good example of the fact that any node in a tree can be identified by giving a path from the root. In nature, plants and rivers have a tree-like structure.

In practice, there are several equivalent ways to characterise the concept of tree, and each one can be used as a definition. Below we introduce the three definitions most commonly found in the literature.

Definition 1.13 (Trees)   *A tree can be alternatively defined as:* (A) *a connected acyclic graph;* (B) *a connected graph with $K = N - 1$ links;* (C) *an acyclic graph with $K = N - 1$ links.*

The three definitions can be proven to be equivalent. Here, we shall only prove that definition (A) implies definition (B), i.e. that $(A) \Rightarrow (B)$, while we will defer the discussion of $(B) \Rightarrow (C)$ and $(C) \Rightarrow (A)$ to Problem 1.4(a). In order to show that definition (A) implies definition (B), we first need to prove the following three propositions, which also show other interesting properties of trees.

Proposition 1.1   *Let $G = (\mathcal{N}, \mathcal{L})$ be a tree, i.e. by using definition (A), a connected acyclic graph. Then, for any pair of vertices $x, y \in \mathcal{N}$ there exists one and only one walk that joins x and y.*

Proof   We will provide a proof by contradiction. Let $x, y \in \mathcal{N}$. Since G is connected, there is at least one path that joins x and y. Let us assume that there exist two paths, denoted by $w_1 = (x_0 = x, x_1, x_2, \ldots, x_r = y)$ and $w_2 = (y_0 = x, y_1, y_2, \ldots, y_s = y)$. Let us denote by $u < \min(r, s)$ the largest index for which $x_i = y_i$. The two walks will reconnect for some indices j and j', i.e. it will be

$$x_j = y_{j'}, \quad x_i \neq y_{i'}, \forall i \in \{u + 1, \ldots j - 1\}, i' \in \{u + 1, \ldots j' - 1\}.$$

As shown in Figure 1.10, it follows that there exists a cycle $\mathbb{C} = (x_u, x_{u+1}, \ldots, x_j = y_{j'}, y_{j'-1}, \ldots, y_u = x_u)$, which contradicts the assumption that G is acyclic. $\square$

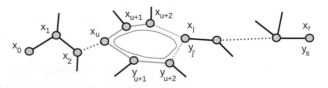

Fig. 1.10   In a tree there cannot exist two different paths that join two nodes, otherwise a cycle would form.

Proposition 1.2 *Let $G = (\mathcal{N}, \mathcal{L})$ be a graph. Suppose that for each pair of distinct nodes of the graph there exists one and only one path joining the nodes. Then G is connected and if we remove any edge $\ell \in \mathcal{L}$, the resulting graph $G - \ell$ will not be connected.*

Proof That G is connected follows immediately from the assumptions. Furthermore, if ℓ is an edge that joins x and y, since there is only one path joining two edges (from proposition 1.1), if we remove it there will be no path joining x and y, and therefore the resulting graph will be disconnected. □

Proposition 1.3 *Let G be a connected graph, such that if $\ell \in \mathcal{L} \Rightarrow G - \ell$ is disconnected. Then G is connected and has $K = N - 1$ links.*

Proof We only need to prove that $K = N - 1$. We will do this by induction on N. For $N = 1$ and $N = 2$ one has respectively $K = 0$ and $K = 1$. Now let G be a graph with $N \geq 3$, and let $x, y \in \mathcal{N}, (x, y) \in \mathcal{L}, x \neq y$. By assumption, $G - (x, y)$ is not connected: it is in fact formed by two connected components, G_1 and G_2, having respectively N_1 and N_2 nodes, with $N = N_1 + N_2$. Because $N_1 < N, N_2 < N$, by induction one has $N_1 = K_1 + 1$ and $N_2 = K_2 + 1$. From $K = K_1 + K_2 + 1$ it follows that

$$N = N_1 + N_1 = K_1 + 1 + K_2 + 1 = K + 1.$$

□

Finally, it is clear that by the successive use of the three propositions above we have proved that definition (A) implies definition (B).

1.5 Graph Theory and the Bridges of Königsberg

As an example of the powerful methods of graph theory, in this section we discuss the theorem proposed by the Swiss mathematician Leonhard Euler in 1736 as a solution to the Königsberg bridge problem. This is an important example of how the abstraction of graph theory can prove useful for solving practical problems. It is also historically significant, since Euler's work on the Königsberg bridges is often regarded as the birth of graph theory. The problem is related to the ancient Prussian city of Königsberg (later, the city was taken over by the USSR and renamed Kaliningrad), traversed by the Pregel river. The city, with its seven bridges, as there were in Euler's time, is graphically shown in the left-hand side of Figure 1.11. The problem to solve is whether or not it is possible to find an optimum stroll that traverses each of the bridges exactly once, and eventually returns to the starting point.

A brute force approach to this problem consists in starting from a side, making an exhaustive list of possible routes, and then checking one by one all the routes. In the case that no route satisfies the requested condition, one has to start again with a different initial point and to repeat the procedure. Of course, such an approach does not provide a general solution to the problem. In fact, if we want to solve the bridges' problem for a different city, we should repeat the enumeration for the case under study. Euler came up with an elegant

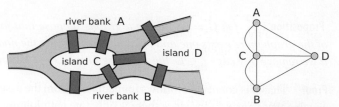

Fig. 1.11 The city of Königsberg at the time of Leonhard Euler (left). The river is coloured light grey, while the seven bridges are dark grey. The associated multigraph, in which the nodes corresponds to river banks and islands, and the links represents bridges (right).

way to answer the question for any given configuration of bridges. First, he introduced the idea of the graph. He recognised that the problem depends only on the set of connections between riverbanks and islands. If we collapse the whole river bank A to a point, and we do the same for river bank B and for the islands C and D, all the relevant information about the city map can, in fact, be encapsulated into a graph with four nodes (river banks and islands) and seven edges (the bridges) shown in right-hand side of Figure 1.11. The graph is actually a multigraph, but this will not affect our discussion. In graph terms, the original problem translates into the following request: "Is it possible to find a circuit (or trail) containing all the graph edges?" Such a circuit (or trail) is technically called an *Eulerian circuit* (*Eulerian trail*).

Definition 1.14 (Eulerian circuits and trails)   *A trail in a graph G containing all the edges is said to be an Eulerian trail in G. Similarly, a circuit in G containing all the edges is said to be an Eulerian circuit in G. A graph is said to be Eulerian if it contains at least one Eulerian circuit, or semi-Eulerian if it contains at least one Eulerian trail.*

Example 1.11   *(Difficulty of an exhaustive search)*   A way to perform an exhaustive search for an Eulerian trail in a graph with N nodes and K edges is to check among the walks of length $l = K$ whether there is one containing all the edges of the graph. If there is such a walk, then it is necessarily a trail, and therefore it is an Eulerian trail. The number of walks of length $l = K$ is thus a measure of the difficulty of an exhaustive search. This number can be calculated exactly in the case of the multigraph in Figure 1.11, although here we will only give an approximate estimate. Let us consider first the simpler case of a complete graph with N nodes, $\mathbb{K}_N$. The total number of walks of length l for such a graph is $N(N - 1)^l$. In fact, we can choose the initial node in N different ways and, at each node, we can choose any of its $N - 1$ edges. In conclusion, to look for Eulerian trails in $\mathbb{K}_N$ we have to check $N(N - 1)^{N(N-1)/2}$ walks. This number is equal to 2916 in a complete graph with $N = 4$ nodes. The same argument applies to a regular graph $\mathbb{R}_{N,k}$, i.e. a graph with N nodes and k links for each node. In such a case we can choose the initial node in N different ways and, at each node, we can choose k edges. Finally, the number of walks of length l is equal to Nk^l, so that if we set $l = K$ we obtain Nk^K walks of length K. By using such a formula with k replaced by $\langle k \rangle = 2K/N$, we can get an estimate for the number of walks of length K in a generic graph with N nodes and K links. This gives 25736 for the graph of

Königsberg, having $\langle k \rangle = 3.5$. This number is of the same order of magnitude as the exact value (see Problem 1.5). Notice that this number grows exponentially with K, so that in a city with a larger number of bridges it can become impossible to explore all the different trips. For instance, in the case of the historical part of Venice, with its 428 bridges, even by assuming a small value $\langle k \rangle = 2$, we get a number of $N \cdot 2^{428}$ walks to check. Thus, an exhaustive search for an Eulerian path over the graph represented by the islands of Venice and its bridges will be far beyond the computational capabilities of modern computers. In fact, even assuming that a computer can check 10^{15} walks per second, it would be able to check about 10^{32} walks in a timespan equal to the age of the universe; this number is much smaller than $N \cdot 2^{428} \approx 7N \cdot 10^{128}$.

After having shown that the problem can be rephrased in terms of a graph, Euler gave a general theorem on the conditions for a graph to be Eulerian.

Theorem 1.1 (Euler theorem)　*A connected graph is Eulerian iff each vertex has even degree. It has a Eulerian trail from vertex i to vertex j, $i \neq j$, iff i and j are the only vertices of odd degree.*

To be more precise, Euler himself actually proved only a necessary condition for the existence of an Eulerian circuit, i.e. he proved that if some nodes have an odd degree, then an Eulerian trail cannot exist. The proof given by Euler can be summarised as follows.

Proof　Suppose that there exists an Euler circuit. This means that each node i is a crossing point, therefore if we denote by p_i the number of times the node i is traversed by the circuit, its degree has to be $k_i = 2p_i$, and therefore it has to be even. If we only assume the existence of an Eulerian trail, then there is no guarantee that the starting point coincides with the ending point, and therefore the degree of such two points may be odd.　□

Euler believed that the converse was also true, i.e. that if all nodes have an even degree then there exists an Eulerian circuit, and he gave some argument about this, but he never rigorously proved the sufficient condition [105]. The proof that the condition that all nodes have even degree is sufficient for the existence of an Eulerian trail appeared more than a century later, and was due to the German mathematician Carl Hierholzer, who published the first characterisation of Eulerian graphs in 1873 [152]. The early history of graph theory, including the work of Euler and Hierholzer, is illustrated in [245].

Here we shall give a complete proof of the Euler theorem based on the concept of partition of a graph into cycles. Consider the set of edges $\mathcal{L}$ of a graph G. We say that a subset $\mathcal{L}_1 \subseteq \mathcal{L}$ is a cycle if there exists a cycle, Z_1, that contains all and only the edges $\mathcal{L}_1$. We say that the set $\mathcal{L}$ is *partitioned* if there exists a certain number s of subsets of $\mathcal{L}$, $\mathcal{L}_1, \mathcal{L}_2, \ldots, \mathcal{L}_s$, such that:

$$\mathcal{L}_i \cap \mathcal{L}_j = \emptyset, \ \forall i,j \in [1,\ldots,s], \quad \cup_{i=1}^{s}\mathcal{L}_i = \mathcal{L}$$

Now we can state the characterisation of Eulerian graphs in the form of equivalence of the following three statements [150]:

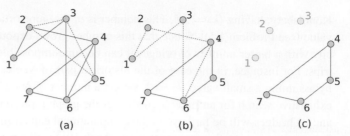

Fig. 1.12 The reduction process to prove that a graph with all nodes with even degree can be partitioned into cycles.

(1) G is an Eulerian graph (i.e. it contains at least one Eulerian circuit)

(2) $\forall i \in \mathcal{N}$, k_i is even

(3) there exists a partition of $\mathcal{L}$ into cycles.

Proof Proof (1) $\implies$ (2). It is clear that the existence of an Eulerian circuit implies that every node i is a crossing point, i.e. every node can be considered both a "starting point" and an "ending point", therefore its degree k_i has to be even. □

Proof Proof (2) $\implies$ (3). From property (2) it follows that in G there exists at least one cycle, Z_1, otherwise G would be a tree, and it would therefore have vertices of degree one (see Section 1.4). If $G \simeq Z_1$,[2] then (3) is proved. If $G \neq Z_1$, let $G_2 = G_1 - \mathcal{L}_1$, i.e. G_2 is the graph obtained from G after removing all edges of Z_1. It is clear that all vertices of G_2 have even degree, because the degree of each node belonging to Z_1 has been decreased by 2. Let G'_2 be the graph obtained from G_2 by eliminating all isolated vertices of G_2. Since all vertices of G'_2 have even degree, this means that G_2 contains at least a cycle, Z_2, and the argument repeats. Proceeding with the reduction, one will at last reach a graph $G'_\ell \simeq Z_\ell$, therefore the set of links $\mathcal{L}$ is partitioned in cycles $\mathcal{L}_1, \mathcal{L}_2, \ldots, \mathcal{L}_\ell$. The procedure is illustrated in Figure 1.12. □

Proof Proof of (3) $\implies$ (1). We now assume that the set of edges $\mathcal{L}$ can be partitioned into a certain number s of cycles, $\mathcal{L}_1, \mathcal{L}_2, \ldots, \mathcal{L}_s$. Let us denote by $Z_1, Z_2, \ldots, Z_s$ the corresponding graphs. If $Z_1 \simeq G$ then (1) is proved. Otherwise, let Z_2 be a cycle with a vertex i in common with Z_1. The circuit that starts in i and passes through all edges of Z_1 and Z_2 contains all edges of Z_1 and Z_2 exactly once. Hence, it is an Eulerian circuit for $Z_1 \cup Z_2$. If $G \simeq Z_1 \cup Z_2$ the assert is proved, otherwise let Z_3 be another cycle with a vertex in common with $Z_1 \cup Z_2$, and so on. By iterating the procedure, one can construct in G an Eulerian circuit. □

The Euler theorem provides a general solution to the bridge problem: the request to pass over every bridge exactly once can be satisfied if and only if the vertices with odd degree are zero (starting and ending point coincide) or two (starting and ending point do not coincide). Now, if we go back to the graph of Königsberg we see that the conditions of the theorem are not verified. Actually, all the four vertices in the graph in Figure 1.11 have

[2] The symbol $\simeq$ indicates that the two graphs are isomorphic. See Section 1.1

an odd degree. Therefore Eulerian circuits and trails are not possible. In the same way, by a simple and fast inspection, we can answer the same question for the city of Venice or for any other city in the world having any number of islands and bridges.

1.6 How to Represent a Graph

Drawing a graph is a certainly a good way to represent it. However, when the number of nodes and links in the graph is large, the picture we get may be useless because the graph can look like an intricate ball of wool. An alternative representation of a graph, which can also prove useful when we need to input a graph into a computer program, can be obtained by using a matrix. Matrices are tables of numbers on which we can perform certain operations. The space of matrices is a vector space, in which, in addition to the usual operations on vector spaces, one defines a matrix product. Here and in Appendices A.4 and A.5 we will recall the basic definitions and operations that we will need in the book. More information can be found in any textbook on linear algebra.

There are different ways to completely describe a graph $G = (\mathcal{N}, \mathcal{L})$ with N nodes and K links by means of a matrix. One possibility is to use the so-called *adjacency* matrix A.

> **Definition 1.15 (Adjacency matrix)**   *The* adjacency *matrix A of a graph is a $N \times N$ square matrix whose entries a_{ij} are either ones or zeros according to the following rule:*
> $$a_{ij} = \begin{cases} 1 & \textit{iff } (i,j) \in \mathcal{L} \\ 0 & \textit{otherwise} \end{cases}$$

In practice, for an undirected graph, entries a_{ij} and a_{ji} are set equal to 1 if there exists the edge (i,j), while they are zero otherwise. Thus, in this case, the adjacency matrix is symmetric. If instead the graph is directed, $a_{ij} = 1$ if there exists an arc from i to j. Notice that in both cases it is common convention to set $a_{ii} = 0, \forall i = 1, \ldots, N$.

Example 1.12   Consider the two graphs in the figure below. The first graph is undirected and has $K = 4$ links, while the second graph is directed and has $K = 7$ arcs. The adjacency

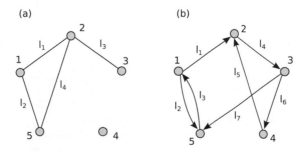

matrices associated with the two graphs are respectively:

$$A_u = \begin{pmatrix} 0 & 1 & 0 & 0 & 1 \\ 1 & 0 & 1 & 0 & 1 \\ 0 & 1 & 0 & 0 & 0 \\ 0 & 0 & 0 & 0 & 0 \\ 1 & 1 & 0 & 0 & 0 \end{pmatrix}, \quad A_d = \begin{pmatrix} 0 & 1 & 0 & 0 & 1 \\ 0 & 0 & 1 & 0 & 0 \\ 0 & 0 & 0 & 1 & 1 \\ 0 & 1 & 0 & 0 & 0 \\ 1 & 0 & 0 & 0 & 0 \end{pmatrix}$$

A_u is symmetric and contains $2K$ non-zero entries. The number of ones in row i, or equivalently in column i, is equal to the degree of vertex i. The adjacency matrix of a directed graph is in general not symmetric. This is the case of A_d. This matrix has K elements different from zero, and the number of ones in row i is equal to the number of outgoing links k_i^{out}, while the number of ones in column i is equal to the number of ingoing links k_i^{in}.

Clearly, Definition 1.15 refers only to undirected and directed graphs without loops or multiple links, which will be our main interest in this section and in the rest of the book. While weighted graphs will be treated in detail in Chapter 10, here we want to add that it is also possible to describe a bipartite graph by means of a slightly different definition of the adjacency matrix than that given above, and which in general is not a square matrix. In fact, a bipartite graph such as that shown in Figure 1.8 can be described by an $N \times V$ *adjacency matrix* A, such that entry $a_{i\alpha}$, with $i = 1, \ldots, N$ and $\alpha = 1, \ldots, V$, is equal to 1 if node i of the first set and node α of the second set are connected, while it is 0 otherwise. Notice that we used Roman and Greek letters to avoid confusion between the two types of nodes. Using this representation of a bipartite graph in terms of an adjacency matrix, we show in the next example how to formally describe a commonly used method to recommend a set of objects to a set of users (see also Box 1.3).

Example 1.13 *(Recommendation systems: collaborative filtering)* Consider a bipartite graph of users and objects such that shown in Figure 1.8, in which the existence of the link between node i and node α denotes that user u_i has selected object o_α. A famous personalised recommendation system, known as *collaborative filtering* (CF), is based on the construction of a $N \times N$ *user similarity matrix* $S = \{s_{ij}\}$. The similarity between two users u_i and u_j can be expressed in terms of the adjacency matrix of the graph as:

$$s_{ij} = \frac{\sum_{\alpha=1}^{V} a_{i\alpha} a_{j\alpha}}{\min\{k_{u_i}, k_{u_j}\}}, \tag{1.1}$$

where $k_{u_i} = \sum_{\alpha=1}^{V} a_{i\alpha}$ is the degree of user u_i, i.e. the number of objects chosen by u_i [324]. Based on the similarity matrix S, we can then construct an $N \times V$ *recommendation matrix* $R = \{r_{i\alpha}\}$. In fact, for any user–object pair u_i, o_α, if u_i has not yet chosen o_α, i.e. if $a_{i\alpha} = 0$, we can define a recommendation score $r_{i\alpha}$ measuring to what extent u_i may like o_α, as:

$$r_{i\alpha} = \frac{\sum_{j=1, j\neq i}^{N} s_{ij} a_{j\alpha}}{\sum_{j=1, j\neq i}^{N} s_{ij}}. \tag{1.2}$$

At the numerator, we sum the similarity between user u_i and all the other users that have chosen object o_α. In practice, we count the number of users that chose object o_α, weighting each of them with the similarity with user u_i. The normalisation at the denominator guarantees that $r_{i\alpha}$ ranges in the interval $[0, 1]$. Finally, in order to recommend items to a user u_i, we need to compute the values of the recommendation score $r_{i\alpha}$ for all objects $o_\alpha, \alpha = 1, \ldots, V$, such that $a_{i\alpha} = 0$. Then, all the non-zero values of $r_{i\alpha}$ are sorted in decreasing order, and the objects in the top of the list are recommended to user u_i.

Let us now come back to the main issue of this section, namely how to represent an undirected or directed graph. An alternative possibility to the adjacency matrix is a $N \times K$ matrix called the *incidence* matrix, in which the rows represent different nodes, while the columns stand for the links.

Definition 1.16 (Incidence matrix) *The incidence matrix B of an undirected graph is an $N \times K$ matrix whose entry b_{ik} is equal to 1 whenever the node i is incident with the link l_k, and is zero otherwise. If the graph is directed, the adopted convention is that the entry b_{ik} of B is equal to 1 if arc k points to node i, it is equal to -1 if the arc leaves node i, and is zero otherwise.*

Example 1.14 *(Incidence matrix)* The incidence matrices respectively associated with the undirected and directed graphs considered in Example 1.12 are:

$$
B_u = \begin{pmatrix} 1 & 1 & 0 & 0 \\ 1 & 0 & 1 & 1 \\ 0 & 0 & 0 & 1 \\ 0 & 0 & 0 & 0 \\ 0 & 1 & 1 & 0 \end{pmatrix}, \quad
B_d = \begin{pmatrix} -1 & 1 & -1 & 0 & 0 & 0 & 0 \\ 1 & 0 & 0 & -1 & 0 & 1 & 0 \\ 0 & 0 & 0 & 1 & -1 & 0 & -1 \\ 0 & 0 & 0 & 0 & 1 & -1 & 0 \\ 0 & -1 & 1 & 0 & 0 & 0 & 1 \end{pmatrix}
$$

Notice that B_u is a 5×4 matrix because the first graph has $N = 5$ nodes and $K = 4$ links, while B_d is a 5×7 matrix because the second graph has $N = 5$ nodes and $K = 7$ arcs. Also, notice that there are only two non-zero entries in each column of an incidence matrix.

Observe now that many elements of the adjacency and of the incidence matrix are zero. In particular, if a graph is sparse, then the adjacency matrix is sparse: for an undirected (directed) graph the number $2K$ (K) of non-zero elements is proportional to N, while the total number of elements of the matrix is N^2. When the adjacency matrix is sparse, it is more convenient to use a different representation, in which only the non-zero elements are stored. The most commonly used representation of sparse adjacency matrices is the so-called *ij-form*, also known as *edge list form*.

Definition 1.17 (Edge list) *The edge list of a graph, also known as ij-form of the adjacency matrix of the graph, consists of two vectors $\mathbf{i}$ and $\mathbf{j}$ of integer numbers storing the positions, i.e. respectively the row and column indices of the ones of the adjacency matrix*

A. Each of the two vectors has M components, with $M = 2K$ for undirected graphs, and $M = K$ for directed graphs.

Notice that storing the indices of the ones of the adjacency matrix is perfectly equivalent to storing the edges, i.e. the ordered pairs (i_k, j_k), $k = 1, \ldots, M$, with $M = 2K$ for undirected graphs, and $M = K$ for directed graphs. The two vectors $\mathbf{i}$ and $\mathbf{j}$ and the total number N of nodes are enough to completely represent a graph whose nodes are identified by integer numbers from 1 to N.

Example 1.15 *(ij-form of the adjacency matrix)* Below we give the *ij*-form of the adjacency matrices, A_u and A_d, of the two graphs in Example 1.12.

$$A_u = (\mathbf{i}, \mathbf{j}), \; \mathbf{i} = \begin{pmatrix} 1 \\ 1 \\ 2 \\ 2 \\ 2 \\ 3 \\ 5 \\ 5 \end{pmatrix}, \; \mathbf{j} = \begin{pmatrix} 2 \\ 5 \\ 1 \\ 3 \\ 5 \\ 2 \\ 1 \\ 2 \end{pmatrix}, \quad A_d = (\mathbf{i}, \mathbf{j}), \; \mathbf{i} = \begin{pmatrix} 1 \\ 1 \\ 2 \\ 3 \\ 3 \\ 4 \\ 5 \end{pmatrix}, \; \mathbf{j} = \begin{pmatrix} 2 \\ 5 \\ 3 \\ 4 \\ 5 \\ 2 \\ 1 \end{pmatrix}$$

For an undirected graph, such a representation is somehow redundant, because (i, j) and (j, i) represent the same edge, and it is enough to store it only once, or, in other words, in a symmetric matrix it is sufficient to store only the non-zero elements (i, j), with, say, $j \leq i$. However, we shall usually adopt the redundant representation, which allows a more uniform treatment of undirected and directed graphs. In a directed graph, the same representation of the adjacency matrix can be used. In this case the number of non-zero elements of the matrix is equal to the number of edges K. Note that the order in which the edges are stored is irrelevant: to change it is equivalent to relabelling the edges, but the graph (and the edge labels as a pair of nodes) is unchanged. We have chosen to store the pairs in such a way that the index i is increasing. This means that we first represent the edges starting from node 1, then the edges starting from node 2, and so on.

The *ij*-form of the adjacency matrix is very flexible, because the order in which the links are stored is not relevant. This form is often used during the construction of a graph which is obtained by adding a new link at a time. However, this form does not reveal straight away the neighbours of a node, so it is not the best choice in tasks such as finding the possible walks on the graph. More suitable graph representations in such cases are the *list of neighbours*, and the *compressed row storage* of the adjacency matrix. A list of neighbours stores for each node i its label, and the labels of the nodes to which i is connected. If the graph is directed we have two possibilities, namely we can list either the out-neighbours or the in-neighbours of each node. The lists of out- and in-neighbours are equivalent, although it can be convenient to choose one or the other according to the problem we face.

Example 1.16 *(List of neighbours)* The lists of neighbours corresponding to the two graphs in Example 1.12 are respectively:

$$
A_u = \begin{pmatrix}
\text{node} & \text{in-neighbours} \\
1 & 2 \quad 5 \\
2 & 1 \quad 3 \quad 5 \\
3 & 2 \\
4 & \\
5 & 1 \quad 2
\end{pmatrix}, \quad
A_d = \begin{pmatrix}
\text{node} & \text{out-neighbours} \\
1 & 2 \quad 5 \\
2 & 3 \\
3 & 4 \quad 5 \\
4 & 2 \\
5 & 1
\end{pmatrix}
$$

Notice that in the case of a directed graph, we have two different possibilities. Namely, for each node i, we can list either the nodes pointed by i, or the nodes pointing to i. In particular, in the matrix A_d given above we have adopted the first choice.

The *compressed row storage*, instead, consists of an array $\mathbf{j}$ of size $2K$ (K for directed graphs) storing the ordered sequence of the neighbours of all nodes, and a second array, $\mathbf{r}$, of size $N + 1$. The k_1 neighbours of node 1 are stored in $\mathbf{j}$ at positions $1, 2, \ldots, k_1$, the k_2 neighbours of node 2 are stored in $\mathbf{j}$ at positions $k_1 + 1, k_1 + 2, \ldots, k_1 + k_2$, and so on. The value of r_i, for $i = 1, 2, \ldots, N$, is the index of $\mathbf{j}$ where the first neighbour of node i is stored, while r_{N+1} is equal to the size of $\mathbf{j}$ plus one ($r_{N+1} = 2K + 1$ for undirected graphs, while $r_{N+1} = K + 1$ for directed graphs). With the definitions given above, we have that $r_{i+1} - r_i$ is equal to the degree of node i if the graph is undirected, or to the out-degree of node i if the graph is directed. For a directed graph, an alternative possibility is to use the *compressed column storage*, where, for each node i, we list all the nodes pointing to i.

Example 1.17 *(Compressed storage)* The vectors $\mathbf{r}$ and $\mathbf{j}$ corresponding to matrix A_u are $\mathbf{r} = (1, 3, 6, 7, 7, 9)$ and $\mathbf{j} = (2, 5, 1, 3, 5, 2, 1, 2)$. Note that the first element of the vector $\mathbf{r}$ is always 1, while the degree k_i of node i is given by $k_i = r_{i+1} - r_i, i = 1, \ldots, N$. Therefore, the $k_1 = 2$ neighbours of node 1 are stored in $\mathbf{j}$ in the positions from $r_1 = 1$ to $r_2 - 1 = 2$, the $k_2 = 3$ neighbours of node 2 are stored in $\mathbf{j}$ in the positions from $r_2 = 3$ to $r_3 - 1 = 5$, and so on. In the case of the directed graph A_d, we have two different lists of neighbours of a node, namely the out-neighbours and the in-neighbours. Consequently the compressed row storage is different from the compressed column storage. The vectors $\mathbf{r}$ and $\mathbf{j}$ corresponding to matrix A_d in the compressed row storage are $\mathbf{r} = (1, 3, 4, 6, 7, 8)$ and $\mathbf{j} = (2, 5, 3, 4, 5, 2, 1)$. Note that the out-degree of node i is $k_i^{\text{out}} = r_{i+1} - r_i$. Instead in the compressed column storage the vectors are $\mathbf{r} = (1, 2, 4, 5, 6, 8)$ and $\mathbf{j} = (5, 1, 4, 2, 3, 1, 3)$, and now $r_{i+1} - r_i$ gives the in-degree k_i^{in} of node i.

In practice, for the implementation of graph algorithms given in the Appendix, we shall make use of either the *ij*-form or the compressed row storage form. The details of the implementation of basic matrix operations in such forms can be found in Appendix A.4.

1.7 What We Have Learned and Further Readings

In this chapter we have introduced the basic language and the fundamental definitions we will use throughout the book. We started from scratch with the very definition of graph, providing the reader with all the different variations of the concept, namely that of *undirected*, *directed*, *weighted* and *bipartite graph*. All of these will be necessary to appropriately describe the richness and variegated nature of real-world networks. We then moved on to the most basic of node properties, the *degree*, and to some properties of pairs of nodes, such as that of *connectedness*, and we also analysed the different ways in which we can explore a graph by visiting its nodes and links. Finally, we briefly discussed how to describe a graph in terms of a matrix or a list of edges. We have also introduced the first data set of a real network of the book, that of children friendships at *Elisa's kindergarten*.

In conclusion, the main message learnt in this chapter is that *graph theory*, a well-developed branch of mathematics with a long tradition and an extended research literature, is the first pillar of complex networks. As an example of the power of graph theory we discussed how Leonhard Euler solved in 1736 the Königsberg bridge problem. This is an important historical example, since it was the first time that a practical problem was solved by transforming it into a graph problem. We will come back to a couple of other useful mathematical results from graph theory in Chapter 3, when we will be studying the properties of large random graphs. However, for a complete introduction to the field of graph theory, we suggest here the books by Douglas B. West [313], Frank Harary [150] and Béla Bollobás [48, 47]. The reader can find more on the history of graph theory in Ref. [245].

Problems

1.1   What Is a Graph?

(a) Consider the geographical map of the USA and draw the graph showing which states have a common border. In what ways do this graph and the one considered in Example 1.2 differ from the graph of social acquaintances in Example 1.1?

(b) The so-called *Four Colour Theorem* demonstrates that any map on a plane can be coloured by using only four colours, so that no two adjacent countries have the same colour [313]. Verify that this is true for the geographical map of Europe and for that of the USA.

(c) The regular solids or regular polyhedra, also known as the Platonic solids, are convex polyhedra with equivalent faces composed of congruent convex regular polygons. There are exactly five such solids, namely the tetrahedron (or triangular pyramid), the hexahedron (or cube), the octahedron, the dodecahedron and the icosahedron, as was proved by Euclid in the last proposition of the Elements.

Consider the connectivity among the vertices of each of the five regular solid polyhedra and construct the associated graphs.

1.2 Directed, Weighted and Bipartite Graphs

(a) Construct the friendship network of your class and show it as a graph, as was done for Elisa's kindergarten network in Box 1.2.

(b) Find an expression of the graph reciprocity r (introduced in Section 1.2) in terms of the adjacency matrix A of the graph.

(c) Consider the adjacency matrix:

$$A = \begin{pmatrix} 0 & 1 & 0 & 0 & 0 \\ 0 & 0 & 1 & 1 & 1 \\ 0 & 0 & 0 & 0 & 0 \\ 0 & 0 & 0 & 0 & 1 \\ 1 & 0 & 0 & 0 & 0 \end{pmatrix} \tag{1.3}$$

Is the corresponding graph directed or undirected? Draw the graph.

(d) Consider the $N \times V$ adjacency matrix with $N = 4$ and $V = 6$:

$$A = \begin{pmatrix} 1 & 0 & 1 & 0 & 0 & 1 \\ 1 & 1 & 1 & 1 & 0 & 0 \\ 0 & 1 & 1 & 0 & 1 & 1 \\ 0 & 0 & 1 & 1 & 1 & 0 \end{pmatrix} \tag{1.4}$$

Draw the corresponding bipartite graph. Construct the $N \times N$ matrix $B = AA^\top$ that describes the projection on the first set of nodes. What is the meaning of the off-diagonal terms and that of the diagonal terms in such a matrix? Express, in terms of matrices A and $A^\top$, the $V \times V$ matrix C that describes the projection of the bipartite graph on the second set of nodes.

1.3 Basic Definitions

(a) Imagine you have organised a party with $N = 5$ people. Some of the people at the party know each other already, so that we focus on the party friendship network. In particular, two of them have one friend each at the party, two of them have two friends each, and one has got three friends. Is this network possible? Why? HINT: Prove that in any graph the number of nodes with odd degree must be even.

(b) Is it possible to have a party at which no two people have the same number of friends? Can you support your answer by a mathematical proof?

(c) Consider a bipartite graph with N nodes of the first type and V nodes of the second type. Prove that $\langle k \rangle_{\mathcal{N}} = \frac{V}{N} \langle k \rangle_{\mathcal{V}}$, where $\langle k \rangle_{\mathcal{N}}$ and $\langle k \rangle_{\mathcal{V}}$ are respectively the average degree of nodes of the first and second type.

(d) Prove that a graph with no odd-length cycles is bipartite.

1.4 Trees

(a) In Section 1.4 we have shown that definition (A) implies definition (B). Prove, now, by contradiction, that definition (B) implies definition (C), i.e. that $(B) \Rightarrow$

(C). Finally, prove that $(C) \Rightarrow (A)$. In this way we have shown the perfect equivalence among the three definitions of a tree (A), (B) and (C), given in Definition 1.13.

(b) Prove the following statement: If an acyclic graph G is a tree, and $(x, y) \notin \mathcal{L}$, $\Rightarrow \exists$ cycle in $G + (x, y)$. This is another possible definition of trees equivalent to definitions (A), (B) and (C) given in Section 1.4.

(c) Evaluate the number of components in a *forest* with N nodes and $K = N - c$ links.

1.5 Graph Theory and the Bridges of Königsberg

(a) The Königsberg bridge problem can be stated in terms of a graph instead of a multigraph. We can in fact construct a graph in which the nodes represent each river bank and island, but also each bridge. Show that, as expected, the obtained graph admits neither Euler circuit nor Euler trail.

(b) Show that the total number of walks of length l in a graph described by an adjacency matrix A is: $\sum_{i,j}(A^l)_{ij}$.

(c) Let us denote by A_l, B_l, C_l, D_l the number of walks of length l starting, respectively, from nodes A, B, C and D, in the multigraph of Figure 1.11. Show that the following iterative rules hold:

$$A_{l+1} = 2C_l + D_l$$
$$B_{l+1} = 2C_l + D_l$$
$$C_{l+1} = 2A_l + 2B_l + D_l$$
$$D_{l+1} = A_l + B_l + C_l$$

with $A_0 = B_0 = C_0 = D_0 = 1$. Use a computer to show that the exact number of walks of length 7 is equal to 32554.

(d) As in Problem 1.7(c), consider the five regular polyhedra, namely the tetrahedron, the cube, the octahedron, the dodecahedron and the icosahedron. Can you trace a path along all of the edges of any of these polyhedra without going over any edge twice?

1.6 How to Represent a Graph

(a) Write the incidence matrix and the list of neighbours for the graph described by the adjacency matrix in Eq. (1.3).

(b) Write the ij-form of the adjacency matrix in Eq. (1.3).

(c) Consider the bipartite graph with $N = 4$ and $V = 6$ corresponding to the $N \times V$ matrix A given in Eq. (1.4), and imagine the graph describes how N users have selected V objects. Suppose, now, you want to recommend objects to users by means of the CF method. Use Eqs. (1.1) and (1.2) to compute the recommendation score for each user.

2 Centrality Measures

Centrality measures allow the key elements in a graph to be identified. The concept of centrality and the first related measures were introduced in the context of *social network analysis*, and more recently have been applied to various other fields. In this chapter we introduce and discuss the centrality measures most commonly used in the literature to characterise and rank the nodes of a network. We will first focus on measures of node centrality based on the node degree, such as the *degree centrality*, the *eigenvector centrality* and the α-*centrality*. We will then consider centrality measures based on shortest paths, such as the *closeness centrality* which is related to the average distance of a node from all the other nodes, or the *betweenness centrality* which counts instead the number of shortest paths a node lies on. As only one possible example of the many potential applications, we introduce a large graph describing a real social system, namely the *movie actor collaboration network*, and we use it to identify the most popular movie stars. In particular, we will rank the nodes according to different types of centralities and we will compare the various *centrality rankings* obtained. We conclude the chapter with a discussion on how to extend the measures of centrality from single nodes to groups of nodes.

2.1 The Importance of Being Central

In addition to the developments in mathematical graph theory, the study of networks has seen important achievements in some specialised contexts, as for instance in the social sciences. *Social networks analysis* originated in the early 1920s, and focuses on relationships among social entities, such as communication and collaboration between members of a group, trades among nations, or economic transactions between corporations [308, 278]. This discipline is based on representing a social system as a graph whose nodes are the social individuals or entities, and whose edges represent social interactions. In Figure 2.1 we report three examples of graphs representing different types of interactions, namely marriages between prominent families in Florence (Italy), joint presences at the river in a group of primates, and contacts between terrorists of the September 2001 attacks. Notice that very diverse systems, such as those reported here, can all be well described in terms of graphs.

Example 2.1 *(Three social networks)* In Figure 2.1 we report the graphs representing interactions in three different social systems. The first graph has $N = 16$ nodes and $K = 20$

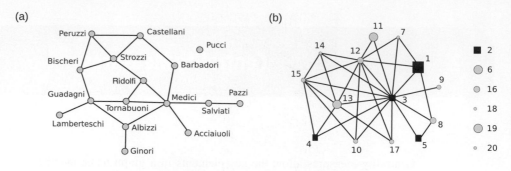

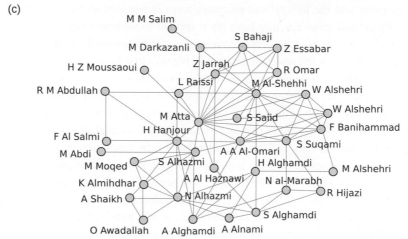

Marital relations between Florentine families (a). Graph of interactions in a group of primates (b). The data set contains also information on the sex (circles represent females) and age (larger size corresponds to older individuals) of each animal. Contact network of hijackers and related terrorists of the September 2001 attacks (c).

edges and describes the marital relations between prominent fifteenth-century Florentine families [308]. Each of the nodes in the graph is a family, and a link between a pair of families exists if a member of one family has married a member of the other. Such a graph is of primary interest for understanding the delicate equilibria between the families. In fact, in the fifteenth century the marital ties were used to solidify political and economical alliances. Hence the graph of marriages is a good proxy to depict the socio-economic relations in Florence in the fifteenth century.

The second graph, with $N = 20$ and $K = 31$, is an example of a social animal interaction network. It represents data recording three months of interactions among a group of 20 primates [108]. The existence or absence of a link was inferred from the count of the number of times each pair of monkeys in a troop was jointly present at the river [108]. This can be seen as an extension to animals of the simple concept that two humans are friends if they go to the pub together! The resulting graph consists of six isolated nodes and a component of 14 nodes.

The third graph with $N = 34$ nodes and $K = 93$ links shows the network of terrorists involved in the September 2001 terrorist attacks on the USA. The nodes represent the 19

Table 2.1 Two socio-economic indicators of the influence of a Florentine family, such as wealth and number of priorates, are correlated with the node degree of the family in the marriage network.

Family	Wealth	Number of priorates	Node degree
Medici	103	53	6
Guadagni	8	21	4
Strozzi	146	74	4
Albizzi	36	65	3
Bischeri	44	12	3
Castellani	20	22	3
Peruzzi	49	42	3
Tornabuoni	48	..	3
Barbadori	55	..	2
Ridolfi	27	38	2
Salviati	10	35	2
Acciaiuoli	10	53	1
Ginori	32	..	1
Lamberteschi	42	0	1
Pazzi	48	..	1
Pucci	3	0	0

hijackers and 15 other associates who were reported to have had direct or indirect interactions with the hijackers. The links in the network stand for contacts reconstructed by using public released information taken from the major newspapers [189].

One of the primary uses of graph theory in social network analysis is the identification of the "most important" actors in a social network. Alex Bavelas and Harold Leavitt, who in the late 1940s conducted a series of experiments at the Group Networks Laboratory at MIT to characterise communication in small groups of people, were the first to notice that important individuals were usually located in strategic locations within the social network [308]. Since these early works, various measures of *structural centrality* have been proposed over the years to describe and measure properties of node location in graphs representing social systems. As a practical example, let us consider the graph of marriages between Florentine families shown in Figure 2.1. For instance, we expect that nodes with a large number of neighbours in the graph play a different role than nodes with just one or a few links. In fact, a family with multiple marital ties, such as the Medici family, has direct relations with many other families, and this can be advantageous also for businesses. In Table 2.1 we report two socio-economic indicators of the importance of each family, together with the associated node degree. In particular, the measure of wealth reported here stands for the net wealth measured in the year 1427, and is coded in thousands of lira, while the number of priorates is the number of seats on the Civic Council held between 1282 and 1344. The families have been sorted in decreasing order of the degree of the corresponding node in the graph. We notice that high-degree nodes often correspond to families with great wealth and with a high number of Civil Count seats. For instance, the only node with six edges corresponds to Medici, the family with the second-greatest wealth, while Strozzi, the family with both the greatest wealth and the largest number of priorates, is one of the two nodes with the second largest value of degree. This is an indication that the degree, as a measure of the structural centrality of a family in the network, is in general positively correlated with the empirical indices of wealth and influence of a family. However, there are also cases, such as that of the Guadagni family, with quite a small wealth but with a large degree in the network of marriages. This is certainly telling us that the degree, although being in general a good proxy of the influence of the family, cannot reproduce all the details. In this chapter we will indeed learn that centrality is a multifaceted concept and that, together with the node degree, there are various other measures to quantify the structural centrality of a node, which captures different ways of being central in the network. But before doing that, we need to return to the concept of graph connectedness.

2.2 Connected Graphs and Irreducible Matrices

A convenient mathematical way to describe the graphs shown in Figure 2.1 is by means of the adjacency matrices. The adjacency matrix of a graph, as we have seen in Section 1.6, is a matrix A whose entry $a_{ij} = 1$ iff there exists the edge (i, j), while $a_{ij} = 0$ otherwise. Before moving to the study of various centrality measures, in this section we discuss an example of how the structural properties of a graph turn into algebraic properties of its adjacency matrix. Namely, we will show that if the graph is connected, its adjacency matrix A is irreducible, while if the graph is unconnected, A is reducible. We start with the definition of reducibility for a general (symmetric or non-symmetric) matrix A.

Definition 2.1 (Reducible-irreducible matrix)   *A matrix A is said to be* reducible *if there exists a $N \times N$ matrix P such that: $P^\top A P = \begin{pmatrix} A_{11} & A_{12} \\ 0 & A_{22} \end{pmatrix}$, with A_{11} and A_{22} square matrices, and where P is a permutation matrix, i.e. a matrix such that each row and each column have exactly one entry equal to 1 and all others 0. Otherwise the matrix is* irreducible.

Notice that reducibility depends only on the sparsity structure of the matrix, i.e. on the location of the non-zero entries of the matrix, not on their values. In particular, if A is the adjacency matrix of a graph, the effect of the permutation matrix P on A corresponds to a permutation of the graph vertex indices $\phi : \mathcal{N} \to \mathcal{N}$. Thus, if the adjacency matrix is reducible, this means that by an appropriate relabelling of the nodes we can put A in a form such that it is evident that some of the nodes are unreachable by other nodes. Let us consider as an example the undirected graph in Figure 2.2. By a visual inspection of the figure, we can immediately conclude that the graph is unconnected. Conversely, this is not evident by simply looking at the adjacency matrix of the graph, which reads:

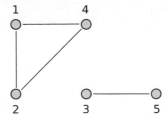

Fig. 2.2　An undirected graph consisting of two components with, respectively, three and two nodes.

$$A = \begin{pmatrix} 0 & 1 & 0 & 1 & 0 \\ 1 & 0 & 0 & 1 & 0 \\ 0 & 0 & 0 & 0 & 1 \\ 1 & 1 & 0 & 0 & 0 \\ 0 & 0 & 1 & 0 & 0 \end{pmatrix}$$

A is a symmetric matrix in this case, because the graph is undirected. But apart from this it has no other visible property from which we can infer that the corresponding graph is unconnected. This is only due to an unfortunate choice of the node labels. In fact, we can find a better labelling of the nodes. If we invert node labels 3 and 4, we get the adjacency matrix:

$$A' = \begin{pmatrix} 0 & 1 & 1 & 0 & 0 \\ 1 & 0 & 1 & 0 & 0 \\ 1 & 1 & 0 & 0 & 0 \\ 0 & 0 & 0 & 0 & 1 \\ 0 & 0 & 0 & 1 & 0 \end{pmatrix}$$

immediately telling us that the group of nodes 1, 2 and 3 does not communicate with the group of nodes 4 and 5. Matrix A' is obtained by switching rows 3 and 4 and columns 3 and 4 of matrix A. Equivalently, we can get A' by left and right multiplying matrix A by the permutation matrix P_{34}:

$$P_{34} = \begin{pmatrix} 1 & 0 & 0 & 0 & 0 \\ 0 & 1 & 0 & 0 & 0 \\ 0 & 0 & 0 & 1 & 0 \\ 0 & 0 & 1 & 0 & 0 \\ 0 & 0 & 0 & 0 & 1 \end{pmatrix},$$

i.e. $A' = P_{34}AP_{34}$. This is equivalent to writing $A' = P_{34}^{\top}AP_{34}$, since we have $P_{34} = P_{34}^{\top}$. Matrix P_{34} is an *elementary permutation matrix*, i.e. a matrix obtained by switching only two rows, or equivalently only two columns, of the identity matrix. Elementary permutation matrices P_{ij} are symmetric and such that $P_{ij}^2 = I$, because if we start with the identity matrix and we switch row i with row j twice we get back the identity matrix. More complex exchanges of indices can be obtained by a composition of elementary matrices P_{ij}, just as any permutation can be obtained by composition of elementary permutations. Notice that, while in the most general case P is not symmetric, it can be proven that $P^{\top} = P^{-1}$ (see Problem 2.2(b)).

Now $A' = \begin{pmatrix} A_{11} & 0 \\ 0 & A_{22} \end{pmatrix}$ is a particular case of the form given in Definition 2.1 with

$A_{12} = \begin{pmatrix} 0 & 0 \\ 0 & 0 \\ 0 & 0 \end{pmatrix}$, which is valid when the adjacency matrix is symmetric, i.e. for undirected graphs. A similar argument holds for directed graphs, with the only differences that A_{12} will in general contain non-zero entries, and A_{11} and A_{22} will in general be non-symmetric. In conclusion, if the adjacency matrix is reducible there are vertices from which one cannot travel to all other vertices. The converse can also be proved, i.e. that if the adjacency matrix of a graph is irreducible, then one can get from any vertex to any other vertex (see Problem 2.3(c)). The latter case, in the realm of undirected (directed) graphs, is called a *connected (strongly connected)* graph. Hence, we can state the following theorem.

> **Theorem 2.1 (Adjacency matrix of connected graphs)**   *An undirected (directed) graph is connected (strongly connected) iff its adjacency matrix A is irreducible.*

While the above discussion makes clear the meaning of irreducibility of the adjacency matrix in terms of graph properties, Definition 2.1 of irreducibility is not very practical in verifying that a given matrix is actually irreducible. See the following example.

Example 2.2   One thing we know for sure is that a matrix with all positive non-diagonal entries, corresponding to a complete graph, is irreducible. Conversely, if we are given the following two matrices:

$$A_B = \begin{pmatrix} 0 & 0 & 1 & 0 \\ 0 & 0 & 0 & 1 \\ 0 & 1 & 0 & 0 \\ 1 & 0 & 0 & 0 \end{pmatrix} \qquad A_C = \begin{pmatrix} 0 & 0 & 1 & 0 \\ 0 & 0 & 1 & 1 \\ 1 & 0 & 0 & 0 \\ 1 & 1 & 0 & 0 \end{pmatrix},$$

it is not immediately obvious that A_B is irreducible while A_C is not. One thing we can do

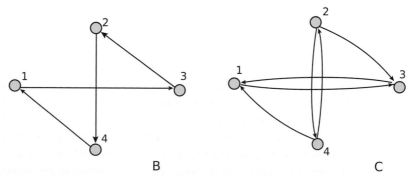

B   C

Fig. 2.3   The two graphs, B and C, respectively corresponding to the adjacency matrices A_B and A_C. The first graph consists of a single strongly connected component, while the second one does not.

is to draw the two corresponding graphs B and C, as in Figure 2.3. Since we have only $N = 4$ nodes, it is easy to recognise that the first graph is strongly connected, while the second is not (for instance, in graph C there is no path from node 1 to node 2). In the case of larger graphs, the visual inspection of the graph is no more of great help. The alternative approach based on Definition 2.1 is useless as well. In fact, we need to keep trying all possible permutations and looking for the zero submatrix A_{21} to appear. For an $N \times N$ matrix there are, of course, $N!$ possible such matrices P. This means only 24 possibilities to explore in our case with $N = 4$ nodes, while there are already over 3 million permutation matrices P to check for $N = 10$, making for a great deal of work.

We thus need a more direct way to determine whether a generic matrix is reducible or irreducible. The following two theorems, valid only for non-negative matrices, are quite helpful in the context of graphs.

Theorem 2.2   *A real non-negative $N \times N$ matrix A is irreducible iff $\forall i, j \in [1, \ldots, N]$, $\exists k = k(i,j) > 0$ such that $(A^k)_{ij} > 0$.*

Proof   Here we give a proof of the theorem based on the correspondence between irreducible matrices and strongly connected graphs. We start from the observation that a non-negative square matrix can be viewed as the adjacency matrix of a weighted graph. Observe that $(A)_{ij} > 0$ iff there is an arc, i.e. a walk of length 1 between node i and node j. Now $(A^2)_{ij} > 0$ iff there is a walk of length 2 from node i to node j. Analogously, $(A^k)_{ij} > 0$ iff there is a walk of length exactly k from i to j. This means that if for all pairs of nodes (i, j), there exists a $k(i,j)$ such that $(A^k)_{ij} > 0$ this means that it is always possible to go from i to j in a finite number of steps, and therefore the weighted graph is connected, and the matrix A is irreducible (see Theorem 2.1). On the other hand, if the non-negative matrix A is irreducible, the corresponding graph is strongly connected, and therefore, for each pair of nodes, there is a walk of a suitable length k that goes from i to j, i.e. $(A^k)_{ij} > 0$.   □

Theorem 2.3   *A real non-negative $N \times N$ matrix A is irreducible iff:*

$$(I + A)^{N-1} > 0. \tag{2.1}$$

Proof   In order to prove the theorem we define $B \equiv (I + A)$ and we show that the (ij) entry of B^m, with $m = 1, 2, \ldots, N - 1$, is different from zero iff there is a walk of length less than or equal to m from node i to node j. Suppose that we start a walk from node i, and we identify the visited nodes by assigning them a non-zero value. The nodes that can be visited starting from node i in one step are those nodes j for which $a_{ij} > 0$. Then the entry $b_{ij} = \delta_{ij} + a_{ij}$ will be non-zero iff it is possible to visit node j from node i in at most one step. This is of course true for node i itself, which is visited in zero steps starting from i. Let us mark these nodes as "visited". This can be done, for example, by using an integer vector and assigning positive values to the nodes marked as "visited" and zero values to the rest of the nodes. Let us denote this vector by $\mathbf{v}$. Then, the vector $\mathbf{u} = B^\top \mathbf{v}$ has non-zero entries

In a directed graph, in-degree and out-degree centrality of node i are respectively defined as:

$$c_i^{D^{in}} = k_i^{in} = \sum_{j=1}^{N} a_{ji}, \qquad c_i^{D^{out}} = k_i^{out} = \sum_{j=1}^{N} a_{ij}, \tag{2.3}$$

where k_i^{in} and k_i^{out} are respectively in- and out-degrees.

The normalised degree centralities C_i^D, $C_i^{D^{in}}$, $C_i^{D^{out}}$ are obtained by dividing the three quantities above by $N - 1$. For instance, in the undirected case we have:

$$C_i^D = \frac{c_i^D}{N - 1} \tag{2.4}$$

The quantity in Eq. (2.2) has the major limitation that comparisons of centrality scores can only be made among the members of the same graph or between graphs of the same size. To overcome this problem, since a given node i can at most be adjacent to $N - 1$ other nodes, it is common to introduce the normalised degree centrality of Eq. (2.4). Such a quantity is independent of the size of the network, and takes values in the range $[0, 1]$.

Example 2.5　The use of the raw degree as in Eq. (2.2) can be misleading, since the degree of a node depends on, among other things, the graph size. A central node with a degree of 25 in a graph of 100 nodes, for example, is not as central as a node with degree of 25 in a graph of 30 nodes, and neither can be easily compared with a central node with a degree 6 in a graph of 10 nodes. The use of the normalised definition in Eq. (2.4), gives a value of centrality respectively equal to 0.25, 0.86, and 0.66, indicating that the node with degree of 25 in a graph of 30 nodes is the most central one.

A degree-based measure of node centrality can be extended beyond direct connections to those at various distances, widening in this way the relevant neighbourhood of a node. For instance, a node i can be assessed for its centrality in terms of the number of nodes that can be reached from it (or can reach it) within a given cut-off distance n. When $n = 1$, this measure is identical to c_i^D. The main problem with such a definition is that, in most cases, the majority of the nodes of a graph (even when the graph is sparse) are at a relatively short path distance. This is the so-called small-world property that will be examined in detail in Chapter 4. Even in a network with, let us say, thousands of nodes, the measure loses significance already for $n = 4$ or $n = 5$, because most, if not all, of the nodes of the graph can be reached at this distance.

A different way to generalise the degree centrality relies on the notion that a node can acquire high centrality either by having a high degree or by being connected to others that themselves are highly central. A simple possibility is to define the centrality c_i of a node i, with $i = 1, 2, \ldots, N$, as the sum of the centralities of its neighbours: $c_i = \sum_{j=1}^{N} a_{ij} c_j$. In practice, we get a set of N linear equations, which can be written in matrix form as

$\mathbf{c} = A\mathbf{c}$ or $(A - I)\mathbf{c} = 0$, where the column vector $\mathbf{c}$ is the unknown centrality score. Being a homogeneous system, this system of equations always admits the trivial solution $\mathbf{c} = 0$, which of course we are not interested in. Moreover, if $\det(A - I) \neq 0$, then the zero solution is the only one. Therefore, these equations would produce a meaningful definition of centrality only under the unlikely condition that $\det(A - I) = 0$, that is in a very limited number of cases. The standard way to overcome this problem is to define the centrality of a node as *proportional* to the sum of the centralities of its neighbours. This is equivalent to multiplying the left-hand side of the equations by a constant λ. This modification was first proposed by Phillip Bonacich back in 1972, and allows a solution to the equations that does not violate the original motivation [52]. We finally have:

$$\lambda c_i^E = \sum_{j=1}^{N} a_{ij} c_j^E \qquad (2.5)$$

where $i = 1, 2, \cdots, N$, and c_i^E is known as the *Bonacich centrality* or alternatively as the *eigenvector centrality* of node i. This is because, in matrix notation, Equations (2.5) can be written as:

$$A\mathbf{c}^E = \lambda \mathbf{c}^E \qquad (2.6)$$

where $\mathbf{c}^E$ is an N-dimensional column vector whose entry i is equal to c_i^E. And this is the familiar problem of finding eigenvalues and eigenvectors: λ is an eigenvalue and $\mathbf{c}^E$ is an eigenvector of the adjacency matrix A.

Of course, the definition in Eq. (2.6) is valid for an undirected graph. If the graph is directed, as for the case of the degree centrality, we have to consider whether we are interested in the ingoing or in the outgoing links [53]. We therefore have two possible definitions, the *in-degree eigenvector centrality* and the *out-degree eigenvector centrality*:

$$\lambda c_i^{E^{\text{in}}} = \sum_{j=1}^{N} a_{ji} c_j^{E^{\text{in}}} \qquad \lambda c_i^{E^{\text{out}}} = \sum_{j=1}^{N} a_{ij} c_j^{E^{\text{out}}} \qquad (2.7)$$

or in vectorial notations:

$$A^{\top} \mathbf{c}^{E^{\text{in}}} = \lambda \mathbf{c}^{E^{\text{in}}} \qquad A\mathbf{c}^{E^{\text{out}}} = \lambda \mathbf{c}^{E^{\text{out}}} \qquad (2.8)$$

The formulae above explain the name given to these centrality measures, but not how to calculate them. In fact, a generic matrix A has N (complex) eigenvalues (counted with the proper multiplicity) $\lambda_1, \ldots, \lambda_N$, and, if it is diagonalisable, it has a set of N linearly independent eigenvectors, that we shall denote by $\mathbf{u}_1, \ldots, \mathbf{u}_N$. For more details about the computation of the eigenvalues and eigenvector of a matrix, see Appendix A.5.1. The question remains as to whether there is a desirable solution. It is intended that λ in Eq. (2.5) should be positive, and that also each c_i^E should be positive. Moreover, there should be just one solution and we should not be faced with an arbitrary choice among conflicting eigenvectors.

To overcome this problem we shall make use of the theoretical results of Perron and Frobenius for non-negative matrices which tell us that, in the case of a connected graph, we can measure centrality by the *eigenvector corresponding to the leading eigenvalue* because its components are all positive real numbers. Let us begin by stating the famous theorem by Oskar Perron [256] and Georg Frobenius [125][1].

Theorem 2.4 (Perron–Frobenius for irreducible matrices)   *If A is an N × N, non-negative, irreducible matrix, then:*

- *one of its eigenvalues is positive and greater than or equal to (in absolute value) all other eigenvalues;*
- *such eigenvalue is a simple root of the characteristic equation of A;*
- *there is a positive eigenvector corresponding to that eigenvalue.*

In practice, for an undirected connected or a directed strongly connected graph, the Perron–Frobenius Theorem 2.4 guarantees the existence of an eigenvector with all positive components. Such an eigenvector, let us call it $\mathbf{u}_1$, is the eigenvector associated with the leading eigenvalue of the adjacency matrix A. We can then therefore give the following definitions:

Definition 2.3 (Bonacich eigenvector centrality)   *In an undirected connected graph, the eigenvector centrality c_i^E of node i is defined as:*

$$c_i^E = u_{1,i} \tag{2.9}$$

where $u_{1,i}$ is the ith component of $\mathbf{u}_1$, the eigenvector associated with the leading eigenvalue λ_1 of the adjacency matrix A, namely the vector that satisfies: $A\mathbf{u}_1 = \lambda_1 \mathbf{u}_1$. In a directed strongly connected graph, the in- and the out-degree eigenvector centralities $c_i^{E^{in}}$ and $c_i^{E^{out}}$ of node i are respectively equal to the ith component of the eigenvectors associated with the leading eigenvalue λ_1 of matrix $A^\top$, and of matrix A. The normalised eigenvector centrality for an undirected connected graph is defined as $C_i^E = c_i^E / \sum_i c_i^E$. Analogous definitions are used for the normalised in- and the out-degree eigenvector centralities of a directed strongly connected graph.

It is easy to show that, if two nodes of the graph, say i and j, share the same set of neighbours, then we have $u_{1,i} = u_{1,j}$. This means that the two nodes not only have the same degree centrality, but they also have exactly the same eigenvector centrality. In the following example we show that two such nodes have in fact equal entries for the corresponding components of all eigenvectors, i.e. $u_{\beta,i} = u_{\beta,j}$ for all $\beta = 1, \ldots, N$.

---

[1]   In 1907 Perron first proved that positive matrices have a single positive eigenvalue r which is a simple root of the characteristic equation, and exceeds in absolute value all other eigenvalues. The corresponding eigenvector has non-zero components, all with the same sign (therefore they can be chosen to be positive). Later, Frobenius extended this result to irreducible non-negative matrices, of which positive matrices are a particular case. Further investigation led Frobenius to the discovery of remarkable spectral properties of irreducible non-negative matrices, and to the definition of primitive matrix, and their relation to graph theory. A more detailed treatment of the theorems by Perron and Frobenius can be found, for example, in the classical books by Gantmacher [126] or by Varga [301].

Example 2.6 *(Spectral coalescence)*   Let us consider a directed graph such that nodes 1 and 2 have exactly the same outgoing neighbours. This means that rows 1 and 2 of the adjacency matrix are the same: $a_{1j} = a_{2j}$, $j = 1, \ldots, N$. The first consequence of this is that one eigenvalue of the matrix is zero, and the corresponding left eigenvector is given by $\mathbf{v}^\top = (1, -1, 0, \ldots, 0)$. Another consequence is that the first two components of all right eigenvectors corresponding to the non-zero eigenvalues are equal. In fact, we have

$$\sum_j a_{1j} \mathbf{u}_{\beta,j} = \lambda_\beta \mathbf{u}_{\beta,1}$$

$$\sum_j a_{2j} \mathbf{u}_{\beta,j} = \lambda_\beta \mathbf{u}_{\beta,2}$$

for any $\beta = 1, \ldots, N$. By subtracting the second equation from the first one, we get:

$$0 = \lambda_\beta (\mathbf{u}_{\beta,1} - \mathbf{u}_{\beta,2})$$

Therefore, if $\lambda_\beta \neq 0$, one has $\mathbf{u}_{\beta,1} = \mathbf{u}_{\beta,2}$. The same property holds for the left eigenvectors if the two nodes 1 and 2 have exactly the same ingoing neighbours.

The previous example shows that nodes with the same set of neighbours all have the same degree centrality and also the same eigenvector centrality. In general, we expect the value of eigenvector centrality of a node to be positively correlated to its degree centrality. However, there are cases when the node rankings induced by the two centrality measures may differ, as shown in the following example, where we have made use of the *power method* described in Appendix A.5 to compute the eigenvector centralities for two of the networks in Figure 2.1.

Example 2.7 *(Computation of the eigenvector centrality)*   The terrorist network reported in Figure 2.1 is connected. Therefore, Perron–Frobenius Theorem 2.4 guarantees that the eigenvector centrality is well defined in this case, since all the components of the eigenvector associated with the largest eigenvalue are positive. We used the power method described in Appendix A.5 to find the largest eigenvalue $\lambda_1 = 7.5988$, and the associated positive eigenvector $\mathbf{u}_1$. From the components of $\mathbf{u}_1$ we can extract the eigenvector centralities $C_i^E, i = 1, \ldots, 34$ of the corresponding nodes. The ten nodes with the highest scores are reported and ranked according to their value of C^E (shown in parenthesis, together with the node degree). We have: Mohamed Atta (0.4168, 16); Marwan Al-Shehhi (0.3921, 14); Abdul Aziz Al-Omari (0.2914, 9); Satam Suqami (0.2490, 8); Fayez Banihammad (0.2415, 7); Wail Alshehri (0.2361, 6); Ziad Jarrah (0.2182, 7); Waleed Alshehri (0.2031, 6); Said Bahaji (0.2020, 6); Hani Hanjour (0.1966, 10). Although high values of C^E correspond in general to high degrees, and vice versa, there are some notable deviations. For instance, among the ten nodes with the largest C^E there are nodes with only six links, while Hani Hanjour, who is the third most important node by degree, is only tenth in the list by eigenvector centrality. Notice that for the case of the terrorist network, the power method converged because the second largest eigenvalue, is equal to -3.7247, and therefore, in absolute value, $\lambda_1 > |\lambda_2|$ (see Appendix A.5).

Analogously, we have computed the eigenvector centralities of the Florentine families. We get $\lambda_1 = 3.2561$, and the associated eigenvector $\mathbf{u}_1$ produces the following ranking: Medici (0.1229, 6); Strozzi (0.1016, 4); Ridolfi (0.0975, 2); Tornabuoni (0.0930, 3); Guadagni (0.0825, 4); Bischeri (0.0807, 3); Peruzzi (0.0787, 3); Castellani (0.0740, 3); Albizzi (0.0696, 3); Barbadori (0.0604, 2); Salviati (0.0417, 2); Acciaiuoli (0.0377, 1); Lamberteschi (0.0254, 1); Ginori (0.0214, 1); Pazzi (0.0128, 1); Pucci (0.0, 0). By comparing C_i^E with the degree k_i we notice that the Ridolfi family, who is third by eigenvector centrality, has a low degree equal to 2. Since the graph of Florentine families is not connected, the Perron–Frobenius Theorem 2.4 is not valid. Despite that, the eigenvector $\mathbf{u}_1$ gives reasonable values of centrality. The only price to pay is that the isolated node corresponding to the Pucci family has $C^E = 0$. Again, the power method converged because we have: $3.2561 = \lambda_1 > |\lambda_2| = |-2.6958|$.

Notice that the Perron–Frobenius theorem for irreducible matrices does not guarantee strict inequality between the largest and the second largest eigenvalue (in absolute value). This is a necessary condition for the convergence of the power method, as discussed in Appendix A.5. Two simple examples, of an undirected connected graph, and of a directed strongly connected graph, for which the power method does not converge, are analysed respectively in Problems 2.3(a) and (b). If we want to make use of the power method, we need to ensure the existence of a single eigenvalue with maximum absolute value. An additional hypothesis is thus required: the adjacency matrix has to be *primitive*.

Definition 2.4 (Primitive matrix) *An $N \times N$ matrix A is said to be primitive if there exists some integer exponent $k > 0$ for which $A^k > 0$; that is, if there exists some integer exponent $k > 0$ such that all the entries of the matrix A^k are positive.*

In practice, to prove that a matrix A is primitive we have to check whether all the entries of A are positive. If this is not true, then we check whether all the entries of A^2 are positive. Again, if this is not true we pass to A^3 and so on until we find a value of k such that $A^k > 0$. Fortunately we do not have to check the sign of matrix A^n for infinitely large n. In fact, it has been proven by Wielandt in 1950 that, if A is primitive, then the power n which should have all positive entries is less than or equal to $N^2 - 2N + 2$ [315]. The Perron–Frobenius theorem can now be formulated for primitive matrices as follows:

Theorem 2.5 (Perron–Frobenius for primitive matrices) *If A is an $N \times N$, non-negative, primitive matrix, then:*

- *one of its eigenvalues is positive and greater than all other eigenvalues (in absolute value);*
- *that eigenvalue is a simple root of the characteristic equation of A;*
- *there is a positive eigenvector corresponding to that eigenvalue.*

Summing up, the Perron–Frobenius Theorem 2.4 for irreducible matrices assures that the eigenvector centrality is well-defined for a connected (strongly connected) graph, since the adjacency matrix has a single positive eigenvector corresponding to the maximum eigenvalue. Furthermore, if the adjacency matrix is primitive, the Perron–Frobenius Theorem 2.5 for primitive matrices guarantees that the power method for the determination of the eigenvector corresponding to the largest positive eigenvalue converges.

Example 2.8   Let us consider again the networks of Figure 2.1. The adjacency matrix A corresponding to the terrorist network is primitive. In fact, it easy to prove that $A^k > 0$ with $k = 5$. Now, Perron–Frobenius Theorem 2.5 helps us understand why the power method converged in Example 2.7. Instead, the adjacency matrix for the Florentine family and for the primate interaction network are not primitive, because they contain lines of zero, corresponding to single isolated nodes. More generally, when a graph is not connected, the adjacency matrix is not primitive. In fact, it is possible to prove that a primitive matrix is irreducible. See Problem 2.3(d).

Can we extend the concept of eigenvector centrality to the case of an undirected graph that is not connected, or to the case of a directed graph that is not strongly connected? First of all, what happens if we compute the eigenvector associated with the largest eigenvalue of the adjacency matrix of an unconnected graph? We have already done that in Example 2.7, when we have computed the eigenvector centrality for the Florentine families network, which contains an isolated nodes. In that particular case we have found a zero eigenvector centrality for the isolated node. There are, however, cases where the computation of the eigenvector centrality in unconnected networks produces inconsistent results, as in the following Example [53].

Example 2.9   Consider the directed network shown in the figure. The graph is not strongly connected. Nevertheless, we can try to evaluate the Bonacich centrality by computing the

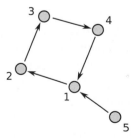

eigenvectors of the adjacency matrix of the graph. Let us consider the case of the in-degree eigenvector centrality. The associated eigenvector problem to solve is $A^\top \mathbf{u} = \lambda \mathbf{u}$, namely:

$$\begin{pmatrix} 0 & 0 & 0 & 1 & 1 \\ 1 & 0 & 0 & 0 & 0 \\ 0 & 1 & 0 & 0 & 0 \\ 0 & 0 & 1 & 0 & 0 \\ 0 & 0 & 0 & 0 & 0 \end{pmatrix} \begin{pmatrix} u_1 \\ u_2 \\ u_3 \\ u_4 \\ u_5 \end{pmatrix} = \lambda \begin{pmatrix} u_1 \\ u_2 \\ u_3 \\ u_4 \\ u_5 \end{pmatrix}$$

We find five different eigenvalues, namely $\lambda = 0, \pm 1, \pm i$. If we look at the eigenvector associated with the eigenvalue $\lambda = 1$, we find that the component corresponding to node 5 is equal to 0, while the other components are all equal. This implies that 1, 2, 3 and 4 have the same Bonacich centrality, a result that is certainly not desirable. In fact, node 1 is clearly different from nodes 2, 3 and 4, because it receives an extra link from node 5, and has a larger in-degree than all the other three nodes. However, node 5 has no in-going links and thus no centrality to contribute to 1. This is certainly a drawback of using the eigenvectors of the adjacency matrix of an unconnected graph to measure the centrality of the graph nodes.

One way out of the problem illustrated in Example 2.9 is to add an intrinsic centrality $\mathbf{e}$ to each of the nodes of the graph. We can therefore replace Eq. (2.8) with:

$$\mathbf{c} = \alpha A^\top \mathbf{c} + \mathbf{e}$$

where $\mathbf{e}$ is a vector with all equal components that, without loss of generality, we can assume to be equal to $\mathbf{1}$. Moreover, thanks to the addition of vector $\mathbf{1}$, we are not forced anymore to fix α to take a precise value, but we can leave it as a free parameter to tune. Such a generalisation of the Bonacich centrality is known as α-centrality [53].

Definition 2.5 (α-centrality)   *In a graph (undirected or directed) the α-centrality c_i^α of node i is defined as the ith component of vector:*

$$c^\alpha = \left(I - \alpha A^\top \right)^{-1} \mathbf{1} \qquad (2.10)$$

where I is the identity matrix and $0 < \alpha \leq 1$. The normalised α-centrality is defined as $C_i^\alpha = c_i^\alpha / \sum_i c_i^\alpha$.

The value of the parameter α reflects the relative importance of the graphs structure, namely the adjacency matrix, versus the exogenous factor corresponding to the addition of the vector $\mathbf{1}$, in the determination of the centrality scores. Of course, we are not considering the case $\alpha = 0$ because it would produces equal scores for all nodes, no matter the structure of the graph. Also the usual convention is to consider $\alpha \leq 1$, and to eventually avoid values of α such that the matrix $I - \alpha A^\top$ is not invertible. The intuitive meaning of the α-centrality will be clear in Chapter 8, when we will introduce another measure of centrality based on the number of walks of different length between two nodes.

Example 2.10   Let us now see how the α-centrality works for the graph in Example 2.9, the same case where we ran into troubles with the Bonacich centrality. For such a graph, we have to avoid the value $\alpha = 1$, when matrix $I - \alpha A^\top$ is not invertible. For $0 < \alpha < 1$ the solution to Eq. (2.10) reads:

$$\mathbf{c}^{\alpha} = \frac{1}{1 - \alpha^4} \begin{pmatrix} 1 + 2\alpha + \alpha^2 + \alpha^3 \\ 1 + \alpha + 2\alpha^2 + \alpha^3 \\ 1 + \alpha + \alpha^2 + 2\alpha^3 \\ 1 + \alpha + \alpha^2 + \alpha^3 + \alpha^4 \\ 1 \end{pmatrix}$$

The order of centrality produced this time is in agreement with what commonly expected. It is in fact easy to prove that $c_1 > c_2 > c_3 > c_4 > c_5$ for any value $0 \le \alpha < 1$. The α-centrality is a good option when we want to deal with unconnected graphs. In the most general case it is also possible that the ranking of the nodes produced by the α-centrality changes with the value of α, as we will see in Problem 2.3(g).

2.4 Measures Based on Shortest Paths

In the previous section we considered centrality measures depending on the node degree. We now turn our attention to another class of measures of centrality, those based on paths between pairs of nodes. We will see three different examples of such measures, such as the *closeness centrality*, based on the *lengths* of the shortest paths from a vertex; the *betweenness centrality*, counting the *number* of shortest paths a vertex lies on; and the *delta centrality*, based both on the number and on the length of shortest paths. In all cases, we need to be able to find shortest paths in a graph. In Appendix A.6 we describe the so-called *Breadth-First Search* (BFS), an optimal algorithm to compute the shortest paths from *a given source node, i, to all the other nodes* of an undirected or a directed graph. The BFS algorithm stores the lengths of the shortest paths, and also keeps track of the precise sequence of nodes along the shortest paths from i to each of the other nodes of the graph. One interesting and useful feature of the algorithm is that, for pairs of nodes i and j such that the shortest path is degenerate, i.e. there is more than one path from i to j with the same minimum length, the algorithm stores all of them.

In Appendix A.6 we show that the time complexity of BFS is $O(N + K)$ for unweighted graphs, and $O(N + K \log K)$ for weighted graphs. Sometimes, we have to compute the shortest paths for each pair of nodes. This problem can be solved by iterating the BFS algorithm over the N starting nodes of the graph, which has time complexity $O(N(N + K))$ for unweighted graphs, and $O(N(N + K \log K))$ for weighted graphs. In a sparse graph this corresponds respectively to time $O(N^2)$ and $O(N^2 \log N)$. We represent the obtained length of shortest paths in the so-called *distance matrix* $\mathcal{D} = \{d_{ij}\}$. The entry d_{ij} of $\mathcal{D}$ gives the distance from node i to node j when $i \ne j$. Usually one sets $d_{ii} = 0 \; \forall i$. Moreover, one usually assumes that $d_{ij} = \infty$ if the two nodes i and j are not connected, even if for practical reasons all the algorithms set $d_{ij} = N$ for pairs of unconnected nodes.

Closeness Centrality

The first measure is based on the idea that an individual that is close to the other individuals of a social network is central because he can quickly interact with them. The simplest way to quantify the centrality of a node is therefore to consider the sum of the geodesic distances from that node to all other nodes in the graph. A node with a low sum distance is on average close to the other nodes. We can then introduce the following centrality measure [123]:

Definition 2.6 (Closeness centrality) *In a connected graph, the closeness centrality of node i is defined as the reciprocal of the sum of distances from i to the other nodes:*

$$c_i^C = \frac{1}{\sum_{j=1}^{N} d_{ij}} \tag{2.11}$$

The normalised quantity, defined as:

$$C_i^C = (N-1)c_i^C \tag{2.12}$$

takes values in the range $[0, 1]$.

To understand the normalisation, consider that in the best possible case $d_{ij} = 1 \forall j \neq i$. Therefore the largest possible value of c_i^C is equal to $1/(N-1)$, and we have to consider the quantity $(N-1)c_i^C$ to fall within the interval $[0, 1]$. Usually the definition of closeness is given for connected graphs only. In fact, in unconnected graphs, there are pairs of nodes that cannot be joined by a path. For such pairs of nodes, let us say nodes i and j, the geodesic length d_{ij} is not a finite number. In order to extend Definition 2.6 also to graphs with more than one component, whenever there is no path between nodes i and j we need to artificially assume d_{ij} is equal to a finite value, for instance equal to N, a distance just greater than the maximum possible distance between two nodes.

Example 2.11 Let us evaluate closeness centrality for the small graph with $N = 7$ nodes shown in the figure. Suppose we want to compare the centrality of node 1 to that of node 4. To calculate c_1^C we need the distances between node 1 and all the other nodes of the graph. It is easy to see that $d_{12} = d_{13} = 1$, $d_{14} = d_{15} = 2$, $d_{16} = 3$ and $d_{17} = 4$. We therefore have $c_1^C = 1/13$ and $C_1^C = (N-1)c_1^C = 6/13$. For node 4, we have instead

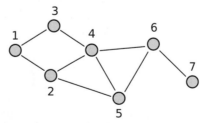

$d_{42} = d_{43} = d_{46} = 1$ and $d_{41} = d_{45} = d_{47} = 2$. Hence, $c_4^C = 1/9$ and $C_1^C = 2/3$. In conclusion, node 4 is more central than node 1, as is also clear from a visual inspection of the graph.

Let us consider the graph in Figure 2.4. This is a tree with $N = 16$ nodes, and is therefore a particularly simple case where to evaluate distances and compare to the degree centrality

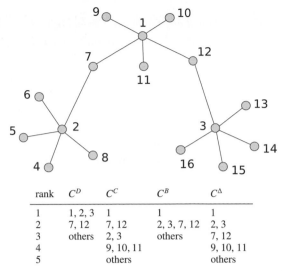

rank	C^D	C^C	C^B	C^Δ
1	1, 2, 3	1	1	1
2	7, 12	7, 12	2, 3, 7, 12	2, 3
3	others	2, 3	others	7, 12
4		9, 10, 11		9, 10, 11
5		others		others

Fig. 2.4 A tree with $N = 16$ nodes, and the ranking of its nodes obtained according to various centrality measures, namely degree, closeness, betweenness and delta centrality.

various measures based on shortest paths. The nodes with the highest degree centrality in the tree are nodes 1, 2 and 3. These three nodes have the same value of C^D, equal to 0.33. Closeness centrality is instead able to differentiate the three nodes. In fact node 1 has a higher closeness than nodes 2 and 3. Moreover, the two nodes 7 and 12 are less central than node 1, but more central than either 2 and 3. This is shown in the associated table, where nodes are rank-ordered in decreasing order of centrality according to degree, closeness and two other centrality measures we are going to define in this section.

Betweenness Centrality

It is frequently the case that the interactions between two non-adjacent individuals depend on the other individuals, especially on those on the paths between the two. Therefore, whoever is in the middle can have a certain strategic control and influence on the others. Betweenness centralities try to capture this idea: a node in a graph is central if it lies between many of the nodes. The simplest measure of betweenness was introduced in 1977 by Linton Freeman. Such a measure is based on the simplification that communication travels *just* along shortest paths, and is evaluated by counting the number of shortest paths between pairs of vertices running through a given node [122, 123].

Definition 2.7 (Betweenness centrality) *In a connected graph, the shortest-path betweenness centrality of node i is defined as:*

$$c_i^B = \sum_{\substack{j=1 \\ j \neq i}}^{N} \sum_{\substack{k=1 \\ k \neq i,j}}^{N} \frac{n_{jk}(i)}{n_{jk}} \tag{2.13}$$

> where n_{jk} is the number of geodesics from node j to node k, whereas $n_{jk}(i)$ is the number
> of geodesics from node j to k, containing node i. The normalised quantity, defined as:
>
> $$C_i^B = \frac{c_i^B}{(N-1)(N-2)} \tag{2.14}$$
>
> takes values in the range $[0, 1]$.

The indices j and k in the double summation must be different, and different from i. Notice also that the number of shortest paths from j to k, n_{jk}, can be larger than one, since the shortest path can be degenerate. Finally, the factor $(N-1)(N-2)$ in Definition 2.14 is the maximum possible number of geodesics going through a node: in this way, the quantity C_i^B assumes its maximum value 1 when node i belongs to all geodesics between pairs of vertices. The above definition can be extended to arbitrary graphs, assuming that the ratio $n_{jk}(i)/n_{jk}$ is equal to zero if j and k are not connected.

Example 2.12   It can be very useful to work out the calculation of the betweenness in a simple case, such as the graph considered in Example 2.11. Now the calculation is a bit more complicated than that of the closeness centrality. In order to calculate the betweenness centrality of node i we need, in fact, to consider the shortest paths between all pairs of nodes, except node i. Moreover, we are not just interested in the lengths of the shortest paths, but on checking whether the shortest paths pass by node i or not. Suppose we want to evaluate c_4^B. The graph is undirected, so we can consider only $(N-1)(N-2)/2 = 15$ pairs of nodes, because we do not need to repeat the same pair twice. Let us start with all the shortest paths from node 1. The shortest paths to nodes 2, 3 and 5 do not pass by node 4, so their contributions to the summations in Eq. (2.13) is zero. Conversely, the pair $(1, 6)$ contributes a term equal to 2/3, since of the three shortest paths between nodes 1 and 6, two pass by node 4. Same thing for the pair $(1, 7)$. Moving to the shortest paths from node 2 to the other graph nodes, we notice that only the pairs $(2, 3)$, $(2, 6)$ and $(2, 7)$ contribute a non-zero term, equal to 1/2 in all three cases, because it is always one shortest path over two passing from node 4. We finally have $c_4^B = 2/3 + 2/3 + 1/2 + 1/2 + 1/2 + 1/2 + 1 + 1 = 16/3$ and $C_4^B = 16/45$, indicating that about 1/3 of the shortest paths make use of node 4.

In Figure 2.4 we report the results obtained by evaluating the node betweenness for the tree with $N = 16$ nodes. The betweenness centrality ranks first node 1, then four nodes, namely 2, 3, 7, 12, and finally all the remaining nodes, hence providing a different ranking than that based on the degree. The network considered here is small and has no cycles, so that shortest paths are unique and can be found even without a computer. In Appendix A.7 we discuss in detail fast numerical algorithms that allow the shortest-path betweenness centrality of all the nodes in a graph to be calculated in time $O(N(K + N))$ for unweighted graphs, and in time $O(N(N + K * logK))$ for weighted graphs. In Section 2.5 we will make use of such algorithms to compute betweenness in much larger networks than that in Figure 2.4.

In most of the cases, communication in a social network does not travel only through geodesic paths. The definition of betweenness can be then modified accordingly, depending on the kind of flow that is assumed to be more appropriate to the network considered. Various measures of betweenness, such as random-walk betweenness or flow betweenness, can be found in the literature. Some are listed in the following example.

Example 2.13   *(Beyond shortest-paths)*   *Random-walk betweenness*, as indicated by the name, is based on random walks. Such a measure is more suited than shortest-path betweenness in all cases in which we want to mimic something propagating on a network with no global knowledge of the network. In fact, in absence of information, at each step, a walker chooses where to go at random from all the possibilities. The random walk betweenness of node i is equal to the number of times that a random walker passes through i in its journey, averaged over a large number of realisations of the random walk [240]. *Flow betweenness* is instead based on the idea that each edge of the graph is like a pipe, and that the quantity of interest moves as a liquid [124]. Finally, the so-called *node load* introduced in Refs. [132, 130] is based on considering the transportation of information data packets over the network. To calculate the load, one assumes that each node sends a unit packet to every other node. The packets are transferred from the source to the target only along the shortest paths between them, and are divided evenly upon encountering any branching point. To be precise, let $\ell_i^{s \to t}$ be the number of packets sent from s (source) to t (target) that pass through node i. Then the load of a node i, ℓ_i, is the accumulated sum of $\ell_i^{s \to t}$ for all s and t, $\ell_i = \sum_s \sum_{t \neq s} \ell_i^{s \to t}$. An example is shown in the following figure. The load at each node due to a unit packet transfer from

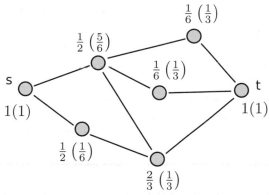

the node s to the node t, is reported. The quantity in parentheses is the corresponding value of the load due to the packet from t to s, $\ell_i^{t \to s}$. Notice that, in general, $\ell_i^{s \to t} \neq \ell_i^{t \to s}$, while the number $n_{jk}(i)$ of shortest paths between j and k containing i used in Definition 2.13 does not depend on the order of the two nodes i and j.

Delta Centrality

The definition of centrality we consider here is based on the idea that the importance of an individual can be measured by its contribution to the performance of the whole network.

Vito Latora and Massimo Marchiori have proposed to measure the so-called delta centrality of a node of graph G, by comparing the performance $P[G]$ of graph G to the performance $P[G']$ of graph G', obtained by deactivating the node from G [199].

Definition 2.8 (Delta centrality) *In a graph G, the delta centrality of node i is defined as:*

$$C_i^\Delta = \frac{(\Delta P)_i}{P} = \frac{P[G] - P[G']}{P[G]} \tag{2.15}$$

where P is a generic quantity measuring the performance of the graph, and $(\Delta P)_i$ is the variation of the performance under the deactivation of node i, i.e. the removal from the graph of the edges incident in node i. The performance P must satisfy the requirement $(\Delta P)_i \geq 0$ for any node i of the graph.

More precisely, in the definition above, G indicates the original graph, while G' is the graph with N nodes and $K - k_i$ edges obtained by removing from G the k_i edges incident in node i. Notice that the quantity C_i^Δ is by definition normalised to take values in the interval $[0,1]$. Of course, the crucial point in Eq. (2.15) is how to measure the performance P of a graph. Since there are several possible ways to define the performance P of a network, there are consequently several different measures of delta centrality C^Δ. Eventually, the meaning and the value of the resulting delta centrality will depend on the choice of P. The simplest possibility is to take $P[G] \equiv K$, where K is the number of edges in the graph G. With such a choice we get $C_i^\Delta = k_i/K$, and the delta centrality C_i^Δ differs from the degree centrality C_i^D only because of a different normalisation factor. Another interesting possibility is to set the performance $P[G]$ in Eq. (2.15) equal to the so-called *efficiency* $E[G]$ of graph G [202, 203].

Definition 2.9 (Efficiency) *Given two nodes i and j of a graph G, we assume that the efficiency ϵ_{ij} in the communication between i and j is equal to the inverse of their distance d_{ij}, namely $\epsilon_{ij} = 1/d_{ij}$. We then define the graph efficiency $E \equiv E[G]$ as:*

$$E = \frac{1}{N(N-1)} \sum_{i=1}^{N} \sum_{\substack{j=1 \\ j \neq i}}^{N} \frac{1}{d_{ij}} \tag{2.16}$$

that is the average of ϵ_{ij} over all pairs of nodes i and j, with $i \neq j$.

The quantity E has the nice property of being finite even for unconnected graphs, with no artificial assumption on the distance of two unconnected nodes. In fact if i and j are not connected we can set $d_{ij} = \infty$, which implies that the efficiency in the communication ϵ_{ij} of the two nodes is equal to zero. And this zero value can be taken into account with no problems in Eq. (2.16) in the calculation of the average quantity E measuring the efficiency of the graph. The definition of delta centrality we adopt from now on is that based on the efficiency. With such a measure of performance, the delta centrality of node i corresponds to the relative drop in the network efficiency caused by the deactivation of i from G. Consequently, the definition of C^Δ will depend both on the number and on the length of shortest paths. In fact, when a node is deactivated, the shortest paths will change: the length of some of them will increase and, eventually, some pairs of nodes will become disconnected. It is immediately clear that C^Δ is positively correlated to the degree centrality C^D, and also to the other two measures of centrality based on shortest paths, namely C^C and C^B, defined in this section. In fact, the delta centrality of node i depends on k_i, since the efficiency $E[G']$ is smaller when a larger number of edges is removed from the graph. C_i^Δ is related to C_i^C since the efficiency of a graph is related to the inverse of $\sum_i \sum_j d_{ij}$. Finally C_i^Δ, similarly to C_i^B, depends on the number of geodesics passing by i, but it also depends on the lengths of the new geodesics, the alternative paths of communication in the graph G' in which node i is deactivated. In this sense, the delta centrality is more sensitive than C_i^B and C_i^C, since both measures are uninfluenced by the rearrangement of the shortest paths due to the removal of the links of node i.

In the tree of Figure 2.4, delta centrality and closeness centrality have the best resolution, being able to differentiate the nodes into five different levels of centrality. The only difference between C^Δ and C^C is the inversion of the ranking of two pairs of nodes. In fact, the closeness centrality ranks the pair of nodes 7 and 12 second, while the delta centrality ranks the pair 2, 3 second, because these two nodes are the centres of two stars of five nodes each.

Comparing Centrality Rankings

Since there are many different ways to evaluate node centrality and to rank, accordingly, the nodes of a network by their importance, it can be very handy to have a quantitative way to determine the similarity of two centrality orderings. This is indeed the same problem you face every time you need to compare two or more *ranked lists* (see Box 2.1). A ranked list is any ordered set of objects, which can be products, individuals, tennis players, football teams, or the nodes of a network as in our case. Although a ranked list is often obtained from the observed values of a given variable which measures a quality the objects possess to a varying extent, we are mainly interested in comparing the rankings produced by two or more variables independently by the actual values taken by the variables. Quantities such as the Spearman's or the Kendall's coefficients allow us to do exactly this [176]. These are in fact *rank correlation coefficients*, i.e. correlation coefficients that depend only on the ranks of the variables, and not on their observed values.

Let us denote as $r_\bullet$ and $q_\bullet$ the two rankings we want to compare, e.g. those produced respectively by two centrality measures, C^X and C^Y. For each node i, r_i is a number indicating the position of node i in the ranking induced by C^X, while q_i is the position according to the other centrality measure C^Y. For instance, values $r_i = 1$ and $q_i = 3$ mean that node i is the most central node in the network according to C^X, while it is only the third in the list, according to C^Y. Notice that usually the rank $r_\bullet$ induced by a centrality C^X is defined to be natural number only if there is no degeneracy, i.e. if there are no two nodes with the same value of centrality. Conversely, if there are u nodes all with the same ℓth largest value of centrality C^X, their rank will be set to $r_i = \frac{1}{u} \left(\sum_{p=\ell}^{\ell+u-1} p \right)$, i.e. to their average position. For example, in the ranking induced by the degree in Figure 2.4, nodes 1, 2 and 3 all have the highest value of C^D. Hence $r_1 = r_2 = r_3 = (1 + 2 + 3)/3 = 2$.

Box 2.1   **Comparing Ranked Lists**

How to quantify the similarity or differences between two ranked lists is a general problem that recurs every time you have two or more ways to order a set of objects, i.e. to assign them the labels "first", "second", "third". Examples of applications are the comparison of two network centrality measures, or of two different recommendation systems (see Box 1.3). In statistics, a *rank correlation coefficient* is an index that measures the degree of similarity between two ranked lists, and can therefore be used to assess the significance of the relation between them. An increasing rank correlation coefficient implies increasing agreement between two rankings. Usually, the coefficient takes values in the range [1, 1], and is equal to 1 if the agreement between the two rankings is perfect (i.e. the two rankings are the same), or is equal to 0 if the rankings are completely independent, while it is 1 if one ranking is the reverse of the other. Popular rank correlation coefficients include Spearman's and Kendall's [176]. The *Spearman's rank correlation coefficient* is nothing other than a Pearson correlation coefficient (see Box 7.2) applied to the rank variables. The *Kendall's rank correlation coefficient* is instead a quantity based on the number of inversions in one ranking compared with another.

The *Spearman's rank correlation coefficient* $\rho_{X,Y}$ of the two rankings is defined as [290]:

$$\rho_{X,Y} = \frac{\sum_i (r_i - \bar{r})(q_i - \bar{q})}{\sqrt{\sum_i (r_i - \bar{r})^2 \sum_i (q_i - \bar{q})^2}} \tag{2.17}$$

where r_i and q_i are the ranks of node i according to the two centrality measures we are considering, while $\bar{r}$ and $\bar{q}$ are the averages over all nodes. Notice that the Spearman's coefficient is simply the Pearson correlation coefficient applied to the two rank variables $r_\bullet$ and $q_\bullet$ (see Box 7.2). As any correlation coefficient, the value of the Spearman's coefficients varies in $[-1, 1]$, with $\rho_{X,Y} = 1$ if the two rankings are identical, and $\rho_{X,Y} = -1$ if they are completely inverted.

The definition of the *Kendall's rank correlation coefficient* $\tau_{X,Y}$ is a bit more complicated. We need in fact to count the number of inversions in one ranking compared to the other. Let us consider two nodes i and j. We say that the pair (i, j) is concordant with respect to the two rankings $r_\bullet$ and $q_\bullet$ if the ordering of the two nodes agree, i.e. if both $r_i > r_j$ and $q_i > q_j$, or both $r_j > r_i$ and $q_j > q_i$. If, instead, $r_i > r_j$ and $q_i < q_j$, or $r_j > r_i$ and $q_j < q_i$, then the pair is said discordant. Finally, if it happens that either $r_i = r_j$ or $q_i = q_j$, then the pair is neither concordant nor discordant. The Kendall's $\tau_{X,Y}$ coefficient quantifies the similarity between two rankings by looking at concordant and discordant pairs [175]:

$$\tau_{X,Y} = \frac{n_c - n_d}{\sqrt{(n_0 - n_X)(n_0 - n_Y)}} \tag{2.18}$$

where n_c is the number of concordant pairs, n_d is the number of discordant pairs, and $n_0 = N(N-1)/2$ is the total possible number of pairs in a set of N elements. The terms n_X and n_Y account for the presence of rank degeneracies. In particular, let us suppose that the first ranking has M_X tied groups, i.e. M_X sets of elements such as all the elements in one of these sets have the same rank. If we call u_g the number of nodes in the gth tied group, then n_X is defined as:

Table 2.2 Spearman's and Kendall's rank correlation coefficients between the four different centrality rankings of the nodes of the graph in Figure 2.4.

ρ	C^D	C^C	C^B	C^Δ	τ	C^D	C^C	C^B	C^Δ
C^D	1.00	0.84	0.98	0.88	C^D	1.00	0.73	0.95	0.84
C^C	0.84	1.00	0.87	0.97	C^C	0.73	1.00	0.82	0.91
C^B	0.98	0.87	1.00	0.87	C^B	0.95	0.82	1.00	0.82
C^Δ	0.88	0.97	0.87	1.00	C^Δ	0.84	0.91	0.82	1.00

$$n_X = \sum_{g=1}^{M_X} \frac{1}{2} u_g (u_g - 1)$$

Similarly, n_Y is defined as follows:

$$n_Y = \sum_{g=1}^{M_Y} \frac{1}{2} v_g (v_g - 1)$$

where we have made the assumption that the second ranking has M_Y tied groups, and that the gth tied group has v_g elements. Similarly to the Spearman's, also the Kendall's τ coefficient takes values in $[-1, 1]$, but in general can give slightly different results.

We have calculated both correlation coefficients in order to compare the four different rankings obtained for the tree with $N = 16$ nodes in Figure 2.4. As expected, all the rankings are strongly positively correlated. In particular, results reported in Table 2.2 indicate that degree and betweenness centrality produce the two most similar rankings, followed by the pair's closeness and delta centrality, while degree and closeness produce the two least similar rankings. This finding is supported both by the values of ρ and by those of τ.

We now conclude the section with an application of the measures of centrality introduced so far, and by their comparison, in the case of two of the real-world networks shown in Figure 2.1. On the left-hand side of Table 2.3 we report C^D, C^C, C^B and C^Δ for each monkey, together with its age and sex. All the measures assign the first score to monkey 3, second score to monkey 12, and third score to monkeys 13 and 15. The six isolated monkeys are the least central nodes according to C^Δ, C^D and C^C. The betweenness centrality C^B assigns a zero score to 14 nodes, the six isolated monkeys, and nodes 4,5,7,9,10,11,14,17. In fact, the latter nodes do not contribute to the shortest paths, although they have a degree equal to or larger than one (for instance, node 14 has four neighbours). Notice that in this particular case the rankings of the 20 nodes produced by C^D, C^C and C^Δ are the same. This is confirmed by the values of $\rho = \tau = 1$ for any pair of these three centralities. Conversely, when computing the rank correlations coefficients between any of these three rankings and the that produced by C^B we obtain $\rho = 0.78$ and $\tau = 0.72$. On the right-hand side of the same table, we report the ten nodes with the highest C^Δ in the terrorist network, and we rank them according to their delta centrality score. The degree of each node is also reported. Delta centrality assigns the highest score to Mohamed Atta, who was on flight

Table 2.3 Comparisons of different centrality measures in the primate interaction network and in the terrorist network reported in Figure 2.1.

	Age group	Sex	C^D	C^C	C^B	C^Δ
1	14–16	M	0.211	0.134	0.006	0.139
2	10–13	M	0.000	0.050	0.000	0.000
3	10–13	M	0.684	0.143	0.260	0.375
4	7–9	M	0.158	0.133	0.000	0.131
5	7–9	M	0.105	0.132	0.000	0.123
6	14–16	F	0.000	0.050	0.000	0.000
7	4–5	F	0.158	0.133	0.000	0.131
8	10–13	F	0.158	0.133	0.003	0.131
9	7–9	F	0.053	0.131	0.000	0.115
10	7–9	F	0.158	0.133	0.000	0.131
11	14–16	F	0.105	0.132	0.000	0.123
12	10–13	F	0.474	0.139	0.060	0.180
13	14–16	F	0.316	0.136	0.011	0.156
14	4–5	F	0.211	0.134	0.000	0.139
15	7–9	F	0.316	0.136	0.011	0.156
16	10–13	F	0.000	0.050	0.000	0.000
17	7–9	F	0.158	0.133	0.000	0.131
18	4–5	F	0.000	0.050	0.000	0.000
19	14–16	F	0.000	0.050	0.000	0.000
20	4–5	F	0.000	0.050	0.000	0.000

rank	name	C_i^Δ	c_i^D
1	Mohamed Atta	0.150	16
2	Salem Alhazmi	0.112	8
3	Hani Hanjour	0.098	10
4	Mamoun Darkazanli	0.091	4
5	Marwan Al-Shehhi	0.091	14
6	Nawaf Alhazmi	0.086	9
7	Hamza Alghamdi	0.080	7
8	Satam Suqami	0.077	8
9	Abdul Aziz Al-Omari	0.075	9
10	Fayez Banihammad	0.067	7

AA-11 crashed into the World Trade Center North, and is also the terrorist with the largest number of direct contacts with other terrorists ($k = 16$). An extremely interesting result is the presence, among the relevant nodes of the network, of individuals with a small degree, such as Mamoun Darkazanli, who has only four ties but is the fourth in the list of C^Δ. Centrality analysis on terrorist organisation networks can contribute to criminal activity prevention. In fact, if some knowledge is available on terrorist organisation networks, the individuation and targeting of the central nodes of the network can be used to disrupt the organisation.

2.5 Movie Actors

It is now time to have fun. The *Internet Movie Database* (*IMDb*) at www.imdb.com is a wonderful data set containing detailed information on all movies ever made. This is the site you may like to look at when you want to choose a film to watch. The site also provides the names of all the actors playing in a given movie. Hence, it is possible to use *IMDb* to construct a bipartite graph where the first set of nodes represents actors, the other set represents movies, and there is a link between an actor and a movie if the actor has played in the movie. Here we will be focusing on the *movie actor collaboration network*, which is obtained as a projection of the original bipartite network on the set of movie actors. The details of this network are illustrated in the DATA SET Box 2.2.

DATA SET 2: The Movie Actor Collaboration Network

In the network of collaborations between actors, the nodes represent movie actors, and two actors who have played in at least one film together are joined by an edge, as sketched in the figure. For each pair of actors we do not consider the number of films they have in common, but only whether they have acted together, so that the obtained undirected network does not contain multiple links.

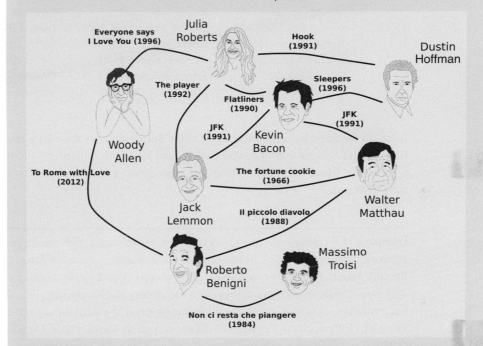

The network provided in this data set has $N = 248243$ nodes and $K = 8302734$ links. It has been generated from data available at the *Internet Movie Database (IMDb)* as of April 1997, and was studied for the first time by Duncan Watts and Steven Strogatz in a paper published in Nature in 1998 [311].

The DATA SET 2 can be downloaded from the book webpage, and we can use it to spot the real stars of the fabulous world of movie actors by applying the various centrality measures defined in the previous sections of this chapter. In particular, for each node of the movie collaboration network, we have computed degree, closeness and betweenness centrality. The list of the top ten actors according to each of the three centrality measures is reported in Table 2.4. In the three lists we can find names of famous actors of different periods and nationality such as Robert Mitchum and Gene Hackman (USA), Max von Sydow (Sweden), Marcello Mastroianni (Italy), Fernando Rey (Spain), Christophe Lee (UK), Donald Sutherland (Canada). Some of them won important prices. For instance, Anthony Quinn won two Oscars, respectively, in 1953 and in 1957. In the following example we

Table 2.4 The ten actors with respectively the highest C^D, C^C and C^B are reported and ranked according to their centrality scores. The number reported adjacent to each actors's name is the degree. The names of actors appearing in more than one top ten are in bold face.

ranking	C^D	C^C	C^B
1	Irving Bacon 3080	Rod Steiger 1599	**Max von Sydow** 1512
2	**John Carradine** 3027	**Christopher Lee** 2340	**Christopher Lee** 2340
3	James Flavin 2788	**Donald Pleasence** 1825	**John Carradine** 3027
4	Lee Phelps 2749	Gene Hackman 2307	Marcello Mastroianni 1617
5	Stanley Andrews 2717	Robert Mitchum 2037	Fernando Rey 1347
6	George Chandler 2657	**Donald Sutherland** 2029	**Donald Pleasence** 1825
7	Byron Foulger 2651	Sean Connery 1802	Savely Kramarov 510
8	Emmett Vogan 2581	**Anthony Quinn** 1948	Viveca Lindfors 1024
9	Pierre Watkin 2579	**John Carradine** 3027	**Donald Sutherland** 2029
10	Selmer Jackson 2554	**Max von Sydow** 1512	**Anthony Quinn** 1948

focus on the three top actors according to the three centrality measures, respectively Irving Bacon, Rod Steiger and Max von Sydow, and we try to understand the reasons for their high scores. More information on all the other actors and their short bios can be found on the *IMDb* website.

Example 2.14   (*Hollywood stars*)   Irving Bacon, the actor with the largest degree in the network, was born in 1893 in Missouri and was a minor character actor who appeared in literally hundreds of films, where he often played friendly and servile parts of mailmen, milkmen, clerks, chauffeurs, taxi drivers, bartenders and handymen. This also explains the huge number (3080) of different collaborators. Over the years he also briefly appeared in some of Hollywood's most beloved classics, such as Capra's "Mr. Deeds Goes to Town" (1936), and was also spotted on popular '50s and '60s TV programs.

Rod Steiger, born in 1925 in Westhampton, New York, began his acting career in theatre and on live television in the early 1950s. His breakthrough role came in 1954, with the classic "On the Waterfront", which earned him an Oscar nomination as Best Actor. Since then, he has been a constant presence on the screen. In 1964, he received his second Oscar nomination for "The Pawnbroker", while in 1967 he played what is considered his greatest role, Chief of Police Bill Gillespie in "In the Heat of the Night" (1967), opposite Sidney Poitier. Steiger received the Best Actor Oscar for his role in that film. In the final part of his career he moved away from big Hollywood pictures, participating instead in foreign productions and independent movies, which might explain his high closeness centrality in the collaboration network. Steiger has a star on the Hollywood Walk of Fame, at 7080 Hollywood Boulevard.

Max von Sydow, born in Sweden in 1929, studied at the Royal Dramatic Theatre's acting school (1948–1951) in Stockholm and started his career as a theatre actor. His work in movies by Ingmar Bergman made him well-known internationally, and he started to get offers from abroad. His career overseas began with "The Greatest Story Ever Told" (1965) and "Hawaii" (1966). Since then, his career includes very different kinds of characters, like Emperor Ming in "Flash Gordon" (1980) or the artist Frederick in "Hannah and Her Sisters" (1986). In 1987 he made his directing debut with "Katinka" (1988). He has become one of Sweden's most admired and professional actors. The top score of Max von Sydow in the betweenness centrality list is probably related to his presence in important movies both in Sweden and in USA, and therefore to its bridging of the European and the American actor communities.

By comparing the results in Table 2.4, the first thing we notice is the small overlap between the three rankings, especially of the ranking based on degree centrality with those based on closeness and on betweenness. Only one actor, John Carradine, from the top ten based on the degree centrality, also appears in both C^C and C^B top tens, while five actors are present in both the C^C and C^B rankings. This result is a clear indication that the three scores measure different ways of being central. It is striking that an actor such as Irving Bacon, the first in the C^D list, who has acted with more than three thousands different actors, is not in the top ten according to closeness and betweenness. Conversely, Max von Sydow is the first actor in the C^B list despite the fact he only has a degree of 1512. Analogously, Donald Sutherland scoring the ninth position by C^B and the sixth position by C^C, has only 2029 first neighbours on the graph.

Together with the names in the top ten, it is also interesting to have an idea about how centrality is distributed over the nodes of the graph. A concept related to that of centrality, which has been thoroughly explored in the context of social network analysis, is that of graph *centralisation*. While centrality has to do with the identification of the key nodes of a graph, centralisation focuses instead on the overall organisation of the network. Generally speaking, an index of centralisation is a single number measuring whether, and how much, the centrality is distributed in a centralised or decentralised way in a given graph. A commonly adopted quantity to quantify the centralisation of a graph, with respect to a given centrality C, is defined as follows [308].

Definition 2.10 (Graph centralisation)   *The centralisation $H^C[G]$ of a graph G with N nodes, with respect to the centrality measure C, is given by:*

$$H^C[G] = \frac{\sum_{i=1}^{N} (C_{i*}[G] - C_i[G])}{\max_{G'} \sum_{i=1}^{N} (C_{i*}[G'] - C_i[G'])} \tag{2.19}$$

where $C_i[G]$ is the value of centrality of node i in graph G, for the centrality index of interest C, and i^* is the node with the highest centrality.

The definition above fully applies to all of the centrality measures we have introduced in this chapter. The reference point in Eq. (2.19) is the most central node in the graph. In fact, $H^C(G)$ measures to what extent the centrality of the most central node exceeds the centrality of the other nodes. The normalisation factor at the denominator is the maximum possible value of the numerator over all graphs G' with the same number of nodes as G, which is obtained in the case of a star with N nodes, so that the graph centralisation $H^C[G]$ ranges in the interval $[0, 1]$ and in particular equals 1 if the graph G is a star. The nice thing about the centralisation is that it produces a single number associated with a graph. The drawbacks are that one can often lose important details, and graphs with completely different structures can give rise to exactly the same value of $H^C[G]$.

For the above reasons, we prefer to adopt here a different way to investigate the overall distribution of centrality in a graph, which is to count the number of nodes $N(C)$ having centrality values in a given interval around the centrality value C and to plot the resulting centrality histogram. How to select the binning to produce such histograms depends on the type of centrality under investigation, and can be a tricky issue. For instance, it is in general easier to deal with the degree centrality because the degree of a node is an integer number, while centralities such as the closeness and the betweenness present more complications since they derive from node scores that are real numbers. We will come back to discuss degree distributions and this point more in detail in Section 3.2 and in Section 5.2. In Figure 2.5 we report the histograms we have obtained for the degree and the betweenness centrality of the movie actor collaboration network. In the case of the degree, shown in panel (a), we observe a plateau for values of C^D smaller than 3×10^{-4}, corresponding to a degree smaller than about 100. This means that we have about the same number of actors with a degree in the range 10–100. When the degree is larger than 100 we see instead that the number of nodes decreases with increasing C^D as a power law, $N(C^D) \sim (C^D)^{-\gamma}$, with a positive value of the exponent γ. In fact, for such values of the degree we notice that the histogram shows a linear behaviour in the double logarithmic scale we have adopted. We will be devoting two entire chapters of this book, namely Chapters 5 and 6, to focus on *power-law degree distributions* and on the role of the exponent γ, given the importance of such distributions and their ubiquity not only in social systems, but more in general in nature (see in particular Box 5.2 and Box 5.3). Panel (b) of Figure 2.5 shows how the betweenness centrality is distributed. Also in this case we observe a slowly decaying histogram with a linear behaviour in a double logarithmic scale.

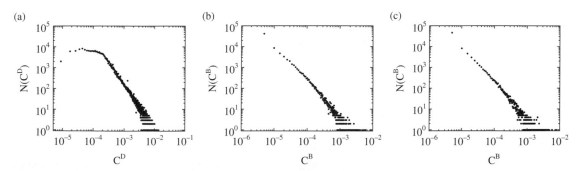

Fig. 2.5   Number of nodes with a given degree (a), betweenness (b), and approximate betweenness (c) centrality in the movie actor collaboration network.

Table 2.5 Top ten as a function of the percentage x of shortest paths used to evaluate the betweenness.

ranking	$x = 50\%$	$x = 80\%$	$x = 90\%$	$x = 100\%$
1	M. von Sydow	M. von Sydow	M. von Sydow	M. von Sydow
2	C. Lee	C. Lee	C. Lee	C. Lee
3	F. Rey	J. Carradine	J. Carradine	J. Carradine
4	D. Pleasence	D. Pleasence	M. Mastroianni	M. Mastroianni
5	D. Sutherland	F. Rey	F. Rey	F. Rey
6	J. Carradine	M. Mastroianni	D. Pleasence	D. Pleasence
7	A. Quinn	A. Quinn	D. Sutherland	S. Kramarov
8	G. Hackman	F. Rey	A. Quinn	V. Lindfors
9	R. Steiger	R. Mitchum	R. Steiger	D. Sutherland
10	M. Mastroianni	R. Steiger	R. Mitchum	A. Quinn

While the degree centrality is extremely easy to compute, the calculation of the betweenness distribution can require a long time, since we have to solve the all-shortest-paths problem. To save computer time, we can evaluate only the contribution of a certain percentage of source – target shortest paths to the betweenness of a node. Namely, we have selected a percentage x of N nodes. For each of these nodes we computed the shortest paths to all other nodes in the graph with the BFS algorithm of Appendix A.6. We finally use only these shortest paths to approximate the value of betweenness of a node i ($i = 1, \ldots, N$). The approximate distribution obtained with $x = 0.1$ is shown in panel (c) of Figure 2.5, and is practically indistinguishable from the exact one reported in panel (b). This means that only 10 per cent of the shortest paths are sufficient to extract the main properties of the betweenness distributions, hence reducing by a factor of ten the numerical efforts. Of course, when we need more details, for instance when we want the precise ranking of the first ten actors, we must consider large values of x. In Table 2.5 we show that, by using the shortest paths originating from half of the nodes, we get eight out of the ten names in the exact top ten (indicated in Table as $x = 100\%$). However, the order is only respected for the first two actors, namely Max von Sydow and Christopher Lee. To improve things we have to increase x even further. With 90 per cent of the shortest paths we reproduce in the exact order the first six actors, while in the top ten there are still two spurious names, Rod Steiger and Robert Mitchum. Such names were present in the closeness centrality top ten, though they are not among the first ten actors by betweenness centrality, when the betweenness is evaluated exactly.

Centrality is not only a useful concept for social networks. It can find interesting applications to very different systems. Think for instance of the city where you live. You certainly know what are the central locations in your city, and can elaborate on the ingredients that make an area central and successful. Usually, central places are more accessible, more visible, have a greater probability of offering a larger diversity of opportunities and goods, and are more popular in terms of people and potential customers. This is why central locations are also more expensive in terms of real estate values and tend to be socially selective. However, despite urban planners and designers having spent a long time trying to understand

and manage centrality in cities, it is still an open question how to define centrality in a precise and quantitative way. Recently, it has been shown that measures of graph centrality, as those discussed in this chapter, can provide a useful help as they correlate with popularity in terms of pedestrian and vehicular flows, micro-economic vitality and social livability. See for instance Refs. [153, 167, 273, 259, 89, 88, 260] and Problem 2.5(c). The map of a city can in fact be turned into a graph whose nodes are intersections and edges are streets. What we get in this way is a graph where the nodes are embedded in the Euclidean plane, and the edges represent real physical connections with their associated length. This is an example of a particular type of graphs known as *spatial graphs*. We will come back to the peculiar properties of spatial graphs in Chapter 8.

2.6 Group Centrality

Let us go back to the networks we started from at the beginning of the chapter, namely to the relations between Florentine families and to the primate interaction network reported in Figure 2.1. Suppose we now want to find the relevant groups of alliances among Florentine families. Or, we are asked to look into whether the youngest are more important than the eldest in the primate network, or the females are more central than the males. Indeed, the concept of centrality can be extended to measure the importance of a group of individuals as a whole, and different types of *group centrality* have been introduced precisely to answer such questions. Measures of group centrality can turn to be very useful in practice when we need to form a team of people that is maximally central in a given social network. They can also contribute to the so-called *inverse centrality* problem, a different but related problem (that we will not treat here in detail) consisting in how to change the structure of a given network in order to obtain any assigned centrality score ranking for its nodes [244].

Most of the node measures we have introduced in the previous sections can be easily extended to quantify the centrality of groups of nodes. However, in most cases the generalisation is not unique. This is the reason why, here, we will closely follow the generalisations proposed by Martin Everett and Steve Borgatti in Ref. [108], which are today widely known and adopted. Before providing a list of the definitions of group centrality measures, let us mention that by group of nodes we mean any subset g consisting of $n \geq 1$ of the N nodes of graph G, independently from whether the n nodes in g are connected or not.

Definition 2.11 (Group degree centrality)   *In an undirected graph, the degree centrality c_g^D of group g is defined as the number of non-group nodes that are connected to group members. Multiple links to the same node are counted only once. A normalised quantity C_g^D ranging in $[0, 1]$ is obtained by dividing the group degree centrality c_g^D by the number $N - n$ of non-group nodes.*

Definition 2.12 (Group closeness centrality)   *In a connected graph, the closeness centrality c_g^C of group g is defined as the sum of the distances from g to all nodes outside the group. A normalised quantity C_g^C ranging in $[0, 1]$ is obtained by dividing the distance score c_g^C by the number $N - n$ of non-group nodes.*

The definition of group closeness given above deliberately leaves unspecified how the distance from a group g to an outside vertex j is defined. To be more precise, consider the set $D = \{d_{ij}\}_{i \in g, j \notin g}$ of all distances between any node i in group g and any node j outside g. We have a certain freedom in how to define the distance from group g to node j based on set D. For instance, the distance between group g and node j can be defined as either the maximum distance in D, the minimum in D or the mean of the values in D. According to the way we evaluate such a distance, we will get slightly different definitions of closeness centrality, except in the trivial case of a group g consisting of a single node, where the three definitions of distances are identical, and the group centrality reduces to the node centrality.

Definition 2.13 (Group betweenness centrality) *In a connected graph, the shortest-path betweenness centrality of group g is defined as:*

$$c_g^B = \sum_{\substack{j=1 \\ j \notin g}}^{N} \sum_{\substack{k=1 \\ k \neq j, k \notin g}}^{N} \frac{n_{jk}(g)}{n_{jk}};$$

(2.20)

where n_{jk} is the number of geodesics from node j to node k, whereas $n_{jk}(g)$ is the number of geodesics from node j to k, passing through the members of group g. A normalised quantity C_g^B ranging in $[0, 1]$ is obtained as: $C_g^B = c_g^B / [(N - n)(N - n - 1)]$.

The group betweenness centrality given above measures the proportion of geodesics connecting pairs of non-group members that pass through the group. In practice, such centrality can be computed by making use of the standard shortest path algorithms of Appendix A.6 in the following way. We need first to count the number of geodesics, n_{jk}, in graph G between every pair i and j of non-group nodes. In order to obtain the quantities $n_{jk}(g)$ for all $j, k \notin g$, we can then repeat the calculation on the graph obtained from G by removing all the links between group members (i.e. between pairs of nodes in g). We can finally sum the ratios $n_{jk}(g)/n_{jk}$ as in Eq. (2.20) to get the betweenness centrality of group g.

Definition 2.14 (Group delta centrality) *In a graph G, the delta centrality C_g^Δ of group g is defined as the relative drop in the graph efficiency E caused by the deactivation of the nodes in g:*

$$C_g^\Delta = \frac{(\Delta E)_g}{E} = \frac{E[G] - E[G']}{P[G]}$$

(2.21)

where G' is the graph obtained by removing from G the edges incident in nodes belonging to g, and the efficiency E is defined as in Eq. (2.16).

As a possible application of the centrality measures defined in this section we report an analysis of groups in the primate interaction network of Figure 2.1(b). We have considered four different age groups and the two sex groups. Namely, group 1 contains the five monkeys of age 14–16, group 2 the five monkeys of age 10–13, group 3 the six monkeys of age 7–9 and group 4 the four monkeys of age 4–5. Group 5 comprises the 5 males and group 6 the 15 females. The values of centrality obtained for primates in the six groups are

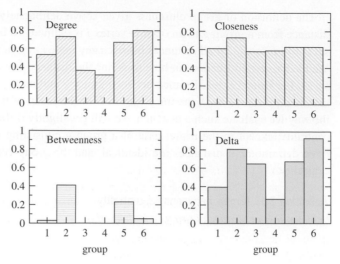

Fig. 2.6   Group degree, closeness, betweenness and delta centrality for six different groups, respectively: (1) age 14–16, (2) age 10–13, (3) age 7–9, (4) age 4–5, (5) males, (6) females.

reported in Figure 2.6. As illustrated in the figure, group 2, containing the 10–13-year-old animals is the most central one among the four age groups, and this is true for each of the four measures of centrality we have computed. Notice that group 2 is also the group containing monkey 3, which is also the most central node at the individual level. The four age groups in decreasing order of importance are: 2, 1, 3, 4 for the degree centrality, 2, 1, 4, 3 for the closeness centrality, 2, 1, 3–4 for the betweenness centrality, which gives a score equal to zero to the two youngest groups, and 2, 3, 1, 4 for the delta centrality. Notice that C^Δ is the only measure assigning the second position to the 7–9-year-olds, while the other three measures attribute the second position to the 14–16-year-olds. Also, all the measures except closeness centrality assign the last position to group 4 (age 4–5). Finally, among the sex groups, the most central one according to both C^D and C^Δ is the female group, which is also the largest group. Notice however that the centrality of a group is not always positively correlated with the number of nodes in the group. For instance the males are more central than the females according to C^B, while C^C attributes the same score to the two groups.

2.7 What We Have Learned and Further Readings

Social network analysis is the discipline that provides quantitative methods and tools to understand social systems by representing and studying them as networks. It is an important component of sociology and is, together with graph theory, one of the pillars of complex networks. Social network analysis provides in fact general and useful ideas such as those of centrality, discussed here, which also have interesting applications outside the realm of social systems. In this chapter we learnt that the *structural centrality* of a node in a network can be measured in many different ways. In particular, there are measures, such as the *degree* and the *closeness*, quantifying how close a node is to the other nodes on the graph. Other centrality indices measure instead how much a node is in *between* other pairs of nodes. Finally, we have introduced *delta centrality*, measuring the importance of a node by the decrease in the network performance when the node is deactivated. We have also briefly discussed the concept of *time complexity* and introduced the first graph algorithms of this book. Namely, we presented two methods to visit a graph, the so-called *Breadth-First Search* (BFS) and the *Depth-First Search* (DFS) algorithm, and a method to compute the largest eigenvalue of a matrix and the corresponding eigenvector, known as the *power method*. In this chapter we have also introduced the first large data set of the book, the *movie actor collaboration network*, which allowed us to compare the most central actors according to various centralities, and also to have a look at how centrality is distributed among the nodes of a large system. We will come back to centrality distributions, with particular focus on the degree, in Chapter 5.

Centrality is only one of the many powerful ideas developed in the context of social network analysis. A comprehensive discussion of the standard techniques to study the structure of a social system can be found in the book by Stanley Wasserman and Katherine Faust [308] and in the book by John Scott [278]. For more information on how the structure of the network influences the dynamics of social interactions we suggest the book by Matthew Jackson [163].

Problems

2.1 The Importance of Being Central

How good is the correlation between node degrees and attributes of the Florentine families shown in Table 2.1? Can you quantify it?

2.2 Connected Graphs and Irreducible Matrices

(a) Let P be the elementary *permutation matrix* obtained by switching rows i and j of the $N \times N$ identity matrix I. Show that the multiplication of matrix P by a vector $\mathbf{v}$ has the effect of switching entries i and j of $\mathbf{v}$. Show also that the left multiplication of matrix P by an $N \times M$ matrix A has the effect of switching rows i and j of matrix A, while right multiplication of $M \times N$ matrix B by the same matrix has the effect of switching columns i and j.

(b) Prove that for any permutation matrix P we have $P^{\top} = P^{-1}$. HINT: any permutation matrix can be written as a product of elementary permutation matrices.

(c) When discussing graph connectedness in Section 2.2 we considered transformations of the adjacency matrix such as $A' = P^{\top}AP$, with P being a permutation matrix. In some particular cases it can happen that $A' = A$ for some $P \neq I$ because the graph has symmetries. Indeed, permutation matrices are also very useful in the context of graph symmetries and automorphisms. We have seen in Definition 1.3 that a *graph automorphism* of G is a permutation of the vertices of graph G that preserves the adjacency relation of G. This can be expressed in mathematical terms by saying that a permutation matrix P is an automorphism of G iff we have $A = P^{\top}AP$ or, analogously, that the permutation matrix P commutes with the adjacency matrix: $PA = AP$. Verify that this is true for all the automorphisms of graph $\mathbb{C}_4$ shown in Figure 1.5.

2.3 Degree and Eigenvector Centrality

(a) Consider an undirected quadrilateral. Verify that the corresponding adjacency matrix is irreducible, and show that in this case the power method for the computation of the maximum eigenvalue does not converge if we start from $x_0 = (1, 0, 0, 0)^{\top}$. Is the adjacency matrix primitive?

(b) Consider a directed triangle with only three arcs, going from node i to node $i+1$ mod 3, $i = 0, 1, 2$. Verify that the corresponding adjacency matrix is irreducible. Show that in this case the power method for the computation of the maximum eigenvalue does not converge for a generic initial condition.

(c) It is possible to associate a directed graph with any non-negative square matrix $A \in \mathbb{R}^{N \times N}$. In fact we say that there is an arc between node i and node j iff $a_{ij} \neq 0$. Prove that a non-negative square matrix A is irreducible iff the corresponding graph G is strongly connected.

(d) Prove that a primitive matrix is irreducible. This is equivalent to proving that reducibility implies non-primitivity. Notice that, if A is an upper (or lower) block triangular matrix, then A^k, with $k \geq 0$, is also an upper (or lower) block triangular matrix.

(e) Consider Elisa's kindergarten network, introduced in Box 1.2. Use a computer to show that the network is strongly connected. By means of the power method, compute the in- and out-degree eigenvector centrality, and rank the nodes accordingly. The two obtained rankings are different from the ones obtained based, respectively, on k^{out} or on k^{in}. Comment on that (in order to compare rankings, evaluate the Spearman's or the Kendall's rank correlation coefficient defined in Section 2.4).

(f) Compute the eigenvector centrality for the graph shown in the figure. Notice that there are three pairs of nodes, namely node 2 and 3, node 5 and 6, and nodes 7 and 4, with the same centrality score. Notice also that the graph has three

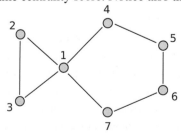

automorphisms (permutations of the vertices that preserve the adjacency relation of the graph): $(1, 2, 3, 4, 5, 6, 7) \rightarrow (1, 3, 2, 4, 5, 6, 7)$, $(1, 2, 3, 4, 5, 6, 7) \rightarrow$

(1, 2, 3, 7, 6, 5, 4), and $(1, 2, 3, 4, 5, 6, 7) \rightarrow (1, 3, 2, 7, 6, 5, 4)$. Prove that, if there is an automorphism switching node i and j, then i and j have the same eigenvector centrality. Make use of the relation $A = P^{\top}AP$ and the fact that the largest eigenvalue is non-degenerate.

(g) Calculate the normalised α-centrality of the nodes of the graph shown in the figure. HINT: Make use of the fact that the infinite sum of the powers of a matrix $\sum_{l=0}^{\infty} \alpha^l A^l$ converges to $(I - \alpha A)^{-1}$ when $|\alpha| < 1/\rho(A)$, where $\rho(A)$ is the *spectral radius* of the matrix A, i.e. the largest eigenvalue in absolute value of A. Notice also that, for the given graph, we have $(A^{\top})^l = 0$ when $l > 3$.

2.4 Measures Based on Shortest Paths

(a) Shortest path lengths on a (unweighted) graph can be computed in the following

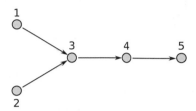

way. Two nodes i and j have distance 1 if $a_{ij} = 1$. If $a_{ij} = 0$ we want to look at the entry $(A^2)_{ij}$ of A^2. If this is different from zero then i and j have distance 2, otherwise we go to check $(A^3)_{ij}$, and so on. Implement and check this algorithm to compute all shortest path lengths in the graph in Figure 2.1(a). Compare the results of this algorithm with those of the BFS algorithm described in Appendix A.6.

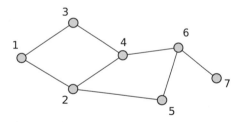

(b) Consider the graph shown in the figure. Calculate the closeness centrality of nodes 1 and 4. Calculate the betweenness centrality of nodes 4.

2.5 Movie Actors

(a) Let A be the $N \times V$ adjacency matrix describing the *IMDb* bipartite network of actors and movies, with $a_{i\alpha} = 1$ if actor i has played in movie α. Prove that the adjacency matrix of the graph of collaborations between actors can be obtained by constructing the $N \times N$ matrix $B = AA^{\top}$, and by setting equal to 0 all the diagonal terms of B, and equal to 1 all non-zero off-diagonal terms.

(b) Evaluate by means of Eq. (2.19) the degree centralisation for the movie actor collaboration network, and compare the number you obtain with the centralisation of other networks of your choice.

(c) Use network centrality to find the central places of a city. Namely, compute node degree, closeness and betweenness centrality for the *urban street networks* in DATA SET 8. Show the results graphically. Then, evaluate shortest paths by taking into account the real length of the traversed streets in meters, instead of counting just the number of links (see Appendix A.6). Recompute then closeness and betweenness based on shortest path lengths in meters, and compare the results with those obtained by considering only step distances (see Refs. [89, 88]).

2.6 Group Centrality

That given in Definition 2.11 is not the unique way to extend the degree in order to measure the centrality of a group of nodes. Can you think of alternative ways to define group degree centrality? You can find source of inspiration in Ref. [108].

Random Graphs

The term *random graph* refers to the disordered nature of the arrangement of links between different nodes. The systematic study of random graphs was initiated by Erdős and Rényi in the late 1950s with the original purpose of studying theoretically, by means of probabilistic methods, the properties of graphs as a function of the increasing number of random connections. In this chapter we introduce the two random graph models proposed by Erdős and Rényi, and we show how many of their average properties can be derived exactly. We focus our attention on the shape of the *degree distributions* and on how the average properties of a random graph change as we increase the number of links. In particular, we study the critical thresholds for the appearance of *small subgraphs*, and for the emergence of a *giant connected component* or of a *single connected component*. As a practical example we compute the component order distribution in a set of large real *networks of scientific paper coauthorships* and we compare the results with random graphs having the same number of nodes and links. We finally derive an analytical expression for the *characteristic path length*, the average distance between nodes, in random graphs.

3.1 Erdős and Rényi (ER) Models

A random graph is a graph in which the edges are randomly distributed. In the late 1950s, two Hungarian mathematicians, Paul Erdős and Alfréd Rényi came up with a formalism for random graphs that would change traditional graph theory, and led to modern graph theory. Up to that point, graph theory was mainly combinatorics. We have seen one typical argument in Section 1.5. The new idea was to add probabilistic reasoning together with combinatorics. In practice, the idea was to consider not a single graph, but the ensemble of all the possible graphs with some fixed properties (for instance with N nodes and K links), and then use probability theory to derive the properties of the ensemble. We will show below how we can get useful information from this approach. Erdős and Rényi introduced two closely related models to generate ensembles of random graphs with a given number N of nodes, that we will henceforth call *Erdős and Rényi (ER) random graphs* [49, 100, 101]. We emphasise that, in order to give precise mathematical definitions, we need to define: a space whose elements are graphs, and also a probability measure on that space. Except when mentioned otherwise, in what follows we shall deal with undirected graphs.

Definition 3.1 (ER model A: uniform random graphs) *Let $0 \leq K \leq M$, where $M = N(N-1)/2$. The model, denoted as $G_{N,K}^{ER}$, consists in the ensemble of graphs with N nodes generated by connecting K randomly selected pairs of nodes, uniformly among the M possible pairs. Each graph $G = (\mathcal{N}, \mathcal{L})$ with $|\mathcal{N}| = N$ and $K = |\mathcal{L}|$ is assigned the same probability.*

In practice, to construct a graph of the ensemble, one considers N nodes and picks at random the first edge among the M possible edges, so that all these edges are equiprobable. Then one chooses the second edge at random with a uniform probability among the remaining $M-1$ possible edges, and continues this process until the K edges are fixed. A given graph is only one outcome of the many possible realisations, an element of the statistical ensemble of all possible combinations of connections. For the complete description of $G_{N,K}^{ER}$ we need to describe the entire statistical ensemble of possible realisations, that is, in matricial representation, the ensemble of adjacency matrices. How many different graphs with N nodes and K edges are there in the ensemble? This number is equal to the number of ways we can select K objects among M possible ones, namely: $\dfrac{M!}{K!\,(M-K)!}$. This quantity is known as the *binomial coefficient*, and is usually denoted as C_M^K or as $\begin{pmatrix} M \\ K \end{pmatrix}$, where the last symbol is pronounced "M choose K" [143, 272] (see Box 3.1).

Box 3.1   **Binomial Coefficient**

In a pizza restaurant the customer can choose two toppings among ten possible ones. How many different kinds of pizza can we get? The first topping can be chosen in 10 different ways, and the second one in 9 different ways. This would give a total of 90 pairs. Notice, however, that the choice mushrooms and olives is the same as olives and mushrooms. Therefore the number of different kinds of pizzas offered by the restaurant is $45 = 90/2$. More generally, suppose we have n objects and we want to choose k of them. The number of ordered k-tuples is given by:

$$n(n-1)(n-2)\ldots(n-k+1) = \frac{n!}{(n-k)!}$$

By dividing this number by the number of permutations of k objects, that is equal to $k!$, we finally obtain the number of possible selections of k of the n objects:

$$C_n^k \equiv \begin{pmatrix} n \\ k \end{pmatrix} = \frac{n!}{k!\,(n-k)!}$$

The quantity C_n^k is called the binomial coefficient, since it appears in the formula that gives the nth power (with n non-negative integer) of the sum of two real or complex numbers a and b:

$$(a+b)^n = \sum_{k=0}^{n} C_n^k a^k b^{n-k} \tag{3.2}$$

known as the *binomial identity*.

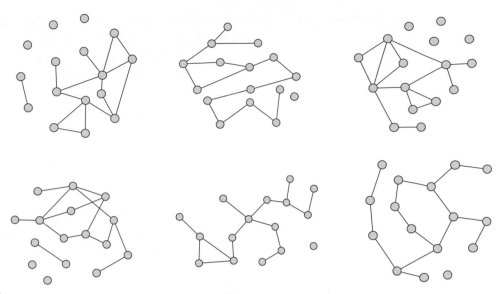

Fig. 3.1 Six different realisations of model A with $N = 16$ and $K = 15$.

The *uniform random graphs* ensemble generated by model A contains $C_M^K = \begin{pmatrix} M \\ K \end{pmatrix}$ graphs, and each of them is selected with a probability of $1/C_M^K$. Therefore, the probability P_G of finding a given graph $G \equiv (\mathcal{N}, \mathcal{L})$ in the model $G_{N,K}^{ER}$ is equal to $1/C_M^K$ if $|\mathcal{N}| = N$ and $|\mathcal{L}| = K$, and is zero otherwise, namely:

$$P_G = \begin{cases} 1/C_M^K & \text{iff } |\mathcal{N}| = N \text{ and } |\mathcal{L}| = K \\ 0 & \text{otherwise} \end{cases} \qquad (3.3)$$

As an example, in Figure 3.1 we show six realisations of the $C_{120}^{15} \approx 4.73 \times 10^{18}$ possible random graphs with $N = 16$ nodes and $K = 15$ links. The graphs have been generated by using the *algorithm for sampling ER random graphs* of model A introduced in Appendix A.10.1.

Definition 3.2 (ER model B: binomial random graphs) *Let $0 \le p \le 1$. The model, denoted as $G_{N,p}^{ER}$, consists in the ensemble of graphs with N nodes obtained by connecting each pair of nodes with a probability p. The probability P_G associated with a graph $G = (\mathcal{N}, \mathcal{L})$ with $|\mathcal{N}| = N$ and $|\mathcal{L}| = K$ is $P_G = p^K(1-p)^{M-K}$, where $M = N(N-1)/2$.*

The ensemble of graphs defined by model B is certainly different from that of model A. In Figure 3.2 we plot six graphs sampled from model B with $N = 16$ and $p = 0.125$. This time we have used the *algorithm for sampling ER random graphs* explained in Appendix A.10.2. We have appropriately chosen the value of p, namely $pM = 15$, so that the graphs in $G_{16,p}^{ER}$ have on average the same number of edges as the graphs in Figure 3.1. However, while all the graphs in the first ensemble have 15 edges, the number of links in the second ensemble

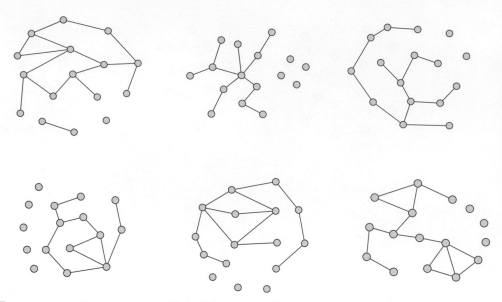

Fig. 3.2 Six different realisations of model B, with $N = 16$ and $p = 0.125$.

can vary. For instance, the six graphs shown in Figure 3.2 have respectively $K_1 = 16$, $K_2 = 11$, $K_3 = 13$, $K_4 = 12$, $K_5 = 14$ and $K_6 = 15$. In principle, when we consider $G^{\mathrm{ER}}_{16,p}$, we can also sample the complete graph with $N = 16$ nodes, or a graph with no links at all. The complete graph is very unlikely, unless $p \simeq 1$, while a graph with no links is likely to appear only when $p \simeq 0$. As we shall see, model B does indeed produce all possible graphs with N nodes, although not all of them with the same probability.

Example 3.1 *(Sampling probability)*   Consider the graph G in the figure. The probability P_G of obtaining such a graph by sampling from $G^{\mathrm{ER}}_{3,p}$ is equal to the probability that there is a link between nodes 1 and 2, a link between nodes 2 and 3, and no link between nodes 1 and 3.

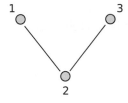

Because the existence (or non-existence) of a link is independent from the existence of the other links, we have:

$$P_G = \mathrm{Prob}\left[\{\exists(1,2)\} \cup \{\exists(2,3)\} \cup \{\nexists(1,3)\}\right] =$$
$$= \mathrm{Prob}\left[\{\exists(1,2)\}\right] \times \mathrm{Prob}\left[\{\exists(2,3)\}\right] \times \mathrm{Prob}\left[\{\nexists(1,3)\}\right] = p \cdot p \cdot (1-p).$$

Let us consider now the general case. We want to find out the probability P_G that a given graph $G = (\mathcal{N}, \mathcal{L})$ with $|\mathcal{N}| = N$ nodes and adjacency matrix $\{a_{ij}\}$ is sampled from $G_{N,p}^{\mathrm{ER}}$. By repeating the argument above we get:

$$P_G = p^{(\sum_{i<j} a_{ij})}(1-p)^{M-(\sum_{i<j} a_{ij})} = p^K(1-p)^{M-K}$$

where $K = \sum_{i<j} a_{ij}$ is the number of links in G. From this formula we can easily calculate that the probabilities of having an empty or a complete graph with N nodes are respectively equal to $(1-p)^M$ and p^M.

Summing up, the probability of finding a graph $G \equiv (\mathcal{N}, \mathcal{L})$ in the model $G_{N,p}^{\mathrm{ER}}$ is:

$$P_G = \begin{cases} p^K(1-p)^{M-K} & \text{iff } |\mathcal{N}| = N \\ 0 & \text{otherwise} \end{cases} \tag{3.4}$$

where $|\mathcal{L}| = K$. Notice that such a probability only depends on the number of links in the graph, provided that the graph has N nodes, and not on other details of its topology. From this observation we can calculate the probability $P(K)$ that a graph sampled from $G_{N,p}^{\mathrm{ER}}$ has exactly K links. This is equal to the product of the probability of sampling a graph with K links, times the number C_M^K of graphs with K links:

$$P(K) = \binom{M}{K} p^K(1-p)^{M-K}. \tag{3.5}$$

The quantity is correctly normalised as a probability. In fact, we have:

$$\sum_{K=0}^{M} P(K) = 1 \tag{3.6}$$

which is a particular case of formula (3.2) with $a = p$, $b = 1 - p$, $a + b = 1$. Model B is usually called *binomial random graphs* model because the distribution of the total number of links $P(K)$ in Eq. (3.5) is equal to $\mathrm{Bin}(M, p, K)$, where the latter function is known in statistics as the *binomial distribution* (See Box 3.2).

We shall return to binomial distributions in the next section, where we will deal with the distribution of degrees in random graphs. Here, from the distribution $P(K)$ we can calculate that the average number of edges $\overline{K}$ in graphs of model B is equal to pM. In fact, we have:

$$\overline{K} = E[K] = \sum_{K=0}^{M} K P(K) = \sum_{K=0}^{M} \frac{M!}{(K-1)!(M-K)!} p^K(1-p)^{M-K} =$$

$$= pM \sum_{j=0}^{M-1} \frac{(M-1)!}{j!(M-1-j)!} p^j(1-p)^{M-1-j} = pM \tag{3.9}$$

where $E[K]$ denotes the expectation value of K over the population of random graphs of model B. This result helps us to elucidate the relation between model A and model B, and explains our choice of p in Figure 3.2. In particular, in a graph with $N = 16$ nodes as that considered in Figure 3.2 we can have $M = 120$ possible links. By fixing

Box 3.2

Binomial Distribution

In statistics, the binomial distribution is valid for processes where there are two mutually exclusive possibilities. The binomial distribution is the discrete probability distribution of the number k of successes in a sequence of n independent yes/no experiments, each of which yields success with probability p:

$$\text{Bin}(n, p, k) \equiv \binom{n}{k} p^k (1 - p)^{n-k} \tag{3.8}$$

where k can take the values $1, 2, 3, \ldots, n$. A simple example is tossing ten coins and counting the number of heads. The distribution of this random number is a binomial distribution with $n = 10$ and $p = 1/2$ (if the coin is not biased). As another example, assume 5 per cent of a very large population to be green-eyed. You pick 100 people randomly. The number of green-eyed people you pick is a random variable which follows a binomial distribution with $n = 100$ and $p = 0.05$.

$p = 0.125$ in model B, then the random graphs we sample have, on average, $pM = 15$ links, which is equal to the number of links in the graphs sampled by model A in Figure 3.1.

So, the idea is very simple. The probability that there is an edge between a given pair of nodes in model A is equal to K/M, while the average number of edges in graphs of model B is equal to pM. Hence, if we have model A with a given K, we can construct a similar ensemble by choosing $p = K/M$ in model B. Likewise, if we are given model B with a fixed value of p, the corresponding model A has $K = pM$. As observed previously, there are some differences in the graphs we obtain. First, in model A the total number of links is fixed, while in model B it is a random variable. In addition to this, the links are independent in model B, while in model A there is necessarily some dependence of an edge being chosen, based on previous choices. This is because, since from the very definition of model A, we have decided to avoid loops and multiple edges. If we allow loops and multiple edges in model A, then we would get a model with no correlations. The price to pay is that the model would generate multigraphs.

The differences between models A and B are less and less relevant as the order of the graph increases, as illustrated in the following two examples.

Example 3.2   *(Correlations in model A)*   Let us denote by (i, j) and (r, s) two different pairs of nodes. In model B, the probability that there exists an edge connecting nodes i and j, does not depend on the existence of an edge between nodes r and s, while this is not the case in model A. More formally, the probability of the existence of edge (i, j) in model A is

$$\text{Prob}\left[(i, j) \in \mathcal{L}\right] = \frac{K}{M},$$

while the probability of the existence of edge (i, j), conditional on the existence of edge (r, s), is

$$\text{Prob}\left[(i, j) \in \mathcal{L} | (r, s) \in \mathcal{L}\right] = \frac{K - 1}{M - 1}.$$

The two probabilities are different, although their ratio tends to 1 as the order of the graph increases. To see this, let us take the limit $N \to \infty$, by keeping constant the ratio $\langle k \rangle = 2K/N$. We have:

$$\frac{K}{M} \frac{M-1}{K-1} = \frac{N\langle k \rangle}{N\langle k \rangle - 2} \frac{N-1-2/N}{N-1} \to 1 \quad \text{as} \quad N \to \infty.$$

In practice this means that, although the presence of a link is correlated to the existence of the other links in model A, such correlations can be neglected in large graphs.

Example 3.3 *(Fluctuations of K in model B)* It is not difficult to prove, by using the properties of the binomial distribution in Eq. (3.5), that the variance in the total number of edges in model B is given by (see Problem 3.1(b)):

$$\sigma_K^2 = E[(K - E[K])^2] = \overline{(K - \overline{K})^2} = \overline{K^2} - \overline{K}^2 = Mp(1-p) \tag{3.10}$$

In conclusion, the ratio between the standard deviation σ_K and the average number of links $\overline{K} = pM$ is given by:

$$\frac{\sigma_K}{\overline{K}} = \sqrt{\frac{(1-p)}{pM}}$$

Since $M \propto N^2$, if we assume that $p(N) = cN^{-1}$ (so to keep constant the average node degree $\langle k \rangle$), the ratio above tends to zero as $N \to \infty$. This proves that, in large graphs, the fluctuations in the value of K of model B can be neglected.

With arguments similar to those in the previous two examples, it can be proven that in the limit of large N the two models are practically indistinguishable [313]. Usually $G_{N,p}^{\text{ER}}$ is used for theoretical analysis, because the fact that edges are independent makes calculations easier. Conversely, model A is preferred when we actually need to sample random graphs on a computer.

Example 3.4 *(Fitting a real network with a random graph)* Suppose you are given a real-world network G with N nodes and adjacency matrix $\{a_{ij}\}$ and you want to model it with a binomial random graph ensemble. What is the ensemble of graphs $G_{N,p}^{\text{ER}}$ that best approximates the real network? We can infer the best value of the parameter p from maximum likelihood considerations. The probability, or likelihood, P that the network G belongs to the ensemble $G_{N,p}^{\text{ER}}$ is given by Eq. (3.4) and only depends on the number of links in the network $K = \sum_{i<j} a_{ij}$. We want to find out the value of p that maximises P_G. It is useful to work with the logarithm of P_G, the so-called log-likelihood $\mathcal{L}(p)$ that the network G belongs to the ensemble:

$$\mathcal{L}(p) = \log P_G(p) = K \log p + [M - K] \log(1-p)$$

Maximising the log-likelihood with respect to p, i.e. solving the equation $d\mathcal{L}(p)/dp = 0$, we find that the best choice of the parameter p to model network G is, as expected, $p = K/M = 2K/(N(N-1))$, where N and K are number of nodes and links in G.

3.2 Degree Distribution

The nice property of random graphs is that they can be studied analytically. Coming back to the topics of graph centrality and centralisation of the previous chapter, in this section we will start by investigating the properties of random graphs by looking at how the degree is distributed among their nodes.

The degree distribution of ER random graphs can be easily derived analytically in model B in the following way. If p is the probability that there exists an edge between two generic vertices, the probability that a specific node i has degree k_i equal to k is given by the following expression:

$$\text{Prob}_{k_i=k} = C_{N-1}^k p^k (1-p)^{N-1-k} \quad 0 \le k \le N-1. \tag{3.11}$$

In fact, p^k is the probability for the existence of k edges, $(1-p)^{N-1-k}$ is the probability for the absence of the remaining $N-1-k$ edges, and $C_{N-1}^k = \begin{pmatrix} N-1 \\ k \end{pmatrix}$ is the number of different ways of selecting the end points of the k edges among the $N-1$ possible. The probability distribution in Eq. (3.11) has the same functional form of Eq. (3.5), i.e. it is a binomial distribution.

From $\text{Prob}_{k_i=k}$, we can now find an expression for the degree distribution p_k in an ensemble of random graphs. By degree distribution of a given graph we mean the probability that a node selected at random among the N nodes of the graph has exactly k edges.

> **Definition 3.3 (Degree distribution)** *The degree distribution of an undirected graph is defined as:*
>
> $$p_k = N_k/N \tag{3.12}$$
>
> *where N is the total number of nodes in the graph, and N_k is the number of nodes with exactly k links.*

We can extend this definition to the case of an ensemble of graphs. by introducing the degree distribution as:

$$p_k = \overline{N}_k/N.$$

where $\overline{N}_k$ denotes the expectation value of the quantity N_k over the ensemble of random graphs. The expectation of the number of nodes with exactly k links is given by the sum of the probabilities that each of the nodes in the graph has k links, namely:

$$\overline{N}_k = \sum_{i=1}^{N} \text{Prob}_{k_i=k}.$$

Since all the nodes in a random graph are statistically equivalent, $\text{Prob}_{k_i=k}$ is the same $\forall i$, and we have:

$$\overline{N}_k = N \, \text{Prob}_{k_i=k}$$

which implies that p_k has the same distribution as $\text{Prob}_{k_i=k}$. This means that the following statement holds.

The degree distribution in a random graph is a binomial distribution:

$$p_k = \binom{N-1}{k} p^k (1-p)^{N-1-k} = \text{Bin}(N-1, p, k) \quad k = 0, 1, 2, \ldots, N-1 \quad (3.13)$$

By using the properties of the binomial distribution, we can compute the average degree of a randomly chosen node in a random graph, and the standard deviation around this quantity [1]:

$$\langle k \rangle = \sum_{k=0}^{N-1} k p_k = p(N-1) \tag{3.14}$$

$$\sigma_k = \sqrt{\langle (k - \langle k \rangle)^2 \rangle} = \sqrt{\langle k^2 \rangle - \langle k \rangle^2} = \sqrt{p(1-p)(N-1)}. \tag{3.15}$$

In Figure 3.3 we check the validity of Eq. (3.13). In order to do so, we have implemented in computer code the *algorithm for sampling ER random graphs* of model B described in Appendix A.10.2. We then sampled one random graph from model B with $N = 10000$ nodes and $p = 0.0015$, and we computed the number of nodes N_k with degree k. In the figure we report the ratio N_k/N (filled circles) together with the analytical expression (solid line) obtained from Eq. (3.13) with $N = 10000$ and $p = 0.0015$. The agreement between solid line and circles is very good. Unfortunately, there is a problem with formula (3.13). When N is large, which is often the case we are interested in, the non-negligible values of p_k are obtained as the product of the binomial coefficient

$$C_N^k = \frac{(N-1)!}{k! \, (N-1-k)!} = \prod_{j=0}^{k-1} \frac{N-j}{k-j}$$

which can be very large, times a term $p^k q^{N-k-1}$, which is very small. The binomial coefficient may produce an overflow error. To avoid this problem, we can use Stirling's formula to approximate the factorial: $\ln N! \approx N(\ln N - 1)$ [210]. In particular, it is possible to transform equation (3.13) into a simpler expression.

In the limits $N \to \infty$ and $p \to 0$, with Np kept constant, the binomial distribution is well approximated by the so-called *Poisson distribution*:

$$p_k = \text{Pois}(z, k) = e^{-z} \frac{z^k}{k!} \quad k = 0, 1, 2, \ldots \tag{3.16}$$

where $z = Np$.

---

[1] Usually by the symbol $\langle x \rangle$ we mean the average of a property x over the nodes of a graph, while we have used the symbol $\bar{x}$ in this chapter to indicate the average of x over the ensemble of graphs.

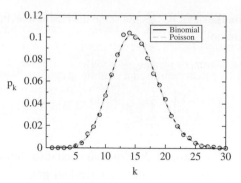

Fig. 3.3 The degree distribution (circles) that results from one realisation of a random graph with $N = 10000$ nodes and $p = 0.0015$ is compared to the corresponding binomial (solid line) and Poisson distributions (dashed line).

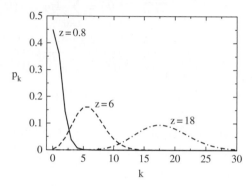

Fig. 3.4 The Poisson distribution of Eq. (3.17) is plotted as a function of k, for three different values of the average degree z.

Remember that $0! = 1$ by convention. This expression is normalised as a probability distribution. In fact, we can prove that it sums to 1:

$$\sum_{k=0}^{\infty} p_k = e^{-z} \sum_{k=0}^{\infty} \frac{z^k}{k!} = e^{-z} e^{z} = 1. \tag{3.17}$$

Here we made use of the fact that the infinite sum $1 + z + \frac{z^2}{2!} + \frac{z^3}{3!} + \ldots$ is equal to the exponential function e^z. Summing up, for large and sparse graphs the degree distribution of a random graph is well approximated by a Poisson distribution. As shown in Box 3.3, the positive parameter z is equal to the average node degree $\langle k \rangle$. The Poisson distribution is more tractable than the binomial distribution since we only need to compute the factorial of a number, k, that for our need is usually much smaller than N. For instance, in Figure 3.3 we have also reported (as dashed line) the Poisson distribution (3.16) with $z = pN = 15$. In practice, the curve we get is indistinguishable from the binomial, and the largest factorial we needed to compute was 30!.

In Figure 3.4 we plot the Poisson distribution as a function of k for different values of the average degree z. As shown in the figure, the distributions are peaked around the average degree, and they all have rather quickly decaying tails. In fact, the tails decay considerably quicker than any exponential (see Problem 3.2(c)). This is a relevant point when it comes to

Box 3.3	Moments of the Poisson Distribution

The moments of a Poisson degree distribution $\text{Pois}(z, k)$ can be derived by simple algebra. For the first and second order moments we have, respectively:

$$\langle k \rangle = \sum_{k=0}^{\infty} k p_k = \sum_{k=0}^{\infty} k \frac{z^k e^{-z}}{k!} = \sum_{k=1}^{\infty} k \frac{z^k e^{-z}}{k!} = z \sum_{k=1}^{\infty} \frac{z^{k-1} e^{-z}}{(k-1)!} = z$$

$$\langle k^2 \rangle = \sum_{k=0}^{\infty} k^2 p_k = \sum_{k=0}^{\infty} k^2 \frac{z^k e^{-z}}{k!} = \sum_{k=1}^{\infty} k^2 \frac{z^k e^{-z}}{k!} = z \sum_{k=1}^{\infty} k \frac{z^{k-1} e^{-z}}{(k-1)!}$$

$$= z \sum_{j=0}^{\infty} (j+1) \frac{z^j e^{-z}}{j!} = z(z+1) = z^2 + z$$

$$\sigma_k = \sqrt{\langle k^2 \rangle - \langle k \rangle^2} = \sqrt{z} \qquad\qquad (3.19)$$

where in the fourth summation of the second equation we performed the substitution $j = k - 1$. An alternative way to derive these relations is to differentiate, respectively once or twice, both sides of equation $\sum_{k=0}^{\infty} p_k = 1$ with respect to z.

model real-world networks. As we shall see, real systems often exhibit a power-law decay in the degree distribution. Such feature cannot be reproduced by Erdős and Rényi random graphs, whose distributions decay much faster. We will see in Chapter 5 how to generalise random graphs to incorporate arbitrary non-Poisson degree distributions.

3.3 Trees, Cycles and Complete Subgraphs

What are the salient features of a typical random graph with N nodes and K edges? Does it contain cycles or small complete subgraphs? How many? Is the graph connected? How do these properties depend on N and K? To answer these questions, let us consider the ensemble generated by model B, and study it as a function of the parameter p. In Figure 3.5 we show four realisations of a random graph with $N = 40$ and different values of p. When $p = 0$ we have N isolated nodes. For connection probabilities different from zero and small, the edges we add are isolated. As p, and with it the number of edges, increases, two edges can attach at a common node forming a tree of order 3. For larger values of p we expect the appearance of cycles and larger trees. For even larger values of p, the components start to merge together forming bigger components, until all nodes belong to a single connected component.

In a series of seminal papers, Erdős and Rényi proved that many important properties of random graphs, such as a random graph contains a tree or a complete graph of order 4, or the graph is connected or has a large connected component, appear suddenly. This means that, while a random graph with N nodes and a certain number of

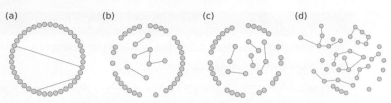

(a)   (b)   (c)   (d)

Fig. 3.5 The typical graph of the ensemble $G_{40,p}^{\mathrm{ER}}$ for four different values of p, namely $p = 2 \times 10^{-3}, 5 \times 10^{-3}, 8 \times 10^{-3}, 3 \times 10^{-2}$.

Box 3.4   **Phase Transitions and Critical Phenomena**

By *phase transitions* we usually intend the transformations of a system from one state of matter (phase) to another one. Phase transitions are common in many branches of physics. Examples are the boiling (transition from the liquid to the gaseous phase) of water at a transition temperature of $100\,^\circ C$ or the melting of ice (transition from solid to liquid) at $0\,^\circ C$ (at the atmospheric pressure). Under ordinary conditions these changes are *first-order phase transitions*, meaning transitions involving heat transfer. For instance, a non-zero quantity of heat, the so-called *latent heat*, is absorbed by ice while it melts at the transition temperature. Such heat is necessary to radically change the crystal organisation of the H_2O molecules of the ice into the more disorganised structure of the liquid. Although the phase changes of H_2O are normally first-order, there are special cases in which the liquid-gas transition is instead a *continuous phase transition*, i.e. it does not involve latent heat. This happens at the so-called *critical point*, defined by a critical temperature $T_c = 374\,^\circ C$ and a critical density $\rho_c = 0.323\, g\, cm^{-3}$. The typical example of continuous phase transitions is the change in the magnetic properties of metals such as iron at a critical temperature known as the Curie temperature. Above the critical temperature $T_c = 374\,^\circ C$ iron is paramagnetic, i.e. it is not magnetised in the absence of an external magnetic field. Below T_c the material is ferromagnetic, i.e. it is magnetised even in the absence of an external field. In this case the temperature plays the role of *control parameter*, while the magnetisation is the so-called *order parameter* and measures the degree of order in the system. The order parameter is non-zero below T_c, and zero above T_c. For an introduction to *critical phenomena*, i.e. the behaviour of systems close to their critical points, we suggest the book by H. Eugene Stanley [292], one of the first on the topic, and a more recent textbook [38].

---

edges is unlikely to have a given property at hand, a random graph with a few more edges is very likely to have that property. Such changes in the properties of a graph are usually called *phase transitions* by analogy to the change of phase observed in thermodynamic systems (see Box 3.4). In this and in the next section we discuss a series of theorems on phase transitions in random graphs. The theorems deal with the asymptotic properties of graph ensembles for infinitely large values of N. We need therefore to introduce a sequence of graph ensembles, one ensemble for each value of N. To do so, we assume the probability p to be an arbitrary function $p(N)$ of the number of nodes N, and we consider the sequence of graph ensembles $G_{N,p(N)}^{\mathrm{ER}}$ as a function of N. Then, we study the property of such a sequence as $N \to \infty$. The theorems state that, for a given function $p(N)$, either almost every graph in the ensemble $G_{N,p(N)}^{\mathrm{ER}}$ has some property or, conversely, almost no graph has it. The term "almost every graph" has the following crucial meaning.

Definition 3.4 (Almost every graph)   *Almost every graph (a.e.g.) has the property Q means that the probability that a graph in the ensemble has the property Q tends to 1 as N → ∞.*

This implies that, for any given property Q, there is an associated critical probability function $p_c(N)$. If $p(N)$ decays slower than $p_c(N)$ as $N \to \infty$, then *a.e.g.* in $G_{N,p(N)}^{\mathrm{ER}}$ has the property Q. If, instead, $p(N)$ decays faster than $p_c(N)$ as $N \to \infty$, then *a.e.g.* having a connection probability $p(N)$, fails to have Q[2].

The first property we study is the appearance of subgraphs, such as trees of a given length, or cycles. As we shall see, their appearance in random graphs depends on the scaling of the probability p as $N \to \infty$. We state a general theorem below, that determines the critical probability $p_c(N)$ marking the appearance of arbitrary connected subgraphs consisting of n nodes and l edges.

Theorem 3.1 (Appearance and number of subgraphs)   *Let $n \geq 2$, $n - 1 \leq l \leq \binom{n}{2}$ and let $F \equiv F_{n,l}$ be a connected graph with n nodes and l edges.*

- *If $p(N)/N^{-n/l} \to 0$ then a.e.g. in the ensemble $G_{N,p(N)}^{\mathrm{ER}}$ does not contain F, while if $p(N)/N^{-n/l} \to \infty$ then a.e.g. in the ensemble contains F.*
- *Let c be a positive constant and set $p(N) = cN^{-n/l}$. Let us denote with $n_F \equiv n_G(F_{n,l})$ the number of F-subgraphs in graph G. For the distribution of subgraph numbers $Prob(n_F = r)$ we have:*

$$\lim_{N \to \infty} P(n_F = r) = e^{-\lambda} \frac{\lambda^r}{r!} \qquad r = 0, 1, 2, \ldots \qquad (3.20)$$

where $\lambda = c^l / a_F$, with a_F being the number of elements in the automorphism group of F.

The first part of the theorem gives us the critical probability function $p_c(N) = cN^{-n/l}$ that marks the appearance of subgraph $F_{n,l}$. The second part tells how many times $F_{n,l}$ appears in G. While the rigorous proof of the theorem can be found in Refs. [49, 47], we present here a simple argument to evaluate the average number of times a given graph appears as a subgraph in the random graph ensemble $G_{N,p}^{\mathrm{ER}}$. Let us consider a graph $F_{n,l}$, with n nodes and l links. The n nodes can be chosen from the total number of nodes N of $G_{N,p}^{\mathrm{ER}}$ in C_N^n ways. Then, we require that the l links are present, and this will happen with a probability p^l. Therefore, we expect, on average, $C_N^n p^l$ graphs like F in G. In practice, in this way we are underestimating the real number, because we have forgotten to consider that we can permute the n nodes and potentially obtain $n!$ new graphs. As shown in the following example, actually, the correct number of new graphs that we can obtain is $n! / a_F$,

---

[2] To be more precise, $p(N)$ decaying faster than $p_c(N)$ as $N \to \infty$ means that $\lim_{N \to \infty} p(N)/p_c(N) = 0$. Conversely, $p(N)$ decaying slower than $p_c(N)$ as $N \to \infty$ means that $\lim_{N \to \infty} p(N)/p_c(N) = \infty$

where a_F is the number of automorphisms of graph F (see Definition 1.3), because a_F of the permutations will produce the same graph.

Example 3.5   Consider, as subgraph $F_{n,l}$, a tree with $n = 3$ nodes and $l = 2$ links, also known as a *triad* and usually denoted with the symbol $\wedge$. For such a tree we have $a_\wedge = 2$, with the two automorphisms respectively corresponding to the identity and to the transformation that keeps the central node fixed and permutes the two leaves (the two nodes of degree 1). Hence, the ratio $n!/a_\wedge$ will be equal to 3, because there are three different ways of selecting the central node. In conclusion, the average number of times the triad appears as a subgraph in $G_{N,p}^{ER}$ can be written as:

$$\bar{n}_\wedge = 3C_N^3 p^2 = \frac{N(N-1)(N-2)}{2}p^2.$$

Suppose, instead, subgraph $F_{n,l}$, with $n = 4$ and $l = 3$, is the star graph $\mathbb{S}_4$ with a central node and three links shown in Figure 1.4. As discussed in Example 1.3, the number of automorphisms of this graph is $a_{\mathbb{S}_4} = 6$. Hence, the ratio $n!/a_{\mathbb{S}_4}$ will be equal to 4, corresponding to the four different ways of selecting the node of degree equal to 3. In conclusion, the average number of times the star graph $\mathbb{S}_4$ appears as a subgraph in the ensemble $G_{N,p}^{ER}$ can be written as:

$$\bar{n}_{\mathbb{S}_4} = 4C_N^4 p^3 = \frac{N(N-1)(N-2)(N-3)}{3!}p^3.$$

Finally, consider subgraph $F_{n,l}$ with $n = 4$ and $l = 3$ is a triangle. We have again $a_\triangle = 6$, but this time $n!/a_F$ will be equal to 1. Then, the expression:

$$\bar{n}_\triangle = C_N^3 p^3 = \frac{N(N-1)(N-2)}{3!}p^3$$

will give the average number of triangles in the ensemble $G_{N,p}^{ER}$.

Finally, the expected number of times a generic graph $F_{n,l}$ is found as a subgraph of $G_{N,p}^{ER}$ can be written as:

$$\bar{n}_F = E[n_F] = C_N^n \frac{n!}{a_F}p^l \approx \frac{N^n p^l}{a_F} \tag{3.21}$$

where the last approximate equality is valid when N is large and n is small. If we consider p as a function of N, Eq. (3.21) gives indication on the critical probability function $p_c(N)$. In fact when $p(N)$ is such that $p(N)/N^{-n/l} \to 0$ as $N \to \infty$, Eq. (3.21) shows that the expected number of subgraphs $E(n_F) \to 0$, i.e. almost none of the random graphs contains subgraph $F_{n,l}$. Conversely, if $p(N)/N^{-n/l} \to \infty$ as $N \to \infty$, the expected number of subgraphs $E(n_F)$ diverges. The marginal case $p(N) = cN^{-n/l}$, with c being a positive constant, is the most interesting one. In such a case, the mean number of subgraphs F is finite and equal to c^l/a_F. This quantity coincides with the constant λ in Eq. (3.20).

In conclusion, the critical probability function $p_c(N)$ for the appearance of a subgraph $F_{n,l}$ is:

$$p_c(N) = cN^{-n/l} \tag{3.22}$$

Of course, Eq. (3.20) of Theorem 3.1 tells us something more. It states that the actual number of such subgraphs, n_F, is distributed according to a Poisson law. Hence the value of n_F in a particular graph of the ensemble can be different from $E(n_F)$, although in most of the cases it will be close to it (see the discussion on the tail of a Poisson distribution in Section 3.2).

Example 3.6 *(Tree, cycles and complete graphs)* From Theorem 3.1, and in particular from Eq. (3.22), it is straightforward to derive the critical probability functions $p_c(N)$ for the appearance of the most notable cases of subgraphs, namely trees, cycles and complete graphs. For instance, a tree of order n has $l = n - 1$ edges. Consequently, the critical probability for the appearance of trees with n nodes is $p_c(N) = cN^{-n/(n-1)}$. Analogously, the critical probability of having cycles of order n is $p_c(N) = cN^{-1}$, since for a cycle $l = n$. As we will see in the following, the scaling $p(N) = cN^{-1}$ is a particularly important case. Finally, the critical probability for the appearance of complete subgraphs of order n is $p_c(N) = cN^{-2/(n-1)}$. In fact, in a complete subgraph of n nodes, there are $l = C_n^2$ links.

The results of Theorem 3.1 are illustrated in Figure 3.6. As seen in the example above, the threshold for the appearance of different subgraphs only depends on the exponent of N in the expression of the critical probability $p_c(N)$. For this reason we consider a sequence of random graph ensembles $G_{N,p(N)}^{ER}$ and we assume that the connection probability $p(N)$ scales as $N^{-\zeta}$, where ζ is a tunable parameter that can take any value between 0 and ∞. In the figure, we show the occurrence of more and more complex structures as ζ decreases. When ζ is larger than 3/2, almost all graphs contain only isolated nodes and edges. When the exponent ζ passes through 3/2, trees of order 3 suddenly appear. When ζ reaches 4/3, trees of order 4 appear, and as ζ approaches 1, the graph contains trees of larger and larger order. However, as long as the exponent $\zeta > 1$, such that the average degree of the graph $\langle k \rangle = pN \to 0$ as $N \to \infty$, the graph is a union of disjoint trees, and cycles are absent. Exactly when the exponent ζ passes through the value 1, at which $\langle k \rangle =$const, the asymptotic probability of cycles of all orders jumps from 0 to 1. Cycles of order 3 can be also viewed as complete subgraphs of order 3. Complete subgraphs of order 4 appear at $\zeta = 2/3$, and as ζ continues to decrease, complete subgraphs of larger and larger order continue to emerge. Notice however that, when the exponent ζ is larger than 1, the average degree of the graph $\langle k \rangle = pN \to \infty$ as $N \to \infty$. In the following section we will consider a phase transition taking place when p scales with N as $p(N) = \langle k \rangle N^{-1}$. In this case, the phase transition is ruled by the precise value of the coefficient $\langle k \rangle$ in front of the scaling

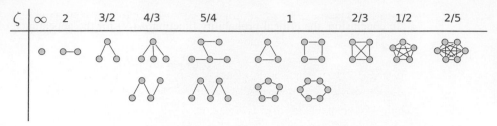

ζ	∞	2	3/2	4/3	5/4	1	2/3	1/2	2/5

Fig. 3.6 Threshold probabilities $p(N) \sim N^{-\zeta}$ for the appearance of different subgraphs in a random graph.

term N^{-1}. This phase transition has to do with the order and the structure of typical graph components, and in particular with the appearance of "large" components.

3.4 Giant Connected Component

In their works, Erdős and Rényi also discovered a phase transition concerning the order of the largest component in the graph, namely the abrupt appearance of a macroscopic component known as a *giant component*. This is probably the most dramatic example of phase transition in random graphs. To define mathematically a giant component we usually need to be able to consider a graph G as a function of its order N.

Definition 3.5 (Giant component) *A* giant component *in a graph G is a component containing a number of vertices which increases with the order of G as some positive power of N.*

Erdős and Rényi considered ensembles of random graphs where p scales with N as $p(N) = \langle k \rangle N^{-1}$, and the value of the average degree $\langle k \rangle$ can be tuned. They proved that a giant component appears at a critical probability function $p_c(N) = \frac{1}{N}$ in model B, or analogously at a critical number of links $K_c(N) = N/2$ in model A. This corresponds to a critical average degree $\langle k \rangle_c = 1$. Thus, there is a striking change of the asymptotic properties of a random graph as the average degree crosses one. This is known as the *double jump*, because the structure of the graphs is significantly different for $\langle k \rangle < 1$, $\langle k \rangle = 1$ and $\langle k \rangle > 1$, hence we have two changes and not one. When $\langle k \rangle < 1$ then the graph contains many small connected components, the maximum order of which scales as $\ln N$. If $\langle k \rangle \geq 1$, a giant component emerges. When $\langle k \rangle = 1$ the order of the largest component scales as $N^{2/3}$, while if $\langle k \rangle > 1$ there is a unique giant component, whose order scales linearly with N, while the second largest component has $\mathcal{O}(\ln N)$ nodes.

The first result on the double jump was obtained by Erdős and Rényi [101], and subsequently, the result was refined by the Hungarian-born British mathematician Béla Bollobás. A precise statement of the theorem is reported below [47]. We need to introduce first some notation that will be useful in the following. For a graph G, let us denote as $g_1, g_2, g_3, \ldots, g_{N_c}$ the N_c connected components sorted from the largest to the smallest, and by s_ℓ, with $\ell = 1, 2, \ldots, N_c$, respectively their order. By definition, we have

$s_1 \geq s_2 \geq s_3 \geq \ldots \geq s_{N_c}$. Furthermore, let us denote as n_s with $s = 1, 2, \ldots, N$ the number of components of order s.

Theorem 3.2 (Appearance of the giant component) *Consider $G_{N,p(N)}^{ER}$ with $p(N) = c/N$, where c is a positive constant. Let $\omega(N) \to \infty$ and $\alpha = c - 1 - \ln c$. Then, for N sufficiently large, the following hold.*

- *If $c < 1$, then for a.e.g. in the ensemble $G_{N,p(N)}^{ER}$ the order of the largest component satisfies:*

$$\left| s_1 - \frac{1}{\alpha} \left(\ln N - \frac{5}{2} \ln \ln N \right) \right| \leq \omega(N).$$

- *If $c = 1$, then there exist two constants $0 < \beta_1 < \beta_2$ such that, for a.e.g. in the ensemble,*

$$\beta_1 N^{2/3} < s_1 < \beta_2 N^{2/3}.$$

- *If $c > 1$, there is a constant $S(c)$, $0 < S(c) < 1$, such that, for a.e.g. in the ensemble,*

$$|s_1 - SN| < \omega(N) N^{1/2}$$

where $S(c)$ is the unique solution of:

$$e^{-cS} = 1 - S. \tag{3.23}$$

Furthermore

$$\left| s_2 - \frac{1}{\alpha} \left(\ln N - \frac{5}{2} \ln \ln N \right) \right| \leq \omega(N).$$

The proof of the theorem can be found in the standard graph theory textbooks [47]. Here, we shall give a simple heuristic argument to calculate the expected order of the giant component [235]. Suppose a graph G has a giant component. Let us indicate as $S = s_1/N$ ($0 < S \leq 1$) the fraction of the graph occupied by the giant component, and as $v = 1 - S$ the fraction of the vertices of the graph that do not belong to the giant component. The quantity v can also be interpreted as the probability that a vertex chosen uniformly at random from the graph is not in the giant component. We can now easily find a self-consistency relation for v. In fact, the probability of a vertex not belonging to the giant component is also equal to the probability that none of the vertex's network neighbours belong to the giant component. This latter is equal to v^k if the vertex has degree k. Averaging this expression over the probability distribution p_k in Eq. (3.16), where z is the average node degree, we find:

$$v = \sum_{k=0}^{\infty} p_k v^k = e^{-z} \sum_{k=0}^{\infty} \frac{(zv)^k}{k!} = e^{-z(1-v)}. \tag{3.24}$$

Hence, we get the following self-consistency relation for S:

$$S = 1 - e^{-zS} \tag{3.25}$$

which is precisely Eq. (3.23) in Theorem 3.2, with c being the average node degree z. This is a transcendental equation and has no closed-form solution. However, it can be solved graphically as in the following example, and admits a solution $S \neq 0$ only if $z > 1$.

Example 3.7 *(Graphical solution)* The transcendental Eq. (3.25) can be solved graphically by plotting the two curves $y_1(S) = S$ and $y_2(S) = 1 - e^{-zS}$ as a function of S, and finding their intersections S^* in the range $[0, 1]$. In fact, being S the fraction of nodes of the graph in the giant component, only a value of S between 0 and 1 has a physical meaning. The two functions $y_1(S)$ and $y_2(S)$ always intersect at $S^* = 0$. However, when $y_2'(S)|_{S=0} < 1$, then $S^* = 0$ is the only intersection. Instead, when $y_2'(S)|_{S=0} > 1$, there is also a second intersection at $S^* \neq 0$.

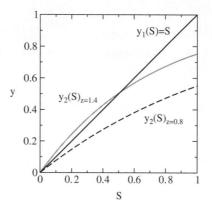

Since $y'(S) = ze^{-zS}$, the solution to Eq. (3.25) depends on the value of the average degree z. For $z \leq 1$ the only non-negative solution is $S^* = 0$, while for $z > 1$ there is also a non-zero solution, which is the order of the giant component. The precise value of the intersection S^* as a function of z, for $z > 1$, can be found numerically, or by means the of the following trick. Although there is no closed-form for the solution $S^*(z)$ to Eq. (3.25), the same equation can be used to express z as a function of S. We get $z(S) = -1/S \ln(1-S)$ that is the curve reported in Figure 3.7.

As shown in the Example 3.7, there is no giant component for $\langle k \rangle \leq 1$. Conversely a giant component can only appear in a random graph if $\langle k \rangle > \langle k_c \rangle = 1$. This result is in agreement with Theorem 3.2. The solution of Eq. (3.25) gives us the order of the giant component as a function of the average degree $z = \langle k \rangle$. This is the solid line reported in Figure 3.7. To check whether the prediction is correct, we have also computed the value of S for a sequence of random graphs sampled from model A. To generate the graphs we have used a computer code based on the *algorithm for sampling ER random graphs* of model A described in Appendix A.10.1. The graphs obtained have $N = 10,000$ nodes and different values of the average degree. The results of the numerical simulations are reported in the figure as triangles, and the agreement with the solid line is perfect. To see

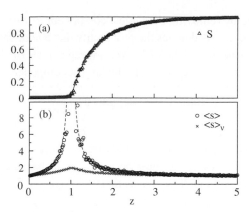

Fig. 3.7 Triangles in panel (a) are the computed values of the giant component $S = s_1/N$, as a function of the average node degree z, obtained from a sequence of random graphs sampled from ER random graph model A, while the smooth curve is the solution to Eq. (3.25). In panel (b) we show, as circles, the computed value of the mean order $\langle s \rangle$ of the finite component to which a randomly chosen node belongs, while the dashed curve is Eq. (3.29). Also for comparison we report the average component order $\langle s \rangle_v$ as crosses.

(a)   (b)

Fig. 3.8 Plot of the largest component g_1 obtained in an ER random graph with $N = 1000$ nodes and with an average degree respectively below, $\langle k \rangle = 0.9$ (a), and above, $\langle k \rangle = 1.1$ (b), the critical value $\langle k \rangle_c = 1$.

by eye the phase transition occurring in the order of the largest component at $\langle k \rangle_c = 1$, we plot in Figure 3.8 the largest connected component obtained in a realisation of a random graph with $N = 1000$ nodes, and for two different values of $\langle k \rangle$, one below the critical threshold, namely $\langle k \rangle = 0.9$, and one above the critical threshold, $\langle k \rangle = 1.1$. The largest component g_1 in the first case has $s_1 = 38$ nodes, while it contains $s_1 = 202$ nodes, i.e. 20 per cent of the graph nodes, in the second case. What is shown is just the outcome of a single realisation of a random graph. However, the value of s_1 found, is not too different from the expected value of 176 nodes, that is obtained by setting $\langle k \rangle = 1.1$ in Eq. (3.25).

Notice from Figure 3.7 that $\langle k \rangle_c$ is also the point at which we observe a divergence in $\langle s \rangle$, namely in the average size of the finite component to which a randomly chosen node belongs. The quantity $\langle s \rangle$ is defined in the following way. Consider the graph $G - g_1$

obtained by removing the largest component g_1 from G, and let $p(g_\ell) = s_\ell/(N - s_1)$ with $\ell = 2, 3, \ldots, N_c$ denotes the probability that a node randomly selected from $G - g_1$ belongs to component g_ℓ. Therefore, the average order of the component to which a randomly selected node, which is not in the largest component, belongs, is given by:

$$\langle s \rangle = \sum_{\ell=2}^{N_c} p(g_\ell) s_\ell = \frac{\sum_{\ell=2}^{N_c} s_\ell^2}{N - s_1} \tag{3.26}$$

This quantity can be obtained directly from the distribution n_s that we have previously defined. However, we must be very careful because n_s is the number of components of order s in a graph. This means that:

$$v_s = \frac{n_s}{\sum_s' n_s} \tag{3.27}$$

is the probability that, by choosing at random with equal probability one component of the graph, excluding the largest one, we get a component of order s. Notice that by the symbol prime in $\sum_s'$ we have indicated a summation over all possible component orders s, which does not include the largest component. In order to calculate $\langle s \rangle$ in Eq. (3.26) we need instead to define a probability different from v_s, namely the probability w_s that, by choosing at random, with uniform probability, a node of the graph, we end up in a component of order s. Taking into account that a component of order s can then be selected in s different ways, the probability w_s can be written in terms of n_s as:

$$w_s = \frac{s n_s}{\sum_s' s n_s} = \frac{s n_s}{N - s_1} \tag{3.28}$$

where the last equality is indeed obtained because in $\sum_s'$ we are not considering the largest component in the summation. Now, it is clear that the average size $\langle s \rangle$ of the component to which a randomly chosen node belongs, given in Eq. (3.26), is nothing other than the first moment of the distribution w_s, namely $\langle s \rangle = \sum_s' s w_s$. To calculate $\langle s \rangle$ in practice, we have performed a so-called *component analysis*, and computed the order of the various components of the random graph as a function of z. Although it is possible to give algebraic characterisation of the various components of a graph using the adjacency matrix, it is much more efficient to detect the various components by algorithms that perform graph exploration. The main algorithms for component analysis (both in the case of undirected and directed graphs) are described in the Appendix A.8. The quantity $\langle s \rangle$ is shown in Figure 3.7 as circles. We notice that as z tends to one from below, the value of $\langle s \rangle$ diverges. Then, it is again finite when $z > 1$. This is so because, by removing the largest component from the definition of the average component order $\langle s \rangle$, we have removed in this case the giant component. As we will see in Chapter 5, it is possible to prove that, in the limit $N \to \infty$, the quantity $\langle s \rangle$ can be written as:

$$\langle s \rangle = \sum_s' s w_s = \frac{1}{1 - z + z S(z)} \tag{3.29}$$

where $S(z)$ is zero for $z \leq 1$ and is given by the solution of Eq. (3.25) for $z > 1$. The quantity $\langle s \rangle$ indeed diverges both as $z \to 1^{-1}$ and $z \to 1^{+1}$. The function in Eq. (3.29)

Box 3.5	The Giant Component of a Real-World Network?

How can we determine if there is a giant component in a real network? Mathematically speaking, the concept of giant component, introduced in Definition 3.5, is related to the way the largest component of a graph scales with the order of the graph. This is perfectly fine when we are dealing with ensembles of random graphs, because we can consider the properties of the ensemble for large values of N. For example, there are some chances that a random graph with $N = 100$ and $K = 40$, hence a graph with average node degree $\langle k \rangle = 0.8$, has a component of order 60. However, we know from Theorem 3.2 that random graphs with $\langle k \rangle = 0.8$ do not have a giant component. This means that if we increase the number of nodes N of the graph, by keeping fixed the average node degree $\langle k \rangle$, the probability of finding a component scaling as a positive power of N will be zero when $N \rightarrow \infty$. Unfortunately, when we deal with a real system, we are usually given only one network, with a fixed number of nodes N. Therefore, it is somewhat arbitrary to decide whether the largest component in a real network is a giant component or not. The only thing we can do is to say that a real network has a *giant component*, if the order of its largest component is a "considerable" fraction of N.

is shown in the figure as a dashed curve, and is in agreement with the numerical simulations. For comparison, in the same figure we have also reported the average order $\langle s \rangle_v$ of the components that is obtained if we select each component with equal probability and not proportionally to its order. The quantity $\langle s \rangle_v$ is obtained as the first moment of the distribution v_s, namely $\langle s \rangle_v = \sum'_s s v_s$, and this is the reason why we have indicated it with the subscript v. Notice that, differently from $\langle s \rangle$ of Eq. (3.26), the quantity $\langle s \rangle_v$ does not diverge in the limit $N \rightarrow \infty$ [239]. The behaviour of S and $\langle s \rangle$ shown in Figure 3.7 will be recognised by those familiar with the theory of phase transitions as a *continuous phase transition* (see Box 3.4). Here, z corresponds to the so-called *control parameter* of the phase transition, while S and $\langle s \rangle$ play respectively the role of the *order parameter* and of the *order-parameter fluctuations* [292, 38].

We finally state a theorem which tells us when a random graph is connected [47].

Theorem 3.3   *Let $\omega(N) \rightarrow \infty$ and set $p_l(N) = (\ln N - \omega(N))/N$ and $p_u(N) = (\ln N + \omega(N))/N$. Then a.e.g. in the ensemble $G_{N,p_l(N)}^{ER}$ is disconnected and a.e.g. in the ensemble $G_{N,p_u(N)}^{ER}$ is connected.*

In practice, Theorem 3.3 tells us that in model B, $p_c(N) = \ln N/N$ is a threshold probability function for the disappearance of isolated vertices. The corresponding threshold in model A is $K_c = \frac{N}{2} \ln N$. The proof of Theorem 3.3 is quite involved [47], and we will not give it here. In the following example we will report instead a simple argument to demonstrate a slightly weaker form of the theorem [313]. Namely, we will assume that $p(N) = c \ln N/N$ with $c > 0$, and we will prove that for $c > 1$ a.e.g. in the sequence has no isolated nodes, while for $c < 1$ a.e.g. in the sequence has isolated nodes. Notice that a connected graph has no isolated nodes, while in general the converse is not true: a graph without isolated nodes might be disconnected. In this sense Theorem 3.3 provides a stronger result. However, it is quite intuitive to understand that removing links at random from a connected graph until it becomes disconnected will first produce, with very large probability, isolated nodes.

Example 3.8   *(Connectedness condition)*   Remember that by n_s we have indicated the number of components of order s. Therefore n_1 is the number of isolated nodes in the graph. We shall show now that, when $p(N) = c \ln N / N$ with $c > 1$, the expectation value $\bar{n}_1$ is equal to zero. If the graph has N nodes, the expectation value of the number of isolated nodes can be written as $\bar{n}_1 = N(1 - p)^{N-1}$, where $(1 - p)^{N-1}$ is the probability that a given node of the graph is isolated. For large N we can write:

$$\bar{n}_1 = N(1 - p)^N = N e^{N \ln(1-p)} = N e^{-Np} e^{-Np^2[1/2 + p/3 + \dots]} = N e^{-Np}$$

where we have expanded $\ln(1 - p)$, and in the last equality we have made use of $Np^2 \to 0$. Now, by plugging the functional form $p(N) = c \ln N / N$ into this expression, we get:

$$\bar{n}_1 = N^{1-c}$$

which tends to 0 for $N \to \infty$, because we have assumed $c > 1$.

3.5 Scientific Collaboration Networks

You may object that the graph of movie actors studied in Chapter 2 is not a good representative example of collaborations in social systems. After all, movie actors have exceptional lives and experiences, so, we expect that their connections will also be governed by uncommon rules. Here, we study another example of collaboration. This time the nodes of the network are scientists, and the relationships are scientific collaborations. One way to define the existence of a scientific collaboration is through scientific publications. We can assume two scientists to be connected if they have coauthored at least one publication. We then obtain scientific coauthorship networks as projections of bipartite networks, in a similar way to our approach to movie actor networks in the previous chapter. In this case, the two sets of nodes represent respectively scientists and scientific publications. Joint scientific publications appear to be a reasonable definition of scientific acquaintances, because people who have been working together will know each other quite well and are more likely to set up an ongoing collaboration. Furthermore, large-scale data related to coauthorships can be easily found on the huge publication records that are now accessible on the Internet. Here, we study a set of coauthorship networks first introduced and analysed in 2001 by Mark Newman, then a physicist at the Santa Fe Institute [234, 233]. The networks are constructed as detailed in the DATA SET Box 3.6, and can be downloaded from the book webpage.

Before moving to the analysis of the seven networks in DATA SET 3, it can be useful to have a better idea of the four databases of articles and preprints they refer to. The first and

DATA SET 3: Coauthorship Networks

In a coauthorship network, the nodes are scientists, and two scientists have a link if they have coauthored at least one publication together, as sketched in the figure for some scientists working in complex networks. The networks provided in this data set are seven unweighted and simple graphs originally constructed by Mark Newman from four publicly available databases of papers respectively in biomedicine, physics, high-energy physics and computer science, namely *Medline*, *ArXiv*, *SPIRES* and *NCSTRL* [234, 233].

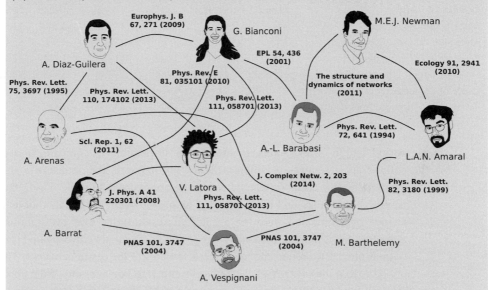

Each network corresponds to papers posted in the five-year period from 1 January 1995 to 31 December 1999, in each of the four databases. In the particular case of ArXiv, together with the entire network, Newman also constructed three partial networks, respectively corresponding to papers posted in the directories astro-ph (astrophysics), cond-mat (condensed matter) and hep-th (high-energy theory). The number of nodes and links and other basic features of the seven networks of this data set are reported in Table 3.1.

largest one, *Medline*, is a database of articles on biomedical research published in refereed journals from 1961 to the present. The physics e-Print archive known as *ArXiv* is instead a website repository of preprints self-submitted by their authors, created by physicist Paul Ginsparg at the Los Alamos National Laboratory (LANL) in New Mexico in 1991, and maintained, from 2001, by the Cornell University Library. The database was originally organised into different specialities within physics, such as astrophysics *astro-ph*, condensed matter *cond-mat* and high-energy physics *hep-th*, and nowadays it includes also sections dedicated to mathematics and computer science. *SPIRES* (Stanford Public Information Retrieval System) is a database of preprints and published papers in high-energy physics, both theoretical and experimental, running from 1974 to the present. Finally, *NCSTRL* (Networked Computer Science Technical Reference Library) is a database of preprints in computer science. Notice that, although the four publication repositories run for much longer periods, all the networks have been constructed considering the same

Table 3.1 Basic properties of scientific coauthorship networks in different disciplines, namely biomedicine, physics and computer science, in the period 1995–1999.

	Medline	ArXiv	astro-ph	cond-mat	hep-th	SPIRES	NCSTRL
Number of nodes (authors) N	1520252	52909	16706	16726	8361	56627	11994
Number of links K	11803060	245300	121251	47594	15751	4573868	20935
Collaborators per author $\langle k \rangle$	15.5	9.27	14.5	5.69	3.77	161	3.49
Number of papers	2163923	98502	22029	22016	19085	66652	13169
Mean papers per author	6.4	5.1	4.8	3.65	4.8	11.6	2.55
Mean authors per paper	3.75	2.53	3.35	2.66	1.99	8.96	2.22
Number of components N_c	89507	4775	1029	1188	1332	4334	2618
Order of largest component s_1	1395693	44337	14845	13861	5835	49002	6396
Order of 2nd largest component s_2	49	18	19	16	24	69	42
Order of 3rd largest component s_3	37	18	15	15	20	23	29

period length of five years. In this way it is possible to compare, in a consistent way, collaboration patterns in different fields. Furthermore, the choice of a short time period ensures that the collaboration networks are approximately static during the study. In fact, names of scientists and the very same nature of scientific collaborations and publishing change over time and, here, we are not interested in the time evolution of the networks, although this is another important aspect that has been also investigated [18, 68].

In Table 3.1 we report the main characteristics of the seven networks. The first thing we notice is that the order N of the networks can vary considerably, from about ten thousand authors for NCSTRL to about a million for Medline. The average degree $\langle k \rangle$ is always smaller than 20, apart from the case of SPIRES, where an author has an average of 161 different coauthors in a period of only five years. This is certainly due to the presence of large experimental collaborations in high-energy physics, which makes this scientific discipline very different from all the others. The table shows other important features of the databases, such as the total number of papers, the average number of papers per author, and the average number of authors per paper. Although these are not directly network quantities, they carry important information on the common collaboration practices in the different disciplines. The biomedicine Medline is the largest database with more than 2 million papers in the considered five-year period, while NCSTRL contains only slightly more than 13,000 papers. The average number of papers per author ranges from around 3 to 7 in all databases, with the only exception of the SPIRES, in which this number is 11.6. The significantly higher number of papers per author in SPIRES could be due to the fact that this is the only database containing both preprints and published papers. The mean number of authors per paper in all subject areas is smaller than 10. Databases with purely theoretical papers, such as hep-th and NCSTRL have on average only two authors per paper, while the averages are higher for databases with both theoretical and experimental papers. For instance, we have 3.75 authors per paper for biomedicine, 3.35 for astrophysics, 2.66 for condensed matter physics, and 8.96 for the high-energy physics database SPIRES. Again, SPIRES is largely different from the other databases, mainly because of the presence of papers written by very large collaborations [234].

By using the algorithm in Appendix A.8 we have performed a *component analysis* to compute the number and order of connected components. The number of components N_c, and the order of the three largest components, s_1, s_2, s_3, is reported in Table 3.1. We notice that in all the graphs considered $s_1 \gg s_2$, i.e. one component is always much larger than the others. This is an indication of the presence in each of the graphs of what we can call a giant component. Indeed, all graphs considered have an average degree $\langle k \rangle$ that is above the critical threshold $\langle k \rangle_c = 1$ for the existence of a giant component in random graphs. In addition to this, three graphs, namely Medline, astro-ph and SPIRES, have an average degree $\langle k \rangle$ larger than the critical threshold $\langle k \rangle_c = \ln N$ for the appearance of a single connected component in a random graph. Consequently, by modelling these three coauthorship networks as random graphs we would expect a single connected component, while this is not what we observe in the real systems. Let us for instance focus on SPIRES, among the seven networks considered, the one with the largest average degree $\langle k \rangle$. SPIRES is also the network with the largest ratio $\langle k \rangle / \ln N = 161/10.94 \approx 14.7 \gg 1$. Nevertheless, SPIRES is made of 4334 components and not of a single one, and its largest component contains only 87 per cent of the graph nodes. At the other extreme, in NCSTRL, the graph with the smallest degree $\langle k \rangle$, the largest component contains about 53 per cent of the nodes of the graphs, while the remaining nodes are organised into 2617 components.

The presence of a large number of components with more than just a few nodes can be better appreciated from a plot of the number n_s of components of order s, as a function of s. In Figure 3.9 we report the results for ArXiv, SPIRES and NCSTRL. For practical reasons the largest component has been removed from the plots. Although there are some fluctuations due to the limited statistics, we notice that the three distributions are slowly decaying and show a linear behaviour in a double logarithmic scale. In the same plots, we have also reported for comparison the average number $\bar{n}_s$ of components of order s in ensembles of random graphs with the same number of nodes and links as in the real networks. We notice that the randomised version of SPIRES is indeed made of a single connected component, not reported in the plot, while the randomised version of ArXiv contains one large component and only a few isolated nodes. Only for NCSTRL do we observe the presence of components other than isolated nodes (and of order up to about ten) in the randomised version of the graph, although the average number of components, $\bar{n}_s$, is a much more rapidly decreasing function of s than in the real network. Average number

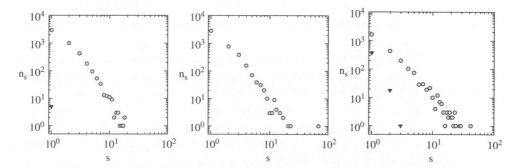

Fig. 3.9 Number of components of order s in three coauthorship networks, namely ArXiv, SPIRES and NCSTRL (circles), and in random graphs with the same number of nodes and links (triangles).

Table 3.2 Average properties of ensembles of random graphs with same order and size of three real scientific coauthorship networks.

	Randomised ArXiv	Randomised SPIRES	Randomised NCSTRL
Number of nodes N	52909	56627	11994
Average degree $\langle k \rangle$	9.27	161	3.49
Average number of components $\overline{N}_c$	6.06 ± 2.36	1	388.3 ± 17.3
Average order of largest component $\overline{s}_1$	52903 ± 2.4	56627	11582 ± 19.3
Average order of second-largest component $\overline{s}_2$	0.99 ± 0.08	–	3.25 ± 0.71
Average of third-largest component $\overline{s}_3$	0.95 ± 0.21	–	2.76 ± 0.55

of components, and average order of the three largest components in random graphs with the same number of nodes and links as in ArXiv, SPIRES and NCSTRL, are reported for comparison in Table 3.2. The averages are performed over ensembles of 1000 random graphs. We observe, indeed, that the randomised version of SPIRES is always made of a single connected component with 56627 nodes, while on average, the randomised version of ArXiv has six components: five isolated nodes and one component with all the remaining nodes. The values $\overline{s}_2 = 0.99$ and $\overline{s}_3 = 0.95$, indicate that 99 per cent of the graphs in the random ensemble corresponding to ArXiv have at least two components, while 95 per cent of them have at least three components. Also the randomised version of NCSTRL has a number of components $\overline{N}_c = 388.3$ which is much smaller than the number of components in the real graph, $N_c = 2618$.

In conclusion, real coauthorship networks are more fragmented than random graphs with the same number of nodes and links. In the next section we define a measure to quantify how well connected a graph is. This will allow us to compare the structure of the largest component in coauthorship networks with that of a random graph.

3.6 Characteristic Path Length

The *distance matrix* $\mathcal{D}$, whose entry d_{ij} is a non-negative integer equal to the length of the shortest paths from node i to node j, contains important information to characterise a graph. In Section 2.4 we used shortest path lengths to define the closeness centrality of a node. Here, we will characterise an entire graph, e.g. a network of scientific collaborations, by computing the typical distance between two nodes. We will compare the results obtained to those expected if the network were a random graph. We start by adding a few more definitions to those already given in Chapter 1.

Definition 3.6 (Graph diameter and radius) *Consider a connected graph. The* eccentricity e_i *of node i is the maximum graph distance between i and any other vertex in the graph:* $e_i = \max_j\{d_{ij}\}$. *The* graph radius R *is the minimum eccentricity:*

$$R = \min_i\{e_i\}, \tag{3.30}$$

while the graph diameter D *is the maximum eccentricity:*

$$D = \max_i\{e_i\} = \max_{i,j}\{d_{ij}\}, \tag{3.31}$$

that is to say the largest possible distance between two nodes in the graph.

Definition 3.7 (Characteristic path length) *Consider a connected graph. The* average distance, *L, is the mean graph distance over all pairs of nodes:*

$$L = \frac{1}{N(N-1)} \sum_{i=1}^{N} \sum_{\substack{j=1 \\ j \neq i}}^{N} d_{ij} \tag{3.32}$$

This quantity is also known as the characteristic path length *of the graph [309, 311].*

The same definitions can be extended to unconnected graphs. However, when a graph is not connected there are pairs of nodes i, j such that $d_{ij} = \infty$. This implies that $L = D = \infty$. In the case of unconnected graphs, it can be therefore more convenient to evaluate L and D of the largest connected component.

In the following example we illustrate how to calculate the eccentricity of a node, and how to derive from this the radius R and the diameter D of a graph.

Example 3.9 *(Node eccentricity)* Consider the two connected graphs with $N = 9$ nodes reported in the figure. Notice that the two graphs differ by exactly one edge. In fact, the edge $(7, 9)$ in the graph shown in panel (a) is replaced by the edge $(2, 9)$ in the graph shown in panel (b). For each graph, we have first computed the distance matrix $\mathcal{D} = \{d_{ij}\}$ by means of the *Breadth-First Search* algorithm introduced in Section 2.4 and described in Appendix A.6. Since the two graphs are connected, all the entries of $\mathcal{D}$ are finite. The eccentricity of a node is the maximum graph distance between that node and any other node in the graph. We can thus obtain the eccentricity e_i of node i as the maximum of the ith row of matrix $\mathcal{D}$. The two resulting vectors $\mathbf{e} = \{e_i\}$ are reported. The values of eccentricity obtained are also shown in parentheses close to each node. The radius R and the diameter D of each graph are given respectively

(a)

Graph with nodes: 9(4), 7(3), 8(3), 6(3), 5(2), 4(4), 2(3), 3(3), 1(4)

$$
\mathcal{D} =
\begin{pmatrix}
0 & 1 & 1 & 2 & 2 & 3 & 3 & 3 & 4 \\
1 & 0 & 2 & 1 & 1 & 2 & 2 & 2 & 3 \\
1 & 2 & 0 & 3 & 1 & 2 & 2 & 2 & 3 \\
2 & 1 & 3 & 0 & 2 & 3 & 3 & 3 & 4 \\
2 & 1 & 1 & 2 & 0 & 1 & 1 & 1 & 2 \\
3 & 2 & 2 & 3 & 1 & 0 & 2 & 2 & 3 \\
3 & 2 & 2 & 3 & 1 & 2 & 0 & 2 & 1 \\
3 & 2 & 2 & 3 & 1 & 2 & 2 & 0 & 3 \\
4 & 3 & 3 & 4 & 2 & 3 & 1 & 3 & 0
\end{pmatrix}
\qquad
e =
\begin{pmatrix}
4 \\ 3 \\ 3 \\ 4 \\ 2 \\ 3 \\ 3 \\ 3 \\ 4
\end{pmatrix}
\qquad
\begin{aligned}
D &= 4 \\[3em]
R &= 2
\end{aligned}
$$

(b)

Graph with nodes: 7(3), 8(3), 6(3), 5(2), 4(3), 9(3), 2(2), 3(3), 1(3)

$$
\mathcal{D} =
\begin{pmatrix}
0 & 1 & 1 & 2 & 2 & 3 & 3 & 3 & 2 \\
1 & 0 & 2 & 1 & 1 & 2 & 2 & 2 & 1 \\
1 & 2 & 0 & 3 & 1 & 2 & 2 & 2 & 3 \\
2 & 1 & 3 & 0 & 2 & 3 & 3 & 3 & 2 \\
2 & 1 & 1 & 2 & 0 & 1 & 1 & 1 & 2 \\
3 & 2 & 2 & 3 & 1 & 0 & 2 & 2 & 3 \\
3 & 2 & 2 & 3 & 1 & 2 & 0 & 2 & 3 \\
3 & 2 & 2 & 3 & 1 & 2 & 2 & 0 & 3 \\
2 & 1 & 3 & 2 & 2 & 3 & 3 & 3 & 0
\end{pmatrix}
\qquad
e =
\begin{pmatrix}
3 \\ 2 \\ 3 \\ 3 \\ 2 \\ 3 \\ 3 \\ 3 \\ 3
\end{pmatrix}
\qquad
\begin{aligned}
D &= 3 \\[3em]
R &= 2
\end{aligned}
$$

by the minimum and maximum component of the corresponding vector $\mathbf{e}$. Note that the two graphs have the same radius $R = 2$, but different values of the diameter, namely $D = 4$ and $D = 3$.

Table 3.3 shows the values of D and L we have obtained for the collaboration networks introduced in the previous section. More precisely, since such networks are unconnected, we have computed the average distance and the diameter of the largest component in each graph. The order, s_1, and the average degree, $\langle k \rangle_{g_1}$, of the largest component are also reported in the table. All the graphs exhibit values of the characteristic path length which are much smaller than the number of nodes in the largest component. It is convenient to compare the values of L obtained for these networks to those of random graphs with the same number of nodes and links. In practice, for any given pair N and K, we can generate an ensemble of random graphs from model A by using the algorithm described in Appendix A.10.1. Then, we can evaluate the ensemble average $\overline{L}$ of the characteristic path length. However, we can save our time, since for random graphs it is possible to derive a simple expression which gives L as a function of N and K.

The characteristic path length of a Erdős-Rényi random graph is:

$$
L^{\mathrm{ER}} \approx \frac{\ln N}{\ln \langle k \rangle} \tag{3.33}
$$

This equation shows that the average node distance in a random graph scales logarithmically with the graph order N. This implies that, in a random graph with $\langle k \rangle = 10$, when the number of vertices increases from 10000 to 100000, L will simply increase from 4 to 5. For the same values of N and $\langle k \rangle$ as in the various scientific collaboration networks considered, we get respectively the values of L^{ER} reported in Table 3.3.

Table 3.3 Diameter D and characteristic path length L of the largest component in various scientific coauthorship networks. The order and average degree of the largest component are also reported. L^{ER} is the prediction of Eq. (3.33) for random graphs with the same number of nodes and links.

Scientific coauthorship network	s_1	$\langle k \rangle_{g_1}$	D	L	L^{ER}
Medline	1395693	16.8	24	4.92	5.01
ArXiv	44337	10.8	19	5.99	4.50
astro-ph	14845	16.1	14	4.80	3.46
cond-mat	13861	6.44	18	6.63	5.12
hep-th	5835	4.74	19	7.03	5.57
SPIRES	49002	186	19	4.14	2.07
NCSTRL	6396	4.96	31	10.48	5.47

A rigorous proof of Eq. (3.33) can be found in Refs. [49, 77, 99]. We can however get the same expression by the following simple argument to estimate the average number, z_m, of nodes at distance m from a given node. We will make use of the fact that a large random graph with a fixed (independent of N) value of $\langle k \rangle$ has, to a good approximation, a *tree-like* structure. In fact, according to Theorem (3.1), at the threshold function $p(N) = cN^{-1}$, the number of cycles is finite and their presence in a large random graph can be neglected. Let us consider then a generic vertex i of the graph. Moving outward from i to its first neighbours, and then to the second neighbours, the situation is that shown in Figure 3.10. On average, starting from i we can reach $z_1 = \langle k \rangle$ nodes in one step. If now we want to calculate z_2, i.e. the average number of nodes at distance two from i, we are interested not in the complete degree of the vertex j reached by following an edge from i, but in the number of edges emerging from j other than the one we arrived along (since the latter edge leads back to vertex i and so does not contribute to the number of second neighbours of i). This number, which we denote by ρ, is one less than the average degree of vertex j:

$$\rho = [(\textit{average degree of first neighbours of } i) - 1] \tag{3.34}$$

Therefore, the average number of nodes at distance two from a generic node i is:

$$z_2 = \rho \cdot \langle k \rangle \tag{3.35}$$

We are in fact sure that there is no overlap between nodes at distance two from i because the graph originating from i is tree-like, as shown in Figure 3.10. At this point we only need to calculate the average degree of the first neighbours of a generic node in a random graph. This quantity, known as the *average nearest neighbours' degree* and indicated as $\langle k_{nn} \rangle$, where the subscript "nn" stands for nearest neighbours, can be written for any graph in terms of its adjacency matrix as follows.

Definition 3.8 (Average nearest neighbours' degree)   *The average degree $k_{nn,i}$ of the nearest neighbours of node i in a graph G is defined as:*

$$k_{nn,i} = \frac{1}{k_i} \sum_{j=1}^{N} a_{ij} k_j. \tag{3.36}$$

3.5 Scientific Collaboration Networks

(a) Consider the scientific coauthorship network from papers posted in the *cond-mat* (condensed matter) directory of *ArXiv* given in DATA SET 3. Construct the so-called *ego-network* of Mark Newman (M.E.J. Newman, node label 759), which is made by Mark Newman, his collaborators, and all the links among them [233]. Draw the ego-network in the usual way: that is, plotting the node corresponding to Mark Newman as the centre of a circle, and all its first neighbours along the circle. How many components does the ego-network have if we remove Mark Newman from it?

(b) The characteristic path length in Eq. (3.32) can also be written as $L = \dfrac{1}{N} \sum_i^N L_i$,

where $L_i = \dfrac{1}{N-1} \sum_{\substack{j=1 \\ j \neq i}}^{N} d_{ij}$. Consider the scientific coauthorship network from papers posted in the *cond-mat* (condensed matter) directory of *ArXiv* given in DATA SET 3. Compute L_i for every node in the largest component of the network, and plot how L_i is distributed.

3.6 Characteristic Path Length

(a) According to current estimates from the United Nations Population Division, there are today about 323 million people leaving in the USA, and 7.4 billion living in the entire world. Can you estimate the characteristic path length of the social network in the USA, and in the entire world, knowing that $\langle k \rangle = 1000$ is a good approximation of the typical number of acquaintances for people in the USA (and assuming the same value of $\langle k \rangle$ for the entire world)?

(b) Compare the definition of the characteristic path length L with that of the efficiency E of a graph given in Eq. (2.16). Evaluate L and E for random graphs with a fixed average degree $\langle k \rangle$ and different number of nodes N. Investigate how L and E scale with N for different values of $\langle k \rangle$. Choose $\langle k \rangle$ in such a way to guarantee that the graphs are connected.

(c) Are you able to use the expression of ρ in Eq. (3.34) to find a condition for the appearance of a giant component in ER random graphs? Notice that ρ gives the average number of edges emerging from the first neighbour of a randomly chosen node other than the one we arrived along. HINT: Use the expression of ρ given in Eq. (3.42) for ER random graphs.

4 Small-World Networks

"It's a small world!" This is the typical expression we use many times in our lives when we discover, for example, that we unexpectedly share a common acquaintance with a stranger we've just met far from home. In this chapter, we show that this happens because social networks have a rather small *characteristic path length*, comparable with that of random graphs with the same number of nodes and links. In addition to this, social networks also have a large *clustering coefficient*, i.e. they contain a large number of *triangles*. As an example of a social network, we will experiment on the collaboration graph of movie actors introduced in Chapter 2. We will show that the *small-world behaviour* also appears in biological systems. For this reason, we will be looking into the *neural network of* C. elegans, the only nervous system that has been completely mapped to date at the level of neurons and synapses. We will then move our focus to the modelling, by introducing and studying both numerically and, when possible, analytically, the *small-world model* originally proposed in 1998 by Watts and Strogatz to construct graphs having both the small-world property and also a high clustering coefficient. This model and the various modified versions of it that have been proposed over the years, are all based on the addition of a few long-range connections to a regular lattice and provide a good intuition about the small-world effect in real systems. In the last section we will try to understand how the individuals of a social network actually discover short paths, even if they just have local knowledge of the network.

4.1 Six Degrees of Separation

Fred Jones of Peoria, sitting in a sidewalk cafe in Tunis, and needing a light for his cigarette, asks the man at the next table for a match. They fall into conversation; the stranger is an Englishman who, it turns out, spent several months in Detroit studying the operation of an interchangeable-bottlecap factory. "I know it's a foolish question," says Jones, "but did you ever by any chance run into a fellow named Ben Arkadian? He's an old friend of mine, manages a chain of supermarket in Detroit. . . " "Arkadian, Arkadian," the Englishman mutters. "Why, upon my soul, I believe I did! Small chap, very energetic, raised merry hell with the factory over a shipment of defective bottlecaps." "No kidding!" Jones exclaims in amazement. "Good lord, it's a small world, isn't it?"

This is the incipit of a paper published in the popular magazine *Psychology Today* by Stanley Milgram, a Harvard social psychologist who conducted a series of experiments in

Table 4.1 Length distribution of the completed chains in Milgram small-world experiment.

Chain length	2	3	4	5	6	7	8	9	10
Number of completed chains	2	4	9	8	11	5	1	2	2

the late 1960s to show that, despite the very large number of people living in the USA and the relatively small number of a person's acquaintances, two individuals chosen at random are very closely connected to one another [287, 298]. In the first of these experiments, Milgram considered a sample of randomly selected people in Omaha, Nebraska. Each of these persons was given the name and address of the same target person, a stockbroker working in Boston and living in Sharon, Massachusetts, and was asked to move a message towards the target person, using only a chain of friends and acquaintances. Each person would be asked to transmit the message to the friend or acquaintance who he thought would be most likely to know the target person. Of 160 chains that started in Nebraska, 44 reached their destination and 126 dropped out. The number of intermediaries needed to reach the target is distributed as shown in Table 4.1. The length of the chains varied from 2 to 10 intermediate acquaintances, with the median of the distribution being equal to 5.5 and its average to 5.43. Rounding up, the experiment suggests that we are all at six steps of distance, the so-called six degrees of separation. Though it was implicit in his work, Milgram never used the phrase "six degrees of separation". This term was introduced for the first time by John Guare, a theatre writer, as the title of his 1990 play [145]. At page 45 of the play Ouisa (Louisa Kittredge), one of the main characters, reflects on the interconnectedness of our world: *"Everybody on this planet is separated by only six other people. Six degrees of separation. Between us and everybody else on this planet. The president of the United States. A gondolier in Venice ...It's not just big names. It's anyone. A native in a rain forest. A Tierra del Fuegan. An Eskimo. I am bound to everyone on this planet by a trail of six people. It's a profound thought ...How every person is a new door, opening up into other worlds. Six degrees of separation between me and everyone else on this planet."* The real story is a little more complex if one looks at the details of Milgram's experiment. One third of the test subjects were from Boston, not Omaha, and one-half of those in Omaha were stockbrokers. Moreover, 25 per cent of the chains moved on to the target by Mr. Jacobs, a clothing merchant in town (a fact which came as a considerable surprise, and even something of a shock for the broker), and 25 per cent arrived at the place of work, a Boston brokerage house, passing through two persons, Mr. Jones and Mr. Brown. The participants were three times as likely to send the letter on to someone of the same sex as to someone of the opposite sex, suggesting that certain kinds of communication are strongly conditioned by sex roles. A large fraction of the letters never reached their destination, because a certain proportion of participants did not cooperate and failed to send on the letter, and were discarded from the distance computation. In addition, the statistics are certainly too poor to make serious claims. However, an average of six steps is, in certain ways, impressive considering the real geographic distance traversed, e.g. almost 3000 km between Nebraska and Massachusetts. Moreover, the letters that reached their destination only provide an upper bound on the distance since there might have been even shorter

Step	Location	Travel	Family	Work	Education	Friends	Cooperative	Other
1	33	16	11	16	3	9	9	3
2	40	11	11	19	4	6	7	2
3	37	8	10	26	6	6	4	3
4	33	6	7	31	8	5	5	5
5	27	3	6	38	12	6	3	5
6	21	3	5	42	15	4	5	5
7	16	3	3	46	19	8	5	0

Table 4.2 Declared reasons for choosing the next recipient at each step of the chain of the email small-world experiment.

routes. Therefore, the results indicate not only that short paths exist between individuals in large social networks, but also that ordinary people can find these short paths. This is not a trivial statement, because people rarely have more than local knowledge about the network. People know who their friends are. They may also know who some of their friends' friends are. But no-one knows the identities of the entire chain of individuals between themselves and an arbitrary target. Nevertheless they are able to find the shortest path to an arbitrary target. We shall come back to this point in Section 4.6.

A group of scientists from the Columbia University performed in 2003 a modern Internet-based version of the Milgram's experiment [95]. More than 60,000 email users participated in the experiment by trying to reach one of 18 target persons from 13 countries by forwarding email messages to direct acquaintances. Targets included a professor at an Ivy League University, an archival inspector in Estonia, a technology consultant in India, a policeman in Australia, and a veterinarian in the Norwegian army. Unfortunately, due to an exponential attenuation of chains as a function of their length, a low chain completion rate was observed: only 384 of 24,163 recorded chains reached their targets. The length distribution of chains arriving at destination is reported in Figure 4.1. The typical length is usually very short, with the average chain length being equal to 4.05. However, this number is misleading because it represents an average only over the completed chains, and shorter chains are more likely to be completed. The authors of the work provided some arguments to correct this effect, and derived that a better estimate of the median ranges from five to seven steps, depending on the separation of source and target, which somehow confirms the results of the Milgram experiment. Participants of the email experiment were also asked why they considered their nominated acquaintance a suitable recipient. It is extremely illuminating to look at the reasons for choosing the next recipient reported in Table 4.2, and how these reasons change at the different steps of a chain. All quantities reported in the table are percentages. The main reasons are: location (means recipient is geographically closer), travel (recipient has travelled to target's region), family (recipient's family originates from target's region), work (recipient has occupation similar to target), education (recipient has similar educational background to target), friends (recipient has many friends), cooperative (recipient is considered likely to continue the chain). Two reasons, geographical proximity of the acquaintance to the target

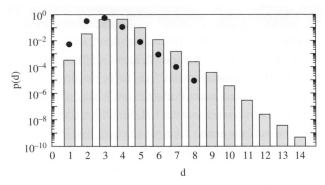

Fig. 4.2
The normalised number $p(d)$ of pairs of actors connected by a shortest path of length d is plotted as a function of d (histogram). The normalised number of actors at distance d from Bacon is also reported for comparison (circles).

Table 4.4 Characteristic path length and clustering coefficient of three real networks (two social networks and a biological one) and of the corresponding ER random graphs with same number of nodes and links.

	N	$\langle k \rangle$	L	L^{ER}	C	C^{ER}
Movie actors	225226	73.71	3.65	2.87	0.79	0.00033
ArXiv	44337	10.79	5.99	4.50	0.62	0.00024
C. elegans	279	16.39	2.44	2.01	0.34	0.06

than or equal to 14 are possible, with most pairs of nodes being at distance 2, 3, 4 or 5. The mean value of the distribution $p(d)$, which coincides with the characteristic path length L of the largest component of the network, is 3.65, which is indeed a very small number for a system with $N = 225226$ nodes. In Table 4.4 we show that the characteristic path length of the network is only slightly larger than the value of L^{ER} for an ensemble of random graphs with the same number of nodes and links. We also report L and the corresponding L^{ER} for the ArXiv coauthorship network introduced in Chapter 3, supporting the idea that all social networks, and not only the movie actors, have characteristic path lengths that are quite small compared to the order of the network.

4.2 The Brain of a Worm

In this section we introduce the first of the biological networks that will be considered in this book. Just as every organism has a genome (that is, the complete set of genes as encoded in its DNA), every organism with a nervous system has also a *connectome*. The connectome of an organism is the complete set of its nerve cells and of the connections among them. It is therefore a network. In 1972 Sydney Brenner, a biologist then at Cambridge University, decided to work out the connections of every cell in the nervous system of the *C. elegans* (or *Caenorhabditis elegans*, if you want to know the extended name), a one-millimetres long, transparent, free-living soil worm of the nematode family, that is commonly found in many parts of the world. He picked this animal because its body has only 959 cells, of which 302 are nerve cells. *C. elegans* is also a hermaphrodite, fertilising itself to produce clones. This means that individual worms are more or less identical. Brenner and his team embedded their worms in blocks of plastic, cut the blocks into thin slices, and then examined each slice through a transmission electron microscope. The resulting pictures allowed them to trace the path of each nerve cell through the entire body of the worm. They also revealed the connections between cells. Over the course of 14 painstaking years, the team managed to map the complete nervous system of the *C. elegans*, a work for which Brenner won a Nobel prize in 2002 [314].

Today we know that the nervous system of the *C. elegans*, in its most common form, comprises 302 neurons grouped into ten ganglia, and located in one of two general units. The first unit is a nerve ring of 20 neural cells at the pharynx. The second unit is composed by a ventral cord, a dorsal cord, and four sublateral processes that run anteriorly and posteriorly from the nerve ring [314, 1]. Connections among neurons are also well known. These can be of two different types, namely synaptic connections and gap junctions. While *synapses* are specialised junctions through which neurons send chemical signals to each other, and also to non-neuronal cells such as those in muscles or glands, *gap junctions* mediate electrical and metabolic coupling between cells. Also notice that, unlike synapses, gap junctions are, in principle bidirectional, i.e. allow impulse transmission in either direction. The *C. elegans* remains to date the only case of a nervous system completely mapped at the level of neurons. For this reason it is a biological system of primary importance, and it can therefore be very interesting to study the structural properties of its network. In DATA SET Box 4.2 we introduce and describe the basic features of the neural network of the *C. elegans* that can be downloaded from the book webpage, and that we will analyse here. Notice that today different brain imaging methods allow more evolved neural systems to be studied, although not at the level of their neurons, but at a more coarse-grained scale of brain areas. Some examples are illustrated in Box 4.3.

Figure 4.3 is a graphical representation of the adjacency matrix of the *C. elegans* neural network. The $N = 279$ nodes of the network are sorted in alphabetical order of their names, and the plot shown is obtained by drawing the two points of coordinates (i, j) and (j, i) whenever there is a link between nodes i and j. As in the previous section we will concentrate here again on how well connected the nodes of a network are, this time for the case of a biological system. As for the other two social networks reported in Table 4.4, we have therefore computed the shortest paths between all the pairs of nodes in the neural network of the *C. elegans*. The value of the characteristic path length of the *C. elegans*, $L = 2.44$, is compared in Table 4.4 with the value expected in an ensemble of random graphs of same order and size. As for the two collaboration networks, also in this case the value of L is indeed quite small, and slightly larger than the corresponding value of L^{ER}. It is interesting to observe that also a biological system shows the *small-world behaviour* we have found in the movie actor collaboration network and in coauthorship networks. What is even more amazing is that networks with the small-world property are found everywhere. Examples include other biological systems, such as food webs and metabolic networks, but

| Box 4.3 | **Neural Networks from Brain Imaging Techniques** |

Almost all organisms with a neural system have a considerably larger number of neurons and connections than the *C. elegans*, as shown in the Table below.

Organism	Neurons	Synapses
C. elegans	3×10^2	8×10^4
Fruit fly	1×10^5	10^7
Honey bee	9×10^5	10^9
Mouse	7×10^7	10^{11}
Rat	2×10^8	5×10^{11}
Human	9×10^{10}	$10^{14} - 10^{15}$

The full mapping of these neural systems is still beyond currently available technology. However, recent developments in brain imaging methods such as *electroencephalography* (EEG), *magnetoencephalography* (MEG), *functional magnetic resonance imaging* (fMRI) and *diffusion tensor imaging* (DTI) have allowed the construction of coarse-grained brain networks of living organisms with about $10^2 - 10^5$ nodes and $10^4 - 10^7$ edges. Notice that each node of these networks does not represent a neuron but an entire area of the brain, usually consisting of thousands to millions neurons, while an edge can indicate the presence of either a physical connection or a significative correlation in the activities of two areas [61, 113, 291, 62, 92]. Several research groups around the world, including among others the groups of Ed Bullmore in Cambridge, Olaf Sporns at Indiana University, Fabio Babiloni in Rome and Mario Chavez in Paris, are systematically applying complex network theory to study such networks in order to reveal how the brain works while we are performing different tasks [29, 112], or to identify structural patterns associated with brain diseases such as schizophrenia [209] and epilepsy [73].

However, since for many real-world systems we only have access to a single network with a given fixed number of nodes N, in order to say that the network has the small-world property we will accept as sufficient requirement that the network has a value of L comparable to that of a random graph with N nodes.

4.3 Clustering Coefficient

We now turn our attention to another important property of social networks: the presence of a large number of triangles. We all know that our friend's friend is very often also our friend. This is because the basic mechanism of growth in a social network is the formation of triangles, i.e. the appearance of links between two individuals with common friends. This property, known in social networks jargon as *transitivity* or *triadic closure*, is usually quantified by evaluating the fraction T of *transitive triads* in the corresponding graph, that is the fraction of *triads* that also close into *triangles*.

Definition 4.2 (Graph transitivity)   A triad *is a tree of three nodes or, equivalently, a graph consisting of a couple of adjacent edges, and the corresponding end nodes. The* transitivity *T of a graph is the fraction of transitive triads, i.e. triads which are subgraphs of triangles [308]:*

$$T = \frac{3 \times (number\ of\ triangles)}{(number\ of\ triads)} = \frac{3n_\triangle}{n_\wedge} \tag{4.1}$$

where $n_\triangle$ is the number of triangles in the graph, and $n_\wedge$ is the number of triads. By definition $0 \le T \le 1$.

The factor 3 in the numerator of Eq. (4.1) compensates for the fact that each triangle in the graph contributes three triads, one centred on each of the three nodes, and ensures that $0 \le T \le 1$, with $T = 1$ for a complete graph.

Example 4.1   Consider the graph with $N = 5$ nodes shown in the figure. To evaluate T we need to count the number of triangles $n_\triangle$ and the number of triads $n_\wedge$ in the graph. The graph contains eight triads, and only one triangle (see Problem 4.3c). The triangle

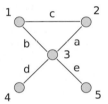

consists of nodes $1, 2, 3$ and links a, b, c. The first triad consists of the two edges c and a of node 2. We can label this triad by listing its two links (c, a) or its three nodes $(1, 2, 3)$. Of course the order of the two links is not important, hence we can indicate the same triad with the symbol (a, c). Analogously, if we list the three nodes to indicate a triad, then what is important is the node that stays in between, while the order of the other two is not important. Hence, $(1, 2, 3)$ and $(3, 2, 1)$ indicate the same triad (c, a). The other seven triads are: (a, b), (b, c), (b, d), (a, d), (b, e), (a, e) and (d, e). Finally, we have $T = 3/8 = 0.375$. Notice that there are three triads associated with each triangle. In our case, the first three triads in the previous list correspond to triangle $1, 2, 3$.

An alternative measure to quantify the presence of triangles in a graph has been introduced more recently. This is the so-called *graph clustering coefficient*, that is the average over all graph nodes of an appropriately defined *node clustering coefficient* [310, 311].

Definition 4.3 (Clustering coefficient)   *The node clustering coefficient, c_i, of node i is defined as:*

$$c_i = \begin{cases} \dfrac{K[G_i]}{k_i(k_i - 1)/2} & for\ k_i \ge 2 \\ 0 & for\ k_i = 0, 1 \end{cases} \tag{4.2}$$

where $K[G_i]$ is the number of links in G_i, the subgraph induced by the neighbours of i.
The graph clustering coefficient, C, is the average of c_i over all the nodes of the graph:

$$C = \langle c \rangle = \frac{1}{N} \sum_{i=1}^{N} c_i \qquad (4.3)$$

By definition $0 \leq c_i \leq 1 \; \forall i$, and $0 \leq C \leq 1$.

The node clustering coefficient c_i expresses how likely it is for two neighbours of node i to be connected. In fact, the value of c_i in Eq. (4.2) is equal to the density of the subgraph G_i induced by the first neighbours of i, i.e. the ratio between the actual number of edges in G_i and their maximum possible number $k_i(k_i - 1)/2$. For this reasons c_i takes values in the interval $[0, 1]$. Notice that $K[G_i]$ is also equal to the actual number $n_i^{\triangle}$ of triangles containing node i, while $k_i(k_i - 1)/2$ is the maximum possible number of triangles centred on node i which, from now on, we will indicate as $n_i^{\wedge}$. Therefore, the node clustering coefficient c_i in Eq. (4.2) can be alternatively seen as the proportion of triads centred in i that close into triangles and, in this sense, the graph clustering coefficient C is the average over nodes of a local version of the graph transitivity defined in Eq. (4.1).

Example 4.2 *(Node clustering coefficient)*   In order to evaluate the clustering coefficient of node i of graph G, we need first to construct the graph G_i induced by the first neighbours of i. In the example shown in the figure, node i has degree $k_i = 4$. The four neighbours of i are labelled as nodes 1, 2, 3, 4, while the edges connecting these nodes to i are shown as dashed lines. Graph G_i has four nodes and the four edges shown as solid lines. Therefore,

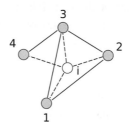

the clustering coefficient of node i is equal to 2/3. In fact, the four nodes of G_i can be connected to each other in six different ways. Of these possibilities, only four are actually realised (the four existing edges) and therefore $c_i = 4/6 = 2/3$.

The subgraph G_i of the neighbours of node i can in general be unconnected. However, this does not affect the definition of c_i. What is instead problematic is a node i which is isolated, or has only one link. In this case, the ratio $2K[G_i]/(k_i(k_i - 1))$ is not defined. We need therefore to set $c_i = 0$, as in Eq. (4.2), when the degree of i is either zero or one.

Example 4.3 *(Expression in terms of the adjacency matrix)* We can rewrite the local clustering coefficient, c_i, of node i as:

$$c_i = \begin{cases} \dfrac{\sum_{l,m} a_{il}a_{lm}a_{mi}}{k_i(k_i - 1)} & \text{for } k_i \geq 2 \\ 0 & \text{for } k_i = 0, 1 \end{cases} \qquad (4.4)$$

Consider the numerator in Eq. (4.2). The number of edges $K[G_i]$ in graph G_i can be written in terms of the adjacency matrix of the graph. In fact $(A^3)_{ij} = \sum_{lm} a_{il}a_{lm}a_{mj}$ is equal to the number of walks of length 3 connecting node i to node j. In particular, by setting $i = j$, the quantity $(A^3)_{ii} = \sum_{l,m} a_{il}a_{lm}a_{mi}$ denotes the number of closed walks of length 3 from node i to itself. This is two times the number of triangles containing node i. The triangle containing node i and the two nodes l and m, is composed by the two edges connected to node i, namely (i, l) and (m, i), and by the edge (l, m) that belongs to graph G_i induced by the first neighbours of i. Since the edge (l, m) appears twice, namely in the closed walk (i, l, m, i) and in the closed walk (i, m, l, i), the number of edges $K[G_i]$ is given by:

$$K[G_i] = \frac{1}{2}(A^3)_{ii} = \frac{1}{2} \sum_{j,m} a_{ij}a_{jm}a_{mi}.$$

We therefore obtain Eq. (4.4) from Eq. (4.2).

Once we have a definition of the clustering coefficient for each node, then the graph clustering coefficient C can be calculated as in Eq. (4.3) as the average of the local clustering coefficient over all graph nodes. Since $0 \leq c_i \leq 1 \; \forall i \in G$, then also the graph clustering C ranges in $[0, 1]$. It can be very interesting to study how the clustering coefficient of a node i depends on other basic node properties, for instance on its degree k_i. In order to do this, we can evaluate the average clustering coefficient $c(k)$ of nodes having degree equal to k as [306, 268, 269, 281]:

$$c(k) = \frac{1}{N_k} \sum_{i=1}^{N} c_i \, \delta_{k_i, k} \qquad (4.5)$$

where N_k is the number of nodes of degree k. We can then study, as in the following example, how $c(k)$ changes as a function of k for some the real-world networks introduced in the previous chapters.

Example 4.4 *(Node clustering coefficient vs. node degree)* The value of the node clustering coefficient can vary a lot from node to node. One way to explore this in the neuronal network of the *C. elegans* introduced in the previous section, and in the two collaboration networks of ArXiv and of movie actors is to plot, as a function of the degree k, the average clustering coefficient $c(k)$ of nodes with degree k defined in Eq. (4.5). The figure below shows that there are nodes (neurons in panel (a), scientists in (b) and actors in (c), respectively) with c_i very close to the maximum value of 1, but also nodes with clustering coefficients as small as 10^{-2}. We also observe that in general $c(k)$ decreases with the degree, i.e. large degree nodes have a clustering coefficient smaller than nodes with only a few links. Since the clustering coefficient of a node is equal to the density of the subgraph induced by its neighbours, this means that, in real-world networks, the density of such subgraphs decreases with their order. Notice that $c(k)$ exhibits a power-law tail for large k. In the case of the largest of the three networks, the collaboration of movie actors, a power law $c(k) \sim k^{-1}$ is reported as dashed line for comparison.

Let us now come back to the definition of the transitivity. At a first glance, the graph clustering coefficient C and the transitivity T look quite similar. In fact, the transitivity is

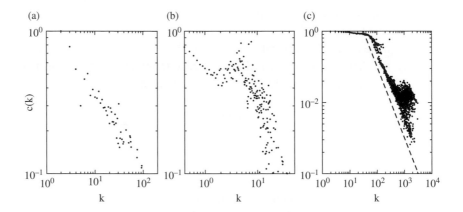

proportional to the total number of triangles in the graph divided by the total number of triads in the graph, while the local clustering coefficient c_i in Eq. (4.2) is equal to the ratio between the number of triangles connected to node i and the number of triads centred on i. However, the similarity between C and T can be misleading: although in many graphs C and T take similar values, there are cases where the two measures can be very different, as shown in the following example.

Example 4.5 *(Clustering coefficient versus transitivity)* The clustering coefficient C of a graph, although apparently similar to T, is indeed a different measure. Consider, for instance, the graph in Example 4.1. The local clustering coefficients of the five nodes are: $c_1 = c_2 = 1$, $c_3 = 1/6$, $c_4 = c_5 = 0$. Thus, the graph has a clustering coefficient equal to $C = 13/30 = 0.433$ which is slightly different from the value of 0.375 we have obtained for T. In some particular cases the differences can be much larger, as in the graph introduced in Ref. [200] and reported in the figure. The graph has $N + 2$ nodes in total: nodes A and B

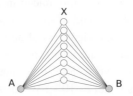

are directly linked, and each of them is also connected to all the other N nodes. The transitivity T of the graph is equal to $3/(N + 2)$, since the graph has $n_\triangle = N$ triangles and $n_\wedge = 3N + N(N - 1)$ triads, namely N triads XAB, N triads XBA, N triads AXB and $N(N - 1)$ triads XAX' or XBX' with $X' \neq X$. Analogously, it is easy to prove that the clustering coefficient C is $\frac{N^2+N+4}{N^2+3N+2}$, since $c_A = c_B = 2/(N + 1)$ and $c_X = 1$. The expressions of T and C looks very different. In particular, in the limiting case of N very large, the value of T converges to 0, which is the minimum possible value for T. Conversely, the clustering coefficient C converges to its maximum possible value 1.

It is not complicated to derive a precise relation between T and the clustering coefficient [58]. From the expression of the clustering coefficient c_i of a node i given in Eq. (4.2), and from the observation that $K[G_i]$ is equal to the actual number $n_i^\triangle$ of triangles containing node i, while $n_i^\wedge = k_i(k_i - 1)/2$ is the number of triads containing a node i, we can in fact write the clustering coefficient of node i (for $k_i > 1$) as:

$$c_i = \frac{n_i^\triangle}{n_i^\wedge}. \tag{4.6}$$

Recalling now the definition of the graph transitivity T in Eq. (4.1), it is possible to express T in terms of the clustering coefficients of the graph nodes. Indeed, from the above equation we have that the total number of triangles, $n_\triangle$, and the total number of triads, $n_\wedge$, can be written as:

$$n_\triangle = \frac{1}{3} \sum_i n_i^\triangle = \frac{1}{3} \sum_i n_i^\wedge c_i$$

$$n_\wedge = \sum_i n_i^\wedge.$$

The factor $1/3$ in the expression of $n_\triangle$ is due to the fact that each individual triangle is counted three times in the summation, given that it belongs to three different nodes. Inserting the expressions above in Eq. (4.1), we can write the transitivity T in terms of the clustering coefficients of the graph nodes:

$$T = \frac{3n_\triangle}{n_\wedge} = \frac{\sum_i n_i^\wedge c_i}{\sum_i n_i^\wedge}. \tag{4.7}$$

Now the relation between T and C is evident. While the graph clustering coefficient C is the arithmetic average of the clustering coefficients of all the nodes, the transitivity T is a weighted average of the clustering coefficients, the weight of each node being equal to the number of triads centred on it. Therefore T gives more weight to high-degree nodes with respect to C. When the node degree distribution, and consequently the distribution of the number of triads at a node, is strongly peaked, then the values of transitivity and graph clustering coefficient will be very similar (exactly the same if k_i does not depend on i). In other cases, the two indices may be very different, as illustrated in Example 4.5.

In the following, we will focus on the clustering coefficient C, since this quantity has recently found a wider use than T in numerical studies and data analysis of different systems. In Table 4.4, together with the values of the characteristic path length for the two collaboration networks and for the neural network of the *C. elegans*, we also report their values of C. Notice that all three networks exhibit a non-zero clustering coefficient, denoting the presence of triangles in their structure. We find $C = 0.34$ in the case of the *C. elegans*, and even larger values of the clustering coefficient, closer to 1 than to 0, in the two collaboration networks. However, a high value of C is not indicative by itself of a strong propensity to close triangles. In fact, triangles can form in a network just by chance, and their number will in general increase with the number of links in the network. A better way to proceed is to compare the value of C of a real network to the value expected in appropriate *network null models*. For each of the three networks we have therefore considered as null model an ensemble of *ER* random graphs with the same number of nodes and links. The corresponding values of the clustering coefficient averaged over the ensemble of random graphs are reported, and indicated as C^{ER}, in Table 4.4. Notice that, in practice, to get these numbers we do not need to sample random graphs from the ER model and then compute numerically their clustering coefficients. We can, in fact, make use of the following result which provides an analytical expression for the clustering coefficient in random graphs (similar to that of Eq. (3.33) for the characteristic path length) in terms of basic graph properties.

> The clustering coefficient of Erdős–Renyí random graphs reads:
>
> $$C^{ER} = p = \langle k \rangle / N \qquad (4.8)$$

The second equality in Eq. (4.8) holds in the usual scaling case, $p(N) = \langle k \rangle N^{-1}$, which corresponds to random graphs with different number of nodes but fixed average degree $\langle k \rangle$, and tells us that in large random graphs the number of triangles can be neglected, since the clustering coefficient tends to 0 for $N \to \infty$. Eq. (4.8) above is not difficult to derive, as illustrated in Example 4.6, and exactly the same expression holds true for the transitivity T^{ER} of random graphs (see Problem 3.3c)).

Example 4.6 *(The clustering coefficient of random graphs)*   Let us consider the ensemble $G_{N,p}^{ER}$ generated by model B of Section 3.1, and calculate the clustering coefficient C as a function of the parameter p. Consider a node i with degree k. The subgraph G_i will contain k nodes and an average of $p \cdot k(k-1)/2$ edges out of the possible maximum number $k(k-1)/2$, since p is the probability of having a link between any two nodes in the graph. Hence $c_i = p$ for any node of the graph, and consequently $C = p$ for a random graph with N nodes and a connection probability p. This means that random graphs with $p(N) \sim N^{-\zeta}$, and $\zeta \in]0, \infty[$, have a vanishing clustering coefficient C in the limit of large system size. This result is in particular true for the case $\zeta = 1$, i.e. when we have $p \propto N^{-1}$ and the node average degree $\langle k \rangle$ is constant, and yields Eq. (4.8).

From Table 4.4 it is now clear that each of the three real networks has a value of C much higher than that of the corresponding null model. This means that the propensity to form

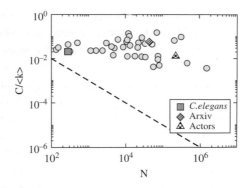

Fig. 4.5 The clustering coefficient of real networks (symbols) is compared to the prediction for ER random graphs given by Eq. (4.8) (dashed line).

triangles in real networks cannot be simply reproduced by a random graph model. This is shown graphically in Figure 4.5 where we consider the same networks as in Figure 4.4 and we plot, in a log–log scale, the ratio between the clustering coefficient and the average degree as a function of the order N. For comparison we report as dashed line the ratio $C^{ER}/\langle k \rangle = N^{-1}$ for ER random graphs, which on a log–log plot corresponds to a straight line of slope -1. We notice that in real networks the quantity $C/\langle k \rangle$ appears to be independent of N, instead of decreasing as N^{-1}. Thus random graphs are inadequate to describe some of the basic properties observed empirically, and we will therefore need to explore other models.

Example 4.7 *(Efficient behaviour of real-world networks)* We have learnt that real world networks have, at the same time, small node distances and a high number of triangles. Hence, we need two different measures, such as L and C, to describe their structure. In 2001, Vito Latora and Massimo Marchiori have proposed a unified framework to characterise the two properties by means of a single measure [202]. The basic idea is to use the network efficiency $E(G)$ of Eq. (2.16) as an alternative to the characteristic path length. We indicate this quantity here as the network *global efficiency* E_{glob}.

	E_{glob}	E_{glob}^{random}	E_{loc}	E_{loc}^{random}
Movie actors	0.37	0.41	0.67	0.00026
ArXiv	0.17	0.21	0.67	0.00024
C. elegans	0.46	0.48	0.47	0.12

If the nodes of a network are at small distances, they can communicate efficiently, and we get a high value of E_{glob}. Now, differently from the characteristic path length, the efficiency has the nice property that it does not diverge for unconnected graphs. Hence, it can also be used to characterise the subgraph (not always connected) of the neighbours of a node. We can then define the local efficiency of a node i as the efficiency $E(G_i)$ of subgraph G_i. To compute such efficiency we obviously need to evaluate node distances in subgraph G_i.

If node i belongs to many triangles, then we expect a high value of the node clustering coefficient c_i and also of the node local efficiency. Finally, we can define the network *local efficiency* E_{loc} as an average of the node local efficiencies over the graph nodes:

$$E_{\text{loc}} = \frac{1}{N} \sum_{i=1}^{N} E(G_i) \tag{4.9}$$

In conclusion, by evaluating a single quantity at a global and at a local scale we can capture the same information provided by L and C. As shown in the table, in social and biological networks E_{glob} is as large as $E_{\text{glob}}^{\text{random}}$, while $E_{\text{loc}} \gg E_{\text{loc}}^{\text{random}}$, meaning that real-world systems are highly efficient both at a global and at a local scale.

The first step to model graphs with a large clustering coefficient is to consider lattices, which are a special class of regular graphs. For example, a triangular lattice, as shown in the left-hand side of Figure 4.6, has $C = 0.4$. We can therefore construct a graph with N nodes and a clustering coefficient $C \approx 0.4$ by taking a suitable finite region of the triangular lattice containing N nodes (see problem 4.4a). In this way, the value of C is essentially independent of N for large enough graphs, and it is similar to the value found for the *C. elegans* in Table 4.4. However, the situation can be very different with other lattices. For instance, the one-dimensional lattice or the two-dimensional square lattice shown in Figure 4.6, have both $C = 0$. This is so because they do not contain triangles. A way to control the number of triangles connected to a given node, and therefore its clustering coefficient, is to connect the node to its second, third, ..., mth neighbours in the lattice. We name such graphs *lattice graphs* and we introduce them by starting from two simple one-dimensional cases, the linear and the circle graphs.

Definition 4.4 (Linear Graph) *Consider a linear lattice of order N, i.e. N nodes uniformly distributed on a finite interval, with each internal node connected to its two first neighbours, and the two extreme nodes connected with their first neighbour. A (N, m)-linear graph, with $1 < m < N/2$, is obtained by additionally linking each node of the linear lattice to its neighbours with lattice distance smaller than or equal to m.*

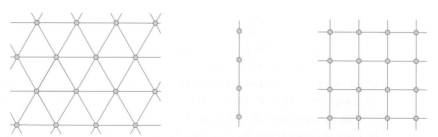

Fig. 4.6   Various examples of lattices: a two-dimensional triangular lattice, a one-dimensional lattice and a two-dimensional square lattice.

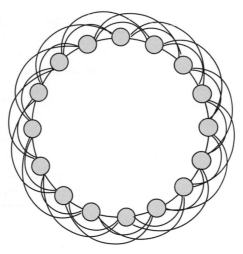

Fig. 4.7   Circle graph with $N = 16$ and $m = 3$

Definition 4.5 (Circle Graph)   *Consider a circular lattice of size N, i.e. N nodes uniformly distributed on a circle, each node connected to its two first neighbours. A (N, m) circle graph, with $1 < m < N/2$, is obtained by additionally linking each node of a circular lattice to the $2m - 2$ closest nodes in the lattice.*

As an example, in Figure 4.7 we show a circle graph with $N = 16$ and $m = 3$. This graph has a total of $K = Nm = 48$ links. In this graph we have a large number of triangles. The exact value of C can be calculated exactly for a (N, m) circle graph, yielding:

$$C^{\text{circle}} = \frac{3(m - 1)}{2(2m - 1)} \tag{4.10}$$

as demonstrated in Example 4.8. In particular, for $m = 1, 2, 3, 4, 5$ we get respectively $C^{\text{circle}} = 0, 1/2, 3/5, 9/14, 12/18$. Increasing the number of links in the graph, the clustering coefficient in the circle graph increases monotonically towards $3/4$ for large m. Thus, circle graphs with $m > 1$ have values of the clustering coefficient as large as those found in real-world networks. For instance, the value of C obtained for $m = 5$, which corresponds to circle graphs with an average degree $\langle k \rangle = 10$, is not very different from that reported in Table 4.4 for the ArXiv collaboration network.

Example 4.8   *(C in circle graphs)*   To calculate C we need to count the number of existing edges between the $2m$ neighbours of a generic site i. In the figure below we show the generic node i, and the subgraph G_i induced by its neighbours in a circle graph with $m = 3$, as an example. An easy way to count the number $K[G_i]$ of edges in G_i avoiding double counting is to scan all the six nodes of G_i clockwise and consider the number of edges connecting each of them to nodes forward in the circle (in clockwise direction). The values reported on each node in the figure represent precisely such a number for $m = 3$. We therefore have a total of nine edges when $m = 3$. For a generic value of m, the total number of existing edges in G_i is:

$$m(m-1) + \sum_{j=1}^{m-1} j = \frac{3m(m-1)}{2}$$

where the first term comes from the nodes to the left of node i (backward nodes), while the second term is due to the nodes to the right of i (forward nodes). Thus the clustering coefficient c_i of node i is:

$$c_i = \frac{3m(m-1)}{2} / \binom{2m}{2} = \frac{3(m-1)}{2(2m-1)}.$$

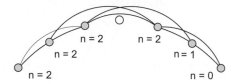

Since the same calculation can be repeated for any vertex of the lattice we get the graph clustering coefficient C of Eq. (4.10).

The above definition of circle graphs can be easily generalised to the D-dimensional case, with $D > 1$.

Definition 4.6 (Square Lattice Graphs) *Consider a D-dimensional square lattice of linear size S with periodic boundary conditions, containing $N = S^D$ nodes. In this way, each node of the network is connected to each $2D$ first neighbours. A (N, m) D-dimensional square lattice graph, with $1 < m < S/2$, is obtained by additionally linking each node of the D-dimensional square lattice to its $2D(m-1)$ closest neighbours along the coordinate directions.*

For instance, in the case $D = 2$, we start with a square lattice as the one reported in the right-hand side of Figure 4.6, which corresponds to the case $m = 1$. In order to create a (N, m) 2-dimensional square lattice with $m > 1$ we then add a link from each node to its $4(m-1)$ closest neighbours on the lattice along the two coordinate directions. For example for $m = 3$, the pattern of links around a generic node will look as in Figure 4.8. A definition similar to Definition 4.6 can be given by removing the periodicity constraints. However, the analysis is in general easier on periodic lattices. In fact, due to translational invariance,

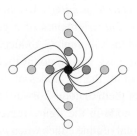

Fig. 4.8   Nearest neighbour structure of a generic node of a bi-dimensional square lattice with $m = 3$. Notice that only nodes along the coordinate directions are linked to the central node.

4.4 The Watts–Strogatz (WS) Model

The problem with regular lattice graphs in reproducing the properties found in real systems, is that they do not exhibit the small-world behaviour. For instance, it can be proven that (see the following example) the characteristic path length in a circle graph is equal to:

$$L^{\text{circle}} \sim \frac{N}{4m} \tag{4.11}$$

i.e. it grows linearly with the order N, rather than logarithmically as in ER random graphs.

Example 4.9 *(L for circle graphs)*   Since all nodes are equivalent we can evaluate L by considering the distances from one node to all the others. For simplicity, let us assume N is an odd number. For symmetry, we only consider the $(N-1)/2$ nodes to the left of node i, and we sort them into groups of nodes having the same distance from i as shown in the figure. The first group has m nodes at distance $d = 1$, the second group has m nodes at distance $d = 2$ from i, and so on. Let us define the quantity:

$$N_g = \frac{N-1}{2m}.$$

If N_g is an integer number, then each of the N_g groups contains m nodes, with distances going from 1 to N_g. Otherwise, there will be $\lfloor N_g \rfloor$ groups with m nodes, and the remaining

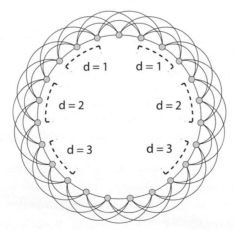

$N - 1 - 2\lfloor N_g \rfloor * m$ nodes at distance $\lfloor N_g \rfloor + 1$. In the simplest case in which N_g is integer, the sum of the distances from node i to the other $(N-1)/2$ nodes is equal to:

$$m \sum_{g=1}^{N_g} g = m\frac{N_g(N_g+1)}{2} = \frac{N-1}{8m}(N-1+2m)$$

dividing this number by $(N-1)/2$ we get the following expression for the characteristic path length:

$$L^{\text{circle}} = \frac{N + 2m - 1}{4m}.$$

Notice that, for $m = 1$, we get a formula for L^{circle}, that is different from that we derived in Example 3.12 for a linear chain without the periodic boundary conditions. If N_g is not an integer, one has to take into account the nodes that are not in the groups (nodes in the red dashed set in the figure). An exact expression for L^{circle} is given by [44]:

$$L^{\text{circle}} = \left(1 - \frac{m}{N-1}\lfloor N_g \rfloor\right)\left(\lfloor N_g \rfloor + 1\right).$$

For large N and for $1 \ll m \ll N$, both expressions reduce to Eq. (4.11).

On the other hand, we have seen that ER random graph models have the small-world property, but their clustering coefficient is negligible. The first model to construct graphs having, at the same time, a high clustering coefficient and a small characteristic path length was proposed by two applied mathematicians at Cornell University, Duncan Watts and Steven Strogatz, in a paper published in *Nature* in 1998 [311]. We will refer to the model as the *Watts and Strogatz small-world model*, or, in short, the WS model. Basically, the idea of Watts and Strogatz is to introduce a certain amount of random connections in a lattice graph. With such a purpose, each of the $K = N \cdot m$ edges of a (N, m) circle graph is visited in turn, and independently randomly rewired with some probability p ($0 \le p \le 1$). In this way the WS model produces an ensemble of graphs with N nodes, K links, and with a variable percentage p of random links. As we will discuss below, a tiny percentage of rewired edges is enough to produce graphs with a small characteristic path length (as small as that of random graphs) still retaining the finite clustering coefficient of lattices.

Definition 4.7 (The WS model)   *Start with a (N, m) circle graph. A (N, m, p) WS graph is constructed by considering each of the links of the circle graph and independently rewiring each of them with some probability p ($0 \le p \le 1$) as follows:*

1 Visit each node along the ring one after the other, moving clockwise.
2 Let i be the current node. Each edge connecting node i to one of its m neighbours in a clockwise sense is rewired with a probability p, or left in place with a probability $1 - p$.
3 Rewiring means shifting the end of the edge, other than that in node i, to a new vertex chosen uniformly at random from the whole lattice, with the constraint than no two vertices can have more than one edge running between them, and no vertex can be connected by an edge to itself.

The procedure to select the edges in the above definition is just a convenient way to scan all the edges of a circle graph avoiding double-counting. The model is illustrated in Figure 4.9, and an algorithm to sample graphs from the model is discussed in Appendix A.11. In the considered case, we start with a circle graph with $N = 16$ nodes and $m = 3$,

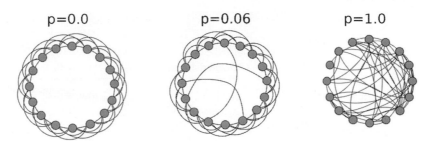

Fig. 4.9 Graphs produced by the WS model with $N = 16$, $m = 3$ and three values of the rewiring probability p.

i.e. $K = 48$ links. Notice that the rewiring procedure keeps the average vertex degree unchanged, while the degree of any particular vertex may change. As p increases from 0 to 1, the typical graph becomes more and more disordered because an increasingly large number of random links is introduced. The random links are usually called *shortcuts* or *long-range connections*. The latter name comes from the following formal definition of *range* of a link:

Definition 4.8 (Range of a link)   *The range of a link l_{ij} is the shortest-path distance in the graph between i and j if the link l_{ij} were deleted.*

Example 4.10   *(Granovetter's strength of weak links)*   Historically, it was Mark Granovetter who first introduced the definition of *range* of a link, although in a slightly different context, and using a different name for it. In 1973, in his very influential paper on "The Strength of Weak Ties", he proposed and tested the idea that *weak ties*, i.e. connections to people who form weaker acquaintance relationships, played a crucial role in bridging parts of the social network otherwise inaccessible through strong friendships [142]. As an empirical example, Granovetter presented results from a labour-market study where he found that people often received important information leading to new jobs not from close friends, but from those with whom they had weak ties. This happens because, although our good friends are in general more motivated to help us with jobs, there is a high overlap in their friendship network. Conversely, weak ties are those giving access to circles different from our own, and as a result of this they provide information different from that which we usually receive, and thus more valuable. Granovetter formalised his idea on the importance of weak ties by introducing the above definition of range of a link, and by considering as *bridges* all edges of range greater than two. He then showed that in most of the cases bridges were made by weak relationships. To understand in full the results of Granovetter we need to consider networks with weights associated with their links, so we will come back to his work in Example 10.8 of Chapter 10 on weighted networks.

In practice, the range of a link is the length of the shortest alternative path between the link endpoints. When p is small, most of the edges in the typical graph produced by the rewiring procedure are still those of the circle graph, and they belong to triangles with large probability. Therefore their range is two. However, there are a few, more precisely

Table 4.5 Characteristic path length and clustering coefficient in the WS model with $N = 1000, m = 5$, and various values of p.

p	Characteristic path length L	Clustering coefficient C
0	50.5	0.667
0.01	8.94	0.648
0.05	5.26	0.576
1	3.27	0.00910

an average of Kp rewired links, having in general a large range. What is particularly interesting in the WS model is that the presence of such a few long-range links, for small p, is
sufficient to bridge different parts of the network and cause a sharp drop in the value of the
characteristic path length, while almost no change is observed in the value of the clustering
coefficient. That is, for small values of p, the WS model produces graphs with a clustering
coefficient $C(p)$ still comparable to C^{circle}, and with a characteristic path length $L(p)$ of the
order of magnitude of L^{ER}. The few shortcuts that make the characteristic path length small
in the model, might represent the few acquaintances one has in other countries or outside
the local professional environment, that explain the short chains found in the experiment
by Milgram.

To have an idea of the structural properties of the networks produced, we have implemented the WS model with $N = 1000$ vertices and $m = 5$. With the choice $m = 5$ and
$N = 1000$ in practice all the graphs obtained after the rewiring are connected also for
$p = 1$, since the condition for the disappearance of isolated vertices in random graphs
$\langle k \rangle = 2m > \ln N$, discussed in Section 3.4 is verified. For increasing rewiring probability
p, we find numerically the values of L and C shown in Table 4.5. The numbers reported are
obtained as averages over 100 different realisations of the rewiring process. Hence, for a
small value of p, the clustering coefficient is similar to the one of the regular lattice, while
the characteristic path length is much shorter. In Figure 4.10 we plot characteristic path

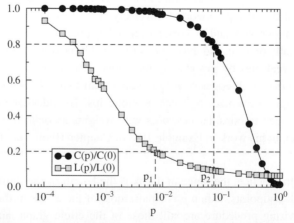

Fig. 4.10 Characteristic path length and clustering coefficient as a function of p in the WS model with $N = 1000$ nodes and $m = 5$. In the interval $[p_1, p_2]$ we have graphs with a small value of L and a large value of C.

length and clustering coefficient as a function of p. The values reported have been respectively normalised to the values $L(0)$ and $C(0)$ for the circle graph. The curves $L(p)/L(0)$ and $C(p)/C(0)$ are again obtained as averages over 100 different realisations of the process. All the graphs have $N = 1000$ vertices and an average vertex degree $\langle k \rangle = 10$ since $m = 5$. We have used a logarithmic horizontal scale to resolve the rapid drop in $L(p)$, corresponding to the onset of the small-world phenomenon. During this drop, $C(p)$ remains almost constant at its value for the regular lattice, indicating that the transition to a small world is almost undetectable at the local level.

Example 4.11   Consider Figure 4.10, and let ϵ be a small positive quantity. Then, let us define a lower bound $p_1(\epsilon)$, such that for $p > p_1$ we have $L(p)/L(0) < \epsilon$, and an upper bound $p_2(\epsilon)$, such that for $p < p_2$ we have $C(p)/C(0) > 1 - \epsilon$. What is interesting is that there are reasonable small values of ϵ for which $p_1(\epsilon) < p_2(\epsilon)$, and the interval $[p_1(\epsilon), p_2(\epsilon)]$ in which we have graphs with high C and small L is of sizeable measure. For instance, let us fix $\epsilon = 0.2$. In a graph with $N = 1000$ nodes and $m = 5$ we have $p_1 = 0.0077$ and $p_2 = 0.074$, as indicated by the dashed lines in the figure. In this case, the region in which the graph exhibits the so-called small-world behaviour and, at the same time, has a large clustering coefficient, spanning one order of magnitude in p.

In practice, there is a sizeable interval $[p_1, p_2]$ in which the WS model produces graphs with a small value of L and a large value of C. The characteristic path length L is the average of the distances d_{ij} over all node pairs i, j. An interesting observation is that, in the small-world regime, not only the average distance $L(p)$ is small, but also most of the pairs of nodes have a distance comparable to $L(p)$. This is shown in Figure 4.11, where we have reported the distribution of distances d_{ij} as a function of p. As expected, for $p \to 0$ (see the case $p = 10^{-4}$ reported in the figure) the number of pairs $N(d)$ at distance d is

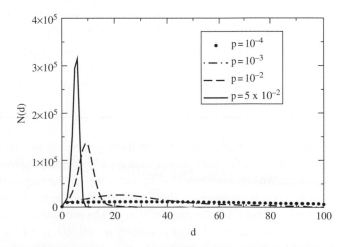

Fig. 4.11   Number of node pairs i and j at distance $d_{ij} = d$ in the WS model with $N = 1000$, $m = 5$ and various values of p.

equal for each value of d between 1 and the largest possible distance between two nodes $N/(2m)$, while the curve $N(d)$ becomes more and more peaked around its mean value $L(p)$ as p grows.

The most interesting feature of the WS model is certainly the rapid decrease produced in the value of L when only a small percentage of the links are rewired. We can now investigate the onset of the small-world behaviour in graphs of different order N.

Example 4.12   *(Scaling with the system order N)*   We report in the figure the curves $L(p)/L(0)$ as a function of p, for graphs of different order. The curves shown correspond to the WS model with $m = 5$ and different values of N, from $N = 100$ to $N = 20000$ (right to left). The results reported are averages over 100 different realisations.

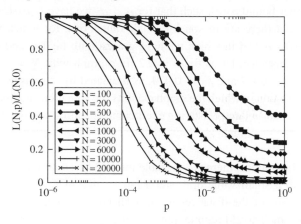

As N becomes larger, we observe two things happening. First of all, the right part of the ratio $L(p)/L(0)$ moves down towards zero as N increases, indicating that $L(p)$, for large values of p, increases less than linearly as a function of N. Remember in fact that $L(0)$, that is L for the circle graph, scales with the graph order as shown in Eq. (4.11). Therefore, a linear growth of $L(p)$ with N would result in a ratio $L(p)/L(0)$ independent of N. We will see below that $L(p)$ for sufficiently large values of p scales indeed as $\ln N$, indicating a small-world behaviour. The second important observation is that the onset of the scaling $L(p) \sim \ln N$ appears for smaller and smaller values of p as N increases, indicating that the transition to the small-world regime might occur for vanishingly small values of p as $N \to \infty$.

The above example suggests that there is a region of values of p in which the characteristic path length scales slower than N, and that this region gets larger and larger as N increases. We can now show that, in such a region, the characteristic path length L increases with $\ln N$, i.e. it has precisely the same scaling in N exhibited by ER random graphs. In Figure 4.12 we plot L as a function of N in graphs produced by the WS model with $m = 5$. We report three different values of p, namely $p = 0.001$, $p = 0.01$ and $p = 1$. As expected, for

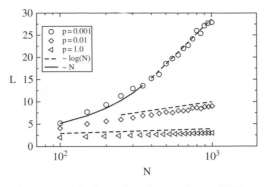

Fig. 4.12 L versus N in the WS model with $m = 5$ and for three values of p, namely $p = 0.001, p = 0.01, p = 1$. Due to the choice of a logarithm scale for the x-axis, the three dashed straight lines correspond to the scaling $L \sim \log N$, while the continuous line corresponds to the scaling $L \sim N$.

$p = 1$, we find that L grows as $\ln(N)/\ln(2m)$, i.e. as in ER random graphs (see Eq. (3.33)). And in fact, the dashed straight line reported close to the points corresponding to $p = 1$ in the figure, has a slope equal to the inverse of the logarithm of $2m$. Notice that we have used a logarithmic scale for the x-axis, so that a behaviour proportional to $\ln N$ gives rise to a straight line. Also the points corresponding to $p = 0.01$ lay on a straight line, indicating the same logarithmic scaling of the characteristic path length. The only difference with respect to the $p = 1$ case is that the proportionality constant, the slope of the dashed straight line, is now larger than before. For even smaller values of the rewiring probability, $p = 0.001$, the logarithm scaling still holds, but only for sufficiently large values of N. Instead, for small values of N, we observe a linear dependence of the characteristic path length L on N, as confirmed by the linear behaviour reported as continuous line. These numerical results suggest that, for each value of p, it is possible to define a threshold N^*, such that the logarithmic behaviour appears only when $N > N^*$. The value of the threshold N^* depends on p. In practice, N^* can be interpreted as the typical size of the regions associated with each ending node of a shortcut. When N is much larger than N^*, the shortcuts become very effective in reducing the average path length, giving rise to the small-world behaviour. In particular, we can show that $L(N) \sim N^* \ln(N/N^*)$ by the following heuristic argument illustrated in Figure 4.13. Let i and j denote two random points on the (N, m, p)-WS graph, with $N = 60$ and $m = 1$, shown in the left part of the figure. The shortest path between i and j is highlighted by the thick lines. In general, a shortest path, as the one shown in the figure, consists in traversing shortcuts and moving on the circle graph from one shortcut to the other. In practice, this is equivalent to considering shortest paths on a new random graph with only N/N^* "macro-nodes", shown by the areas delimited by the lines in the right-hand side of the figure, and Nmp random links. This random graph is constructed from the WS model in the following way. Consider all the end nodes of the shortcuts (shown as circles) and associate with each one of them all the other nodes closest to it. A macro-node of the new random graph is then made by each end node of the shortcut and by all the nodes associated to it. If N^* is the typical size of macro-nodes, the new random graph has N/N^* macro-nodes, each containing N^* nodes of the original graph. The

(a)   (b)

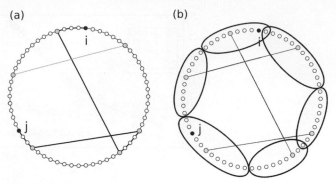

A (N, m, p)-WS graph, with $N = 60$ nodes and $m = 1$. The shortest path between two nodes i and j is highlighted (left). The associated "macro-node" random graph, where nodes of the original graph belonging to the same macro-node are delimited by a closed line (right).

links of the random graph are the shortcuts of the WS model. The typical distance between any two macro-nodes on this random graph scales as $\ln(N/N^*)$. This means that we can connect two macro-nodes in an average of $\ln(N/N^*)$ steps. In conclusion, we can connect i and j on the original graph in an average of $(N^*/m) \cdot \ln(N/N^*)$ steps, since we need to traverse $\ln(N/N^*)$ shortcuts, and in order to reach each shortcut we need to perform N^*/m steps on the lattice on average.

More in general, it is possible to show that the characteristic path length in the WS model obeys the following *scaling law* as a function of N [27, 21]:

$$L(N) \sim N^* \cdot F\left(\frac{N}{N^*}\right) \tag{4.12}$$

where N^* is a function of p, and $F(x)$ is a function of x with the asymptotic behaviours $F(x) \sim x$ for $x \ll 1$, and $F(x) \sim \ln x$ for $x \gg 1$. The precise form of $F(x)$ can be extracted numerically, and has also been derived analytically under various kinds of approximations [239]. The behaviour of $F(x)$ describes the transition from a lattice-like behaviour to a small-world one:

$$L(N) \sim \begin{cases} N & \text{for } N \ll N^* \\ N^* \ln(\frac{N}{N^*}) & \text{for } N \gg N^*. \end{cases} \tag{4.13}$$

That is, for a fixed value of p, the transition between the scaling $L(N) \sim N$ and the small-world behaviour $L(N) \sim \ln N$ takes place at a threshold value of the graph order N^*, which of course depends on p. As said before, N^* can be approximated as the typical size of the regions between shortcuts. Thus, N^* can be evaluated as the total number of nodes N, divided by two times the average number Nmp of shortcuts:

$$N^* = \frac{1}{2mp}. \tag{4.14}$$

This relation, together with Eq. (4.13), tells us that for a given p, if the characteristic size of the regions between shortcuts, $(2mp)^{-1}$, is much smaller than the graph order N, the shortcuts connect distant region of the graph and produce the small-world effect. Or, alternatively, at fixed N, we have the small-world behaviour for $p \gg 1/(2mN)$. Thus, we can define the critical probability function:

$$p_c(N) = (2mN)^{-1}, \tag{4.15}$$

which tells us how strong the disorder has to be, in a graph of order N, in order to induce a transition from large-world to small-world. Eq. (4.15) indicates that there is a critical threshold function $p_c(N)$ for the appearance of the small world behaviour, and not a finite critical value, independent of N. This behaviour is in all similar to those observed in Chapter 3, when we studied the appearance of various properties in models of random graphs.

4.5 Variations to the Theme

The graph produced by the WS model for $p = 1$ are not exactly equivalent to an ensemble of ER random graphs. For instance, the degree of any vertex in the model is forced to be larger than or equal to m. The exact degree distribution is derived in the following example. Nevertheless, the characteristic path length and the clustering coefficient in WS model with $p = 1$ have the same order of magnitude as in ER random graphs.

Example 4.13   *(Degree distribution in the WS model)*   The figure below shows, as symbols, the degree distribution p_k obtained numerically for a system with $N = 1000$, $m = 5$ and three values of p, namely $p = 0.2, 0.4, 1$. An analytical expression that fits the numerical results can be derived in the following way [21]. Since m of the initial $2m$ connections of each vertex are left untouched by the construction, the degree of vertex i can be written as $k_i = m + n_i$, with $n_i \geq 0$. The quantity n_i can then again be divided in two parts: $n_i^1 \leq m$ links have been left in place, each one with a probability $1 - p$, the other $n_i^2 = n_i - n_i^1 = k_i - m - n_i^1$ links have been reconnected from other nodes towards i, each one with a probability p/N. We then have $P_1(n_i^1) = \binom{m}{n_i^1}(1 - p)^{n_i^1} p^{m - n_i^1}$, and $P_2(n_i^2) = \frac{(mp)^{n_i^2}}{n_i^2!} \exp(-mp)$ valid for large N. Finally, if we indicate as k the degree of a generic vertex, we can write the distribution of k as:

$$P_p(k) = \sum_{n=0}^{\min(k-m,m)} \binom{m}{n}(1 - p)^n p^{m-n} \frac{(mp)^{k-m-n}}{(k - m - n)!} e^{-pm} \quad \text{for } k \geq m$$

where we are summing over the possible number of links $n = n_i^1$ left in place, while $k - m - n = n_i^2$ is the number of links arriving from other nodes. We report in the figure, as dashed lines, the analytical expressions obtained for $m = 3$ and the three values of p studied numerically. The analytical expressions are in good agreement with the points obtained numerically. We notice that, as p grows, the distribution becomes broader. In particular, in the limit $p \rightarrow 1$, the analytical expression above reduces to:

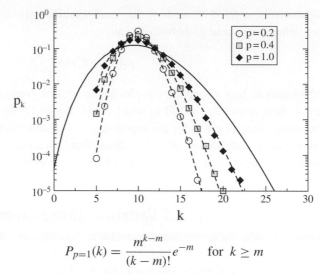

$$P_{p=1}(k) = \frac{m^{k-m}}{(k-m)!}e^{-m} \quad \text{for } k \geq m$$

which is a Poisson distribution for the variable $k-m$, with average equal to m. This is different from the degree distribution of an ER random graph with same average degree, reported as a continuous line.

It is possible to modify the rewiring process of the WS model so that one gets exactly ER random graphs in the limit $p=1$. We refer to this model as the *modified Watts–Strogatz (mWS) model*, which is in practice obtained by rewiring both ends of each selected edge.

Definition 4.9 (The modified WS model)   *Start with a (N,m) circle graph. A (N,m,p) modified WS graph is constructed by considering each of the links of the circle graph and independently rewiring each of them with some probability p $(0 \leq p \leq 1)$ as follows:*

1 *Visit each node along the ring one after the other, clockwise.*
2 *Let i be the current node. Each edge connecting node i to one of its m neighbours clockwise, is considered and rewired with a probability p, or left in place with a probability $1-p$.*
3 *Rewiring means replacing the edge by a new edge connecting two nodes chosen uniformly at random from the whole lattice, with the constraint than no two vertices can have more than one edge running between them, and no vertex can be connected by an edge to itself.*

We have implemented the modified WS model with the same parameters used for the WS model, namely $N=1000$ vertices and $m=5$. The results in Table 4.6 are averages over 100 different realisations. Comparing these numbers to the values obtained by the rewiring process of the original WS model, it turns out that there are no large differences in the characteristic path length and in the clustering coefficient of the two models.

In both this and the original WS procedure, there is a finite probability for large values of p to produce unconnected graphs, especially if m is small. The existence of small

Table 4.6 Characteristic path length and clustering coefficient in the modified WS model with $N = 1000, m = 5$, and various values of p.

p	Characteristic path length L	Clustering coefficient C
0	50.5	0.667
0.01	8.91	0.652
0.05	5.21	0.595
1	3.24	0.0106

disconnected subgraphs makes the characteristic path length poorly defined, since we may find pairs of vertices without a path joining them. And assuming that, in such a case, the path length is infinite does not solve the problem. What is usually done in numerical simulation is to neglect small components, defining the characteristic path length only for the largest connected part. While this can be acceptable for numerical studies, for analytic work, having to deal with poorly defined quantities is not very satisfactory. To avoid disconnecting the network, Mark Newman and Duncan Watts have proposed adding new edges between randomly selected pairs of vertices, instead of rewiring existing edges [228, 239]. We indicate this as the *Newman and Watts (NW)* model.

Definition 4.10 (The NW model) *Start with a (N, m) circle graph. A (N, m, p) NW graph is constructed by adding, with some probability p $(0 \le p \le 1)$, a new edge for each of the edges of the circle graph. New edges connect pairs of nodes chosen uniformly at random from the whole lattice.*

To make this model as similar as possible to the Watts and Strogatz model, a shortcut is added with a probability p for each edge of the original lattice, so that there are again, on average, mNp shortcuts, as in the WS model. Since these shortcuts are added to the existing links, the average node degree is a function of p, and is equal to $2m(1 + p)$. Furthermore, in the NW model, it is possible to have more than one edge between a pair of vertices and to have edges connecting a vertex to itself. The model is equivalent to the WS model for small p, and does not produce unconnected graphs when p becomes comparable to 1. For the usual values $N = 1000$ and $m = 5$ we get numerically the results in Table 4.7. Again there are no big differences with respect to the previous models.

In all models considered so far, the target nodes for the rewired links are chosen at random on the circle, with uniform probability. As a consequence, the length l of the rewired links, defined in terms of the lattice distance, is characterised by a uniform distribution over all possible ones, since we can get with equal probability links of length $l = m+1, m+2, \ldots$ Since long-range connections usually have high costs in real systems, it may be more reasonable also to consider models in which the number of rewired links of a given length is a decreasing function of the length. One such model was proposed in 2006 at the École Polytechnique Fédérale de Lausanne by two physicists, Thomas Petermann and Paolo De Los Rios [258].

Table 4.7 Characteristic path length and clustering coefficient in the NW model with $N = 1000, m = 5$, and various values of p.

p	Characteristic path length L	Clustering coefficient C
0	50.5	0.667
0.01	8.89	0.655
0.05	5.10	0.610
1	2.70	0.180

Definition 4.11 (The Petermann–De Los Rios model)   *Start with a (N, m) circle graph. Each node i is assigned an extra link with a probability p $(0 \leq p \leq 1)$. The ending node j is chosen according to the distribution:*

$$q(l) = cl^{-\beta}, \quad l = 1, 2, 3, \dots \tag{4.16}$$

where l is the distance on the original circle between node i and node j, $\beta \geq 0$, and c is a normalisation constant. Multiple links are rejected.

The model has two control parameters, namely p and β. The value of p determines the number of links, since on average we have pN additional links on the circle. This is because a new link is added with a probability p to *each node*, and not *for each link* of the circle graph as in the NW model. However, the main difference with respect to the previous models comes with the introduction of parameter β. In fact, the end-point of the additional links are not chosen at random uniformly, but according to a power-law distribution $q(l) \sim l^{-\beta}$ of the distance l, with an exponent $\beta \geq 0$. This is sketched in Figure 4.14 for three different values of β, with the width of the links originating from a generic node i (in red) representing the probability of their occurrence. When $\beta = 0$ the shortcuts are completely random and we get back to a standard small-world model. When $\beta \neq 0$ the additional connection from node i is biased towards nodes j closer to i on the circular lattice, with larger values of β meaning shorter lengths of the additional links.

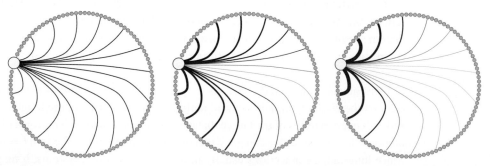

Fig. 4.14   Graphs produced by the Petermann–De Los Rios model consist in linear lattices where each node is connected to its $2m$ closest nodes in the lattice ($m = 1$ in the case shown) and is assigned, with a probability p, an extra link. The length of the extra link is tuned by the exponent β in Eq. (4.16) as sketched here for three different values of β (in increasing order) starting from $\beta = 0$.

Clearly, a certain number of real shortcuts, i.e. links connecting distant nodes on the original lattice, are needed for a small-world behaviour to emerge. And in fact, it is found that the exponent β has to be smaller than a critical value β_c, otherwise we cannot observe the transition to the small-world regime [258]. More precisely, we need to consider separately the two cases $\beta < 1$ and $\beta > 1$.

When $\beta < 1$ the behaviour of the model is in all ways similar to that of the standard case $\beta = 0$, with two possible regimes: small-world and non-small-world. This behaviour can again be described by the scaling of the characteristic path length as in Eq. (4.13), with a critical size N^* inversely proportional to p as in Eq. (4.14), or with a critical probability p_c inversely proportional to N as in Eq. (4.15).

Conversely, when $\beta > 1$, things are qualitatively different from the standard case $\beta = 0$. It is in fact possible to prove that:

$$p_c(N) \sim N^{\beta-2}. \tag{4.17}$$

Notice that, at variance with the case $\beta < 1$, the scaling of the critical probability here depends on the value of β. This means that there is a critical value:

$$\beta_c = 2 \tag{4.18}$$

such that, for $1 < \beta < \beta_c$, it is still possible to observe the transition to a small-world behaviour for a sufficiently large value of p. Instead, when $\beta > \beta_c = 2$, there is no transition to small-world behaviour in the thermodynamic limit, no matter how large the value of p.

This model can be easily extended to higher dimensions, in the sense that we can consider D-dimensional square lattices with the addition of random links [258]. And it can be very convenient to be able to tune the typical length of the additional links based on the position of the nodes on a D-dimensional square lattice. For instance, as we will see in the next section, the case $D = 2$ will be used to represent a social system where the individuals are embedded in a two-dimensional geographical area and are preferentially linked to other nodes that are geographically close.

Definition 4.12 (The Petermann–De Los Rios model in D dimensions)   *Consider a D-dimensional square lattice of linear size S with periodic boundary conditions, with $N = S^D$ nodes. Each node i is assigned an extra link with a probability p ($0 \leq p \leq 1$). The ending node j is chosen according to the probability distribution:*

$$q(l) = cl^{-\beta}, \quad l = 1, 2, 3, \ldots \tag{4.19}$$

where l is the distance on the original lattice between node i and node j, β is the tuning parameter, and c is a normalisation constant. Multiple links are rejected.

The limit $p = 0$ of the model is a regular lattice in D dimensions, having a finite clustering coefficient and a characteristic path length proportional to $N^{1/D}$ and not to $\ln N$ (see Problem 4.4c). For $p \neq 0$ we have a certain percentage of long-range links, and, for any fixed values of p, the structure of the resulting graph will depend on the value of the tuning exponent β. Notice that we have not specified the range of variation of β in Definition 4.12.

This is because, only when $D = 1$, the value $\beta = 0$ corresponds to choosing all the nodes with the same probability. For $D > 1$ it is possible to consider also negative values of β, as we shall see in the following.

As in the one-dimensional model, a certain amount of links connecting distant nodes on the original lattice is necessary to observe a small-world behaviour. In fact, when $\beta < 1$ the model has the same behaviour as if the additional links were chosen at random with equal probability. For $\beta > 1$ it is instead possible to prove that the critical probability depends both on the parameter β and on the dimension D [258]:

$$p_c(N) \sim N^{\frac{\beta-1}{D}-1} \tag{4.20}$$

which means that in D dimension the critical value β_c, such that for $\beta > \beta_c$ there is no transition to small-world behaviour no matter how large the value of p, is equal to:

$$\beta_c = D + 1. \tag{4.21}$$

In the following example we provide an argument to understand this result.

Example 4.14 *(The role of β)* Following Ref. [258] we can study the transition to a small-world behaviour of the model in Definition 4.12 by an argument similar to the one that led us to Eq. (4.14) and Eq. (4.15). Let us in fact assume that, for a fixed value of p and β, we can have a small-world behaviour in a system whose size N is much larger than N^*, the typical size of the region associated with each ending node of a shortcut. As for the WS model, N^* can be evaluated as the total number of nodes N divided by two times the average number of shortcuts. This time, however, not all the added links give rise to shortcuts because, depending on the value of β, some link may connect two nodes that are very close on the lattice. Therefore, the average number of shortcuts present can be obtained as the number of added links, Np, times the probability Q that each of these links is really a shortcut, i.e. that it spans the lattice, being effectively able to shrink distances in the graph. We can quantify Q as the probability that the length of the link is of the order of its maximum possible value $S/2$. We can formally write this as an integral $Q = \int_{(1-\epsilon)S/2}^{S/2} q(l)dl = c \int_{(1-\epsilon)S/2}^{S/2} l^{-\beta}dl$, where $\epsilon < 1$ is a small but finite number. It is crucial now to determine the normalisation constant c. In order to do this, we need to impose the normalisation condition $c \int_1^{S/2} l^{-\beta} = 1$, which gives $c = \frac{1-\beta}{(S/2)^{1-\beta}-1}$ for $\beta \neq 1$. We can then write:

$$Q = \int_{(1-\epsilon)S/2}^{S/2} q(l)dl = \frac{(S/2)^{1-\beta}}{(S/2)^{1-\beta} - 1}(1 - \beta)\epsilon + O(\epsilon^2).$$

The probability Q that an added link is a shortcut is therefore a finite quantity, approximately equal to $(1 - \beta)\epsilon$, if $\beta < 1$. Instead, if $\beta > 1$, Q goes to zero as $S^{1-\beta}$ when the system size S increases. Finally, we have that, for a given value of p and β, the transition to a small-world behaviour takes place at:

$$N^* = \frac{N}{2NpQ} \sim \begin{cases} [2p(1 - \beta)\epsilon]^{-1} & \text{if } \beta < 1 \\ (S/2)^{\beta-1}[2p(\beta - 1)\epsilon]^{-1} = N^{(\beta-1)/D}[2^\beta p(\beta - 1)\epsilon]^{-1} & \text{if } \beta > 1. \end{cases}$$

Notice that N^* is finite for $\beta < 1$, while for $\beta > 1$ it diverges as the graph size increases. Therefore, for $\beta < 1$ we can certainly have a transition to a small world, while for $\beta > 1$ the situation is more complicated and depends on the value of β and on the dimensionality D. Following the same argument adopted for the WS model, the transition from a behaviour $L \sim N^{1/D}$ to a behaviour $L \sim (N^*)^{1/D} \ln(N/N^*)$ takes place when $N \sim N^*$. For a given value of N we can therefore obtain the critical probability, by solving the previous relations for p with $N* = N$:

$$p_c(N) \sim \begin{cases} [2N(1-\beta)\epsilon]^{-1} & \text{if } \beta < 1 \\ N^{-1}(S/2)^{\beta-1}[2(\beta-1)\epsilon]^{-1} \sim N^{\frac{\beta-1}{D}-1} & \text{if } \beta > 1. \end{cases}$$

Hence, for $\beta < 1$ we have the typical scaling $p_c(N) \sim N^{-1}$ found for the WS model in Eq. (4.15). For $\beta > 1$ we have instead proven the validity of Eq. (4.20).

Summing up, if we introduce the exponent β to modulate the length of the additional links as in the Petermann–De Los Rios model, we can have a transition to a small-world behaviour, i.e. a characteristic path length L which depends only logarithmically on the network order, $L(N) \sim (N^*)^{1/D} \ln(N/N^*)$, only if β is smaller than a certain critical value. Such a critical value β_c depends on the dimensionality of the system D as in Eq. (4.21). In particular we have $\beta_c = 2$ in a one-dimensional system, and $\beta_c = 3$ in a two-dimensional system.

One important remark is now necessary. We have already said that in $D = 1$, choosing $\beta = 0$, i.e. a constant distribution $q(l)$, is equivalent to connecting a node i to a node j chosen at random on the circle with a uniform probability. However things are different when $D > 1$. Let us consider for instance the case of a square lattice in $D = 2$ dimensions, and denote by Ω_S the subset of nodes at distance $d \leq S/2$ from a given node, say node i. It is easy to prove that Ω_S contains exactly $S^2/2 + S$ nodes. In Figure 4.15 we represent the

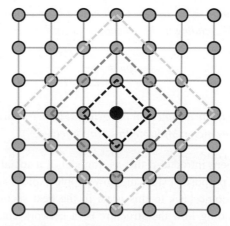

Fig. 4.15   Subset of nodes respectively at distance $d = 1, 2, 3$ from a node i (black node) in a two-dimensional square lattice with $S = 6$.

case $S = 6$ in which Ω_S contains 24 nodes, 4 at distance 1 from i, 8 at distance 2, and 12 at distance 3. What is the probability that a randomly selected node on the lattice has a graph distance l from node i? Such probability can be directly computed by counting the number of nodes at distance l, which is equal to $4l$, and dividing by the total number of nodes distinct from the origin, which is $S^2/2 + S$. The probability of finding a node at distance l is then $4l/(S^2/2 + S)$. Therefore, if we connect node i to a node j chosen at random with a uniform probability, on a two-dimensional lattice we obtain a distribution $q(l)$ of the link lengths l which is not uniform, but equal to:

$$q(l) = \frac{4l}{S^2/2 + S}.$$

In practice, by selecting nodes with equal probability, we get more connections to nodes at large distance from i, simply because in two dimensions the number of nodes at distance l increases linearly with l. This means that in $D = 2$ dimensions the model in Definition 4.12 reduces to the completely random case for $\beta = -1$, and not for $\beta = 0$. Using the same argument in D dimensions, one obtains that, by selecting nodes with equal probability in a D-dimensional lattice, we get a distribution of link lengths which depends on the value of D:

$$q(l) \propto l^{D-1}.$$

This geometric effect is due to the change of coordinates, when dealing with random variables. Consequently, in D dimensions we will get the completely random case when we adopt a power-law distribution of lengths $q(l) = cl^{-\beta}$, with an exponent $\beta = 1 - D$. This means that, while in the one-dimensional case the interesting range of values of β to explore is $\beta \geq 0$, in D dimensions we will have to explore $\beta \geq 1 - D$.

In D dimensions it can therefore be more natural to consider a model defined in a slightly different way, which was proposed for the first time by Jon Kleinberg, a computer scientist working at Cornell University [180].

Definition 4.13 (Kleinberg network model)   *Consider a D-dimensional square lattice of linear size S with $N = S^D$ nodes. Each node i is assigned an extra link with a probability p ($0 \leq p \leq 1$). The ending node j is chosen according to the probability distribution:*

$$p_{ij} = \frac{d_{ij}^{-\alpha}}{\sum_{\substack{j=1 \\ j \neq i}}^{N} d_{ij}^{-\alpha}} \quad j = 1, 2, \ldots, N \tag{4.22}$$

where d_{ij} is the distance on the original lattice between node i and node j, and $\alpha \geq 0$ is the tuning parameter. Multiple links are rejected.

Kleinberg introduced this as a model of a social system embedded in geography, with the main purpose of studying how people explore and find short paths in a social network. We will come back to this issue in the next section. Here, we are instead interested in understanding how the structure of a network changes by tuning the length of random links. Kleinberg model is in all similar to the Petermann–De Los Rios model introduced in Definition 4.12, with the only difference that the stochastic variable in the Petermann

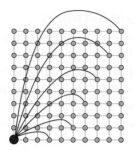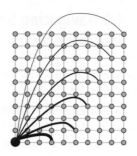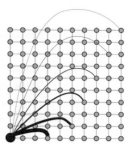

Fig. 4.16   Graphs produced by the Kleinberg model in $D = 2$ dimensions consist in two-dimensional square lattices where each node is assigned a shortcut link with a probability p. The probability p_{ij} that a shortcut goes from node i (black node in the left-bottom corner) to a node j depends on the distance d_{ij} between i and j on the original lattice as in Eq. (4.22). We respectively show the case $\alpha = 0$ where the shortcuts are completely random, and other two cases with $\alpha > 0$.

and De Los Rios model is the link length l, while it is the ending node j in Kleinberg model. Therefore, the distribution in Eq. (4.19) is normalised over the link lengths l, while Kleinberg normalises p_{ij} by summing over all nodes j different from i. In Figure 4.16 we show three cases of graphs produced by the model in $D = 2$ dimensions, respectively, for $\alpha = 0$, $\alpha = \alpha_1 > 0$, and $\alpha = \alpha_2 > \alpha_1$. The nice thing about the graph model proposed by Kleinberg is that, when $\alpha = 0$, we get to the limiting case in which all nodes are chosen with equal probability, and this is true for any dimension D. However, it is extremely simple to relate this model to the Petermann–De Los Rios model. In fact, if we choose the node j with a probability $p_{ij} = f(d_{ij})$, where d_{ij} represents the distance between nodes i and j on the lattice, then the probability distribution of the link lengths will be proportional to $l^{D-1} \cdot f(l)$ [1]. In particular, if $f(d_{ij}) \sim d_{ij}^{-\alpha}$ as in the Kleinberg model, we get:

$$q(l) \propto l^{D-1-\alpha}.$$

Finally, the relation between the exponent β appearing in Eq. (4.19) of the Petermann–De Los Rios model and the exponent α of the Kleinberg model is:

$$\beta = \alpha - D + 1.$$

We can therefore tune the link lengths by changing the value of α, as we did by changing the value of β in the Petermann–De Los Rios model. As α increases, the long-range connections of a node become more and more clustered in its vicinity. In particular, in terms of α, we expect no small-world behaviour if α is larger than a critical value:

$$\alpha_c = 2D \tag{4.23}$$

corresponding to the critical value of β in Eq. (4.23). This means $\alpha_c = 2$ in one dimension, or $\alpha_c = 4$ in the two-dimensional case

[1] To be more precise, due to periodic boundary conditions, this relation only holds for $l \le S/2$, because the number of links with length $l > S/2$ is equal to $4(S - l)$ instead of $4l$.

4.6 Navigating Small-World Networks

There is another interesting aspect in the Milgram experiment. In fact, the experiment showed not only that short paths exist in social networks, but also that people are indeed able to find them. In the last two sections we presented a series of models to generate small-world graphs. Here, we concentrate instead on the second aspect, namely we assume that short paths exist in a network, and we try to understand how individuals discover and exploit them to efficiently send messages over the network. Of course, if people had a full global knowledge of the network, the shortest path between two nodes could be computed simply, for instance, by breadth-first search. But the astonishing fact is that the participants in the Milgram experiment were indeed able to reach the target efficiently, by only having *local knowledge* of the graph. In his work on navigation in small worlds published in Nature in 2000, Jon Kleinberg, the same author of the network model in Definition 4.13, provided a possible mechanism to explain this apparent puzzle [180]. Since geography strongly shapes our social interactions so that we tend to interact more often with those living or working close to us, Kleinberg modelled a social network as a modified version of a two-dimensional lattice, where each person knows the next-door neighbours, but has also non-zero probability to know someone who lives far away. Then, his main idea was to consider the *geographical space* also as the underlying substrate guiding a search in a social system. In order to mimic how a letter navigated the social system towards a target node in the Milgram's experiment, Kleinberg proposed a *decentralised navigation model* where the current letter holder has only local information on the structure of the network, and passes the letter to one of its local or long-range contacts, the one that is geographically closest to the target. In this way the message is sent closer and closer geographically to the target, until the target is reached. Here is the precise definition of the navigation model.

Definition 4.14 (Kleinberg navigation model)   *Consider a graph produced by the Kleinberg network model in Definition 4.13, with $D = 2$. Given a source node s and a target node t on the graph, a message is passed from one node to one of its neighbours, based on the following three assumptions:*

1 *each node in the graph knows the location of the target t on the lattice;*
2 *each node in the graph knows its own location on the lattice, and also the location of all its neighbours (on the graph);*
3 *the node currently holding the message chooses to pass it to a neighbour (on the graph) that is as close to the target t as possible in terms of lattice distances.*

The navigation starts at node s and ends when the message arrives at node t.

Of course, there are two crucial points in Kleinberg's idea: the decentralised navigation model uses only local information, and the probability for two nodes to be connected in the network decreases with their geographical distance. To quantify the performance of a search on a given network, Kleinberg considered the *expected delivery time*, T, defined as the number of steps needed to pass the message from a source s to a target t, averaged over all the possible pairs s and t chosen uniformly at random among all pairs of nodes. He found that the value of T depends crucially on the structure of the network, namely on the value of the exponent α tuning the length of the random links in the network model of Definition 4.13.

In order to understand this, let us first consider the simplest possible case, namely $\alpha = 0$. Since the pN random links added to the lattice are, in this case, uniformly distributed as in the original WS model, shortcuts appear in the system, and the shortest distance between every pair of nodes in the graph is proportional to $\ln N$. However, there is no way for the individuals living in a social network with $\alpha = 0$ to find the target quickly with only local information on the network, such as in the navigation model of Definition 4.14. In mathematical terms, Kleinberg was in fact able to prove that, when $\alpha = 0$, the expected delivery time T of the navigation model is at least proportional to $S^{2/3}$. This means that the Milgram experiment would have failed in a social network with completely random long-range links, since the short paths would have been there, but it would have been impossible for the people in the network to find and exploit them to deliver the letter to the target. Intuitively, this happens because a node holding the message can only exploit its own long-range link, while it does not know the features of the long-range links of its neighbours. This will certainly be useful in the first few steps to get closer to the target, but then the long-range links will very often be useless to reach the target because they will generally point to other nodes far away from the target. In fact, in the navigation system of Definition 4.14, a shortcut is "useful" if it points to nodes closer to the target in terms of lattice distances. Ideally, the best network to navigate would be the one in which the fraction of useful shortcuts is roughly constant along the navigation, and this is obtained by tuning the value of α and assigning a suitable distribution of shortcut lengths.

In the case $\alpha \neq 0$, Kleinberg proved the following two theorems providing, respectively, lower bounds for the expected delivery time T for $0 \leq \alpha < 2$ and for $\alpha > 2$, and an upper bound when $\alpha = 2$ [180, 178].

Theorem 4.1 (Kleinberg theorem 1)   *Let $\alpha = 2$. There is a constant c_2, independent of S, so that the expected delivery time T of the decentralised algorithm is at most $c_2(\ln S)^2$.*

Theorem 4.2 (Kleinberg theorem 2)   *Let $0 \leq \alpha < 2$. There is a constant c_α, depending on α, but independent of S so that the expected delivery time T of any decentralised algorithm is at least $c_\alpha S^{(2-\alpha)/3}$.*

Let $\alpha > 2$. There is a constant c_α, depending on α, but independent of S so that the expected delivery time T of any decentralised algorithm is at least $c_\alpha S^{(\alpha-2)/(\alpha-1)}$.

The theorems demonstrate that α has a crucial role in the effective searching for targets in the graph. In practice, in a navigation with local information, it is possible to achieve a rapid delivery time only for $\alpha = 2$. Exactly at this very special value of α we have a logarithm scaling with the system size, $T \sim (\ln S)^2$. What happens is the following. Minimising the graph characteristic path length as a function of α is not necessarily the same as minimising the delivery time of a decentralised search. In fact, even when short paths exist, it could be impossible to find them and actually use them, with only local information. On the other hand, a certain amount of the minimisation of the path lengths could be sacrificed in order to provide some structural clues that can be used to guide the message towards the target. As α increases from 0 to 2, the long-range connections become shorter, making them less useful for the transmission of the message at large distances on the lattice in one step. However, at the same time, the graph becomes more easy to navigate and the delivery time T decreases, since the decentralised algorithm can take advantage of the correlation between the length of the shortcuts and the geographic structure of the network. When $\alpha > 2$, the delivery time gets worse again as α continues to increase because short paths actually start becoming too rare in the graph.

Summing up, the best structure to navigate is a graph with $\alpha = 2$. In other words, when the probability of friendship with an individual at distance d on the lattice decreases like the square of the distance, we have a small world in which the paths are not only there, but also can be found quickly by individuals with only a local knowledge of the network. As illustrated in the following example, a power law with an exponent $\alpha = 2$ has indeed some special properties of scale-invariance in $D = 2$ dimensions.

Example 4.15 When the probability to link a node to another node at distance d on a 2D lattice decreases as d^{-2}, it is possible to prove that the probability of having a link of any given length is the same at all scales [177]. To see this, let us compute the probability that a link spans a lattice distance in the range $[d, fd]$, where f is a factor larger than 1. It is easy to prove that, in a two-dimensional space, such a probability is proportional to:

$$\int_d^{fd} r^{-\alpha} r \, dr = \begin{cases} \frac{f^{2-\alpha}-1}{2-\alpha} d^{2-\alpha} & \text{for } \alpha \neq 2 \\ \ln f & \text{for } \alpha = 2 \end{cases}$$

for any large distance d, and for any factor $f > 1$. Therefore, when $\alpha = 2$ this probability is independent of distance d. This means that, when the link probability decreases according to an inverse-square law in two dimensions, then on average people have the same number of friends at any scale. In practice, the number of links of length 1–10 will be equal to those of length 10–100, to those of length 100–1000, and so on. It is exactly this property that allows the message to descend gradually through the various distance scales, finding ways to get significantly closer to the target at each step. This argument is in agreement with the intuition that, when $\alpha = 2$, the fraction of "useful" shortcuts is roughly constant along the navigation.

The exact solution of the decentralised navigation model is today available [70, 71]. These findings complement the bounds established by Kleinberg and allow to study more precisely the dependence of the delivery time on the graph order, for various values of the exponent α. Namely, we have:

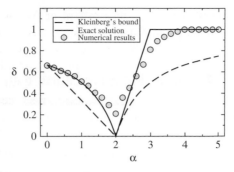

Fig. 4.17 The scaling exponent δ of the delivery time $T \sim S^\delta$ for a decentralised navigation in a 2D Kleinberg graph model is plotted as a function of the parameter α tuning the shortcut probability in Eq. (4.22). Numerical results (circles) are compared with the lower bounds found by Kleinberg (dashed line) and with the exact solution presented in Refs. [70, 71] (solid line). Notice that, when $\alpha > \alpha_c = 4$, we are outside the small-world regime.

$$
T \sim \begin{cases}
S^{\frac{2-\alpha}{3-\alpha}}/p^{\frac{1}{3-\alpha}} & \text{for } 0 \le \alpha < 2 \\
(\log S)^2/p & \text{for } \alpha = 2 \\
S^{\alpha-2}/p & \text{for } 2 < \alpha < 3 \\
S/(p \ln S) & \text{for } \alpha = 3 \\
S/[1 + c_\alpha p] & \text{for } 3 < \alpha < 4.
\end{cases}
\tag{4.24}
$$

These results are summarised in Figure 4.17, where the exponent δ of the scaling $T \sim S^\delta$ of the expected delivery time T versus the lattice linear size S, is plotted as a function of α, and compared with numerical results obtained from simulations of the model.

Although we have considered here only the case $D = 2$, the results we have presented can be generalised to lattice graphs in any dimension D. The main conclusion is that, in a system with D dimensions, the best Kleinberg network model to navigate is the one in which the number of long-range links decays with distance with an exponent $\alpha = D$. In fact, a decentralised navigation is able to achieve an expected delivery time T that is polynomial in the logarithm of the total number of nodes only when $\alpha = D$.

After Kleinberg published his results, various empirical studies have indeed confirmed that real social networks are organised in a such a way that the probability of having a link of a certain distance follows an inverse-square law. For instance, in 2005 David Liben-Nowell and collaborators used data from online social networks to test the predictions of Kleinberg small-world navigation model [207]. In particular, they studied the friendship network of half a million people who reported hometown locations and lists of friends on the public blogging site LiveJournal. In order to deal with the fact that real human population densities are highly nonuniform, Liben-Nowell and collaborators had to define the distance between two people in a slightly different way. They investigated how the probability that a person i forms a link to a person j depends on the number of people who are closer to i than j, rather than on the geographical distance between i and j. And with such a definition of distance, they showed that the data are in good agreement with a generalisation of the inverse-square law. This finding is certainly surprising, and it is today still an intriguing open question to understand why and how real social networks have organised themselves in a pattern of friendships over distance that is optimal for forwarding messages to distant targets.

4.7 What We Have Learned and Further Readings

In this chapter we have learnt that real-world networks exhibit the *small-world behaviour*; that is, they can have very small characteristic path lengths L, as small as the characteristic path lengths of random graphs. In mathematical terms, the typical distance between nodes *scales as the logarithm of the number of nodes*, $L \sim \ln N$, the same scaling found in the previous chapter for ER random graphs. However, while in random graphs cycles can be neglected so that by following two different paths from a node we have access to different nodes; this is not true in real-world networks, which have instead a large number of cycles. To quantify this property we have focused on the shortest possible cycles, the triangles, and we have introduced a novel measure, the so-called node *clustering coefficient*, that counts the number of triangles a node belongs to. In this way we have shown that, indeed, real-world networks have small characteristic path lengths and, at the same time, large clustering coefficients. This discovery opens up to the need of introducing novel models. We have therefore shown, by studying the *Watts and Strogatz model*, that we can reproduce both properties of real-world networks by rewiring at random a few of the links of regular lattices. We can, in fact, easily construct regular lattices with a high number of triangles. And, although regular lattices have large characteristic path lengths, the addition of just a few random links, that work as shortcuts, produces the small-world behaviour. For an historical introduction to small-world models and their applications in a remarkable variety of fields, we recommend the book by Duncan Watts, one of the two researchers who first proposed these ideas [309]. For a complete mathematical treatment of clustering in complex networks see the work by M. Ángeles Serrano and Marián Boguñá [280].

In this chapter we have also introduced the first database of a biological system, the *neural network of the* C. elegans. For readers who are further interested in the network analysis of the *C. elegans*, we suggest a review paper that recently appeared in the journal *PLoS Computational Biology* [303]. Particular attention to the distribution of the nodes of the network of the *C. elegans* in space and time can be found in the paper by Sreedevi Varier and Marcus Kaiser [302], and in the paper by Vincenzo Nicosia, Petra Vértes et al. [243]. For a more general introduction to brain networks extracted from different neuroimaging techniques, such as electroencephalography and functional magnetic resonance imaging in humans, we recommend the review papers by Ed Bullmore and Olaf Sporns [61, 62].

Problems

4.1   Six Degrees of Separation

   (a) Set up an experiment similar to that by Milgram at different scales, for instance in your city (or in your university). What you need first is to find someone available to be the target of the experiment. Of course, as target you can use one of your friends, but then you have to be careful to not use other friends of yours as sources of the letter. If you do so, you have biased the experiment because you have chosen sources and target at distance two (you are the common acquaintance!).

(b) Compute the distribution of Mastroianni's and Clooney's numbers for the movie actor collaboration graph introduced in Chapter 2. Compute average and standard deviation of the two distributions. Marcello Mastroianni and George Clooney respectively correspond to node 62821 and node 8530 in the data set.

(c) Perform a component analysis of the movie actor collaboration graph. Focus your attention on order and number of links in components other than the largest one.

(d) If a randomly chosen actor can be linked to Kevin Bacon in an average of three steps (see Table 4.3), does this mean that two randomly chosen actors in the largest component can be linked by a path through Kevin Bacon in an average of six steps?

4.2 The Brain of a Worm

Compute the clustering coefficient c for each of the nodes of the neural network of the *C. elegans* in DATA SET Box 4.2. Construct and plot the distribution of the node clustering coefficient $p(c)$.

4.3 Clustering Coefficient

(a) Calculate the number of triangles $n_\triangle$ in a complete graph with N nodes.

(b) Can the number of triangles in a graph be larger than the number of edges? Are you able to find a graph with more triangles than edges?

(c) Prove that the number of triads in a undirected graph is equal to half the number of non-zero off-diagonal entries of matrix A^2, where A denotes the adjacency matrix. Apply this to the graph in Example 4.1 and verify that the number is exactly eight.

(d) Find an expression for the graph transitivity T in terms of the elements of the adjacency matrix. Notice that the number of triangles in a graph is related to the trace of matrix A^3 (see Eq. (8.10) in Chapter 8).

(e) In 1992 Ronald Stuart Burt proposed a theory on the benefits associated with occupying *brokerage* positions between otherwise disconnected individuals in a network [63]. To quantify the presence of structural holes in the neighbourhood of a node, he proposed to calculate the so-called effective size S_i of a node i:

$$S_i = \sum_{j \in \mathcal{N}_i} \left[1 - \sum_\ell p_{i\ell} m_{j\ell} \right]$$

where $\mathcal{N}_i$ is the set of neighbours of i, and the quantities $p_{i\ell}$ and $m_{j\ell}$ reduce for an undirected graph to [201]:

$$p_{i\ell} = \frac{a_{i\ell}}{k_i} \qquad m_{j\ell} = a_{j\ell}.$$

The effective size of node i ranges from its smallest value equal to 1, when node i belongs to a clique, to a maximum value equal to the node degree k_i, when there are no links (j, ℓ) connecting any two neighbours j and ℓ of i, that is in the very special case in which i is the centre of a star graph. In general, the more redundant the neighbours of i are, the smaller the value of S_i is, and vice versa. There is a simple relation between the effective size of a node i, S_i, and its clustering coefficient C_i [201]. Are you able to derive it?

4.4 The Watts–Strogatz (WS) Model

(a) Prove that a triangular lattice has a clustering coefficient $C = 0.4$. Prove that for any triangular tessellation of the plane, the regular triangular lattice is the one with the smallest clustering coefficient.

(b) Compute the clustering coefficient $C(N, m)$ of a linear graph with N nodes and generic m. Show that in the limit for large N, $C(N, m)$ converges to C^{circle} given in Eq. (4.10).

(c) Consider a (N, m) D-dimensional square lattice network (see Definition 4.6). Prove that the expression (4.10) for the clustering coefficient of a 1D lattice generalises, in this case, to:

$$C^{\text{reg}} = \frac{3(m - 1)}{2(2mD - 1)}.$$

Prove, also, that the characteristic path length scales with the number of nodes in the lattice as:

$$L^{\text{reg}} \sim N^{1/D}.$$

4.5 Variations to the Theme

(a) Consider a (N, m) circle graph. Write down the expression of the transitivity T, defined in Eq. (4.1), for such a graph in terms of m and N. Does this expression coincide with that of the graph clustering coefficient in Eq. (4.10)?

(b) Due to the fact that no link of the underlying circle graph is rewired, analytical calculations in the NW model (see definition 4.10) are much easier than in the original WS model. Find that the transitivity T of (N, m, p) NW graphs can be written as [239]:

$$T(p) = \frac{3(m - 1)}{2(2m - 1) + 8mp + 4mp^2}$$

as a function of the probability p to add shortcuts.

4.6 Navigating Small-World Networks

Consider a D-dimensional square lattice of linear size S, and show that the Kleinberg decentralised algorithm has a delivery time T that scales as $T \sim S$.

5 Generalised Random Graphs

Many real networks, such as the network of hyperlinks in the *World Wide Web*, have a degree distribution p_k that follows a power law for large values of the degree k. This means that, although most nodes have a small number of links, the probability of finding nodes with a very large degree is not negligible. Both Erdős–Rényi random graphs models and the small-world models studied in the previous chapter produce instead networks with a degree distribution peaked at the average degree $\langle k \rangle$ and rapidly decreasing for large values of k. In this chapter we show how to generalise the Erdős–Rényi models to reproduce the degree distributions of real-world networks. Namely, we will discuss the so-called *configuration model*, a method to generate random graphs with any form of p_k. We will then derive the *Molloy and Reed criterion*, which allows us to determine whether random graphs with a given p_k have a giant component or not. We will also work out some expressions for the clustering coefficient and for the characteristic path length in terms of the average degree $\langle k \rangle$ and of the second moment $\langle k^2 \rangle$ of the degree distribution. Specifically, we will use all these results to study the properties of random graphs with *power-law degree distributions* $p_k \sim k^{-\gamma}$, with $2 < \gamma \leq 3$. The final section of the chapter is devoted to *probability generating functions* and introduces an elegant and powerful mathematical method to obtain results on random graphs with any degree distribution p_k in a compact and direct way.

5.1 The World Wide Web

In the previous two chapters we have shown that various real-world networks have short characteristic path lengths and high clustering coefficients. However, we have not examined yet how the degree is distributed among the nodes of these networks. As an example, we will study here the degree distribution of the *World Wide Web* (WWW), a network that has become fundamental in the everyday life of billions of people as one of the main sources of information, as a platform for the diffusion and exchange of knowledge and opinions, and more recently as the location where an increasing proportion of our social activities take place. The vertices of this network are *webpages* (i.e. documents written in the HTML language, each associated with a different Uniform Resource Locator, or URL, used to identify and retrieve a webpage), while the directed links represent *hyperlinks* pointing from one document to another. Each webpage is hosted at a specific *website*, identified by a certain domain name. For instance, the URL `http://en.wikipedia.org/wiki/Complex_network` identifies one of the few millions of webpages (namely the webpage `wiki/Complex_network`) of the English version of Wikipedia, a website which hosts the largest free-access encyclopedia in the world and is associated with the domain name `en.wikipedia.org`.

The WWW is a directed network: given two webpages i and j, there can be a hyperlink pointing from i to j, a hyperlink pointing from j to i, or both. With billions of nodes, the WWW is certainly one of the largest mapped network available on the market. And it is a perfect example of a self-organised technological system. The growth of the WWW is in fact totally unregulated: any individual, company or institution is free to create new websites with an unlimited number of documents and links. Moreover, existing websites keep evolving with the addition and deletion of webpages and links. We can easily get a picture of how fast the Web is growing from the statistics on the total number of websites in the entire World Wide Web. In Figure 5.1 we report the evolution of this number from 1991 to 2013. Starting by the first webpage put online by Tim Berners-Lee in August 1991, the WWW counted ten thousand websites only three years later, when the first search engine Yahoo! was launched. Today it consists of more than 600 million websites. Despite the accurate statistics on the number of websites, obtaining the complete map of the WWW at the level of webpages is still an open challenge. There are, however, various partial network samples that we will consider here, whose size is large enough to allow a reliable statistical analysis of their degree distribution. The first database that we will study is the network of all the webpages of the University of Notre Dame. Such a network, was mapped in 1999 by Réka Albert, Hawoong Jeong, and Albert-László Barabási, a group of physicists then at the University of Notre Dame [5]. This is a network of great historical importance because it is the first example of a systematic study of degree distributions in the Web. We will also consider other three more recent samples of the Web, respectively indicated as Stanford, BerkStan and Google. All the four networks are described in DATA SET Box 5.1 and available from the book webpage.

Before studying the degree distributions of the networks of the WWW described in Box 5.1 this is the right time to go back to the concepts of connectedness and components

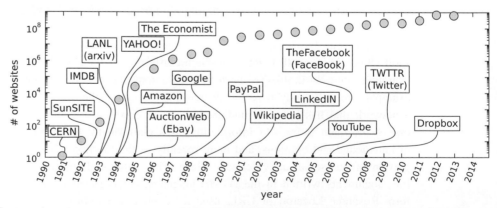

Fig. 5.1 The growth of the World Wide Web over the years from its initial creation at CERN in 1991. The number of websites is plotted versus time, together with the year of appearance of some well-known websites.

Box 5.1   DATA SET 5: Network Samples of the World Wide Web

In the network of the World Wide Web, the nodes represent webpages and the arcs stand for directed links pointing from one page to another. In the figure below we show, as an example, some of the hyperlinks connecting a few pages of Wikipedia.

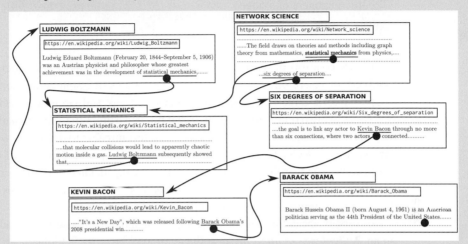

In this data set we provide four different network samples of the World Wide Web. The first network refers to the entire domain of the University of Notre Dame (nd.edu), and was constructed in 1999 by Réka Albert, Hawoong Jeong, and Albert–László Barabási [5]. The Notre Dame World Wide Web network has $N = 325,729$ nodes and $K = 1,469,680$ arcs which corresponds to an average degree $\langle k^{out} \rangle = \langle k^{in} \rangle = 4.51$. It is important to notice that, while all the N nodes in this network have at least one incoming arc, only $136,934$ of them have out-degree different from 0. The other three networks of the World Wide Web, respectively based on the entire domain of the University of Stanford (stanford.edu), on the webpages both from the University of California Berkeley (berkely.edu) and the University of Stanford, and on a sample of the Web released by Google in the 2002 programming contest, were constructed and analysed by Jure Leskovec and collaborators [206]. The three networks have respectively $N = 281903, 685230, 875713$ nodes and $K = 2312497, 7600595, 5105039$ arcs.

introduced in Chapter 1. As with any directed graph, the World Wide Web has in fact a richer component structure than that of an undirected graph, which is therefore worth exploring. We know from Definition 1.12 that the components of a directed graph can be of two different types, namely weakly and strongly connected. We now introduce the notion of components *associated with a node* of a directed graph. In this case, we need to define four different concepts.

Definition 5.1 (Node components in directed graphs)   *The* weakly connected component of *a node i of a directed graph G, denoted as WCC(i), is the set of vertices that can be reached from i ignoring the directed nature of the edges altogether.*

The strongly connected component of a node *i, denoted as SCC(i), is the set of vertices from which vertex i can be reached and which can be reached from i.*
The out-component of node *i, denoted as OUT(i), is the set of vertices which can be reached from i.*
The in-component of a node *i, denoted as IN(i), is the set of vertices from which i can be reached.*

The two concepts of weakly and strongly connected components associated with a node *i* defined above are directly related to the definitions of graph components introduced in Chapter 1 for a directed graph. This is because weak and strong connectedness between two nodes are two reflexive, symmetric and transitive properties, i.e. in mathematical terms, they are *equivalence relations*. Consequently, it is possible to define the weakly connected components (WCCs), and the strongly connected components (SCCs) of a graph in terms of the weakly and strongly connected components associated with the nodes of the graph. In particular, the weakly (strongly) connected component of a node coincides with one of the weakly (strongly) connected components of the graph.

Conversely, the definitions of out-component and in-component of a node *i* are not based on an *equivalence relation* because the symmetry property is violated. In fact $i \in \text{OUT}(j)$ does not imply $j \in \text{OUT}(i)$, and analogously $i \in \text{IN}(j)$ does not imply $j \in \text{IN}(i)$. This means that out- and in-components are concepts associated only with nodes and cannot be extended to the entire graph. In practice, we cannot partition a graph into a disjoint set of in- or out-components, while it is possible to partition a graph into a disjoint set of weakly or strongly connected components. However, it is possible to define the out- and in-components of a set of nodes, for instance the out- and in-components of a strongly connected component. Also, the in- and out-components of the nodes of a directed graph can be used to define the strongly connected components of the graph. From the above definitions, we observe that $i \in \text{OUT}(j)$ iff $j \in \text{IN}(i)$. Furthermore, we notice that *i* and *j* are strongly connected iff $j \in \text{OUT}(i)$ and, at the same time, $i \in \text{OUT}(j)$. Or equivalently, iff $j \in \text{OUT}(i)$ and $j \in \text{IN}(i)$. Therefore the strongly connected component of node *i*, which is one of the graph strongly connected components, can be obtained as the intersection of $\text{IN}(i)$ and $\text{OUT}(i)$.

Going back to the World Wide Web, the network is made of many (disjoint) weakly connected components as sketched in Figure 5.2. Each weakly connected component can contain one or more strongly connected components. Let us focus on the internal structure of one of the WCCs of the World Wide Web, supposing, for simplicity, that the WCC contains only one SCC as shown in Figure 5.2. By definition, every node of the WCC is reachable from every other node, provided that the links are treated as bidirectional. Instead, the SCC consists of all those nodes in the WCC that are also strongly connected. All the nodes reachable from the SCC are referred to as the OUT *component* of the SCC, while the nodes from which the SCC is reachable are referred to as its IN *component*. Therefore, the SCC is the intersection of the IN and the OUT component. There are also nodes in the WCC which are neither in the OUT nor in the IN-component of the SCC. These nodes are referred to as the *tendrils* of the SCC. And the whole structure around any

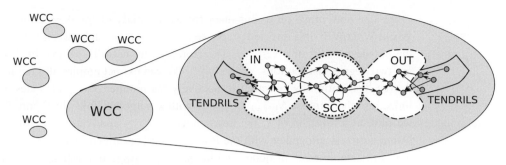

Fig. 5.2 The World Wide Web consists of many weakly connected components (WCCs), each of them containing one or more strongly connected components (SCCs). In the simplest case in which a WCC contains only one SCC, the nodes of the WCC can be partitioned into the SCC, its IN- (dotted line) and OUT-(dashed line) components, and the tendrils. This is known as the *bow tie structure* of the Web.

SCC of the network is known, for obvious reasons, as the *bow tie structure of the Web* [59]. In the following example we focus specifically on the study of components in the network samples introduced in Box 5.1.

Example 5.1 *(Component analysis of the WWW)* The algorithm for the discovery of connected components in undirected graphs introduced in Chapter 3 and described in Appendix A.8.1 can be straightforwardly used to detect the weakly connected components of a directed graph. In fact, it is sufficient to feed the algorithm with the undirected version of the graph to obtain as output the set of weakly connected components of the original directed graph. The detection of strongly connected components requires a slightly more sophisticated algorithm, as detailed in Appendix A.8.2. In the table below we report, for each of the four WWW samples, the number of weakly and strongly connected components, and the characteristics of the largest weakly connected component (LWCC) and of its largest strongly connected component (LSCC).

Network	N	# WCC	# SCC	LWCC	LSCC	IN	OUT	tendrils
Notre Dame	329729	1	203608	325729	53968	53968	325729	0
Stanford	281903	365	29913	255265	150532	183317	218048	4432
BerkStan	685230	676	109405	654782	334857	494566	459831	35242
Google	875713	2746	371764	855802	434818	615720	600493	74407

Notice that, with the only exception of Notre Dame, all the networks have more than one weakly connected component, each of them consisting of one or more strongly connected components. The typical bow-tie structure of the WWW is evident in Stanford, BerkStan and Google, whose largest strongly connected component LSCC has non-trivial IN and OUT components and tendrils (their respective sizes are reported in the last three columns of the table).

We can now move on to our main focus, the study of the degree distributions of the World Wide Web. In a directed graph we have to distinguish between the in-degree k_i^{in} and the out-degree k_i^{out} of a vertex i. Therefore, in any network sample of the WWW, there are two degree distributions: p_k^{in} and p_k^{out}. We can extract empirically the in- and the out-degree distribution by counting respectively the number $N_{k^{in}=k}$ of nodes with k ingoing links, and the number $N_{k^{out}=k}$ of nodes with k outgoing links. In Figure 5.3 (a) and (b) we plot, as a function of k, the histograms of $N_{k^{in}=k}$ and $N_{k^{out}=k}$ obtained for the Notre Dame network, altogether with the Poisson distribution expected in a random graph with the same total number of nodes and links. The message in the figure is clear. The degree distributions of the WWW are totally different from that of a random graph. This certainly came as a big surprise when it was first observed by the three physicists of Notre Dame University. The network of the WWW is the result of billions of agents acting independently, so there were no a priori reasons to find such large deviations from a random graph. And the differences from a Poisson distribution are indeed so large that the real degree distribution of the Web cannot be shown entirely in the same scale as that of the Poisson distribution. In fact, in the WWW there are a few nodes with an extremely large number of links. For instance, the five nodes with the largest k^{out} have respectively 3444, 2660, 2647, 1892, 1735 links, while the five nodes with the largest k^{in} have 10721, 7618, 7026, 4299, 4279 links. These numbers are much larger than those expected for random graphs. For instance, the probability of finding a node with 3444 links in a randomised version of the WWW graph is $8.36 \cdot 10^{-8438}$. This means that the expected number of nodes with 3444 links in a random graph is $2.72 \cdot 10^{-8432}$, practically zero. To observe entirely the long tail of the distribution in a single plot it is much better to use a log–log scale, as done in Figure 5.3 (c) and (d). We are now plotting directly the degree distributions, which are obtained by dividing by N the number of nodes with a certain degree: $p_k^{in} = N_{k^{in}=k}/N$ and $p_k^{out} = N_{k^{out}=k}/N$[1]. The fact that the points stay on a straight line in a log–log plot means that we can write:

$$\log p_k = a + b \log k \tag{5.1}$$

where a and b are respectively intercept with the vertical axis and slope of the straight line. This is equivalent to say that the probability p_k of having a node with degree k is a *power law*:

$$p_k = c k^{-\gamma} \tag{5.2}$$

where $c = 10^a$ and $\gamma = -b > 0$. See Box 5.2 for more details on power-law functions and their properties. Of course, here we need to assume $k \geq 1$, since for $k = 0$ the expression in Eq. (5.2) diverges. We can also try to fit the points in Figure 5.3 (c) and (d) with a straight line in order to compute the value of the power-law exponent γ in the two cases. However, we postpone this to Section 5.2 where we will explicitly deal with the best methods to extract the exponent of a power law from a data set. What we want

---

[1] Notice that, here, and in all cases in which we deal with real networks rather than mathematical models, p_k denotes the probability that a randomly chosen node *in one particular graph* has k links. This is different from the original definition of degree distribution, Eq. 3.12, given in Chapter 3, where p_k is calculated from the $\overline{N}_k$, the expectation number of N_k taken *over an ensemble of graphs*.

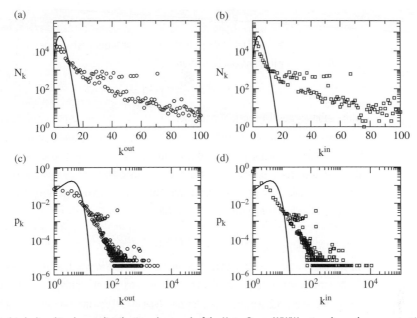

Fig. 5.3 The out- (circles) and in-degree distributions (squares) of the Notre Dame WWW network are shown respectively in a log–linear (panel a and b) and in a log–log scale (panel c and d) together with the corresponding Poisson distributions (solid lines).

to stress now is that for the WWW graph we have found that both p_k^{out} and p_k^{in} follow a power-law behaviour over about four orders of magnitude. Power-law degree distributions are remarkably different from the degree distributions of ER random graphs or of the WS model, so that graphs with power-law degree distributions have been given a special name. They are called *scale-free graphs*, because of the main property of power laws, that of being *scale-invariant* or, in other words, *scale-free*. See Box 5.2 and the following example.

Example 5.2   It is simple to verify that a power law $f(x) = cx^{-\gamma}$ satisfies the scale-invariance condition of Eq. (5.6), with multiplicative factor $a(\lambda) = \lambda^{-\gamma}$. But it is also possible to prove that power laws are the only functions satisfying Eq. (5.6). In our derivation we will follow Ref. [242]. We can first get rid of the multiplicative factor $a(\lambda)$ by setting $x = 1$ in Eq. (5.6). We get $a(\lambda) = f(\lambda)/f(1)$, and then we can rewrite Eq. (5.6) as:

$$f(\lambda x) = \frac{f(\lambda)f(x)}{f(1)}.$$

Since this expression is valid for any scale factor λ, assuming f is a differentiable function and taking the derivative of both sides with respect to λ, we get:

$$xf'(\lambda x) = \frac{f'(\lambda)f(x)}{f(1)}$$

A power law is a functional relationship between two quantities, where one quantity varies as a power of the other. A power-law function can be written as:

$$f(x) = cx^{-\gamma} \qquad (5.5)$$

with c arbitrary positive constant, and where $\gamma \in \mathbb{R}$ is called the exponent of the power law. Usually, x can only take non-negative values, so that $f(x) : \mathbb{R}^+ \to \mathbb{R}^+$. Furthermore, when we want to consider power-law probability distributions, we need to assume $\gamma > 1$ in order to guarantee the proper normalisation. The main property of power laws is that they are *scale-invariant*, or *scale-free*, meaning that they are the same at any scale, they are independent from the scale at which we observe it. In mathematical terms, this means that the function $f(x)$ in Eq. (5.5) satisfies:

$$f(\lambda x) = a(\lambda) \cdot f(x) \qquad (5.6)$$

for any scale factor $\lambda > 0$. That is, if we increase the scale by which we measure the quantity x by a factor λ, the form of the function $f(x)$ does not change, apart from a multiplicative factor $a(\lambda)$. And indeed, as shown in Example 5.2, a power law is the only type of function that satisfies Eq. (5.6).

where f' indicates the derivative of f with respect to its argument. By setting $\lambda = 1$ we obtain the first-order differential equation:

$$xf'(x) = -\gamma f(x)$$

where $\gamma = -f'(1)/f(1)$. This differential equation can be solved by separation of variables, yielding:

$$\ln f(x) = -\gamma \ln x + C$$

where the integration constant C is equal to $\ln f(1)$. Finally, taking exponentials of both sides we get:

$$f(x) = f(1) x^{-\gamma}$$

which is a power law.

Scale-free functions are ubiquitous in nature and in made-man systems. In fact, the distributions of a wide variety of physical, social and biological phenomena can be well approximated by a power law over a wide range of magnitudes [242, 80]. Some notable examples are illustrated in Box 5.3. Power law also have an important role in statistical physics because of their connections to phase transitions and fractals [292, 111, 64]. In particular, in the case of the WWW, a scale-free degree distribution implies that there are nodes with all scales of degree. The probability of finding documents with a large number of outgoing links is significant, and the network connectivity is dominated by highly connected webpages. Notice, for instance, that the node with the highest in-degree $k_i^{\text{in}} = 10721$ is pointed to by more than 3 per cent of the other nodes in the network.

Gutenberg–Richter, Zipf's and Pareto's Laws

Distributions with a power-law form have been discovered over the years in the most varied of contexts. For instance, in seismology, if we characterise the strength of an earthquake by its amplitude A, the number $N(A)$ of registered earthquakes of amplitude greater or equal to A follows a power-law distribution $N(A) \sim A^{-b}$, with an exponent $b \simeq 1$. This is known as the *Gutenberg–Richter law* [148]. A similar expression, although in the slightly different form of a ranked-distribution, was empirically found by the American linguist George Kingsley Zipf in his studies on word occurrences in different languages. The so-called *Zipf's law* states that, if the words of a text written in any language are ranked from the most to the least frequent, the number of times N a word appears is inversely proportional to its rank r, namely $N(r) \sim r^{-1}$ [325]. Thus, the most frequent word will occur approximately twice as often as the second most frequent word, three times as often as the third most frequent word, etc. Zipf's law is also valid for many other rankings unrelated to language, such as for the population ranks of cities or for the income ranks of individuals in various countries. In the latter case of wealth distributions, the power law is better known as *Pareto's law*, after the Italian economist who first discovered the "80–20 rule" which says that 20 per cent of the population controls 80 per cent of the wealth [252]. Finally, notice that plotting the probability distribution $p(x)$ of observing a certain value x, or plotting the value $x(r)$ as a function of the rank r, are completely equivalent. Consequently, if $p(x)$ is a power law with a given exponent, also $x(r)$ is a power law with an exponent that can be related to the first one [64].

Similarly, for incoming links, the probability of finding very popular addresses, to which a large number of other documents point, is non-negligible. Notice that, while the owner of each webpage has complete freedom in choosing the number of links on a document and the addresses to which the links point, the power law observed in the in-degree distribution is highly non-trivial being the result of the self-organised behaviour of all the owners.

The study of the World Wide Web as a graph is not only fascinating in its own right, but also yields valuable insights into algorithms for searching information and items on the Web as we discuss in the following example.

Example 5.3 *(Hubs and authorities)*   In Section 2.3 we ran into two types of centrality generalising the concepts of in- and out-degree of a node in a directed graph. The in- and the out-degree eigenvector centrality, $c_i^{E^{\text{in}}}$ and $c_i^{E^{\text{out}}}$, of a node i are defined respectively in terms of left and right eigenvectors of matrix A corresponding to its largest eigenvalue λ_1, namely $A^\top \mathbf{c}^{E^{\text{in}}} = \lambda_1 \mathbf{c}^{E^{\text{in}}}$ and $A \mathbf{c}^{E^{\text{out}}} = \lambda_1 \mathbf{c}^{E^{\text{out}}}$. In the World Wide Web, a node with high in-degree eigenvector centrality corresponds to a webpage pointed by many other pages with a high in-degree. Conversely, a webpage with high out-degree eigenvector centrality is a webpage pointing to many pages with many outgoing links. This is shown in panels (a) and (b) of the figure below. However, when we seek information on the Web, two other definitions of eigenvector centrality turn out to be more appropriate than these. We are in fact interested in finding good resource lists, the so-called information *hubs*, linking in

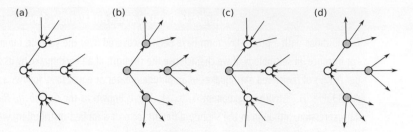

a correlated way to a collection of pages containing relevant information on a common topic, the so-called *authorities*. In 1999, the same Jon Kleinberg of the small-world model of Section 4.5, introduced an algorithmic definition of *hubs* and *authorities* in a directed network [179]. The main idea is that hubs and authorities have to be defined concurrently as shown in panels (c) and (d). Although a good hub may not contain directly relevant information on a given topic, it can certainly tell you where to find such information. We can then define a good hub (filled circle in panel c) as a node that points to a large number of good authorities (open circles). This means that, in order to define hubs, we need to define at the same time authorities, and this can be done in terms of the hubs. A good authority, that is likely to contain relevant information, is a node as the empty circle in panel (d), which is pointed to by many highly ranked hubs (filled circle). Summing up, Kleinberg proposed to associate with each node i a hub centrality score c_i^H and an authority centrality score c_i^A defined self-consistently in the following way:

$$\nu c_i^A = \sum_j a_{ji} c_j^H \qquad \mu c_i^H = \sum_j a_{ij} c_j^A.$$

The previous equations in vectorial notation, read:

$$A^\top \mathbf{c}^H = \nu \mathbf{c}^A \qquad A\mathbf{c}^A = \mu \mathbf{c}^H. \tag{5.7}$$

By eliminating c^H from the first equation we get:

$$A^\top A\mathbf{c}^A = \lambda \mathbf{c}^A \tag{5.8}$$

where we have set $\nu\mu = \lambda$. Analogously, for the hub centrality we have:

$$AA^\top \mathbf{c}^H = \lambda \mathbf{c}^H \tag{5.9}$$

so that the primary difference between what we did in Section 2.3 is the replacement of the adjacency matrix A and $A^\top$ by the symmetric products $AA^\top$ and $A^\top A$. Notice that matrices $AA^\top$ and $A^\top A$ may be reducible even when the original matrix A is irreducible. For example a directed cycle of order N is strongly connected, so its adjacency matrix A is irreducible, while the corresponding symmetric matrices $AA^\top$ and $A^\top A$ are reducible. This might be a problem for the validity of the hypothesis of the Perron–Frobenius Theorem 2.4. Therefore, the best way to proceed is to perform a component analysis of the weighted undirected graph defined by $AA^\top$ or by $A^\top A$, and to evaluate hub and authority scores only for nodes belonging to the largest connected component.

5.2 Power-Law Degree Distributions

We have seen that power laws are very particular distribution functions that appear in many situations. In this section we discuss some of the mathematical properties of power-law degree distributions which will be useful later on, and we also focus on the numerical methods to extract the exponent of a power law from real data.

Let us consider the expression in Eq. (5.2). The most important property of a degree distribution is that it has to be normalisable. The constant c in Eq. (5.2) is then fixed by the normalisation condition:

$$1 = p_0 + \sum_{k=1}^{\infty} p_k = p_0 + c \sum_{k=1}^{\infty} k^{-\gamma}$$

where $p_0 = N_0/N$ is the fraction of isolated nodes in the network. Since nodes with $k = 0$ cannot be included in the power-law expression, we neglect them from the beginning by setting $p_0 = 0$, so that p_k is the probability of finding nodes with k links on the population of nodes with at least one link. The series can be summed for $\gamma > 1$ and is equal to the Riemann ζ-function[2]. Rearranging, we find that $c = 1/\zeta(\gamma)$ and we finally have a properly normalised degree distribution:

$$p_k = \frac{k^{-\gamma}}{\sum_{k=1}^{\infty} k^{-\gamma}} = \frac{k^{-\gamma}}{\zeta(\gamma)} \quad \gamma > 1 \tag{5.10}$$

that is valid for $k = 1, 2, \ldots$, while $p_0 = 0$.

Quantities of great importance to characterise a distribution are the mth order moments:

$$\langle k^m \rangle = \sum_{k=1}^{\infty} k^m p_k \tag{5.11}$$

where m is a positive integer: When $m = 1$ we get the average degree, while the second moment is the mean square. For a power law with exponent γ as that in Eq. (5.10) the value of the mth order moment depends on the values of m and γ. In particular, it is easy to prove, by using the definition of the Riemann's ζ-function, that the mth order moment is infinite if $\gamma < m + 1$, while it is finite and equal to:

$$\langle k^m \rangle = \frac{\zeta(\gamma - m)}{\zeta(\gamma)} \quad \text{if} \quad \gamma > m + 1. \tag{5.12}$$

In particular we have:

$$\langle k \rangle = \sum_{k=1}^{\infty} k p_k = \frac{\zeta(\gamma - 1)}{\zeta(\gamma)} \quad \text{if} \quad \gamma > 2$$

$$\langle k^2 \rangle = \sum_{k=1}^{\infty} k^2 p_k = \frac{\zeta(\gamma - 2)}{\zeta(\gamma)} \quad \text{if} \quad \gamma > 3. \tag{5.13}$$

[2] Historically, Leonhard Euler considered the above series in 1740 for integer values of γ. Later Pafnuty Chebyshev extended the definition to real values of γ, and then Bernhard Riemann considered the series for complex values of γ and published in 1859 a relation between its zeros and the distribution of prime numbers.

When one deals with power-law probability distributions of discrete variables, the mathematics can be hard and, as we have seen, involves special functions. However, in large networks, k can be considered as a positive continuous variable, and the degree distribution can be well approximated by a function of a real variable k:

$$p(k) = ck^{-\gamma} \quad k \in \Re, k > 0, \ \gamma > 0. \tag{5.14}$$

Now, $p(k) \, dk$ represents the probability of having a value from k to $k + dk$. In this approximation, usually known as the *continuous-k approximation*, as we show below, special functions are replaced by more tractable integrals.

First of all notice that, since $\gamma > 0$, the integral $\int_0^\infty ck^{-\gamma} dk$ diverges at the origin, i.e. the function $p(k)$ is not normalisable. Hence, as in the case of the discrete degree distributions, we must assume that the power law in Eq. (5.14) holds from some minimum value $k_{min} > 0$. Usually, a good choice is $k_{min} = 1$ as in the discrete case. For generality, we assume that k_{min} can be any positive value. The constant c in Eq. (5.14) is then given by the normalisation condition:

$$1 = \int_{k_{min}}^\infty p(k)dk = c \int_{k_{min}}^\infty k^{-\gamma} dk = \frac{c}{1-\gamma} \left[k^{-\gamma+1} \right]_{k_{min}}^\infty.$$

This expression only makes sense if $\gamma > 1$, since otherwise the right-hand side of the equation would diverge at infinity. We thus obtain $c = (\gamma - 1)k_{min}^{\gamma-1}$, and the correct normalised expression for the power law is the following:

$$p(k) = \frac{\gamma - 1}{k_{min}^{1-\gamma}} k^{-\gamma} \quad \gamma > 1. \tag{5.15}$$

Now, the expressions for the various moments of a power-law degree distribution are easily obtained by solving the integrals:

$$\langle k^m \rangle = \int_{k_{min}}^\infty k^m p(k)dk \tag{5.16}$$

for $m = 1, 2, \dots$ As detailed in the following example, we can prove that the moments of order m, for $\gamma > m + 1$, are finite quantities equal to:

$$\langle k^m \rangle = \frac{\gamma - 1}{\gamma - 1 - m} k_{min}^m. \tag{5.17}$$

Conversely, when $\gamma \le m+1$, moments of order m are infinite in an infinitely large network, as shown in the following example.

Example 5.4 *(Moments of a power law in the continuous-k approximation)*   By substituting Eq. (5.15) in the Eq. (5.16), and solving the integral, we get:

$$\langle k^m \rangle = c \int_{k_{min}}^\infty k^{-\gamma+m}dk = \begin{cases} \frac{c}{m+1-\gamma} \left[k^{m+1-\gamma} \right]_{k_{min}}^\infty & \text{for } \gamma \neq m + 1 \\ c \left[\ln k \right]_{k_{min}}^\infty & \text{for } \gamma = m + 1 \end{cases}$$

where m is a positive integer. The terms in the brackets depend in general on the values of the moment order, m, and of the degree exponent, γ. If $\gamma \leq m + 1$, they diverge at infinity. Thus, moments of order m are infinite in an infinitely large network when $\gamma \leq m + 1$, while, for $\gamma > m + 1$, they are finite and equal to Eq. (5.17).

This means that the average of a power-law degree distribution is finite and well defined only for $\gamma > 2$, and is given by the expression:

$$\langle k \rangle = \frac{\gamma - 1}{\gamma - 2} k_{\min}. \tag{5.18}$$

The second moment is finite and well-defined only for $\gamma > 3$, and is given by the expression:

$$\langle k^2 \rangle = \frac{\gamma - 1}{\gamma - 3} k_{\min}^2. \tag{5.19}$$

We will now show that the World Wide Web and many other real-world networks exhibit power-law degree distributions with exponents $2 < \gamma < 3$. This implies that we have to face with networks with large fluctuations in the node degrees, namely networks whose average degree $\langle k \rangle$ is finite, while the second moment $\langle k^2 \rangle$ diverges with the number of nodes N in the system.

How to Fit a Power-Law Degree Distribution?

From the formulas above we understand that the behaviour of the moments depends crucially on the precise value of γ. Identifying a power-law behaviour and extracting the exponent γ can be a very tricky issue. The standard strategy we have already seen of making a simple histogram and plotting it on a double-log scale to see if it looks straight is, in most cases, a poor way to proceed [242]. Let us consider the case of the out-degree distribution of the WWW reported in Figure 5.3(c). We notice that the distribution is very fluctuating, especially for large values of k. In this region there are, in fact, only a few nodes, if any, for each value of k. Hence, the statistical fluctuations in the bin counts are large and this appears on the plot as a curve p_k with a noisy tail. However, we cannot simply ignore the data for large values of k. The tail of the curve is important to assess whether the distribution is a power law, to identify the range of validity of the power law, and eventually also to extract a correct value of the exponent. In order to reduce the fluctuations in the tail we can use larger and larger bins as k increases. A common procedure is the so-called *logarithmic binning*, in which the ratio a between the size of two successive bins is constant. For instance, we can set $a = 2$, so that the first bin contains nodes of degree 1, the second bin contains nodes with $k = 2$ and $k = 3$, the third bin contains nodes with $k = 4, 5, 6, 7$ and so on. In general, the pth bin will contain nodes with degree $k \in [2^{p-1}, 2^p - 1]$. The height of the histogram must be the number of nodes contained in each bin divided by the width of the bin they fall in, so that the area of each rectangle of the histogram is equal to the number of nodes contained in each bin. This means the bins

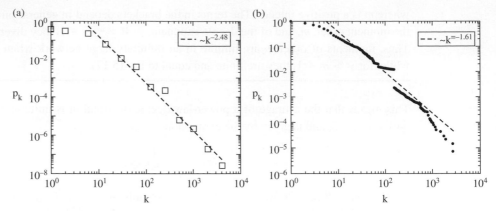

Fig. 5.4   Out-degree distribution of Notre Dame WWW network. We report in (a) the histogram constructed by using a logarithmic binning, and in (b) the cumulative distribution.

in the tail of the distribution get more nodes than they would if bin sizes were fixed, and this reduces the statistical fluctuations in the tail.

In Figure 5.4 (a), we plot the out-degree distribution of the WWW obtained with a logarithmic binning with $a = 2$. The logarithmic binning has also the nice side-effect that the points representing the individual bins appear equally spaced in the log plot. The straight-line power-law form of the histogram is now much clearer, and it extends for about a decade further than in Figure 5.3(c). Even with the logarithmic binning, as shown in the following example, there is still some noise in the tail.

Example 5.5 *(Choice of the optimal binning)*   Suppose the bottom of the lowest bin is at $k_{\min}$ and the ratio of the widths of successive bins is a. Then the nth bin extends from $x_{n-1} = k_{\min}a^{n-1}$ to $x_n = k_{\min}a^n$, and the expected number of samples falling in this interval is:

$$\int_{x_{n-1}}^{x_n} p(k)dk = c \int_{x_{n-1}}^{x_n} k^{-\gamma} dk$$

$$= c \frac{a^{\gamma-1} - 1}{\gamma - 1}(k_{\min}a^n)^{1-\gamma}.$$

Thus, as long as $\gamma > 1$ which, as shown in Table 5.1, is true for most of the degree distribution of real networks, the number of samples per bin goes down as n increases as $a^{(1-\gamma)n}$. Consequently, the bins in the tail will have more statistical noise than those that precede them.

We can further reduce the fluctuations in the tails with a different binning in which the ratio of the width of successive bins is not necessarily constant. Such an *ideal binning* would indeed be obtained by imposing that the expected number of samples in each bin is *exactly* the same. Mathematically, this is equivalent to imposing that:

$$\int_{x_{n-1}}^{x_n} p(k)dk = \frac{1}{N_b}$$

where N_b is the number of bins. We thus obtain $(\frac{x_{n-1}}{k_{\min}})^{1-\gamma} - (\frac{x_n}{k_{\min}})^{1-\gamma} = \frac{1}{N_b}$ which gives:

$$x_n = k_{\min}(1 - \frac{n}{N_b})^{-\frac{1}{\gamma-1}}. \tag{5.20}$$

This choice is the one that minimises the overall fluctuations for the same number of bins and samples. However, the problem with this choice is that the two sides of a bin are in general not integer numbers.

Another, and in many ways a superior, method of plotting the data is to calculate a *cumulative distribution function*. Instead of plotting a simple histogram $p_k = N_k/N$ of the data, we make a plot of the so-called cumulative probability P_k, that is the probability that the degree has a value greater than or equal to k:

$$P_k = \sum_{k'=k}^{\infty} p_{k'} \quad k = 1, 2, \ldots \tag{5.21}$$

By definition, for all cumulative distributions $P_{k_{\min}} = 1$, since all nodes have a degree larger or equal to $k_{\min}$. In practice, to construct P_k from the data, we need to count the total number of nodes $N_{\geq k}$, with degree at least k: $N_{\geq k} = \sum_{k' \geq k} N_{k'}$, $k = 1, 2, \ldots$. This can be readily done after sorting all the nodes in increasing order of their degrees, as shown in the following example.

Example 5.6 Consider the graph of Florentine families introduced in Section 2.1 and shown in Figure 2.1. By sorting the 16 nodes in increasing order of degree, we get the

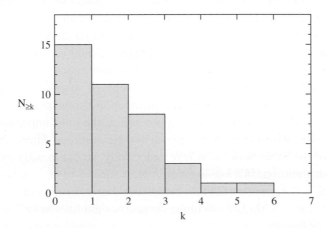

following list: 0, 1, 1, 1, 1, 2, 2, 2, 3, 3, 3, 3, 3, 4, 4, 6, which means that there are four nodes with one link, three nodes of degree 2, five nodes of degree 3, two nodes of degree 4, and one node of degree 6. Therefore, we have $N_0 = 1, N_1 = 4, N_2 = 3, N_3 = 5, N_4 = 2, N_6 = 1$. Neglecting the isolated node for obvious reasons, we can construct the histogram of $N_{\geq k}$ versus k by drawing an horizontal line of height equal to 15 up to a value of k equal to 1. Then the height of the horizontal line is decreased by $N_1 = 4$ units. In a similar way, the height of the horizontal line is decreased at $k = 2$ by $N_2 = 3$ units, and so on. We finally obtain the histogram below.

Notice that, if the distribution p_k is a power law with an exponent γ, as in Eq. (5.10) or in Eq. (5.14), then the cumulative distribution P_k also follows a power law $P_k \sim k^{-\gamma'}$, with an exponent $\gamma' = \gamma - 1$. This is very simple to see in the continuous-k approximation, as shown in the following example.

Example 5.7 *(Cumulative distribution of a continuous variable)* Suppose we have a power-law distribution function $p(k) = ck^{-\gamma}$ of the continuous variable k. The corresponding cumulative distribution is defined as:

$$P(k) = c \int_k^\infty k'^{-\gamma} dk' = \frac{c}{\gamma - 1} k^{-(\gamma-1)}. \tag{5.22}$$

Thus if $p(k)$ is a power law, the cumulative distribution function $P(k)$ also follows a power law, but with a different exponent $\gamma - 1$, which is 1 less than the original exponent. Thus, if we plot $P(k)$ on logarithmic scales we should again get a straight line, but with a shallower slope.

By definition, P_k is well defined for every value of k and so it can be plotted directly with no need of binning. This avoids all questions about the bin size. Moreover, plotting cumulative distributions means a better use of all the information contained in the data. In fact, by binning data, and considering together all degrees within a given range, we throw away all the details contained in the individual values of the samples within each bin. In Figure 5.4(b) we report the cumulative distribution we have obtained for the out-degree of the WWW. Again we notice a well-behaved straight line, but with a shallower slope than before. Plotting the cumulative distribution also helps to visually assess whether a network is scale-free. In the case of a power law, the most important thing to do is to extract the value of the exponent γ. This can be done in several different ways. For instance, in the case of the network of the WWW under study we can directly try to fit the slope of the degree distributions in a log–log plot. A least-squares fit of the straight line in Figure 5.4(a) gives $\gamma = 2.48$, while a fit of the cumulative degree distribution in Figure 5.4(b) gives $\gamma' = 1.61$, i.e. $\gamma = 2.61$. The two fitting curves (straight lines in the log–log scale) are also plotted in the figures.

An alternative and simpler way for extracting the exponent is based on the following formula:

$$\gamma = 1 + N\left[\sum_{i=1}^{N} \ln \frac{k_i}{k_{min} - \frac{1}{2}}\right]^{-1}. \tag{5.23}$$

which gives γ in terms of the degrees of all the nodes, k_i, $i = 1 \ldots N$, and of the value of the minimum degree, k_{min}. An estimate of the expected statistical error σ on the value of γ of Eq. (5.23) is given by:

$$\sigma = \sqrt{N}\left[\sum_{i=1}^{N} \ln \frac{k_i}{k_{min} - \frac{1}{2}}\right]^{-1} = \frac{\gamma - 1}{\sqrt{N}}. \tag{5.24}$$

In many practical situations in which the power law does not hold all the way down to k_{min}, the same formulas can be used by replacing k_{min} with the smallest degree at which we still observe a straight line in the log–log plot of the degree distribution, and by replacing N with the number of nodes with degree larger or equal than such value. A rigorous derivation of Eq. (5.23) and Eq. (5.24) can be found in Refs. [242, 80]. However, a simpler interpretation of Eq. (5.23) can be obtained in the continuous-k approximation, as shown in following example.

Example 5.8   Consider a set of N real values k_i, $i = 1 \ldots N$. The probability that these values were generated by sampling N times from a power-law distribution as that in Eq. (5.15) is equal to:

$$\text{Prob}\,[k_1, k_2, \ldots, k_N] = \prod_{i=1}^{N} p(k_i) = \prod_{i=1}^{N}\left(\frac{\gamma - 1}{k_{min}^{1-\gamma}} k_i^{-\gamma}\right).$$

This quantity is called the *likelihood* of the data set $\{k_1, k_2, \ldots, k_N\}$, since it quantifies how probable its occurrence is. Now, the idea is to consider γ as a tuning parameter, and to find the value of γ that best fits the data, as that which maximises the likelihood. Being a product of many terms, it is convenient to work with the logarithm of the likelihood:

$$\ln \text{Prob}\,[k_1, k_2, \ldots, k_N] = \sum_{i=1}^{N}\left[\ln(\gamma - 1) - \ln k_{min} - \gamma \ln \frac{k_i}{k_{min}}\right]$$

$$= N\ln(\gamma - 1) - N \ln k_{min} - \gamma \sum_{i=1}^{N} \ln \frac{k_i}{k_{min}}.$$

Since the logarithm is a monotonic increasing function, maximising the likelihood with respect to γ is the same as maximising the log likelihood. Hence, by solving the equation $\frac{d \ln Prob}{d\gamma} = 0$ with respect to γ, we find:

$$\gamma = 1 + N\left[\sum_{i=1}^{N} \ln \frac{k_i}{k_{min}}\right]^{-1}.$$

that is the equivalent of Eq. (5.23) in the case of continuous distribution. The derivation of Eq. (5.23) treating k as a discrete variable is in all analogous to the one above, with the complications due to the calculation of the derivative of special functions.

We notice that, in general, the value of the exponent given by Eq. (5.23) can vary a lot depending on the choice of k_{min}. For instance, for the distribution of out-degree in the Notre Dame WWW network we obtain $\gamma_{out} = 2.1$, $\gamma_{out} = 2.26$ and $\gamma_{out} = 2.56$ when we set k_{min} respectively equal to 10, 20 and 30. Statistical hypothesis tests can then be used to determine the best value of k_{min}. In particular, it is possible to show that the exponent of the best power-law fit of that distribution according to a Goodness-of-Fit test based on the Kolmogorov–Smirnov statistics [80] is $\gamma = 2.06 \pm 0.01$, which is obtained by setting $k_{min} = 8$. In this case we have $N \simeq 37000$ nodes whose out-degree is larger than k_{min}, so that the associated statistical error computed by means of Eq. (5.24) is $\sigma \simeq 0.01$.

In Figure 5.5 we report other examples of scale-free degree distributions that have been found over the years in real-world networks. Together with the cumulative in-degree distribution of Notre Dame, shown in panel (a), we report in panel (b) the in-degree distribution of the largest WWW network sample in DATA SET 5, namely that of Google WWW. In this case, we observe a straight line with a constant slope in the usual double-logarithmic scale over a range of k^{in} going from 10 to about 1000. This means that the cumulative distribution follows a power law for part of its range, while, above a certain value of k, it deviates from the power law and falls off quickly towards zero. We observe similar deviations in the movie actor collaboration network and in cond-mat coauthorship network, reported respectively in panels (c) and (d). These deviations from an ideal power-law behaviour are certainly due to the finite order of real networks, and to constraints limiting the addition of new links, such as the limited capacity of nodes of acquiring and maintaining links [9]. For instance, each link in a co-authorship network represents a scientific collaboration, which is the result of the time and energy the two co-authors have dedicated to a common research activity. And since the number of hours per day a scientist can dedicate

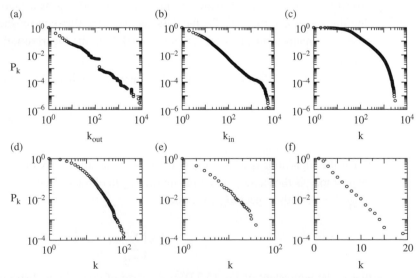

Fig. 5.5 Cumulative degree distributions for a number of different networks, namely: (a) Notre Dame WWW (in-degree), (b) Google WWW (in-degree), (c) movie actor collaborations, (d) cond-mat coauthorships, (e) yeast protein interaction network and (f) the US power grid.

Table 5.1 Power-law exponents characterising the degree distribution of real-world scale-free networks.

Network	N	$\langle k \rangle$	γ	γ_{out}	γ_{in}
Notre Dame WWW	325729	4.51	–	2.06	1.91
Stanford WWW	281903	8.20	–	2.28	2.03
Berkley-Stanford WWW	685230	11.2	–	2.09	2.00
Google WWW	875713	5.83	–	3.63	2.57
AltaVista WWW from Ref. [59]	$2 \cdot 10^8$	7.5	–	2.72	2.1
Internet AS	11174	4.19	2.08	–	–
Internet routers	190914	6.36	2.54	–	–
Movie actor collaboration	225226	73.71	2.24	–	–
Cond-mat coauthorship	16726	5.69	3.57	–	–
Medline coauthorship	1520252	15.5	2.91	–	–
Sexual contacts from Ref.[208]	2810		2.6	–	–
Metabolic interactions from Ref.[166]	778	7.4	2.2	–	–
Protein interactions from Ref.[165]	1870	2.39	2.4	–	–

to research are finite, also the total number of collaborations he can initiate and maintain is limited. As a consequence, it is quite unusual for a researcher to have more than a few dozen co-authors during his entire career, and there exists only a few rare cases of scientists who have collaborated with one hundred or more people during their careers. In panel (e) we consider an example from biology, namely the protein interaction network of the yeast *Saccharomyces cerevisiae*. The cumulative distribution shown here was obtained for a network of $N = 1870$ nodes, representing proteins, connected by $K = 2240$ links representing physical interactions between the proteins and shows a power law with a rapid drop for large values of k [165]. Of course, it is important to say that not all networks in the real world are scale-free. Some networks, such as the power grid of the USA reported in panel (f), have a purely exponential degree distribution [9]. Notice that in this plot, differently from all the other cases, we have used a log–linear scale. And the straight line in the log–linear scale indicates that the degree distribution is exponential.

As stated before, the best way to characterise a power-law distribution is to extract the value of its exponent. In Table 5.1 we report the values of the out- and of the in-degree exponents obtained for all the four WWW network samples in DATA SET 5. We notice that, with the exception of γ_{out} for Google WWW, all the other values of the exponents are in the range $(2, 3]$. Such range of values was confirmed by a larger study of the WWW performed by Andrei Broder and collaborators who studied samples of 200 million pages from two AltaVista crawls [59]. The values of γ obtained are also reported in the table. Scale-free degree distributions are found in other technological networks, but also in social and biological systems. The other examples shown in Table 5.1 refer respectively to the Internet, a network that we will study in detail in Chapter 7, to a list of social networks, including the collaboration networks introduced in the previous chapters and a network of sexual contacts in Sweden studied in Ref. [208], and to two biological networks respectively describing chemical interactions between metabolites [166], and the physical interactions between proteins whose cumulative degree distribution was reported in panel (e) of Figure 5.5 [165]. For each network we report the order of the graph N, the average degree $\langle k \rangle$, and the power-law exponent γ. Again, we notice that all the cases considered, with the only exception of the cond-mat coauthorship network, have exponents γ larger than 2, and smaller than 3. This is a particularly important range of values. In fact, as seen in Example 5.4, the second order moment $\langle k^2 \rangle$, and consequently the variance $\sigma^2 = \langle k^2 \rangle - \langle k \rangle^2$, of an infinitely large network with a power-law degree distribution with $2 < \gamma \leq 3$ is infinite, while the average degree $\langle k \rangle$ of such network is finite and given by Eq. (5.17). In a few words this means that the average degree $\langle k \rangle$ alone is not able to characterise a scale-free network, because the fluctuations around the mean value are unbounded. As we will see in Sections 5.4 and 5.5, this will have dramatic consequences on the network properties.

The Natural Cut-Off

We need to keep in mind that real-world networks are always finite size entities. Hence, in real systems, power-law distributions can only extend up to a maximum value of k. It is therefore important to understand how the moments of the distribution scale as a function of the maximum degree, k_{max}. Consider a finite network with smallest degree k_{min} and largest degree k_{max}. By imposing a finite cut-off on the largest degree we avoid the divergence at infinity in the expression (5.16) for the moments of the distribution. In fact, the mth order moment now reads:

$$\langle k^m \rangle = \int_{k_{min}}^{k_{max}} k^m p(k) dk = \begin{cases} \frac{c}{m+1-\gamma}(k_{max}^{m+1-\gamma} - k_{min}^{m+1-\gamma}) & \text{for } \gamma \neq m+1 \\ \\ c \ln(k_{max}/k_{min}) & \text{for } \gamma = m+1 \end{cases} \tag{5.25}$$

which is a finite quantity depending on the value of k_{max}. Now, if $\gamma > m+1$, the mth moment converges to a constant value as k_{max} increases. Instead, when $\gamma < m+1$, the mth moment grows as $k_{max}^{m+1-\gamma}$, while, when $\gamma = m+1$, it grows as $\ln(k_{max})$. In particular, if we set $m = 1$, we find that the average degree is finite for $\gamma > 2$, while it grows as a function of k_{max} as:

$$\langle k \rangle \sim \begin{cases} k_{max}^{2-\gamma} & \text{for } \gamma < 2 \\ \\ \ln(k_{max}) & \text{for } \gamma = 2. \end{cases} \tag{5.26}$$

Analogously, the mean square is finite for $\gamma > 3$, while it grows as:

$$\langle k^2 \rangle \sim \begin{cases} k_{max}^{3-\gamma} & \text{for } \gamma < 3 \\ \\ \ln(k_{max}) & \text{for } \gamma = 3. \end{cases} \tag{5.27}$$

As we have seen at the beginning of this section, all the real-world networks considered exhibit power laws with a value of the exponent $2 < \gamma \leq 3$. Consequently, we are in the case in which the average degree is a finite constant, while the second moment of the distribution grows with the upper degree k_{max} as in Eq. (5.27).

It is possible to estimate how k_{max} depends on the network order by using the following condition:

$$N \int_{k_{max}}^{\infty} p(k)dk \simeq 1. \qquad (5.28)$$

This relation indicates that the number of vertices of degree greater than k_{max} is of the order of 1. Assuming that a power law of the form (5.15) with exponent $\gamma > 2$ holds for $k > k_{min}$, it is easy to prove that the condition above gives (see Problem 5.2a):

$$k_{max} \simeq k_{min}N^{1/(\gamma-1)} \qquad (5.29)$$

where k_{min} is the lower bound of the power-law region of the degree distribution. Expression (5.29) is known as the *natural cut-off* of the network.

Example 5.9   We can use Eq. (5.29) to get a fast but rough approximation of the γ exponent of a given scale-free network of N nodes in the case we know the value of the maximum degree from empirical data. In fact, by inverting Eq. (5.29) we get an expression for γ as a function of k_{min}, k_{max} and N, namely $\gamma \approx 1 + \ln N / \ln(k_{max}/k_{min})$. In the case of the WWW, the largest observed degree is $k_{max} = 3444$. If we use the previous formula with $k_{min} = 1$ and N equal to the number $N_{\geq 1} = 136,934$ of nodes with degree larger or equal to 1, we get an exponent $\gamma = 2.5$. If instead we consider only the tail of the distribution, by fixing $k_{min} = 8$ (the value we found before which yields the best power-law fit) and using $N = N_{\geq 8} = 37,538$, we get an exponent 2.7.

5.3 The Configuration Model

The ER random graph models introduced in Chapter 3 produce Poisson degree distributions. A Poisson distribution cannot reproduce by any means the fat tails observed in many real networks that we have reported in the previous section. This might sound as a limitation when we want to model real-world networks as random graphs. However, it is not a real problem, since ER models can be easily extended to consider random graphs in which the degree of each node is arbitrarily assigned to take a precise value, or random graphs in which the probability of connecting each pair of nodes is arbitrarily assigned. These two models are respectively the generalisation of ER models A and B, and will allow us to generate graphs with a given arbitrary degree distribution p_k. We first discuss a model, known in the mathematical literature as the *configuration model*, to describe ensembles of random graphs with N nodes, K edges, and a given *degree sequence*. The model was introduced by Edward Bender and Rodney Canfield in 1978 [31] and refined by Belà Bollobás [48], and is an extension of ER model A.

Definition 5.2 (Configuration model) *Let $N > 0$, $K > 0$ and let $\mathcal{K} = \{k_1, k_2, \ldots, k_N\}$ be a degree sequence, i.e. a set of N non-negative integers, such that $\sum_i k_i = 2K$. The configuration model denoted as $G_{N,\mathcal{K}}^{\text{conf}}$ consists in the ensemble of all graphs of N nodes and K edges, in which vertex i has the specified degree k_i, with $i = 1, 2, \ldots, N$, each graph being assigned the same probability.*

In practice, a simple way to generate all the graphs in the ensemble defined by a given degree sequence $\mathcal{K} = \{k_1, k_2, \ldots, k_N\}$ consists in assigning to each node i a number of *half-edges*, also known as *stubs*, equal to its degree k_i. A graph in the ensemble is then formed by matching, at random with uniform probability, pairs of half-edges together, until all K edges of the graph are formed. There are, however, a couple of subtleties that have to be taken into account, as shown by means of the particular simple case of a graph with $N = 5$ nodes and with degree sequence $\mathcal{K} = \{k_1 = 2, k_2 = 2, k_3 = 2, k_4 = 1, k_5 = 1\}$ considered Figure 5.6 and in Example 5.10. It is not difficult to realise that, with such a degree sequence, there are 23 different graphs, more precisely, the seven labelled graphs in the first two groups in the first row, and the 16 labelled multigraphs in the remaining groups of Figure 5.6. Notice, that we have implicitly assumed that the nodes are distinguishable, so it is crucial to work with labelled graphs. Things would be different if we were interested in unlabelled graphs. For instance, the first six graphs in the first row would be lumped into the same unlabelled graph, the last three multigraphs in the second row would be lumped into another unlabelled graph, and so on. In the figure, all labelled graphs corresponding to the same unlabelled graph are placed in the same rectangle. In the following example we focus on a procedure to generate in a systematic way the graphs in Figure 5.6. We will see that each of the 23 graphs can be obtained in one, or more than one, way(s), by matching the stubs of the nodes. The number of matchings generating each graph is reported in the figure above each rectangle, since it turns out to be equal for every graphs in the rectangle.

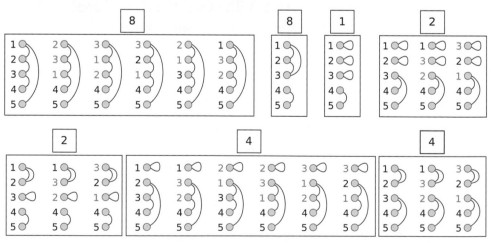

Fig. 5.6 The 23 different graphs with $N = 5$ nodes and $K = 4$ links in the configuration model ensemble $G_{N,\mathcal{K}}^{\text{conf}}$ with degree sequence $\mathcal{K} = \{k_1 = 2, k_2 = 2, k_3 = 2, k_4 = 1, k_5 = 1\}$.

Example 5.10   In order to generate ensemble $G_{N,\mathcal{K}}^{\mathrm{conf}}$ shown in Figure 5.6, with $N = 5$ and degree sequence $\mathcal{K} = \{k_1 = 2, k_2 = 2, k_3 = 2, k_4 = 1, k_5 = 1\}$, we can imagine associating two stubs with nodes 1, 2 and 3, and one stub with node 4 and node 5, and then consider all the possible matchings of the stubs. Of course, each of the 23 labelled graphs can be obtained, in principle, with more than one matching of the stubs. Let us consider, for

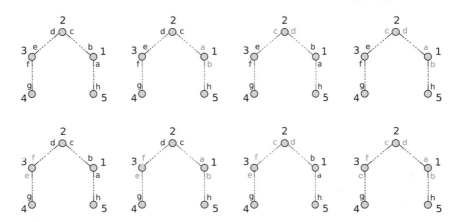

instance, the first graph in Figure 5.6. This can be obtained by eight different matchings, all shown in the figure below, where we denoted by a and b the stubs of node 1, by c and d the stubs of node 2, by e and f the stubs of node 3, and by g and h respectively the stubs of node 4 and 5. Observe that all the possible matchings can be produced from the first one by switching the order of the stubs coming out from the same node (in each graph of the figure we have indicated in grey the pairs of stubs which have been switched). Since we have three nodes with two stubs, and for each pair of stubs there are two possible orders, the total number of matchings producing the same labelled graph is $2 \times 2 \times 2 = 8$. Analogously, we can prove that each one of the other six graphs in the first and second rectangle can be generated by eight different matchings. The number of matchings generating each graph is reported in the small squares in Figure 5.6. Finally, we have a total of 105 different matchings.

As illustrated in the previous example, the procedure to match stubs at random generates all graphs with a specified degree sequence. We can easily understand that the number of different matchings, M^{conf}, only depends on $2K$, the total number of stubs we have to match. Namely, we have:

$$M^{\mathrm{conf}} = (2K - 1)!! \tag{5.30}$$

different matchings, where the double factorial is defined as $n!! = n \cdot (n - 2) \cdot (n - 4) \cdot \ldots \cdot 3 \cdot 1$ if n is odd, or as $n!! = n \cdot (n - 2) \cdot (n - 4) \cdot \ldots \cdot 4 \cdot 2$ if n is even. This is because, if we have $2K$ stubs, the first one can be matched to one of the other $2K - 1$ stubs. After the first matching is done, the second stub can be connected to one of the $2K - 3$

remaining stubs and so on. In the case considered in the previous example $(2K-1)!! = 7!! = 105$.

However, the fact that all matchings have the same probability of being produced does not mean that all the labelled graphs will appear with the same probability. For instance, each of the seven graphs in the first two rectangles of Figure 5.6 will be sampled with a probability of $8/105$. Conversely the graph in third rectangle will be produced with a probability of $1/105$, all the graphs in the fourth rectangle with a probability of $2/105$, and so on. Notice that the seven graphs in the first two rectangles do not contain loops or multiple edges, while the last 16 do, and the difference between graphs and multigraphs is crucial for the choice of a good method to generate the graphs of the model. It is indeed possible to prove that each graph with degree sequence $\mathcal{K} = \{k_1, k_2, \ldots, k_N\}$ and with no loops and no multiple edges can be generated by $\Pi_{i=1}^{N}(k_i!) = k_1! k_2! \ldots k_N!$ matchings, since $k_i!$ are the permutations of the k_i half-edges of node i. This number is a constant for a given degree sequence, so it is the same for all graphs with no loops and no multiple edges. In the case in Figure 5.6, each of the first seven graphs (those in the first two rectangles) can indeed be generated in $\Pi_{i=1}^{N}(k_i!) = 2 \cdot 2 \cdot 2 = 8$ ways. Instead, the number of matchings generating a given multigraph is smaller than $\Pi_{i=1}^{N}(k_i!)$, and will depend on the particular multigraph under study. In fact, the switch of the two stubs in a loop does not correspond to a different matching, and the simultaneous switching of two pairs of stubs in a multiple edge will correspond to the same matching. The formula for the exact number of matchings generating a given multigraph can be found in Ref [239]. Here we do not need such a number. In fact, from the discussion above, it follows that a possible way to uniformly sample graphs with an assigned degree sequence is to use an algorithm that implements the matching procedure described above, and rejects all matchings producing multigraphs (see Appendix A.12). Since all graphs are produced by the same number of possible matchings, a uniform distribution in the matchings will result in a uniform distribution of graphs.

The alternative possibility is that of sampling uniformly among all possible stub matchings, not rejecting multigraphs. For practical reasons, this is the approach that is implicitly adopted in the analytical treatments of the configuration model, such as those considered in the following section. Strictly speaking, this will not produce all possible graphs and multigraphs exactly with a uniform probability. However, as we shall see later in this section, self-edges and multiple edges can be neglected when N is large.

Let us calculate how many links we have on average between a given pair of nodes in a model with loops and multiple edges. Consider two different nodes i and j, having, respectively, degrees k_i and k_j. The expected number of links between i and j, $\bar{e}_{ij}^{\text{conf}}$, can be computed as the ratio between the number of stub matchings in which i and j are connected, and the total number of possible stub matchings, M^{conf}. The first number is given by $k_i k_j (2K-3)!!$. This is because there are $k_i k_j$ different ways of connecting the k_i stubs of i with the k_j stubs j, while the remaining $2K-2$ stubs give rise to a total of $(2K-2-1)!!$ possible matchings, see Eq. (5.30). We finally have

$$\bar{e}_{ij}^{\text{conf}} = \frac{k_i k_j (2K-3)!!}{(2K-1)!!} = \frac{k_i k_j}{(2K-1)}. \tag{5.31}$$

The situation is different in the case of $i = j$. In this case the number of possible ways in which the stubs of node i can be connected to form a loop is equal to the number of distinct pairs of stubs $k_i(k_i - 1)/2$. Repeating the argument above, we have

$$\bar{e}_{ii}^{\text{conf}} = \frac{k_i(k_i - 1)}{2(2K - 1)}. \tag{5.32}$$

Notice that for large graphs $\bar{e}_{ij}^{\text{conf}}$ is usually a very small number, and in practice can be interpreted as the probability that a link between i and j occurs. The same thing can be said for $\bar{e}_{ii}^{\text{conf}}$.

Example 5.11 Let us consider random graphs with the same degree sequence as in Figure 5.6. The expected number of links between two nodes, for instance nodes 1 and 2, in $G_{N,\mathcal{K}}^{\text{conf}}$ can be calculated by means of Eq. (5.32). We therefore obtain:

$$\bar{e}_{12}^{\text{conf}} = \frac{k_1 k_2}{2K - 1} = \frac{4}{7}.$$

Since we have a small number of nodes, we can check the correctness of our result by the direct enumeration of the number of configurations. We have in fact seen that there are 105 different stub matchings. From Figure 5.6 we can count that there are 48 matchings in which nodes 1 and 2 have one link, and six matchings in which the two nodes have a double link. The total number of links connecting nodes 1 and 2 in all configurations is therefore $48 + 2 \times 6 = 60$, therefore the expected number of edges connecting nodes 1 and 2 is $60/105 = 4/7$, in agreement with the analytical prediction.

As shown in the following example, the expected number of links in the configuration model satisfies the following normalisation condition:

$$\sum_i \sum_{j \leq i} \bar{e}_{ij}^{\text{conf}} = K \tag{5.33}$$

where K is the total number of links in the network.

Example 5.12 To show that the expected number of links between two nodes i and j is normalised as in Eq. (5.33) it is necessary to consider loops separately from the other links. We have:

$$\sum_i \sum_{j \leq i} \bar{e}_{ij}^{\text{conf}} = \sum_i \sum_{j < i} \bar{e}_{ij}^{\text{conf}} + \sum_i \bar{e}_{ii}^{\text{conf}} = \sum_i \sum_{j < i} \frac{k_i k_j}{2K - 1} + \sum_i \frac{k_i^2 - k_i}{4K - 2}$$

$$= \frac{1}{4K - 2} \sum_{i \neq j} k_i k_j + \frac{1}{4K - 2} \sum_i (k_i^2 - k_i)$$

$$= \frac{1}{4K - 2} \sum_{ij} k_i k_j - \frac{1}{4K - 2} \sum_i k_i = \frac{(2K)^2}{4K - 2} - \frac{2K}{4K - 2} = K$$

We are now ready to evaluate the average number of loops and multiple edges in the configuration model. The average number of loops can be immediately evaluated from the quantity $\bar{e}_{ii}^{\text{conf}}$. In fact the expected number of links $\bar{e}_{ii}^{\text{conf}}$ gives the average number of loops belonging to node i. To find the expected number of loops $\bar{n}^{\text{loops}}$ in a graph of the configuration model we have to sum this quantity over all nodes. Hence we have:

$$\bar{n}^{\text{loops}} = \sum_i \bar{e}_{ii} = \frac{1}{2(2K-1)} \sum_i k_i(k_i - 1) \approx \frac{1}{2} \frac{\langle k^2 \rangle - \langle k \rangle}{\langle k \rangle} \tag{5.34}$$

where we have approximated $2(2K-1)$ with $4K$ and we have divided and multiplied by N.

The expected number of multiple edges can be obtained by a combinatorial argument similar to the one used to derive $\bar{e}_{ii}^{\text{conf}}$. In fact, the number of stub matchings leading to a double edge between nodes i and j is equal to

$$\frac{k_i k_j (k_i - 1)(k_j - 1)}{2} \cdot (2K - 5)!\,!$$

The first factor represents the number of different ways of forming two edges between the two nodes: once we have one of the $k_i k_j$ possible links joining nodes i and j, the number of available stubs at each node is reduced by one, and therefore we have to multiply $k_i k_j$ by $(k_i - 1)(k_j - 1)/2$, where the factor 2 is needed to avoid double counting of the double link. The factor $(2K - 5)!!$ represents the number of different matchings of the remaining $2K - 4$ stubs. Dividing by the total number of stub matchings $(2K - 1)!!$, we have

$$\bar{n}_{ij}^{\text{multiple}} = \frac{k_i(k_i - 1)k_j(k_j - 1)}{2(2K-1)(2K-3)}. \tag{5.35}$$

Finally, the expected number of multiple edges in the graph is

$$\bar{n}^{\text{multiple}} = \frac{1}{2} \sum_{ij} \bar{n}_{ij}^{\text{multiple}} = \frac{1}{2} \sum_{ij} \frac{k_i(k_i - 1)k_j(k_j - 1)}{2(2K-1)(2K-3)}$$

$$\approx \frac{1}{4(2K)^2} \sum_i k_i(k_i - 1) \sum_j k_j(k_j - 1)$$

$$= \frac{1}{4} \left(\frac{\langle k^2 \rangle - \langle k \rangle}{\langle k \rangle} \right)^2. \tag{5.36}$$

Example 5.13 Let us consider again random graphs with the same degree sequence as in Figure 5.6. The number of multiple links between two nodes, for instance nodes 1 and 2, can be calculated using Eq. (5.35). We get:

$$\bar{n}_{12}^{\text{multiple}} = \frac{4}{2 \cdot 7 \cdot 5} = \frac{2}{35}.$$

We can check this result by direct enumeration. From Figure 5.6 we see that there are six matchings in which nodes 1 and 2 have multiple (double) links, therefore the expected number of multiedges between nodes 1 and 2 is $6/105 = 2/35$, in agreement with the analytical prediction.

From expressions (5.36) we can observe that the expected number of multiple edges depends on the ratio $\langle k^2 \rangle / \langle k \rangle$. Thus, the number of multiple edges is finite if $\langle k^2 \rangle / \langle k \rangle$ is finite as $N \to \infty$. In this case, the density of multiple edges vanishes as $1/N$ and therefore multiple edges can be neglected in large networks. Exactly the same can be said about loops. The situation is different when $\langle k^2 \rangle / \langle k \rangle$ diverges with N, as for scale-free networks with an exponent $\gamma \in (2,3)$. Here, we have to check whether $(\langle k^2 \rangle / \langle k \rangle)^2$ increases faster or slower than N (see Example 5.20).

The configuration model is a generalisation of the $G_{N,K}^{\mathrm{ER}}$ random graph model in which, not only the total number of links, but also the degree of each node is fixed. Analogously, we can construct generalisations of ER random graph model B, $G_{N,p}^{\mathrm{ER}}$. One possibility is to consider graphs in which we assign the value p_{ij} to the probability of connecting each pair of nodes i and j. Alternatively we can fix $E = \{e_{ij}\}$, a set of real numbers representing the expected number of links between different pairs of nodes. In particular, since we want the graph to be random, we have to factorise this quantity into a term depending on i and a term depending on j. We name the following model the configuration model B.

Definition 5.3 (Configuration model B)   *Let $N > 0$ and let $\mathcal{K}' = \{d_1, d_2, \ldots, d_N\}$ be a set of N non-negative real numbers. Let us define the quantity M as $\sum_i d_i = 2M$. The model, denoted as $G_{N,\mathcal{K}'}^{\mathrm{conf}}$, consists in the ensemble of all graphs of N nodes in which the expected number of links between different pairs of nodes is given by:*

$$
e_{ij} = \begin{cases} \dfrac{d_i d_j}{2M} & \text{if } j \neq i \\[2ex] \dfrac{d_i^2}{4M} & \text{if } j = i. \end{cases} \tag{5.37}
$$

It is very simple to see that in this version of the configuration model the degree of each node can fluctuate around an average value. In fact, the expectation value of the degree of node i is given by:

$$
\bar{k}_i = \sum_{j \neq i} e_{ij} + 2e_{ii} = \sum_j \frac{d_i d_j}{2M} = d_i.
$$

Also, the total number of links is not fixed, but can fluctuate from graph to graph around an average value equal to M. We have in fact:

$$
\overline{K} = \frac{1}{2} \sum_i \bar{k}_i = \frac{1}{2} \sum_i d_i = M.
$$

Furthermore, it is possible to prove that the probability that node i has a particular degree k in a graph of the ensemble, follows a Poisson distribution around d_i [239]. Consequently, the probability $P(K)$ that a graph sampled from $G_{N,\mathcal{K}'}^{\mathrm{conf}}$ has exactly K links is a Poisson distribution with average equal to M. This result generalises what we have seen for ER random graph model B. Notice, however, that there is a slight difference between the interpretation of e_{ij} in $G_{N,\mathcal{K}'}^{\mathrm{conf}}$ and that of p in ER model B. The quantity e_{ij} is an expected number of links, and it may give rise to more than one link between two nodes, while p is a probability. The reason for this difference is that, in the ER random graph model, we consider all the $\binom{N}{2}$

pairs of nodes and assign a uniform probability p that there is a link between them: by construction, there can be only one link between two nodes, and the quantity p denotes at the same time the expectation number of edges between the two nodes, and the probability of finding an edge between them.

Example 5.14 A possible variant of ER model B, which is indeed a particular case of $G_{N,\mathcal{K}'}^{\text{conf}}$ can be constructed as follows. Consider N^2 pairs of nodes (i, j), obtained by sampling i and j uniformly and independently from $1, \ldots, N$. Each time, nodes i and j are connected by a link with probability $p/2$. In this way, the quantity p will have the meaning of the expected number of links between the pair (i, j), not the probability of having a link. More precisely, the expected number of links between nodes i and j will be p if $i \neq j$, and $p/2$ if $i = j$. This model is the same as $G_{N,\mathcal{K}'}^{\text{conf}}$, with $e_{ij} = p$ if $i \neq j$ and $e_{ij} = p/2$ if $i = j$, which means $\mathcal{K}' = \{d, d, \ldots, d\}$ with $d = \sqrt{2Mp}$.

5.4 Random Graphs with Arbitrary Degree Distribution

Let us now turn from abstract models to our practical problem. In Sections 5.1 and 5.2 we have extracted empirically the degree distribution $p_k = N_k/N$ of various real networks, such as the WWW, by counting the number N_k of nodes with k links. Our need is thus to construct an ensemble of graphs with an assigned p_k, rather than with an assigned degree sequence $\mathcal{K}$. Since in the case we are interested in the quantities $N_k = p_k N$ are integer numbers for any $k = 0, 1, 2, \ldots$, one way to sample graphs from an assigned degree distribution p_k is first to extract a random degree sequence $\mathcal{K}$ from the set $\{N_k, k = 0, 1, \ldots\}$, and then sample the corresponding configuration model A. Obviously, there are many degree sequences corresponding to the same set $\{N_k\}$. We can therefore adopt the following procedure. First, we can construct the so-called canonical degree sequence, by assigning degree $k = 0$ to the N_0 nodes with label from 1 to N_0, degree $k = 1$ to the N_1 nodes with labels from $N_0 + 1$ to $N_0 + N_1$, and so on. All the other degree sequences can then be constructed from the canonical one by a random permutation of the labels of the nodes. More precisely, we can prove that for a given $\{N_k\}$, there is a total of

$$\frac{N!}{\prod_i N_{k_i}!}$$

distinct degree sequences. The numerator in the formula is the number of permutations of the labels of the N nodes, while the denominator represents the number of permutations giving rise to the same degree sequence. Because each degree sequence appears exactly $\prod_i N_{k_i}!$ times in the $N!$ permutations of the nodes, sampling uniformly among all the $N!$ permutations is equivalent to sampling uniformly among all the possible degree sequences. Notice that most of the average quantities of all configuration models corresponding to the same N_k are the same, and therefore it is sufficient to sample one specific configuration model from a given N_k, and compute such a quantity for that ensemble.

When we want to sample graphs with a given functional form for the p_k, as for instance the power law p_k in Eq. (5.15), then $\overline{N}_k = Np_k$ is in general not an integer number, and therefore it has to be interpreted as an average number of nodes with degree k. In practice, what we can do in this case is to sample, for each node i, $i = 1, \ldots, N$, a degree k_i with probability p_{k_i}. Once a degree sequence $\mathcal{K} = \{k_1, \ldots, k_N\}$ has been constructed in this way, we can sample the corresponding configuration model A and calculate the desired average properties. This procedure can be repeated to improve the statistics of the ensemble averages.

The Molloy and Reed Criterion

As seen in Chapter 3, random graphs are a good playground for analytical approaches. In the Nineties Michael Molloy and Bruce Reed proved a series of theorems that generalise the findings of Erdős and Rényi to the case of random graphs with an arbitrary degree distribution [222, 223]. One of the main results found is a condition on the probability distribution p_k for the existence of a giant component, which is known as the *Molloy and Reed criterion* and can be stated as follows:

Theorem 5.1 (Molloy and Reed)   *Given a sequence of non-negative real numbers $p_0, p_1, \ldots,$ which sum to 1, consider the ensemble of random graphs with N nodes and $\{p_k\}$ as degree distribution. If*

$$Q(\{p_k\}) \equiv \sum_{k \geq 1} k(k-2)p_k = \langle k^2 \rangle - 2\langle k \rangle > 0 \tag{5.38}$$

then a.e.g. in the ensemble contains a giant component, while if $Q(\{p_k\}) < 0$ then all components are small.

A more precise statement of the theorem originally given by Molloy and Reed deals with sequences $N_k(N)$, $k = 0, 1, \ldots$, and requires also conditions on the maximum degree $k_{\max}$ in the network. A more general proof, without such restrictions, can be found in the book by Rick Durrett [99].

A simple way to understand the result of the theorem is to estimate the average number, z_m, of nodes at distance m from a given node [231]. We will thus follow a similar argument to the one adopted in Section 3.6 for ER random graphs. The argument relies on the fact that random graphs have to have a good approximation of a *tree-like* structure, independently of their degree distribution. For the moment we will assume this to be valid for any degree distribution. There are however a number of caveats, and we will come back to this point when we discuss below the clustering coefficient of random graphs with assigned p_k. Let us consider a generic vertex i of the graph. On average, starting from i we can reach $z_1 = \langle k \rangle$ nodes in one step. Following the same argument as in Section 3.6, the average number z_2 of nodes at distance two from i is equal to $\langle k \rangle$ times ρ, where $\rho = \langle k_{\mathrm{nn}} \rangle - 1$. Thus ρ is equal to one less than the average number of nodes $\langle k_{\mathrm{nn}} \rangle$ we can reach from a first neighbour of *i.* To find an expression for ρ in random graphs with assigned degree distribution p_k we need to recall Eq. (3.40) from Section 3.6. In fact, in a random graph the average degree $\langle k_{nn} \rangle$ of the first neighbours of a node is equal to the average degree of nodes to which an edge leads. Using Eq. (3.38) and Eq. (3.39) we get $\langle k_{nn} \rangle = \langle k^2 \rangle / \langle k \rangle$. However, this time we cannot use the expression $\langle k^2 \rangle = \langle k \rangle^2 + \langle k \rangle$ which is valid for ER random graphs only and gives $\langle k_{nn} \rangle = \langle k \rangle + 1$. The following example shows that, in scale-free random graphs with an exponent $2 < \gamma \leq 3$, differently from ER random graphs, the value of the average degree of the first neighbours of a randomly chosen node $\langle k_{nn} \rangle$ can be much larger than $\langle k \rangle$.

Example 5.15 *(Your friends are more social than you!)* Suppose that a network of $N = 10000$ individuals and the friendship relations among them is well described by a scale-free random graph with $k_{min} = 1$, $k_{max} = 1200$, and a degree exponent $\gamma = 2.2$. We want to calculate and compare the average degree of a randomly chosen node to the average degree $\langle k_{nn} \rangle$ of the first neighbours of a randomly chosen node. From Eq. (5.25) we have:

$$\langle k \rangle = \frac{c}{2 - \gamma} \left(k_{max}^{2-\gamma} - k_{min}^{2-\gamma} \right) = -\frac{c}{0.2} \left(1200^{-0.2} - 1 \right) \simeq 3.79c$$

and

$$\langle k^2 \rangle = \frac{c}{3 - \gamma} \left(k_{max}^{3-\gamma} - k_{min}^{3-\gamma} \right) = \frac{c}{0.8} \left(1200^{0.8} - 1 \right) \simeq 363.29c$$

where $c \simeq (\gamma - 1) k_{min}^{\gamma-1} = 1.2$. Therefore, the average degree of the network is $\langle k \rangle \simeq 4.5$, while the average degree of the first neighbours of a randomly chosen node is $\langle k_{nn} \rangle = \langle k^2 \rangle / \langle k \rangle \simeq 96.9$, which is a much larger number than 4.5. This proves, that in a scale-free network, your friends have more friends than you do! Again, the reason for this apparent paradox is very simple. You are more likely to be friends with someone who has more friends than with someone who has fewer friends. If you are still not convinced of this, do the following experiment. List all of your friends. Then, ask each one of them how many friends they have got. Independently of who you are, your age, sex or the number of friends you have, you will find that your friends on average have more friends than you do [115]. The result of living in a scale-free network can be more depressing than this. In fact, the same argument applies to your partners, and predicts that your lover has had more lovers than you have.

We finally have:

$$\rho = [(\textit{average degree of first neighbours of } i) - 1] = \langle k_{nn} \rangle - 1$$
$$= \frac{\langle k^2 \rangle}{\langle k \rangle} - 1 = \frac{\langle k^2 \rangle - \langle k \rangle}{\langle k \rangle}. \tag{5.39}$$

Multiplying this by the mean degree of i, $z_1 = \langle k \rangle$, we find that the average number of nodes at distance two from a node i is:

$$z_2 = \rho \cdot \langle k \rangle = \frac{\langle k^2 \rangle - \langle k \rangle}{\langle k \rangle} \langle k \rangle = \langle k^2 \rangle - \langle k \rangle. \tag{5.40}$$

We can extend this calculation to further neighbours also, and calculate z_m for $m = 3, 4, \ldots$. The average number of edges leading from each second neighbour, other than the one we arrived along, is also given by Eq. (5.39), and indeed this is true at any distance m away from vertex i. Thus, the average number of neighbours at distance m is:

$$z_m = \rho z_{m-1}$$

This relation can be iterated, and we finally get the following result.

In random graphs with an arbitrary degree distribution p_k, the average number of nodes reached by a node i in m steps is:

$$z_m = \rho^{m-1} z_1 \tag{5.41}$$

where $\rho = z_2/z_1 = \langle k^2 \rangle / \langle k \rangle - 1$.

Finally, from this equation together with $z_1 = \langle k \rangle$, we can derive the Molloy and Reed criterion. In fact, depending on whether ρ is greater than 1 or not, the expression 5.41 will either diverge or converge exponentially as m becomes large. This implies that the average total number of neighbours of vertex i at all distances is finite for $\rho < 1$, or infinite, in the limit of infinite N, for $\rho > 1$. If the number of nodes that can be reached from a generic node of the graph is finite, then clearly there can be no giant component in the graph. Conversely, if it is infinite, then there must be a giant component. Thus the graph shows a phase transition when $\rho = 1$. The condition $\rho > 1$ is perfectly equivalent to the Molloy and Reed condition of Eq. (5.38), namely $\langle k^2 \rangle / \langle k \rangle > 2$. The following example shows that, for Erdős–Rényi random graphs, the Molloy and Reed criterion reduces to the condition $\langle k \rangle > \langle k \rangle_c = 1$ we derived in Chapter 3.

Example 5.16 Consider Erdős–Rényi random graphs. For a Poisson degree distribution we have $\langle k \rangle = z$ and $\langle k^2 \rangle = z^2 + z$ (see Eqs. 3.19). Consequently, we get $z_1 = \langle k \rangle$, $z_2 = \langle k \rangle^2$, $z_3 = \langle k \rangle^3$, and so on. In general we have $z_m = \langle k \rangle^m$, i.e. the mean number of m-neighbours of a generic vertex in an ER random graph is just the mth power of the average degree. In this case the Molloy and Reed criterion given in Eq. (5.38) reduces to $z^2 - z = \langle k \rangle^2 - \langle k \rangle > 0$. This is equivalent to say that we can observe a giant component in ER random graphs only when $\langle k \rangle > \langle k \rangle_c$, that is when the average degree is larger than a critical value $\langle k \rangle_c = 1$. This is in agreement with the result derived in Section 3.4.

Characteristic Path Length

An expression for the characteristic path length of random graphs with a generic p_k can be obtained by an argument similar to that used in Section 3.6. Let us consider a connected graph, or assume that the Molloy and Reed criterion is satisfied, and focus on the giant component. The total number of nodes *within distance m* from a node i is equal to: $z_1 + z_2 + \ldots + z_m = z_1 + \rho z_1 + \rho^2 z_1 + \ldots + \rho^{m-1} z_1$. Such a number is dominated by the number of nodes exactly *at distance m*, especially if we are well above the transition, so that $\rho = z_2/z_1 \gg 1$. However, in a finite network, as m increases, the number of nodes at distance m will at a certain point stop to increase, and it will rapidly decrease when m approaches the diameter of the graph. We thus expect that most of the nodes will be at a distance around the average distance L. As in the case of ER random graphs, we can therefore estimate L as the value of m such that the total number of vertices at distance m from a given vertex i in the giant component is of the order of N, namely from:

$$N \approx z_L = \rho^{L-1} z_1. \tag{5.42}$$

The only difference with respect to the case of ER random graphs is that now the quantity ρ is equal to $\langle k^2 \rangle / \langle k \rangle - 1$, and does not simply reduce to $\langle k \rangle$. By taking the logarithm of the left- and right-hand sides of Equation (5.42), we obtain the following result.

Well above the transition, the characteristic path length of random graphs with an arbitrary degree distribution p_k reads:

$$L^{\text{conf}} \approx \frac{\ln(N/\langle k \rangle)}{\ln\left(\frac{\langle k^2 \rangle - \langle k \rangle}{\langle k \rangle}\right)} + 1. \tag{5.43}$$

A more rigorous derivation of the characteristic path length in graphs with a given p_k can be found in Ref. [76].

Example 5.17   In the particular case of ER random graphs, Eq. (5.43) reduces to the expression derived in Section 3.6. In fact, by substituting $\langle k \rangle = z$ and $\langle k^2 \rangle = z^2 + z = \langle k \rangle^2 + \langle k \rangle$, we get:

$$L^{\text{ER}} \approx \frac{\ln(N/\langle k \rangle)}{\ln \langle k \rangle} + 1 = \frac{\ln N}{\ln \langle k \rangle}$$

which is equal to Eq. (3.33).

Clustering Coefficient

The various definitions of graph clustering coefficient and transitivity defined in Section 4.3 are different measures of the probability that two neighbours of a node are also neighbours of each other. Here, to evaluate C for the ensemble of graphs with a given p_k, we calculate the probability of having a link between two neighbours of a given node, averaged over the graphs of the ensemble. This can be done by the usual combinatorial argument based on counting matchings of node stubs to produce different graphs with the same degree sequence.

Consider a node i that is connected to two nodes j and l. Suppose j and l have respectively degrees equal to k_j and k_l. We are thus assuming that four of the stubs of the graph have already been matched, and node i is the centre of a triad. Now, the probability that there is also a link between j and l, which closes the triad into a triangle, can be written as the number of matchings that give rise to a triangle divided by the total number of matchings:

$$\frac{(k_j - 1)(k_l - 1)(2K - 7)!!}{(2K - 5)!!} = \frac{(k_j - 1)(k_l - 1)}{2K - 5}.$$

The first two factors in the numerator of the first expression account for the number of different ways of forming an edge between j and l. Since both nodes j and l are connected to i, the number of available links at each node is reduced by one. Once the link between j and l is formed, then we have three links already fixed (six stubs have already been matched), and the factor $(2K - 7)!!$ in formula represents indeed the number of different matchings of the remaining $2K - 6$ stubs. Finally, we have to divide the number of matchings giving rise to the triangle by the total number of matchings $(2K - 5)!!$ of $2K - 4$ stubs (two links, that from i to j and that from i to l are already fixed). The clustering coefficient C^{conf} of the ensemble of graphs with a given p_k can be obtained by averaging over all the possible degrees of the nodes j and l. We get:

$$C^{\mathrm{conf}} = \sum_{k_j} \sum_{k_l} q_{k_j} q_{k_l} \frac{(k_j - 1) \cdot (k_l - 1)}{N \langle k \rangle} = \frac{1}{N \langle k \rangle} \left[\sum_{k} (k - 1) q_k \right]^2$$

where the factor $q_{k_j} \cdot q_{k_l}$ is indeed the probability that j and l have degree respectively equal to k_j and k_l, and where we have approximated $2K - 5$ as $2K = \langle k \rangle N$. Finally, recalling that $q_k = k p_k / \langle k \rangle$, we get the following result [231].

The clustering coefficient of random graphs with an arbitrary degree distribution p_k reads:

$$C^{\mathrm{conf}} = \frac{\langle k \rangle}{N} \left[\frac{\langle k^2 \rangle - \langle k \rangle}{\langle k \rangle^2} \right]^2. \tag{5.44}$$

That is, the clustering coefficient of a random graph with an arbitrary degree distribution is equal to the value of the clustering coefficient of an ER random graphs, $\langle k \rangle / N$, times an extra factor that can be quite large, since its leading term goes as the square of $\langle k^2 \rangle / \langle k \rangle^2$. Consequently, if $\langle k^2 \rangle$ is finite, then C still vanishes for $N \rightarrow \infty$. In this case, we can neglect triangles (also the same thing can be proven for longer cycles), and the graph has to a good approximation a *tree-like structure*. This is precisely the property we have used to derive the number of neighbours at distance m from a given node in Eq. (5.41), and the Molloy and Reed criterion. However, the value of the clustering coefficient in Eq. (5.44) may be not negligible for highly skewed degree distributions and finite graph sizes as those observed empirically. We will come back to this point in the next section, when we will discuss graphs with power-law degree distributions. The example below shows that random graphs with a given degree distribution can indeed reproduce the finite values of clustering observed in real systems.

Example 5.18   *(The clustering coefficient of the World Wide Web)*   We can use Eq. (4.3) to compute the clustering coefficient of some WWW samples introduced in Section 5.1. If we ignore the directed nature of the links, we get $C = 0.23$ and $C = 0.59$, respectively for Notre Dame and for Stanford. As shown in the table below, ER random graphs with same number of nodes N and average degree $\langle k \rangle$ would have a much smaller clustering coefficient. Certainly, if we want to use random graphs as *network null models* of the WWW, we should utilise random graphs with degree distributions similar to those empirically observed and reported in Figure 5.5. If we extract the first two moments

	N	$\langle k \rangle$	$\langle k^2 \rangle$	C^{ER}	C^{conf}	C
Notre Dame WWW	325729	9.02	2406	0.000028	0.024	0.23
Stanford WWW	281903	16.40	28.24	0.000058	0.64	0.59
Internet AS	11174	4.19	1113	0.00037	1.5	0.30
Movie actor collaboration	225226	73.71	23940	0.00033	0.006	0.79
Cond-mat coauthorship	16726	5.69	73.57	0.00034	0.0015	0.62

of the degree distribution (notice that we get different values of $\langle k \rangle$ with respect to those reported in Table 5.1, because we are ignoring here the directionality of the links), and we plug such numbers into Eq. (5.44), we get better estimates of the correct values of C than those obtained by C^{ER}. In the case of Stanford WWW the value of C^{conf} we get is of the same order as C. However, this is a particularly fortunate case, as can be seen from the results of similar calculations for other man-made and social networks reported in the table.

5.5 Scale-Free Random Graphs

Graphs with a power-law degree distribution can be obtained as a special case of random graphs with an arbitrary degree distribution. As explained in Section 5.4, what we need to do is to sample the degree of each node i, $i = 1, \ldots, N$ from the power-law degree distribution in Eq. (5.10). A simple method to sample from a discrete power law $p_k = k^{-\gamma}/\zeta(\gamma)$ with a tunable exponent γ can be found in Appendix A.9.5. Once we have constructed a degree sequence $\mathcal{K} = \{k_1, k_2, \ldots, k_N\}$, we can generate random graphs with N nodes and with a power-law degree distribution using one of the algorithms given in Appendix A.12, as discussed in Section 5.3.

In this section we will explore how the basic properties of scale-free graphs, such as the values of their characteristic path length and clustering coefficient, or the existence of a giant connected component, change as a function of the exponent γ.

Giant Component

The first thing we will focus on is the condition to have a giant component in a scale-free random graph. Namely, we will use the Molloy and Reed criterion in Eq. (5.38) to determine the critical value of γ for the appearance of a giant component. We know from Section 5.2 that, when $2 < \gamma < 3$, the second-order moment of the degree distribution is infinite in an infinitely large system. Hence, the Molloy and Reed criterion is always satisfied in this case. Instead, when $\gamma > 3$, both $\langle k \rangle$ and $\langle k^2 \rangle$ are finite quantities and, by using expressions (5.13), the condition in Eq. (5.38) becomes:

$$\zeta(\gamma - 2) - 2\zeta(\gamma - 1) > 0. \tag{5.45}$$

The solution to this inequality can be obtained numerically, and gives a critical value of the exponent:

$$\gamma_c = 3.47875\ldots \tag{5.46}$$

such that, for $\gamma < \gamma_c$ random graphs with a scale-free distribution $p_k \sim k^{-\gamma}$ have a giant component, while for $\gamma > \gamma_c$ they have no giant component. The following example shows that the study of the threshold for the appearance of a giant component is indeed a case where we get a wrong result if we treat the degree k as a real number.

Example 5.19 *(Continuous-k approximation)* Let us estimate the threshold for the appearance of a giant component by approximating the node degree k as a real variable. Again, as in the discrete k case, the Molloy and Reed criterion (5.38) is always satisfied for $2 < \gamma \leq 3$ because $\langle k^2 \rangle$ diverges while $\langle k \rangle$ is finite. For $\gamma > 3$, also the second-order moment is finite and we need to use the expressions for $\langle k \rangle$ and $\langle k^2 \rangle$ given in Eqs. (5.18) and (5.19). Plugging such expressions into the Molloy and Reed condition, and assuming $k_{min} = 1$, we get the inequality $\gamma - 2 > 2(\gamma - 3)$. Thus, we obtain that the graph has a giant component for values of the exponent γ smaller than the critical value $\gamma_c = 4$. Such a critical value is appreciably different from the exact result reported in Eq. (5.46).

Characteristic Path Length

Concerning the characteristic path length L, Reuven Cohen and Shlomo Havlin have proven that scale-free graphs with $2 < \gamma < 3$ are *ultrasmall*, i.e. their diameter scales as $\ln \ln N$, which is a slower scaling than the $\ln N$ valid for ER random graphs [82]. Fan Chung and Lin Yuan Lu have given a more formal mathematical proof of the results obtained by Cohen and Havlin. They have demonstrated that the average distance L in random graphs with a power-law degree distribution scales as $\ln N$ when $\gamma > 3$, while if $2 < \gamma < 3$, L is $\mathcal{O}(\ln \ln N)$, while the diameter is $\mathcal{O}(\ln N)$ [76].

Clustering Coefficient

The *clustering coefficient* C of scale-free random graphs can be obtained by plugging the expressions of the first and second moment of a power-law degree distribution into Eq. (5.44). In the case $\gamma > 3$, both $\langle k \rangle$ and $\langle k^2 \rangle$ are finite, and Eq. (5.44) reduces to the same scaling valid for ER random graphs, namely $C \approx N^{-1}$. Therefore, when $\gamma > 3$, triangles can be neglected in large networks. Let us now consider the more interesting case, $2 < \gamma < 3$, in which the average degree is finite, while the second order moment is infinite. In such a case, Eq. (5.27) tells us that the second moment grows with the largest degree in the network, k_{max}, as in $\langle k^2 \rangle \approx k_{max}^{3-\gamma}$. Using Eq. (5.29) for the scaling of the natural cut-off k_{max} with N, we get $\langle k^2 \rangle \approx N^{(3-\gamma)/(\gamma-1)}$. Plugging this expression into Eq. (5.44), we finally get the scaling of the clustering coefficient of a scale-free random graph [231, 72, 286]:

$$C \approx N^{\frac{7-3\gamma}{\gamma-1}} \quad \text{for } 2 < \gamma < 3. \tag{5.47}$$

Notice that this formula leads to two different behaviours, according to whether the value of the exponent γ is larger or smaller than $7/3 \approx 2.33$. In fact, when $7/3 < \gamma < 3$, the exponent of Eq. (5.47) is negative, and also in this case C tends to zero. However, this time we have $C \approx N^\alpha$, with an exponent $\alpha = (7 - 3\gamma)/(\gamma - 1)$ which depends on the value of γ. In particular, the decrease of C with N is slower than that of ER random graphs, because $-1 < \alpha < 0$. Figure 5.7 shows how the value of the scaling exponent α changes as a function of the degree exponent γ. Summing up, if $\gamma > 7/3$, the clustering coefficient of a scale-free graph tends to zero when $N \to \infty$. In addition to this, if $\gamma > 3$, the scaling with N is exactly the same as that of ER random graphs.

The situation is radically different when $\gamma \leq 7/3$. In fact, at $\gamma = 7/3$, the exponent α is zero, therefore C is either a constant, or changes slower than any power of N. For $2 < \gamma < 7/3$, the exponent α ranges from 0 ($\gamma = 7/3$) to 1 ($\gamma = 2$), and Eq. (5.47) is clearly anomalous, since it leads to a diverging clustering coefficient C for large N while, by definition,

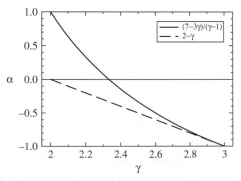

Fig. 5.7 The two exponents $\alpha = \frac{7-3\gamma}{\gamma-1}$ and $\alpha = 2 - \gamma$ governing the scaling $C \approx N^\alpha$ of the clustering coefficient in scale-free random graphs with degree distribution $p_k \sim k^{-\gamma}$, and respectively with or without multiple edges, are reported as a function of γ.

C must be smaller than 1 [72]. This problem is due to the fact that, as shown in Example 5.20, there can be more than one edge between two nodes with a common neighbour.

Example 5.20 *(Number of multiple edges in a scale-free graph)*   Eq. (5.36) gives the expected number of multiple edges, $\overline{n}^{\text{multiple}}$, in a random graph with an arbitrary p_k. Now, if $\langle k^2 \rangle / \langle k \rangle$ is finite as $N \to \infty$, then the density of multiple edges vanishes as $1/N$. However, multiple edges cannot always be neglected in the configuration model. In fact, in networks with power-law degree distributions with exponent $\gamma \in (2, 3)$, the quantity $\langle k^2 \rangle / \langle k \rangle$ diverges when $N \to \infty$, so that we have to check whether the ratio $(\langle k^2 \rangle / \langle k \rangle)^2$ in Eq. (5.36) increases faster or slower than N. Using Eqs. (5.27) and (5.29), respectively for $\langle k^2 \rangle$ and the natural cut-off k_{max}, we get:

$$\left(\frac{\langle k^2 \rangle}{\langle k \rangle} \right)^2 \sim N^{2\frac{3-\gamma}{\gamma-1}}$$

so that the density of multiple links reads:

$$\overline{n}^{\text{multiple}} / N \sim N^{\frac{7-3\gamma}{\gamma-1}}$$

which goes to zero only for $\gamma > 7/3 \approx 2.33$. For $\gamma < 7/3$ the density of multiple links increases with N, and this also leads to a value of the clustering coefficient C which increases with increasing system size, as we found from Eq. (5.47).

To avoid this problem with the expression of the clustering coefficient in Eq. (5.47), we need to consider random graphs without multiple edges. We show below that in a network with no multiple edges, there is a cut-off on the largest possible degree of the network due to the intrinsic structure of the network. This is the so-called *structural cut-off*, and is in general different from the *natural cut-off* discussed in Section 5.2, which derives from the finite order of the network.

The Structural Cut-Off

We have seen in Section 5.2 that, in a scale-free network, the largest degree k_{max} scales as a power of the number of nodes in the network, with an exponent which depends on the degree exponent as in Eq. (5.29). This is known as the network *natural cut-off*. In a network with no multiple edges, there is another cut-off, $k_{\text{max}}^{\text{struct}}$, this time due to the intrinsic structure of the network, which grows as the square root of the number of nodes, namely:

$$k_{\text{max}}^{\text{struct}} \simeq \sqrt{\langle k \rangle N}. \tag{5.48}$$

This is the so-called *structural cut-off*, and was proposed for the first time in 2004 by Marián Boguñá, Romualdo Pastor-Satorras and Alessandro Vespignani [45]. A precise derivation of Eq. (5.48) will be presented in Section 7.3 of the chapter on degree–degree correlations. We propose here a simple argument, based on Eq. (5.31), to find the dependence of $k_{\max}$ on the network order N. Eq. (5.31) gives the expected number of links $\bar{e}_{ij}^{\text{conf}}$ between two nodes i and j in the configuration model as a function of the degrees k_i and k_j of the two nodes. Now, if we do not allow multiple edges between the same pair of nodes in a graph, the quantity $\bar{e}_{ij}^{\text{conf}}$ cannot be larger than 1 for any pair i, j. Suppose $k_{\max}$ is the largest degree in the graph. This means that $k_i \leq k_{\max}, \forall i$. The maximum possible value of $\bar{e}_{ij}^{\text{conf}}$ is thus obtained when $k_i = k_j = k_{\max}$. We can therefore estimate the value of $k_{\max}^{\text{struct}}$ from the condition:

$$\frac{(k_{\max}^{\text{struct}})^2}{2K} \simeq 1$$

which implies that the largest degree scales as the square root of the number of nodes in the network, and indeed gives the expression of the structural cut-off reported in Eq. (5.48) [45]. Notice that the structural cut-off coincides with the natural cut-off when the degree exponent γ of a scale-free network is exactly equal to 3. When $\gamma > 3$ the structural cut-off diverges faster than the natural cut-off, and therefore the latter should be selected as the appropriate one. Conversely, when $\gamma < 3$, the exponent of the natural cut-off is greater than 1/2. As a consequence, scale-free networks with degree exponent $\gamma < 3$ and without multiple edges must have a largest degree that behaves as the structural cut-off and not as the natural cut-off. In the Appendix A.12 on algorithms to sample graph from the configuration model, we discuss a general criterion to decide whether a given degree sequence admits a graph without loops and multiple edges.

Let us now come back to the expression for the clustering coefficient of scale-free networks with $2 < \gamma < 3$. We can still make use of Eq. (5.44) together with Eq. (5.27) for the scaling of $\langle k^2 \rangle$ with the largest degree in the network. However, now, to avoid multiple edges we have to use expression (5.48) for the largest degree $k_{\max}$, and not the natural cut-off. This leads to the scaling of the clustering coefficient of a scale-free random graph with no multiple edges [231, 72, 286]:

$$C \approx N^{2-\gamma} \quad \text{for } 2 < \gamma < 3 \tag{5.49}$$

which implies that, in this case, C tends to zero in the whole range $2 < \gamma < 3$. Moreover, since we have $C \approx N^{\alpha}$, with $\alpha = 2 - \gamma$, the decrease of C with N is slower than that of ER random graphs, because $-1 < \alpha < 0$. Figure 5.7 also reports the scaling exponent α in Eq. (5.49) as a function of the degree exponent γ.

5.6 Probability Generating Functions

Most of the results presented in the previous sections can be obtained by means of the so-called generating functions formalism. This is an elegant and powerful mathematical tool extensively used in probability, which was employed by Mark Newman, Steven Strogatz and Duncan Watts to derive the properties of random graphs with a given degree distribution [238, 231]. A *probability generating function* is an alternative representation of a probability distribution. Let us consider the degree probability distribution $p_k, k = 0, 1, 2, \ldots$ of a graph. The corresponding probability generating function, denoted as $G_p(x)$, is a function of the real variable x, defined as:[3]

$$G_p(x) = p_0 + p_1 x + p_2 x^2 + \ldots = \sum_{k=0}^{\infty} p_k x^k. \qquad (5.50)$$

We say that the function $G_p(x)$ generates the probability distribution p_k because it contains all the information present in the original distribution p_k, which can be expressed in terms of $G_p(x)$. In fact, the Taylor series expansion of $G_p(x)$ reads:

$$G_p(x) = \sum_{k=0}^{\infty} \frac{G_p^{(k)}(0)}{k!} x^k$$

where $G_p^{(k)}(0) = \left[\frac{d^k G_p(x)}{dx^k} \right]_{x=0}$ is the kth derivative of the generating function $G_p(x)$ evaluated at the point $x = 0$[4]. Comparing the two expressions for $G_p(x)$, one obtains:

$$p_k = \frac{G_p^{(k)}(0)}{k!} \quad k = 0, 1, \ldots \qquad (5.51)$$

Hence, we can recover the degree probability distribution p_k from the generating function $G_p(x)$ by simple differentiation.

Analogously, the different-order moments of the distribution p_k can be expressed in terms of the derivatives of the generating function:

$$\langle k^n \rangle = \sum_{k=0}^{\infty} k^n p_k = \left[\left(x \frac{d}{dx} \right)^n G_p(x) \right]_{x=1} \qquad (5.52)$$

as shown in Example 5.21.

Example 5.21 *(Moments of p_k)*   Since the degree distribution p_k is normalised to 1, the generating function $G_p(x)$ must satisfy:

$$G_p(1) = p_0 + p_1 + p_2 + \ldots = \sum_k p_k = 1 \qquad (5.53)$$

which coincides with Eq. (5.52) with $n = 0$.

---

[3] Since the degree k is a non-negative integer number we will deal here with a special but important case of generating functions. A general introduction to generating functions can be found in any textbook on probability. See for instance Ref. [143].

[4] Notice that in the following we will also indicate the first and second derivatives of $G_p(x)$ respectively as $G_p'(x)$ and $G_p''(x)$.

The mean of the distribution p_k can be calculated directly by differentiation of Eq. (5.50). In fact, we have:

$$G'_p(x) = \frac{d}{dx}G_p(x) = p_1 + 2p_2x + 3p_3x^2 + \ldots = \sum_k kp_kx^{k-1}$$

and setting $x = 1$ we get:

$$G'_p(1) = \sum_k kp_k = \langle k \rangle$$

which is equal to Eq. (5.52) with $n = 1$.

Analogously, by considering second and third derivative of the generating function, and by setting $x = 1$, we obtain:

$$G''_p(1) = \sum_k k(k-1)p_k = \langle k^2 \rangle - \langle k \rangle$$

$$G'''_p(1) = \sum_k k(k-1)(k-2)p_k = \langle k^3 \rangle - 3\langle k^2 \rangle + 2\langle k \rangle$$

which are linear combinations of different order moments.

In order to prove Eq. (5.52) let us introduce the following sequence of functions $G_{p,n}(x)$ recursively defined as:

$$G_{p,n}(x) = \sum_{k=0}^{\infty} p_{k,n}x^k = \begin{cases} G_p(x) & \text{if } n = 0 \\ x\dfrac{d}{dx}G_{p,n-1}(x) & \text{if } n > 0 \end{cases}$$

which implies

$$G_{p,n}(x) = \left(x\frac{d}{dx}\right)^n G_p(x).$$

By the above definition of $G_{p,n}(x)$ we have, for $n > 0$,

$$G_{p,n}(x) = \sum_{k=0}^{\infty} p_{k,n}x^k = x\frac{d}{dx}G_{p,n-1}(x) = x\sum_{k=0}^{\infty} k\,p_{k,n-1}x^{k-1} = \sum_{k=0}^{\infty} k\,p_{k,n-1}x^k$$

which means that, $\forall k \geq 0$,

$$p_{k,n} = k\,p_{k,n-1} = k^2\,p_{k,n-2} = \cdots = k^n p_k.$$

Therefore $G_{p,n}(x) = \sum_{k=0}^{\infty} p_{k,n}x^k = \sum_{k=0}^{\infty} k^n p_k x^k$, which implies:

$$\langle k^n \rangle = G_{p,n}(1) = \left[\left(x\frac{d}{dx}\right)^n G_p(x)\right]_{x=1}$$

which means that we can calculate any moment of the distribution by taking a suitable derivative.

These expressions are very practical to use because they allow the moments of p_k to be calculated by performing a series of opportune differentiations of the generating functions, instead of by directly performing the summations in the definition $\langle k^n \rangle = \sum_k k^n p_k$. For instance, the following example shows how to construct the generating function associated with a binomial distribution, and how to use it to derive, in a very simple way, average degree $\langle k \rangle$ and fluctuations in ER random graphs.

Example 5.22   *(Generating function of a binomial distribution)*   When we deal with ER random graphs, it can be useful to construct the generating function of a binomial distribution. By inserting the expression of the degree distribution of Eq. (3.13) into Eq. (5.50), and using the binomial identity of Eq. (3.2), we get:

$$G_p(x) = \sum_{k=0}^{\infty} \binom{N-1}{k} p^k x^k (1-p)^{N-1-k} = (px + 1 - p)^{N-1}. \tag{5.54}$$

Now, by means of Eq. (5.52), we get for $n = 1$:

$$\langle k \rangle = \left[x \frac{d}{dx} G_p(x) \right]_{x=1} = \left[xp(N-1)(px + 1 - p)^{N-2} \right]_{x=1} = p(N-1)$$

which coincides with the average node degree of ER random graphs in Eq. (3.14). Analogously, for $n = 2$:

$$\langle k^2 \rangle = \left[x \frac{d}{dx} \left(x \frac{d}{dx} G_p(x) \right) \right]_{x=1}$$
$$= \left[xp(N-1)(px + 1 - p)^{N-2} + x^2 p^2 (N-1)(N-2)(px + 1 - p)^{N-3} \right]_{x=1}$$

which gives:

$$\langle k^2 \rangle = p(N-1)[p(N-2) + 1]$$

and

$$\sigma_k^2 = \langle k^2 \rangle - \langle k \rangle^2 = p(1-p)(N-1)$$

in agreement with the result in Eq. (3.15).

One of the most important properties of generating functions which will turn out to be very useful for the study of random graphs is the following. Suppose we have *two independent* random variables whose distributions are respectively described by two generic functions p_k and $r_{k'}$. Let us indicate as $G_p(x)$ and $G_r(x)$ the two corresponding generating functions. If we are now interested to calculate how the sum $k + k'$ is distributed we have to perform the so-called *convolution* of p_k and $r_{k'}$, which is a rather complicated calculation [143]. However, as illustrated in Example 5.23, it is possible to prove that the generating function of the sum of the two independent random variables, which we indicate as $G_{p \oplus r}(x)$, can be obtained just as the product of the two generating functions: $G_{p \oplus r}(x) = G_p(x) G_r(x)$. And then, from such a generation function, we can easily derive important properties of the distribution of the sum $k + k'$, such as its moments.

Example 5.23　*(Sum of two independent random variables)*　Let us indicate as $G_p(x)$ and $G_r(x)$ the generating functions associated with the distribution functions p_k and $r_{k'}$ of two independent random variables k and k'. Consider the product of the two generating functions:

$$G_p(x)G_r(x) = \sum_{k=0}^{\infty} p_k x^k \sum_{k'=0}^{\infty} r_{k'} x^{k'}$$
$$= (p_0 + p_1 x + p_2 x^2 + \ldots)(r_0 + r_1 x + r_2 x^2 + \ldots)$$
$$= p_0 r_0 + (p_0 r_1 + p_1 r_0)x + (p_0 r_2 + p_1 r_1 + p_2 r_0)x^2 + \ldots$$
$$= \sum_{m=0}^{\infty} c_m x^m.$$

Now, the latter summation can indeed be seen as the generating function $G_{p\oplus r}(x)$ of the distribution of the sum of the two original random variables, since the coefficients are respectively equal to:

$$c_0 = p_0 r_0 \quad c_1 = p_0 r_1 + p_1 r_0 \quad c_2 = p_0 r_2 + p_1 r_1 + p_2 r_0 \quad \ldots$$

In fact, the coefficient c_0 is the probability that the product of the two variables is equal to zero, and this happens when both variables k and k' are equal to zero. The coefficient c_1 is the probability that the product of the two variables is equal to 1, and this happens when the first variable takes the value 1 and the second is zero, and vice versa, therefore with a probability $p_0 r_1 + p_1 r_0$. In general we can write:

$$c_m = \sum_{k+k'=m} p_k r_{k'}$$

for $m = 0, 1, 2, \ldots$, which are the coefficients of the generating function $G_{p\oplus r}(x)$.

The same argument holds for the *sum of n independent random variables*. For instance, in a random graph, the node degrees are uncorrelated. Therefore, if $G_p(x)$ is the generating function of the degree distribution p_k, the sum of the degrees of n randomly chosen nodes has a distribution which is generated by the function $[G_p(x)]^n$. To see this, we just need to generalise what we did in Example 5.23. We get that the coefficient of x^m in the function $[G_p(x)]^n$ contains one term of the form $p_{k_1} p_{k_2} \ldots p_{k_n}$ for each set $k_1, k_2, \ldots, k_m$ of the degrees of the n nodes such that $\sum_i k_i = m$. But these terms are precisely the probabilities that the n degrees sum to m in every possible way. Hence $[G_p(x)]^n$ is indeed the generating function of the sum of the n degrees. Many of the results on random graphs we will derive in this section rely on such a property.

In order to study properties such as the component size distribution in a random graph, together with the generating function $G_p(x)$ of the degree distribution p_k, we also need to define a second generating function, associated with the degree of nodes reached by following an edge in the graph. We saw in Chapter 3 that the probability q_k of finding a node with degree k by following a link of the graph can be expressed in terms of p_k as in Eq. (3.38). More precisely here, for symmetry reasons, it is better to consider not the number of links of a node we reach by following an edge in the graph, but the number of links other than the one we arrived along. We therefore introduce the new distribution $\tilde{q}_k$ which gives the probability that a node we reach by following an edge in the graph has k links *other than the one we arrived along*. Of course we have:

$$\tilde{q}_k = q_{k+1} = \frac{(k+1)p_{k+1}}{\langle k \rangle}. \tag{5.55}$$

If we indicate as $G_{\tilde{q}}(x)$ the generating function associated with such a distribution, we have:

$$G_{\tilde{q}}(x) = \sum_{k=0}^{\infty} \tilde{q}_k x^k = \frac{\sum_{k=0}^{\infty}(k+1)p_{k+1}x^k}{\langle k \rangle} = \frac{\sum_{k=1}^{\infty} kp_k x^{k-1}}{\langle k \rangle}$$

We finally get:

$$G_{\tilde{q}}(x) = \sum_{k=0}^{\infty} \tilde{q}_k x^k = \frac{G_p'(x)}{\langle k \rangle}, \tag{5.56}$$

where again $G_p'(x)$ denotes the first derivative of $G_p(x)$ with respect to its argument. Thus $G_{\tilde{q}}(x)$ can be easily derived from $G_p(x)$.

Example 5.24 *(Generating functions of a Poisson distribution)*   When we deal with ER random graphs, it can be extremely useful to work with the generating function of Poisson degree distributions. Plugging Eq. (3.16) into Eq. (5.50), we get:

$$G_p(x) = e^{-z} \sum_{k=0}^{\infty} \frac{z^k}{k!} x^k = e^{-z+zx}. \tag{5.57}$$

From this generating function we can easily derive the average degree and the standard deviation of ER random graphs:

$$\langle k \rangle = \left[x \frac{d}{dx} G_p(x) \right]_{x=1} = G_p'(1) = \left[z e^{-z+zx} \right]_{x=1} = z$$

$$\langle k^2 \rangle = \left[x \frac{d}{dx} \left(x \frac{d}{dx} G_p(x) \right) \right]_{x=1} = \left[x G_p'(x) + x^2 G_p''(x) \right]_{x=1} = \langle k \rangle + G_p''(1)$$

$$= z + \left[z^2 e^{-z+zx} \right]_{x=1} = z + z^2$$

$$\sigma_k = \sqrt{\langle k^2 \rangle - \langle k \rangle^2} = \sqrt{z}$$

obtaining the same results we found in Box 3.3. Notice that here, in order to derive $\langle k^2 \rangle$ we have made use of the derivative product rule before the substitution of the expression of G_p, and this allows a simpler calculation than that of Example 5.22.

The generating function $G_{\tilde{q}}(x)$ of the other degree of nodes reached by following an edge is also easily found, from Eq. (5.56):

$$G_{\tilde{q}}(x) = \frac{G'_p(x)}{z} = e^{-z+zx}. \qquad (5.58)$$

Thus, for the case of the Poisson distribution we have

$$G_{\tilde{q}}(x) = G_p(x). \qquad (5.59)$$

This identity is the reason why the properties of ER random graphs are particularly simple to derive analytically. For instance, this identity is closely connected to Eq. (5.41) and to the result we derived in Chapter 3 that the mean number of second neighbours of a vertex in ER random graphs is simply the square of the mean number of first neighbours. In fact, the average number of *other* links of the first neighbours of a node can be derived from $G_{\tilde{q}}(x)$, and turns out to be equal to $\langle k \rangle$ because of the identity in Eq. (5.59). Hence, the average number of second neighbours of a node in ER random graphs is $\langle k \rangle^2$.

Example 5.25   *(Generating functions of a power-law distribution)*   The generating function corresponding to a power law $p_k = k^{-\gamma}/\zeta(\gamma)$ with exponent $\gamma > 1$ can be obtained by plugging Eq. (5.10) into Eq. (5.50):

$$G_p(x) = \sum_{k=1}^{\infty} p_k x^k = \frac{1}{\zeta(\gamma)} \sum_{k=1}^{\infty} k^{-\gamma} x^k.$$

Unfortunately, there is no simple analytical expression for the series appearing on the right-hand side of the above relation; however, this is a well-studied function known as *polylogarithm* of x, and is denoted by $\mathrm{Li}_\gamma(x)$:

$$\mathrm{Li}_\gamma(x) \equiv \sum_{k=1}^{\infty} k^{-\gamma} x^k.$$

Hence the generating function of a power-law distribution can be written as:

$$G_p(x) = \frac{\mathrm{Li}_\gamma(x)}{\zeta(\gamma)}. \qquad (5.60)$$

Observe that $\mathrm{Li}_\gamma(1) = \zeta(\gamma)$ and therefore $G_p(1) = 1$. Also observe that

$$x\frac{d}{dx}\mathrm{Li}_\gamma(x) = x\sum_{k=1}^{\infty} kx^{k-1}k^{-\gamma} = \sum_{k=1}^{\infty} k^{-(\gamma-1)}x^k = \mathrm{Li}_{\gamma-1}(x).$$

We can therefore apply Eq. (5.52) to derive moments of different order of a power-law degree distribution:

$$\langle k^m \rangle = \left[\left(x\frac{d}{dx}\right)^m G_p(x)\right]_{x=1} = \frac{1}{\zeta(\gamma)}\left[\left(x\frac{d}{dx}\right)^m \mathrm{Li}_\gamma(x)\right]_{x=1}$$

$$= \frac{1}{\zeta(\gamma)}\mathrm{Li}_{\gamma-m}(1) = \frac{\zeta(\gamma-m)}{\zeta(\gamma)}$$

which is in agreement with Eq. (5.12).

Component Size Distribution

Having defined the two generating functions $G_p(x)$ and $G_{\tilde{q}}(x)$, we are now ready to explore number and size of components in random graphs with a generic p_k, as we did in Section 3.4 for ER random graphs. We will closely follow the derivations presented in Refs. [238, 231] and in Mark Newman's book on complex networks [239].

Let us first assume to be *below the phase transition*, in the regime in which there is no giant component in the graph. If we indicate as n_s the actual number of components of size s in the graph, the distribution w_s of the sizes of the components to which a randomly chosen node belongs is equal to:

$$w_s = \frac{sn_s}{\sum_s sn_s} = \frac{sn_s}{N} \tag{5.61}$$

since there are s different ways to find a component of size s by selecting a node at random. Notice that this definition is slightly different from the definition of w_s we adopted in Eq. (3.28): since we are below the phase transition, there is no need to exclude the largest component in the normalisation constant. We can now introduce two other generating functions, namely those associated with the component size distributions. Let us indicate as $H_w(x)$ the generating function of the probability distribution w_s:

$$H_w(x) = \sum_{s=1}^{\infty} w_s x^s. \tag{5.62}$$

Notice that the summation over s here starts from $s = 1$, since by definition a randomly chosen node does at least belong to a component with 1 node. The crucial assumption we will use here to derive an expression for w_s is that there is no giant component, and we can neglect cycles in any finite component of connected nodes of a large random graph. In Section 3.6 we discussed this point for ER random graphs. For random graphs with an arbitrary degree distribution, we showed in Section 5.4 that the clustering coefficient tends to zero as $N \to \infty$. And the same idea can be extended to neglect cycles of lengths larger than 3. In fact, the expected number of links between two randomly chosen nodes i and j of degrees k_i and k_j, given in Eq. (5.31), is the same regardless of where the nodes are, and tends to zero as $N \to \infty$. Let us now consider a randomly selected node i. Such a node will have k links with a probability p_k. Suppose, for instance, to have selected a node i with $k = 4$ links, as that shown in Figure 5.8. The component to which i belongs is a tree, and therefore its size is equal to 1 plus the number of all nodes we can reach by following each of the branches, in this case four, originating in i. Or, in other words, the size of the component to which i belongs is equal to 1 plus the size of the four components g_1, g_2, g_3, g_4 we get by removing node i from the graph, as shown in the right-hand side of the figure.

The size distribution of such components is not given by the probability w_s, but rather by another probability distribution, namely the probability ρ_s of ending up into a component of size s by following a randomly chosen link. Let us therefore denote as:

$$H_\rho(x) = \sum_{s=1}^{\infty} \rho_s x^s \tag{5.63}$$

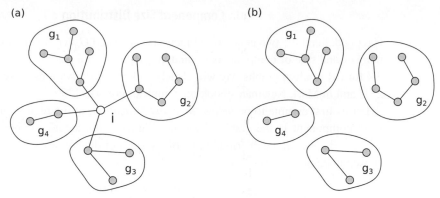

(a)   (b)

Fig. 5.8 The size of the component to which a randomly chosen node i belongs to is equal to 1 plus the size of the components we get by removing node i from the graph. This is shown here for a node i with degree $k_i = 4$.

the generating function associated with ρ_s. Again, the summation starts at $s = 1$, since the components have at least one node. Now, the probability that node i in the left-hand side of Figure 5.8 belongs to a component with s nodes is exactly equal to the probability $\text{Prob}(s_1 + s_2 + s_3 + s_4 = s - 1)$ that the sum of the sizes of the four components shown in the right-hand side of the figure is equal to $s - 1$. Since a node i with k links is selected with a probability p_k, we can therefore write:

$$w_s = \sum_{k=0}^{\infty} p_k \text{Prob}(s_1 + s_2 + \ldots + s_k = s - 1).$$

While it is complicated to get an expression for w_s, we can easily find a simple and useful self-consistent set of equations for the generating functions $H_w(x)$ and $H_\rho(x)$, which also involves the generating function of the degree distributions $G_p(x)$ and $G_{\tilde{q}}(x)$. In fact, by substituting the above expression of w_s in Eq. (5.62), we get:

$$H_w(x) = \sum_{s=1}^{\infty} \sum_{k=0}^{\infty} p_k \text{Prob}(s_1 + s_2 + \ldots + s_k = s - 1)x^s$$

$$= x \sum_{k=0}^{\infty} p_k \sum_{s=1}^{\infty} \text{Prob}(s_1 + s_2 + \ldots + s_k = s - 1)x^{s-1}.$$

The latter summation in this expression is nothing other than the generating function associated with the distribution of the sum of the sizes of the k components reached from i, by following its k links. By using the property discussed in Example 5.23 this can be written as $[H_\rho(x)]^k$, since the distribution of the size of the component reached by following a link is generated by $H_\rho(x)$, and the k sizes are independent random variables. We finally have:

$$H_w(x) = x \sum_{k=0}^{\infty} p_k [H_\rho(x)]^k = x G_p(H_\rho(x)) \qquad (5.64)$$

which relates the generating function $H_w(x)$ to the generating function $H_\rho(x)$. To get an analogous expression for $H_\rho(x)$ we observe that the probability ρ_s that following a randomly chosen link in one direction we can reach a component of size s can be written as:

$$\rho_s = \sum_{k=0}^{\infty} \tilde{q}_k \text{Prob}(s_1 + s_2 + \ldots + s_k = s - 1).$$

In fact, the probability that, following the link, we get into a node with k links, other than the link we arrived along, is equal to $\tilde{q}_k$, and the probability that a node with k links belongs to a component with s nodes is again equal to the probability $\text{Prob}(s_1 + s_2 + \ldots + s_k = s - 1)$ that the sum of the number of nodes reachable through each of the k branches is equal to $s - 1$. Inserting this expression of ρ_s into Eq. (5.63), and using again the multiplication rule to get the generating function associated with the distribution of the sum of independent random variables we get:

$$H_\rho(x) = x \sum_{k=0}^{\infty} \tilde{q}_k [H_\rho(x)]^k = x G_{\tilde{q}}(H_\rho(x)). \tag{5.65}$$

Now, in some particular cases, it is possible to solve exactly the self-consistent Eq. (5.65) for $H_\rho(x)$, and to obtain $H_w(x)$ by substituting the result into Eq. (5.64). We illustrate this procedure in the following example for a case of a graph having only nodes with 0, 1, 2, and 3 links.

Example 5.26   Consider the simple case of a graph in which all vertices have degree 0, 1, 2 or 3, with probabilities p_0, p_1, p_2, p_3. In this case, the two generating functions $G_p(x)$ and $G_{\tilde{q}}(x)$ read:

$$G_p(x) = p_3 x^3 + p_2 x^2 + p_1 x + p_0$$
$$G_{\tilde{q}}(x) = q_2 x^2 + q_1 x + q_0.$$

The average degree of the network is $\langle k \rangle = G_p'(1) = 3p_3 + 2p_2 + p_1$ and, by using Eq. (5.56), we can also express the second generating function in terms of the probabilities p_k as:

$$G_{\tilde{q}}(x) = \frac{G_p'(x)}{\langle k \rangle} = \frac{3p_3 x^2 + 2p_2 x + p_1}{3p_3 + 2p_2 + p_1}.$$

We can now obtain information on the component size distribution w_s by constructing the generating function $H_w(x)$ from Eq. (5.64) and Eq. (5.65). In fact, Eq. (5.65) implies that the generating function $H_\rho(x)$ is a solution of the quadratic equation:

$$q_2 [H_\rho(x)]^2 + \left(q_1 - \frac{1}{x}\right) H_\rho(x) + q_0 = 0,$$

which gives:

$$H_\rho(x) = \frac{-q_1 + \frac{1}{x} \pm \sqrt{\left(q_1 - \frac{1}{x}\right)^2 - 4q_0 q_2}}{2q_2}.$$

Finally, we have a closed-form solution for the function generating the distribution w_s:

$$H_w(x) = x\left(p_3[H_\rho(x)]^3 + p_2[H_\rho(x)]^2 + p_1[H_\rho(x)] + p_0\right)$$

by simply inserting the expression of $H_\rho(x)$ into Eq. (5.64).

Once we have an explicit expression of $H_w(x)$, we have a complete information on the distribution of the component sizes w_s to which a randomly chosen node belongs. For instance, in analogy with Eq. (5.51), by taking the first derivative $H'_w(0)$ we get the number of isolated nodes, by taking the second derivative $H''_w(0)$ we obtain the number of components with two nodes, and so on.

Mean Component Size

Cases such as the one in the Example 5.26, in which closed-form solutions for $H_w(x)$ and $H_\rho(x)$ can be obtained, are more of an exception than a rule. For instance, Eq. (5.65) is transcendental and has no closed-form solution for a Poissonian degree distribution. Therefore, for ER random graphs and in most of the other cases, we have to rely on approximate solutions of Eq. (5.65) and Eq. (5.64). However, even when there is no closed-form solution for $H_w(x)$ and $H_\rho(x)$, we can calculate at almost no price the moments of the distribution w_s which, after all, are the most useful quantities.

Here we show this for the first moment, the average component size. From the definition of $H_w(x)$ in Eq. (5.62) we have:

$$\langle s \rangle = \sum_s s w_s = H'_w(1).$$

By taking the derivative of $H_w(x)$ in Eq. (5.64), we get:

$$H'_w(1) = \left[G_p(H_\rho(x)) + xG'_p(H_\rho(x))H'_\rho(x)\right]_{x=1} = 1 + G'_p(1)H'_\rho(1)$$

where, in the last equality, we have made use of the generating functions normalisation conditions, $G_p(1) = 1$ and $H_\rho(1) = 1$. Analogously, by taking the derivative of $H_\rho(x)$ in Eq. (5.65), we get:

$$H'_\rho(1) = \left[G_{\tilde{q}}(H_\rho(x)) + xG'_{\tilde{q}}(H_\rho(x))H'_\rho(x)\right]_{x=1} = 1 + G'_q(1)H'_\rho(1)$$

which gives:

$$H'_\rho(1) = \frac{1}{1 - G'_{\tilde{q}}(1)}, \tag{5.66}$$

Inserting this expression in the previous equation for $H'_w(1)$, we finally get:

$$\langle s \rangle = 1 + \frac{G'_p(1)}{1 - G'_{\tilde{q}}(1)}. \tag{5.67}$$

In practice, from Eq. (5.64) and Eq. (5.65) we have been able to find a closed-form expression for the average component size below the transition in terms of the generating functions $G_p(x)$ and $G_{\tilde{q}}(x)$. Notice that, when $G'_{\tilde{q}}(1) = 1$, the value of $\langle s \rangle$ diverges, indicating the formation of a giant component.

This result can be rewritten in a more familiar form. In fact, the derivatives of the generating functions $G_p(x)$ and $G_{\tilde{q}}(x)$ can be expressed in terms of the moments of the degree distribution $\langle k \rangle$ and $\langle k^2 \rangle$:

$$G'_p(1) = \sum_k k p_k = \langle k \rangle = z_1 \tag{5.68}$$

$$G'_{\tilde{q}}(1) = \frac{G''_p(1)}{\langle k \rangle} = \frac{\sum_k k(k-1)p_k}{\langle k \rangle} = \frac{\langle k^2 \rangle - \langle k \rangle}{\langle k \rangle} = \frac{z_2}{z_1}. \tag{5.69}$$

Notice that we have also made use of the symbol z_2 defined in Eq. (5.40) to indicate the average number of nodes at distance two from a given node. Substituting into Eq. (5.67) we get an expression for the average component size below the transition:

$$\langle s \rangle = 1 + \frac{z_1^2}{z_1 - z_2}. \tag{5.70}$$

It is now immediate to see that the critical condition for the appearance of a giant component, $z_1 = z_2$, coincides with the *Molloy and Reed criterion* of Eq. (5.38) derived in Section 5.4 with different arguments. The extra information we get here, as a payoff for our efforts in dealing with the mathematics of generating functions, is an expression for $\langle s \rangle$ valid in the whole subcritical regime.

Size of the Giant Component

The previous arguments are valid below the phase transition, where there is no giant component in the graph and all the finite components in the limit of large N are tree-like. If we want to find the size of the giant component, we have to work instead *above the phase transition*, in the regime where the graph has a giant component. To be more specific, let us indicate as s_1 the number of nodes in the largest component of a graph of order N. When $N \to \infty$, the ratio $S = s_1/N$ ($0 < S \le 1$) will represent the fraction of the graph nodes occupied by the giant component. The quantity $v = 1 - S$ will then be the probability that a node, chosen uniformly at random from the graph, is not in the giant component. Now, the main problem is that we cannot use straightforwardly the generating function formalism to calculate the value of S. This is because the giant component does, in general, contain cycles, and therefore all the arguments we have used so far are not valid anymore. The problem can be easily overcome by treating the giant component separately from the other components. Namely, we define the generating function $H_w(x)$ associated with w_s as:

$$H_w(x) = \sum_s' w_s x^s \tag{5.71}$$

where, by the symbol prime in $\sum'_s$ we have indicated a summation over all possible component orders s, excluding the largest component. Analogously, we define:

$$H_\rho(x) = \sum_s{}' \rho_s x^s. \tag{5.72}$$

These definitions are slightly different from the original ones in Eq. (5.62) and Eq. (5.63). The main point with introducing these new quantities is that the finite components are still tree-like even above the transition, so that Eq. (5.64) and Eq. (5.65) hold with the new definitions of $H_w(x)$ and $H_\rho(x)$. The price to pay is that now $H_w(1)$ and $H_\rho(1)$ are no longer equal to 1. We have in fact:

$$H_w(1) = \sum_s{}' w_s = 1 - S \tag{5.73}$$

which means that $H_w(1)$ is equal to the fraction of nodes not in the giant component. And it is precisely this equation which allows us to calculate the size S of the giant component above the transition from the fraction of nodes in the finite components: $S = 1 - H_w(1)$. In fact, by using Eq. (5.64) and Eq. (5.65), we get the system:

$$\begin{cases} S = 1 - G_p(H_\rho(1)) \\ H_\rho(1) = G_{\tilde{q}}(H_\rho(1)). \end{cases} \tag{5.74}$$

Hence, the value of S can be obtained by finding $H_\rho(1)$ from the second equation, and then by plugging it into the first one. Notice that $H_\rho(1)$ has a precise physical meaning. In fact, it represents the probability that a node reached by following a randomly chosen link does not belong to the giant component, just as $H_w(1)$ represents the probability that a randomly chosen node in the graph does not belong to the giant component. The system of Eqs. (5.74) is in general not solvable in closed form, so that we need to find the solution numerically.

Example 5.27 *(Size of the giant component in ER random graphs)* In the case of ER random graphs we can use the expressions for the generating functions of Poisson degree distributions:

$$G_p(x) = G_{\tilde{q}}(x) = e^{-z+zx}$$

Eqs. (5.74) can then be written as:

$$\begin{cases} S = 1 - u \\ u = e^{-z(1-u)} \end{cases}$$

which means that S must satisfy the self-consistency relation:

$$S = 1 - e^{-zS}.$$

This is precisely Eq. (3.25) we derived in Section 3.4.

Based on Eqs. (5.74), we can also find an expression for the average size of finite components above the critical threshold:

$$\langle s \rangle = \frac{\sum_s' s w_s}{\sum_s' w_s} = \frac{\sum_s' s^2 n_s}{N - s_1}$$

that complements Eq. (5.67), which is valid only below the threshold. Notice that the expression of $\langle s \rangle$ above coincides with the definition we adopted in Eq. (3.26) of Section 3.4. By differentiating, respectively, Eq. (5.64) to get an expression for $H_w'(1)$, and Eq. (5.65) to get an expression for $H_\rho'(1)$, we obtain (see Problem 5.6(b)):

$$\langle s \rangle = \frac{\sum_s' s w_s}{\sum_s' w_s} = \frac{H_w'(1)}{H_w(1)} = \frac{1}{H_w(1)} \left[G_p(H_\rho(1)) + \frac{G_p'(H_\rho(1))G_{\tilde{q}}(H_\rho(1))}{1 - G_{\tilde{q}}'(H_\rho(1))} \right].$$

Now, by means of Eq. (5.56) we can write $G_p'(H_\rho(1))$ as $\langle k \rangle \cdot G_{\tilde{q}}(H_\rho(1))$, and by using Eq. (5.73) and Eqs. (5.74) we finally get the following expression for the average size of finite components:

$$\langle s \rangle = 1 + \frac{\langle k \rangle \cdot H_\rho^2(1)}{[1 - S][1 - G_{\tilde{q}}'(H_\rho(1))]}, \tag{5.75}$$

where the values of S and $H_\rho(1)$ are those found from Eqs. (5.74). Notice that the expression in Eq. (5.75) is also valid below the phase transition, when there is no giant component. In fact, if we set $S = 0$ and $H_\rho(1) = 1$ (meaning that the probability that a node reached by following a randomly chosen link does not belong to the giant component is equal to 1) in Eq. (5.75), this equation reduces exactly to Eq. (5.67), since we know that $\langle k \rangle = G_p'(1)$.

We used Eq. (5.74) and Eq. (5.75) to compute, as a function of the exponent γ, the size of the giant component S and the average size of finite components $\langle s \rangle$ for graphs with a power-law degree distribution $p_k \sim k^{-\gamma}$ and $k_{min} = 1$. The results are reported as solid lines respectively in Figure 5.9(a) and in Figure 5.9(b). Notice that S is a monotonically decreasing function of γ in the interval $[2.0, \gamma_c]$, where $\gamma_c = 3.47875\ldots$ is the critical value of the exponent, determined by the Molloy–Reed criterion in Eq. (5.46). Above such a value we expect that $S \to 0$ when $N \to \infty$. Similarly to what we have seen in Figure 3.7, the average size of finite components $\langle s \rangle$ diverges at the phase transition. Interestingly, the theory is in perfect agreement with the values of S and $\langle s \rangle$ obtained numerically in synthetic networks with $N = 10^6$ nodes constructed through the configuration model (reported as symbols in Figure 5.9). Slight discrepancies between Eq. (5.75) and the values of $\langle s \rangle$ measured in synthetic graphs are indeed visible around the phase transition. Such fluctuations are related to the finite value of the maximum degree in finite-size networks.

Example 5.28 *(Mean component size in ER random graphs)*   For simplicity, let us set $u = H_\rho(1)$. As seen in Example 5.27, then u satisfies the system:

$$\begin{cases} S = 1 - u \\ u = e^{-z(1-u)}. \end{cases}$$

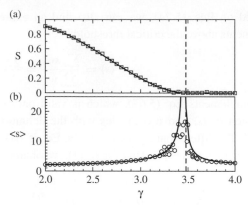

Fig. 5.9 The size S of the giant connected component (a), and the mean size $\langle s \rangle$ of finite components (b) in random graphs with power-law degree distributions $p_k \sim k^{-\gamma}$ as a function of the exponent γ. The theoretical predictions of Eq. (5.74) and Eq. (5.75) are in perfect agreement with the values obtained numerically on realisations of the configuration model with $N = 10^6$ nodes (respectively, squares and circles) The vertical dashed line corresponds to the critical value $\gamma_c = 3.47875\ldots$ of Eq. (5.46), above which no giant component exists.

We thus have:

$$\langle s \rangle = 1 + \frac{zu^2}{[1 - S][1 - G'_{\tilde{q}}(u)]}.$$

Since for a Poisson distribution $G_{\tilde{q}}(x) = e^{-z+zx}$, then $G'_{\tilde{q}}(u) = ze^{-z(1-u)} = zu$. Finally, the average size of finite components in ER random graphs is equal to:

$$\langle s \rangle = 1 + \frac{z(1 - S)^2}{[1 - S][1 - z(1 - S)]} = \frac{1}{1 - z + zS}.$$

And this proves Eq. (3.29), as given in Chapter 3.

5.7 What We Have Learned and Further Readings

In this chapter we have introduced and studied a data set of four network samples of the *World Wide Web*. In particular, we have shown that the *degree distributions* of these networks are *scale free*, i.e. the probability p_k that a network node has degree k is a *power-law* $p_k \sim k^{-\gamma}$ for large values of k, with a value of γ ranging in the interval (2, 3]. This is indeed true both for the out- and the in-degree distributions, and it certainly came as a surprise when this result was first published for the World Wide Web of the University of Notre Dame in 1999. Power-law degree distributions differ a lot from the Poisson degree distributions of random graphs. They indicate that most vertices are sparsely connected, while a few vertices have an extremely large number of links and therefore play a crucial role for the functionality of the World Wide Web. It is possible to show mathematically that, in an infinite system with $p_k \sim k^{-\gamma}$ and $\gamma \in (2, 3]$, the average degree

$\langle k \rangle$ is finite while the second moment $\langle k^2 \rangle$ diverges. This means that in a finite network as the World Wide Web we have to deal with large fluctuations in the node degrees, and is remarkable that the structure of this system has organised in this peculiar way. What is even more surprising is that *scale-free networks* are not just found in the World Wide Web, but are ubiquitous in other man-made systems, and also in social systems and in nature. Although the precise value of the degree exponent γ depends on the details of the network under study, most of the exponents found empirically are larger than 2 and smaller than 3.

Driven by the need to better model real-world systems, in this chapter we have therefore generalised the Erdős–Rényi models to generate random graphs with any shape of degree distribution p_k. In particular, we have studied the so-called *configuration model* in which the degree of each node is assigned to take a precise value but, apart from this constraint, the links are distributed at random. Our main interest was in graphs with power-law degree distributions. We used the so-called *Molloy–Reed criterion* to predict the existence of a giant component for γ smaller than a critical value $\gamma_c \simeq 3.47$, and we found an expression for the characteristic path length and for the clustering coefficient. An important take-home message of this chapter is that, being real-world networks always finite, there is a cut-off on their largest node degree. This is the so-called *natural cut-off* of a network. In addition to this, in scale-free random graphs without multiple links we have also to take into account the so-called *structural cut-off* that is related to the intrinsic structure of the network.

The configuration model is not the only way to produce graphs with power-law degree distributions. There are various methods to construct scale-free networks by arbitrarily assigning each node to take a precise value for a quantity that does not necessarily need to be the degree. For instance, Guido Caldarelli, Andrea Capocci, Paolo De Los Rios and Miguel Muñoz have proposed a *fitness model* in which each node i has an attractiveness or fitness a_i which is a real number sampled from a fitness distribution $\rho(a)$. Then, for every pair of nodes i and j, a link is drawn with a probability $f(a_i, a_j)$, with f being a symmetric function of its arguments. The model generates power-law degree distributions for various forms of the attractiveness function ρ and of the attaching function f, while it produces ER random graph when $f(a_i, a_j) = p$ $\forall i, j$ [65]. Kwang-Il Goh, Byungnam Kahng and Doochul Kim have proposed another model based on similar ideas [133, 134]. In their model, a node i is assigned an attractiveness $a_i = i^{-\alpha}$, where α is a tunable parameter in $[0, 1)$. Then two different nodes i and j are selected with probabilities equal to the normalised attractiveness, respectively equal to $a_i / \sum_l a_l$ and $a_j / \sum_l a_l$, and are connected if there is not already a link between them. This operation is repeated until there are K links in the network. When $\alpha = 0$ the model produces ER random graphs, while for $\alpha \neq 0$ the graphs obtained have power-law degree distributions with exponent $\gamma = 1 + 1/\alpha$. Thus, by varying α in $[0, 1)$ one obtains an exponent γ in the range $2 < \gamma < \infty$. We will come back to fitness models in the context of growing graphs in Chapter 6.

Generalisations to include clustering properties in random graphs have also been explored in the literature. One of the oldest models for random graphs with clustering is the so-called random biased network model introduced by Anatol Rapoport in 1957 [267]. Two more recent methods to construct random graphs with an arbitrary degree distribution and with a tunable clustering coefficient can be found in the paper by Mark Newman [227], and in the paper by M. Ángeles Serrano and Marián Boguñá [281].

Concerning further readings related to this chapter, we suggest the complete review on power laws by Mark Newman [242].

Problems

5.1   The World Wide Web

(a) Show that the scale-free property of Eq. (5.6) does not hold for an exponential function.

(b) Consider hub and authority centrality scores introduced in Example 5.3. Prove that in the case of a cycle of order N, $AA^\top = A^\top A = I_N$, that is the identity matrix in dimension N.

(c) Go back to Figure 5.2. Is it possible to have an arc pointing from a node in the tendrils of the OUT component to a node in the tendrils of the IN component without changing the internal structure of the WCC? And an arc pointing in the opposite direction? Explain your answer.

5.2   Power-Law Degree Distributions

(a) Consider a network with N nodes and a degree distribution $p(k) = ck^{-\gamma}$, $k \in \Re$, $k > 0$, and a degree exponent $\gamma > 2$. Derive Eq. (5.29) for the scaling of the *natural cut-off* $k_{\max}$ as a function of N. Why is the expression found to be valid only for $\gamma > 2$?

(b) Consider a network with N nodes and a power-law degree distribution $p(k) = ck^{-\gamma}$, with exponent $\gamma > 1$. Assume that $k_{\min} = 1$ and $k_{\max} = \min\left(N^{1/(\gamma-1)}, N\right)$. Working in the continuous-k approximation (i.e. treating the degree k as a positive real number), find an expression for $\langle k \rangle$ and $\langle k^2 \rangle$ when $N \to \infty$. Consider separately the three following cases: $\gamma \in (1, 2]$, $\gamma \in (2, 3]$, and $\gamma > 3$.

5.3   The Configuration Model

(a) Are you able to construct graphs with 10 nodes, four nodes with degree 1, three nodes with degree 2, two nodes with degree 3 and one node with degree 4? What are the basic necessary conditions for such a construction? Suppose you want to avoid multigraphs in the construction. HINT: Have a look at Appendix A.12.

(b) Consider the configuration model with $N = 3$ nodes, $K = 3$ edges, and degree sequence $\mathcal{K} = \{k_1 = 2, k_2 = 2, k_3 = 2\}$. Draw and count the number of labelled and unlabelled graphs with such a degree sequence. How many different matchings of node stubs can produce each of these graphs?

(c) What are the expected number of edges and the expected number of multiple edges between two nodes in an ensemble of graphs as that considered in point (a)? Evaluate also the expected number of loops in a graph of the ensemble.

5.4 Random Graphs with Arbitrary Degree Distribution

(a) Given the ensemble of Erdős–Renyí (ER) random graphs $G_{N,p}^{\text{ER}}$ with $N = 1000$ nodes, and where each pair of nodes is connected with a probability p, consider the two cases $p = 0.0005$ and $p = 0.01$. Do the networks in the two ensembles have a giant connected component?

(b) Consider an ensemble of random graphs with N nodes, all having degree equal to m. Write the degree distribution and find for which values of m a.e.g. in the ensemble has a giant connected component in the limit $N \rightarrow \infty$. Write the expression of the clustering coefficient as a function of m.

5.5 Scale-Free Random Graphs

(a) Together with the natural cut-off, in a scale-free network with no multiple edges, there is also the so-called structural cut-off, scaling as $k_{\max} \sim \sqrt{N}$. Answer the same question as in Problem 5.2(b) assuming, this time, $k_{\max} = \min\left(N^{1/(\gamma-1)}, \sqrt{N}\right)$. Again, consider separately the three cases: $\gamma \in (1, 2]$, $\gamma \in (2, 3]$, and $\gamma > 3$.

(b) Does the network considered in Problem 5.5(a) have a giant component for $N \rightarrow \infty$, when $\gamma \in (2, 3]$?

5.6 Probability Generating Functions

(a) Consider the ensemble of ER random graphs $G_{N,p}^{\text{ER}}$. Use the generating function associated with the node degree distribution to show that the sum κ of the degrees of two nodes of a graph is distributed as:

$$p_\kappa = \binom{2N - 2}{\kappa} p^\kappa (1 - p)^{2N-2-\kappa}$$

where N is the number of nodes in the graph and p is the probability of connecting two nodes.

(b) Derive the expression in Eq. (5.75) for $\langle s \rangle$ above the transition starting by the definition: $\langle s \rangle = \sum_s' s w_s / \sum_s' w_s = H_w'(1)/H_w(1)$.

(c) Consider the same ensemble as in Problem 5.4(b). Using the generating function formalism, write down an expression for the average size of components $\langle s \rangle$ below the critical threshold, for the average size of finite components $\langle s \rangle$ above the critical threshold, and for the size of the giant component as a function of the value of m.

6

Models of Growing Graphs

Many of the networks around us continuously grow in time by the addition of new nodes and new links. One typical example is the network of the World Wide Web. As we saw in the previous chapter, millions of new websites have been created over recent years, and the number of hyperlinks among them has also increased enormously over time. *Networks of citations* among scientific papers is another interesting example of growing systems. The size of these networks constantly increases because of the publication of new papers, all arriving with new citations to previously published papers. All the models we have studied so far in the last three chapters deal, instead, with static graphs. For instance, in order to construct random graphs and small-world networks we have always fixed the number N of vertices and then we have randomly connected such vertices, or rewired the existing edges, without modifying N. In this chapter we show that it is possible to reproduce the final structure of a network by modelling its dynamical evolution, i.e. by modelling the continuous addition in time of nodes and links to a graph. In particular, we concentrate on the simplest growth mechanisms able to produce scale-free networks. Hence, we will discuss in detail the *Barabási–Albert model*, in which newly arrived nodes select and link existing nodes with a probability linearly proportional to their degree, the so-called *rich gets richer mechanism*, and various other extensions and modifications of this model. Finally, we will show that scale-free graphs can also be produced in a completely different way by means of growing models based on optimisation principles.

6.1 Citation Networks and the Linear Preferential Attachment

As authors of scientific publications we are all interested in our articles being cited in other papers' bibliographies. Citations are in fact an indication of the impact of a work on the research community and, in general, articles of high quality or broad interest are expected to receive many more citations than articles of low quality or limited interest. This is the reason why citation data are a useful source not only to identify influential publications, but also to find hot research topics, to discover new connections across different fields, and to rank authors and journals [87, 194].

Eugene Garfield was among the first to recognise the importance of the role of bibliographic information in understanding the propagation of scientific thinking. He founded in 1960 the *Institute for Scientific Information* (ISI) where he developed the so-called *Science Citation Index*, the first comprehensive citation index for papers published in academic journals [127]. Today, the Science Citation Index can be searched through subscription-based online platforms such as the *Web of Science* maintained by Thomson Reuters, while similar bibliographic databases, such as *Google Scholar*, are freely available. Thanks to scientific citation indexing services and to the large diffusion of online journals and of electronically available citation data, a new interdisciplinary research field advancing the scientific basis of science and innovation policy with a quantitative data-driven approach, the so-called *Science of Science*, has emerged [151, 106].

Citation networks will be the focus of this section. In a *citation network* each node is a published article, while the arcs represent citations between papers. Notice that, at variance with most of the networks previously considered in this book, in a citation network there is a natural way to order the nodes based on their time of arrival in the network, namely the time of publication of the corresponding articles. It is therefore possible to study how a citation network grows over time. The typical growth process is illustrated in Figure 6.1. The number of nodes increases over time by the addition of new articles. Nearly every new article contains a certain number of references to old ones. So, each new node arrives together with a set of arcs pointing to nodes already in the network. The number of references in the bibliography of a given paper, the out-degree of the corresponding node, is

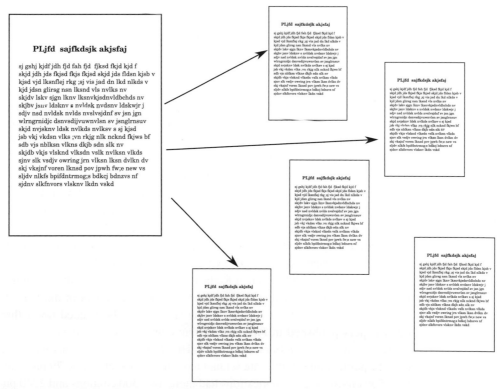

Fig. 6.1   Growth mechanism of citation networks. A new article is added to the network with a set of arcs pointing to some of the previously published articles, namely those in its bibliography.

usually small and depends on the limits imposed by the journal where the paper is published. Conversely, the number of references received by a paper, its in-degree, is the result of the decision of the authors of other papers and can be as large as the number of nodes in the network. Notice that arcs pointing from a node to a newer one are impossible. This is another distinctive feature of citation networks related to the fact that, once a paper is published, its bibliography cannot be updated any more. Hence, reciprocated links, directed triangles and higher-length directed cycles are not allowed. In other words a citation network is a typical example of a *directed acyclic graph* (DAG), i.e. it is a *forest* consisting of directed trees, and as such it needs to be treated [136, 81]. In principle there can be exceptions to this rule. In fact, since the publication of a paper is not an instantaneous process, it is possible that two or more papers that are published almost at the same time are cross-referenced. However, such exceptions are rare and can be neglected.

As an example of a small but complete citation network, we consider here a data set of citations constructed by Olle Persson from 27 years of articles published in *Scientometrics* [257]. Interestingly, *Scientometrics* is an international journal publishing articles precisely on bibliometrics and on the new Science of Science. To be consistent, only citations from articles published in *Scientometrics* to articles published in the same journal in the same period were considered. The network is illustrated in the DATA SET Box 6.1.

From citation plots like the one reported in Box 6.1 we can identify highly influential papers as those nodes corresponding to horizontal lines of dots. In fact, the number of citations of a paper, i.e. the in-degree of the corresponding node in the citation graph, is usually adopted as a natural and simple measure of the popularity and impact of a publication. The five nodes with the highest in-degree in the network correspond to papers number 412, 218, 466, 192 and 6 which have respectively acquired 71, 59, 39, 30 and 29 citations from other papers in the same data set. It is worth noticing that the total number of citations obtained by a paper is not always correlated with its age. For instance, paper number 6 and number 192 have roughly the same number of citations, despite paper number 6 having been published in 1979, six years before paper 192. However, both paper 6 and paper 192 are outperformed by paper 412, which appeared in 1989.

As an example of a larger system where it is possible to get a better statistics, we refer to an empirical study by Sidney Redner, a physicist from Boston University who analysed 110 years of citations for physics articles published in the journals of the American Physical Society (APS) [270]. The data set he considered included important physics journals such as *Physical Review Letters* and *Physical Review A, B, C, D* and *E,* and covered $N = 353,268$ papers and $K = 3,110,839$ citations from the first issues published in July 1893, through June 2003. When Redner first published his study, one thing that emerged from it certainly sounded a bit depressing to the researchers in the field. In fact, he discovered that about 70 per cent of all the articles had been cited fewer than 10 times by other papers published in *Physical Review* journals, and the median over the whole data set was of only six citations per article. Highly cited publications were a large minority: only 11 publications in the data set had more than 1000 citations, 79 publications had more than 500 citations, 237 publications had more than 300 citations and 2340 publications had more than 100 citations. Table 6.1 shows the ten most cited publications we have extracted from an updated and larger version of the same data set (made available by the APS) that

Box 6.1 | **DATA SET 6: The *Scientometrics* Citation Network**

The network provided in this data set has $N = 1655$ nodes, corresponding to all the articles published in the journal *Scientometrics* since 1978 and up to 2004, and $K = 4123$ arcs representing citations among them [257]. Since articles are sorted and labelled in chronological order, we can obtain a nice graphical representation of the adjacency matrix by drawing, as in the figure, the point of coordinate (i, j) in the plane, whenever article i cites article j.

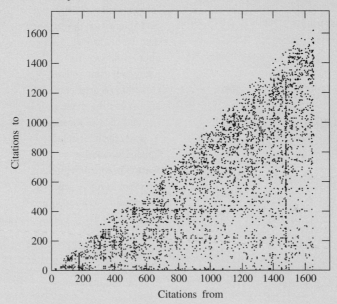

The plot has a typical triangular shape because only points with $i > j$ are possible. The first thing we notice is that the distribution of dots is denser in the region close to the diagonal and sparser towards the lower right-hand side of the plot, indicating that authors tend to give more attention to recent articles than to older ones. Notice also the presence of vertical and horizontal lines of dots. Vertical lines are the typical effect of review articles that cite a large number of earlier articles, and correspond to high-out-degree nodes. An example is article number 1480, which cites 156 of earlier *Scientometrics* papers. Conversely, highly cited articles corresponding to high-in-degree nodes appear as horizontal lines of dots, as in the case of article number 412 which has collected 71 citations since its publication in 1989. It is interesting to notice that only 1059 of the 1655 articles in the data set have cited at least one other paper in *Scientometrics*, thus having a degree larger than zero, and only 1053 of them have received at least one citation from another paper in the same data set.

includes $N = 450,084$ papers published in APS journals until 2009, and $K = 4,710,547$ citations. The paper with the highest number of citations is the condensed matter physics article on "Self-Consistent Equations Including Exchange and Correlation Effects" published by Kohn and Sham in 1965. This is quite a recent paper if you consider that the new data set covers the period 1893–2009, and has been cited 4763 times in 44 years only from articles published in APS journals, an average of more than 100 citations per year. Interestingly, with a total of 3227 citations, corresponding to about 85 citations per year, the same

Table 6.1 The top ten articles published in APS Journals, according to the total number of citations received up to 2009 from other articles in the same data set. For each paper we report title and author names.

Publication	Citations	Title	Authors
PR 140, A1133 (1965)	4763	Self-consistent equations including exchange and correlation effects	Kohn, Sham (KS)
PR 136, B864 (1964)	3712	Inhomogeneous electron gas	Hohenberg, Kohn
PRB 23, 5048 (1981)	3191	Self-interaction correction to density-functional approximations for many-electron systems	Perdew, Zunger
PRL 77, 3865 (1996)	3088	Generalized gradient approximation made simple	Perdew, Burke, Ernzerhof (PBE)
PRL 45, 566 (1980)	2651	Ground state of the electron gas by a stochastic method	Ceperley, Alder (CA)
PRB 13, 5188 (1976)	2569	Special points for Brillouin-zone integrations	Monkhorst, Pack
PRB 54, 11169 (1996)	2387	Efficient iterative schemes for ab initio total-energy calculations using a plane-wave basis set	Kresse, Furthmüller,
PRB 41, 7892 (1990)	1951	Soft self-consistent pseudopotentials in a generalized eigenvalue formalism	Vanderbilt
PRB 43, 1993 (1991)	1950	Efficient pseudopotentials for plane-wave calculations	Troullier, Martins
PR 108, 1175 (1957)	1725	Theory of superconductivity	Bardeen, Cooper, Schrieffer (BCS)

paper was also the top one in the data set analysed by Redner (see Table 1 in Ref. [270]). This means that the average number of citations per year received by this article has further increased in the last six years.

Thanks to the long time span covered by the APS data set, Redner was also able to look at how highly cited articles acquired their citations over time, i.e. how the in-degrees of corresponding nodes in the citation graph changed over time. He found a rich variety of different citation histories. While the typical trend consists as expected in a peak in the number of new citations in the few years after the publication of a paper followed then by a decay, there are also papers that steadily accumulate citations for long time periods. In Example 6.1 we show and discuss the case of a few representative articles.

Example 6.1 *(Popularity over time of physics articles)* As typical examples of the citation histories of highly cited articles, the figure below reports the number of citations received each year by six articles published in the APS journals. The large majority of articles in the data set have citation histories similar to those shown in panel (a) and panel

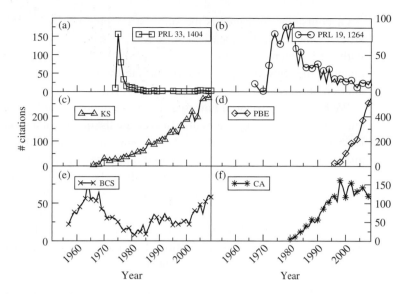

(b), corresponding respectively to "Experimental Observation of a Heavy Particle J" by Aubert et al., and to "A Model of Leptons" by Weinberg. A typical paper receives in fact the majority of its citations in the first few years after its publication, and then the number of new citations per year starts decreasing in time either exponentially as in panel (a), or as a power law as in panel (b). Notwithstanding this decay, the two papers reported have acquired respectively 361 and 1397 citations up to 2009, and they are ranked 430th and 18th in the data set. Climbing the citation ladder up to the top is a completely different story, as evident by the unusual citation timeline of the 1965 paper by Kohn and Sham (KS) in panel (c). Instead of decreasing, the number of citations acquired by this paper has steadily increased over time from about 50 citations per year in the 80s to more than 200 new citations per year in the last decade, an impressive indication of the long-term impact of this work. Some other papers are just an immediate success, like the 1996 *Physical Review Letter* by Perdew, Burke and Ernserhof reported in panel (d), which has made it to the fourth place in the citation ranking in less than 13 years and is currently acquiring more than 600 citations per year only from other papers in APS, making it a likely candidate to become the most cited article in the data set in the next few years.

In some specific cases the citation pattern of a paper is bound to the popularity of a research topic. This happens e.g. to the famous "Theory of Superconductivity" by Bardeen, Cooper and Schrieffer (BCS), whose citation timeline reported in panel (e) closely follows the research activity in superconductivity. The paper was published in 1957, received a very large number of citations in the period 1960–1970, had a minimum of new citations in 1985, the year before the discovery of high-temperature superconductivity, and is now acquiring more and more citations again. There are also cases, such as the one of the 1980 paper by Ceperley and Alder shown in panel (f), where the popularity of a paper increases for a couple of decades and then remain pretty stable for a relatively long period.

For other examples of the rich variety of citation histories and the definition of typical classes of behaviours such as "hot papers", "major discoveries" and "sleeping beauties" (i.e. the revival of old classics), we suggest you to read the beautiful paper by Redner [270].

Of course the in-degree is the simplest possible way to quantify the popularity of an article. Other centrality measures such as those discussed in Chapter 2 [74], or the overlap with the bibliographies of other articles [78] can be used in alternative to the in-degree to quantify the relevance of a scientific article by looking at its role in the structure of the citation network.

We have talked so far of how to measure the importance of an article. Nowadays, article citations are increasingly used also to evaluate the quality of a scholar, or to rank scientific journals. For instance, it is always more common in decisions for appointment, promotion and funding to evaluate a scientist based on his/her citation record. Therefore, in modern science a successful scholar needs not only to be active in publishing, according to the famous motto "publish or perish", but also to produce highly cited papers. The quality of a scientist can be measured by various indicators such as the number of papers published per year, the average number of citations per paper, or, even better, the *h-index* illustrated in Box 6.2, a recently proposed quantity which takes into account both the number of papers and the number of citations of a scientist.

Box 6.2   **Author's Ratings: The h-Index**

The *h*-index was introduced in 2005 by Jorge Eduardo Hirsch, a physicist from UC San Diego, as a tool for determining the quality of theoretical physicists, and has become a widely adopted bibliometric parameter in all research fields [154, 155]. The *h*-index is defined as follows: a scientist with N_p publications has index equal to h if h of the N_p papers have at least h citations each, and the other $(N_p - h)$ papers have $\leq h$ citations each. In practice, the *h*-index of a given author can be determined by searching the *Web of Science* or *Google Scholar* for all articles by the author, and by ranking them by the number of citations, so that article 1 has the most citations. The value of h is the rank of the lowest ranking article whose number of citations is greater or equal than its rank.

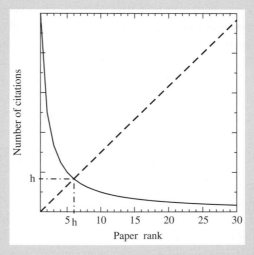

A plot of the number of citations versus paper rank as the one reported in the figure can help understanding the meaning of the *h*-index. The total number of citations is the area under the curve. The intersection of the curve with the $45°$ line determines the value of h.

The *h*-index has become popular because it is simple to compute and also because it has been proven that large *h* and, even more importantly, large variations of *h* over time indicate successful scientists. For instance the *h*-index is highly predictive of whether a scientist has won honours such the Nobel Prize or National Academy memberships [154, 155]. However the index depends on the "academic age" of a scientist, and must be used only to compare scholars working in the same years and in the same fields since citation conventions can show large variations from one field to another [265, 264]. If you are interested to know more about author's rating, for a comparative testing of different measures we suggest you read Ref. [205].

Finally, citation records can also be used to produce a rating of the different scientific journals. The best known index of a journal status is the so-called *Impact Factor* (IF) which reflects the average number of citations to recent articles published in that journal. Details can be found in Box 6.3.

As for the case of the *h*-index, also the validity of the IF has been recently questioned because the dynamics of publications and citations can vary a lot across different fields. Let us consider, for instance, the case of multidisciplinary physics, one of the sub-field of physics defined by ISI. The three highest Impact Factor periodicals in this category in 2013 were *Reviews of Modern Physics* (IF=42.9), *Physics Reports* (IF=22.9) and *Nature Physics* (IF=20.6), while *Physical Review Letters*, one of the historically most respected journals in physics, ranked only sixth and was outperformed by *Reports on Progress in Physics* (IF=15.6) and *Physical Review X* (IF=8.5), a new online multidisciplinary journal launched by APS only in 2011. One of the main problems here is that many of the top IF journals in each subfield specialise in review articles, which are typically highly cited, and hence have anomalously high values of IF, especially when compared with those of original research journals [218]. For this reason, alternative measures making use of the PageRank algorithm [15], or extending the *h*-index to journal rating have been proposed.

After this brief digression on how to quantify the relevance of a publication, the influence of a scientist, or the impact of a journal, we can finally come back to our main interest, that of understanding the basic rules governing the growth of citation networks.

The first important thing to notice is that citations produce networks with scale-free degree distributions. As an example, we report in panel (a) and (c) of Figure 6.2 the in-degree distributions of papers respectively published in *Scientometrics* and in the APS journals. In both cases, the distributions have broad tails that are reasonably well

---

Box 6.3	Journal's Ratings: The Impact Factor

The Impact Factor (IF) is a measure of the popularity of a journal introduced by the Institute of Scientific Information (ISI), the company founded by Eugene Garfield and now part of Thomson Reuters. The IF of a journal is defined as the ratio between the number of citations acquired in a certain year by all the articles published by that journal in the two preceding years and the number of articles published in the same time interval. For instance, the 2013 IF of the journal *Physical Review Letters* is $54382/7037 \simeq 7.73$, where 7037 is the number of papers published in *Physical Review Letters* in 2011 and 2012, while 54382 is the number of citations collectively accumulated by those papers in 2013.

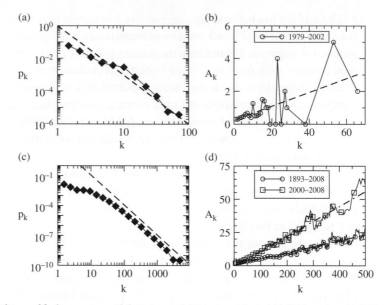

Fig. 6.2 The probability p_k of finding a paper with k citations in (a) *Scientometrics* and (c) APS has a broad tail (the dashed line in each plot corresponds to $p_k \sim k^{-3}$). The annual attachment rate A_k for (b) year 2003 in *Scientometrics* and (d) year 2009 in APS (for papers having accumulated citations respectively in the period 1893–2008 and 2000–2008), increases linearly with k in both cases.

approximated by a power law $p_{k_{in}} \approx k_{in}^{-\gamma_{in}}$. Historically, Derek de Solla Price was the first author to point out, in a work published in 1965, that the in-degree of citation networks are distributed as power laws. And he extracted a value of $\gamma_{in} \approx 2.5 - 3.0$ for the in-degree exponent [289]. De Solla Price's paper has been a source of inspiration for many other studies on citation networks and can be considered the earliest published example in the literature of a scale-free network.

Thanks to the temporal information present in the two data sets under study, we have another important lesson to learn on how citation networks grow over time. In fact, despite the huge diversity in the citation histories of individual papers, we shall see that on average the nodes of the network accumulate new links in proportion to the number of links they already have. This *rich gets richer* mechanism, known today as *linear preferential attachment* will turn out to be a fundamental ingredient in the growth of many real-world systems, not only of citation networks. Let us concentrate on how to extract this information from our data sets. The growth of a network can be characterised by an attachment rate A_k denoting in our case the likelihood that a paper with k citations will be cited by a new article [270, 225]. The attachment rate can be computed in the following way. We first count the number of citations each paper has received up to a certain time t. This allows us to define the in-degree of each paper, and also to compute N_k, the number of papers being cited exactly k times up to t. Then, we count the total number of citations M_k^{cit} accumulated in a time window starting at t to articles having accumulated k links up to time t. Finally, to get A_k, we divide this number by N_k: $A_k = M_k^{cit}/N_k$. Summing up, A_k gives the average number of times a paper with a given k at time t has been cited in the subsequent window. If the duration of the window is one year, then A_k represents the average number of citations a paper with k citations receives in one year. If there is no preferential attachment, A_k will be roughly the same for all values of k. Conversely, the presence of a preferential attachment mechanism will be signalled by the increase of A_k as a function of k. We have computed A_k both for *Scientometrics* and APS citation networks. In the first case, we have considered all the papers published before 2002, and we have evaluated their in-degree k by counting the citations they have received in the period 1979–2002. Finally, we have computed A_k from the citations arriving in a window of one year, namely year 2003. In the case of APS we have constructed the in-degree in two different ways, namely considering citations accumulated respectively in the period 1893–2008 and in the period 2000–2008. In both cases, we have then computed A_k from the citations arriving in year 2009. Results reported in Figure 6.2 (b) and (d) suggest that both in *Scientometrics* and in APS A_k increases with k. In particular, the computation of A_k as a function of k for the larger data set of the *Physical Review* journals reported in Figure 6.2 (d) indicates that A_k is an almost linear function of k. This result suggests that *linear preferential attachment* is the mechanism at work in the growth of citation networks. In practice, a new paper cites old papers with a probability linearly proportional to the number of citations they already have. This is intuitively easy to justify, since the probability of coming across a particular paper while reading the literature is expected somehow to increase with the number of other papers that cite it. However, what is striking is that the dependence of the citation probability on previous citations found and reported in Figure 6.2 is strictly linear. Interestingly enough, the same linear preferential attachment mechanism, with the nodes in the network receiving links from new nodes with a probability linearly proportional to their degree, has also been found in other growing networks such as the World Wide Web.

6.2 The Barabási–Albert (BA) Model

We now turn our attention on how to model the growth of a graph in order to produce scale-free degree distributions. The first model we study was proposed in a very influential paper published in 1999 in the journal *Science* by Albert László Barabási and Réka Albert, two physicists then at Notre Dame University [19]. The *Barabási–Albert (BA) model* mimics the mechanisms observed in the growth of real systems such as the WWW and citation networks. In particular, it assumes that the number of links increases proportionally to the number of nodes, and that the existing nodes of the network receive links from new nodes with a probability linearly proportional to their degree, the so-called *rich gets richer mechanism*. Although the model provides an extremely simplified description of growing networks, as we shall see, it gives rise to scale-free networks with power-law degree distributions with exponent $\gamma = 3$. This is close to the value of γ_{in} found in networks of citations, while other real growing networks, such as the World Wide Web, have distributions characterised by different exponents (see Table 5.1). Nevertheless, the BA model clearly identifies the minimal ingredients in the growth of a scale-free graph, which is the reason for its appeal and success. In the following, we study the properties of the BA

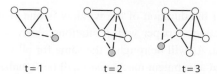

Fig. 6.3   Illustration of the BA growth process for $n_0 = 3$ and $m = 2$. At $t = 0$ we start with a complete graph of three nodes. At each time step a new node n is added, which is connected to two vertices, preferentially to the vertices with high connectivity, according to Eq. (6.1). At each time step, the new node is shaded, and the two new edges are drawn with dashed lines.

model, both analytically and numerically ([19, 20]). A series of more elaborated models originated from the BA model will then be discussed in Sections 6.4 and 6.5.

Definition 6.1 (The BA model)   *Given three positive integer numbers N, n_0 and m (with $m \leq n_0 \ll N$), let n_t and l_t denote the number of nodes and links of the graph produced at time t. At time $t = 0$, one starts with a complete graph with n_0 nodes, labelled as $1, 2, \ldots, n_0$, and $l_0 = \binom{n_0}{2}$ edges. The graph grows, as illustrated in Figure 6.3, by iteratively repeating at time $t = 1, 2, 3, \ldots, N - n_0$ the following two steps:*

1 A new node, labelled by the index n, being $n = n_0 + t$, is added to the graph. The node arrives together with m edges.

2 The m edges link the new node to m different nodes already present in the system.

The probability $\Pi_{n \to i}$ that a new link connects the new node n to node i (with $i = 1, 2, \cdots, n - 1$) at time t is:

$$\Pi_{n \to i} = \frac{k_{i,t-1}}{\sum_{l=1}^{n-1} k_{l,t-1}} = \frac{k_{i,t-1}}{2l_{t-1}} \quad (6.1)$$

where $k_{i,t}$ is the degree of node i at time t.

In the model, the *attachment probability* $\Pi_{n \to i}$ is normalised so that $\sum_{i=1}^{n-1} \Pi_{n \to i} = 1$. Notice that $\Pi_{n \to i}$ in Eq. (6.1) depends only on the degree k_i of node i, and on the total number of links l_{t-1}, i.e. $\Pi_{n \to i} = \Pi(k_{i,t-1}, l_{t-1})$. More precisely it is *linearly* proportional to the degree of node i. This mimics the linear preferential attachment observed in real systems, as for instance in the scientific citation networks. From now on we imply the dependence on time and on the total number of links, and we simply write $\Pi_{n \to i} = \Pi(k_i)$, where

$$\Pi(k_i) = \frac{k_i}{\sum_l k_l}.$$

Notice that, since the probability $\Pi_{n \to i}$ depends only on k_i, the normalisation denominator $\sum_l k_l$ can be written as $\sum_k kN_k$. The expression of $\Pi(k)$ is therefore

$$\Pi(k) = \frac{k}{\sum_k kN_k},$$

where N_k denotes the number of nodes with degree k, with the obvious normalisation $\sum_k N_k \Pi(k) = 1$.

Example 6.2   In Figure 6.2 we have extracted the attachment rate A_k, defined as the average number of times a specific paper with k citations has been cited in the last year, and we have found that A_k is a linear function of k. We now show that the linear preferential attachment of the BA model contained in Eq. (6.1) is indeed consistent with this empirical finding. We can in fact evaluate the attachment rate A_k in the BA model as:

$$A_k = N_{\text{year}} m \Pi(k)$$

where N_{year} is the number of new nodes in the last year, $N_{\text{year}} m$ is the total number of links arriving in the last year, and $\Pi(k) = k / \sum N_k k$ is the probability that each new link attaches to a node of degree k. Since $N_{\text{year}} m$ does not depend on k, a linear dependence of $\Pi(k)$ on k, as that used in Eq. (6.1), implies that also A_k is a linear function of k.

At a given time step t, the growth process of the BA model produces a graph with:

$$n_t = n_0 + t \quad \text{nodes}$$
$$l_t = \binom{n_0}{2} + mt \quad \text{edges.} \tag{6.2}$$

The procedure is iterated up to time $N - n_0$, generating an undirected graph with N nodes and $K = m(N - n_0) + n_0(n_0 - 1)/2$ edges. For large N, this corresponds to a graph with an average degree $\langle k \rangle = 2m$. We will now show, through a series of computer simulations, that the graphs produced have a power-law degree distribution. See Appendix A.13.1 for a description of the algorithms to generate graphs according to the BA model.

In Figure 6.4 we report the results of simulations obtained for different values of N and m. The plot indicates that the growth process evolves into a scale-invariant state in which the probability that a node has k edges follows a power-law $p_k \sim k^{-\gamma}$, with an exponent $\gamma = 3$. Panel (a) shows that the degree distribution is stable already for $N = 10,000$ and does not change sensibly when N increases. Moreover, panel (b) shows that the exponent γ does not depend on the number m of links created by each new node. Notice also that, although we have started the growth with a complete graph of n_0 nodes, the properties of the obtained graphs for large N do not depend on n_0 and on the structure of the initial graph.

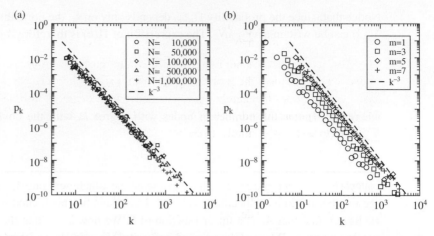

Fig. 6.4   (a) As N increases, the degree distribution of the BA model quickly converges to a power law $p_k \sim k^{-\gamma}$ with $\gamma = 3$. (b) The exponent γ is independent from the number m of edges created when a new node joins the network.

All these results can be explained analytically by writing down and solving the *rate equations*[1] for the average number of nodes with a certain degree. To be more specific, let us indicate as $\bar{n}_{k,t}$ the average number of nodes with degree k ($k \geq m$) present in the graph at time t. Here the average is performed over infinite realisations of the growth process with the same parameters N, n_0 and m. The quantities $\bar{n}_{k,t}$ are real numbers which satisfy the normalisation condition $\sum_k \bar{n}_{k,t} = n_t$ at each time t. Dealing with the time evolution of average quantities has several advantages in the mathematical treatment. In fact, in a single growth process, the number of nodes with k links, $n_{k,t}$, is an integer number. And this number increases or decreases by $0, 1, 2, \ldots$, or m at each time step, each increment or decrement happening with a given probability. The simulation of a single growth can therefore be performed by random sampling, and macroscopic information can be obtained by averaging over several realisations of the process. Alternatively, we can directly obtain information on the averages by writing a set of deterministic equations for the time evolution of the average quantities $\bar{n}_{k,t}$. In fact, when a new node enters the network at time t, the change in the average number of nodes of degree k can be written as:

$$\bar{n}_{k,t} = \bar{n}_{k,t-1} - \text{LOSS} + \text{GAIN} \qquad k \geq m$$

where loss and gain terms are real numbers representing the average changes of $\bar{n}_{k,t-1}$, $k \geq m$, due to the introduction of the new node and of the m new links. Notice that, while $\bar{n}_{k,t}$ are real numbers, both time and node degrees are treated as integer numbers in the equations.

The key point to evaluate loss and gain terms is to notice that, in one realisation of the process, the probability $T_{k,t}$ that a new edge arriving at time t, is linked to *any* node of degree k, is equal to:

[1]   The rate equations are a simple yet powerful method to analyse growing networks. In addition to providing information on the degree distribution, the rate equations can be easily adapted to study correlations between node degrees, and to analyse directed networks. For a complete review on rate equations in growing networks see Ref. [185].

$$T_{k,t} = \sum_{i:k_{i,t-1}=k} \Pi_{n \to i} = \sum_{i:k_{i,t-1}=k} \frac{k_{i,t-1}}{2l_{t-1}} = \frac{kn_{k,t-1}}{2l_{t-1}} = \frac{kn_{k,t-1}}{\sum_k kn_{k,t-1}} \tag{6.3}$$

where $n_{k,t-1}$ denotes the number of nodes with degree k at time $t-1$ in the particular realisation. In practice, the probability $T_{k,t}$ of linking at time t to any node of degree k is equal to $\Pi(k)$ times the number $n_{k,t-1}$ of nodes with degree k at time $t-1$. Notice that T_k is perfectly normalised, in fact $\sum_k T_{k,t} = 1 \ \forall t$. The average $\overline{T}_{k,t}$ over the ensemble of different realisations is simply given by:

$$\overline{T}_{k,t} = \frac{k\overline{n}_{k,t-1}}{2l_{t-1}} \tag{6.4}$$

because the number of links l_{t-1} is equal for all the realisations. We can therefore rewrite l_{t-1} as a sum over node degrees, $2l_{t-1} = \sum_k k\overline{n}_{k,t-1}$, and we finally have:

$$\overline{T}_{k,t} = \frac{k\overline{n}_{k,t-1}}{\sum_k k\overline{n}_{k,t-1}} \tag{6.5}$$

which explicitly shows that $\overline{T}_{k,t}$ is also normalised so that $\sum_k \overline{T}_{k,t} = 1 \ \forall t$. Now, the loss term describes the new edges connecting to nodes with k edges and turning them into nodes with $k+1$ edges, consequently decreasing the number of nodes with k edges. Therefore it can be written as:

$$LOSS = m\overline{T}_{k,t} = m \frac{k\overline{n}_{k,t-1}}{\sum_k k\overline{n}_{k,t-1}} \qquad k \geq m$$

where the factor m comes into play since we add m new links at each time step.

The gain term is made by two contributions, namely:

$$GAIN = m\overline{T}_{k-1}(t) + \delta_{k,m} = m \frac{(k-1)\overline{n}_{k-1,t-1}}{\sum_k k\overline{n}_{k,t-1}} + \delta_{k,m} \qquad k \geq m.$$

The first contribution accounts for the new edges that connect to nodes with $k-1$ edges, thus increasing their degree to k, and is different from zero only for $k > m$. The second contribution, $\delta_{k,m}$ describes the addition, at each time step, of the new node with m edges. Finally, the set of rate equations for the model is:

$$\overline{n}_{k,t} - \overline{n}_{k,t-1} = m \frac{(k-1)\overline{n}_{k-1,t-1} - k\overline{n}_{k,t-1}}{2l_{t-1}} + \delta_{k,m} \quad k \geq m \tag{6.6}$$

where we have made use of the relation $\sum_k k\overline{n}_{k,t-1} = 2l_{t-1}$. Since we are interested in the degree distribution produced for large times, we express the equations in terms of the degree distributions $p_{k,t} = \overline{n}_{k,t}/n_t$ and $p_{k,t-1} = \overline{n}_{k,t-1}/n_{t-1}$. Observing that $n_t = n_{t-1} + 1$, we obtain:

$$n_{t-1}(p_{k,t} - p_{k,t-1}) = -p_{k,t}$$
$$+ \frac{mn_{t-1}}{l_{t-1}} \left(\frac{k-1}{2} p_{k-1,t-1} - \frac{k}{2} p_{k,t-1} \right) + \delta_{k,m} \quad k \geq m. \tag{6.7}$$

Let us assume that $p_{k,t}$ tends to a stationary solution p_k, as $t \to \infty$, in agreement with the numerical simulation in Figure 6.4. In this case the right-hand side will converge to a quantity Q_k independent on t. The problem is that the left-hand side of Eq. (6.7) is an indeterminate form of $\infty \cdot 0$ type. Now we prove that, as $t \to \infty$, the limit of the left-hand side is indeed zero, and therefore we can set $Q_k = 0$. For large time t, by observing that $n_t \approx t$ and $l_t \approx mt$, we obtain:

$$(t-1)(p_{k,t} - p_{k,t-1}) = Q_k \tag{6.8}$$

with

$$Q_k \equiv -p_k + \left(\frac{k-1}{2}p_{k-1} - \frac{k}{2}p_k \right) + \delta_{k,m}$$

This means that, for large time, the asymptotic behaviour of the probability $p_{k,t}$ satisfies the following recurrence relation (in time):

$$p_{k,t} = p_{k,t-1} + \frac{Q_k}{t-1}.$$

This relation leads to a divergence in time, because the right-hand side gives rise to a harmonic series, unless $Q_k \equiv 0$. This implies that the asymptotic distribution p_k satisfies the recurrence relation in k:

$$p_k = \frac{k-1}{2}p_{k-1} - \frac{k}{2}p_k + \delta_{k,m} \qquad k \geq m. \tag{6.9}$$

Example 6.3   By definition, we expect that any degree distribution p_k satisfies the normalisation condition $\sum_k p_k = 1$. We can therefore check if the quantities p_k, with $k \geq m$, in Eqs. (6.9) are properly normalised. By summing p_k over k we get:

$$\sum_{k \geq m} p_k = \sum_{k \geq m} \frac{k-1}{2}p_{k-1} - \sum_{k \geq m} \frac{k}{2}p_k + \sum_{k \geq m} \delta_{k,m}.$$

Now, the first two terms on the right-hand side cancel out, while $\sum_k \delta_{k,m} = 1$. We therefore find the correct normalisation: $\sum_{k \geq m} p_k = 1$.

In order to get p_k we need to solve the following iterative equations:

$$p_k = \begin{cases} \frac{k-1}{k+2}p_{k-1} & \text{for} \quad k > m \\ \frac{2}{m+2} & \text{for} \quad k = m. \end{cases}$$

We get:

$$p_k = \frac{k-1}{k+2}\frac{k-2}{k+1}\frac{k-3}{k}\frac{k-4}{k-1}\frac{k-5}{k-2} \cdots \frac{m+3}{m+6}\frac{m+2}{m+5}\frac{m+1}{m+4}\frac{m}{m+3}p_m$$
$$= \frac{(m+2)(m+1)m}{k(k+1)(k+2)}p_m$$

which finally gives:

$$p_k = \frac{2m(m+1)}{k(k+1)(k+2)} \simeq 2m(m+1)k^{-3}. \tag{6.10}$$

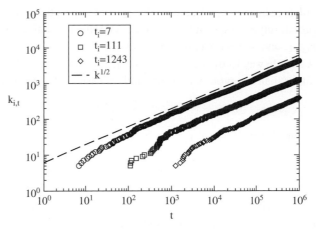

Fig. 6.5 The time evolution of the degree $k_{i,t}$ of a generic node i in the BA model is described by Eq. (6.11), which predicts a power-law increase $k_{i,t} \sim (t/t_i)^{1/2}$ (dashed line), where t_i is the time at which node i joined the network. The plots of $k_{i,t}$ for three different values of i in a typical realisation of the BA model with $N = 10^6$ are in agreement with this scaling.

Expression (6.10) proves that the BA model produces indeed a power-law degree distribution. Moreover, the exponent γ of the power law is independent of the value of m, and is equal to 3, in agreement with what found numerically in Figure 6.4.

The rate equation approach is a powerful and exact method to derive the asymptotic degree distributions in the BA model. However, this technique is not of much use if we want to study how the degree of different nodes increases over time. In Figure 6.5 we consider the degree of two nodes added to the graph at very different times, respectively $t_1 = 5$ and $t_2 = 95$. The plot shows that both their degrees increase in time as power laws. The two straight lines in a double logarithmic scale have the same slope, indicating that the exponent of the power law does not depend on the node considered.

It is in fact possible to prove that the following rule:

$$k_{i,t} = m \left(\frac{t}{t_i} \right)^{1/2}, \qquad t \geq t_i \qquad (6.11)$$

holds for all the nodes in the graph, where t_i is the time at which node i is introduced. This implies that in the BA model older nodes always have higher degree.

We derive Eq. (6.11) in the following example, by making use of the so-called *mean-field* approximation.

Example 6.4 *(Time evolution of node degrees in the mean-field approximation)*   Equation (6.11) can be easily obtained by the following approximate argument [19, 20]. Old nodes increase their degree by acquiring new links at each time step from the newly introduced node n. In practice, this means that, at each time step, the degree k_i of a given node i

either increases by a quantity $\Delta k_i = 1$ or stays constant. The first process happens with a probability $m\Pi(k_i)$, and the second one with a probability $1 - m\Pi(k_i)$. Therefore, we have $k_{i,t+1} = k_{i,t}+1$ with a probability $m\Pi(k_{i,t})$, while $k_{i,t+1} = k_{i,t}$ with probability $1-m\Pi(k_{i,t})$. A simple way to treat this process is to consider the evolution of the ensemble average $\bar{k}_{i,t}$, rather than the evolution of the random variable $k_{i,t}$ itself. This satisfies the equation:

$$\bar{k}_{i,t+1} = \bar{k}_{i,t} + m\Pi(\bar{k}_{i,t}).$$

By approximating $\bar{k}_{i,t+1} - \bar{k}_{i,t} \approx d\bar{k}_i(t)/dt$ we get:

$$\frac{d\bar{k}_i}{dt} = m\Pi(\bar{k}_i) = m\frac{\bar{k}_i}{\sum_j \bar{k}_j}. \tag{6.12}$$

This is known as *mean-field* approximation. The sum in the denominator goes over all nodes in the system except the newly introduced one, thus its value for large times is $\sum_j \bar{k}_j \approx 2mt$, leading to:

$$\frac{d\bar{k}_i}{dt} = \frac{\bar{k}_i}{2t}. \tag{6.13}$$

This differential equation can be easily solved with the initial condition that node i has been introduced at time $t = t_i$, such that $\bar{k}_i(t = t_i) = m$. We get:

$$\int_m^{\bar{k}_i(t)} \frac{dk}{k} = \frac{1}{2} \int_{t_i}^t \frac{dt'}{t'}$$

whose solution $\ln(\bar{k}_i(t)/m) = 1/2 \ln(t/t_i)$ gives the result reported in Eq. (6.11). Strictly speaking, Eq. (6.11) describes the evolution of the ensemble average $\bar{k}_{i,t}$. However, the relative deviations of $k_{i,t}$ from its average $\bar{k}_{i,t}$ become smaller and smaller as time increases. Notice that the mean-field approximation can also be used to derive the degree distribution of the BA model (see Problem 6.2).

Equation (6.11) points out that the degree of all nodes evolves in the same way, namely as a power law of time with an exponent 0.5. The prediction of the mean field is then in good agreement with the numerical results reported in Figure 6.5, and this is a confirmation that the mean-field approximation is indeed working well. This result also shows that the only difference between two nodes is the intercept of the power law, that is the time t_i at which the node has been added to the graph. In practice, in the BA model, two nodes of the final network have a different degree only because they were introduced at two different times, with older nodes being on average those with the largest degrees. This can be unrealistic for some systems, and in Section 6.5 we will see how to modify the BA model in such a way that new nodes can also overcome older ones.

We conclude this section by focusing on characteristic path length and clustering coefficient in the BA model. Concerning the characteristic path length, it can be proven that the networks produced by the model are indeed small worlds: they have an average node distance L which scales logarithmically with N [3, 50]. This is the same scaling exhibited by ER random graphs, although the values of L found in the BA model are always smaller than those of ER random graphs with same N and K. Concerning the clustering coefficient, it has been proven numerically that it vanishes with the system order as $C \sim N^{-0.75}$. Such a scaling is slower than the $C \sim N^{-1}$ holding for ER random graphs, but it is still different from the behaviour observed in real small-world networks and in the small-world models considered in Chapter 4, where instead C is a constant independent of N. To solve this problem, Petter Holme and Beom Jun Kim have proposed in 2002 a modification of the BA model that allows scale-free graphs with a finite clustering coefficient to be constructed. The modification consists in the addition of a mechanism of triangle formation to the standard linear preferential attachment of the BA model. In the *Holme–Kim (HK) model* the graph grows as follows [156].

Definition 6.2 (The HK model)   *Given three positive integer numbers: N, n_0 and m (with $m \leq n_0 \ll N$), and a real number $0 \leq q \leq 1$, the graph grows, starting with a complete graph with n_0 nodes, and iteratively repeating at time $t = 1, 2, 3, \ldots, N - n_0$ the following three steps:*

1 *A new node, labelled by the index n, being $n = n_0 + t$, is added to the graph. The node arrives together with m edges.*
2 *The first new edge links the new node to a node already present in the system. The node i (with $i = 1, 2, \cdots, n - 1$) is chosen with the preferential attachment probability $\Pi_{n \to i}$ of Equation 6.1.*
3 *For each of the remaining $m - 1$ edges, with a probability q create a triangle, i.e. add the edge from n to a randomly chosen neighbour of i^*, where i^* is the node linked to n in the preferential attachment step 2 (if all neighbours of i^* are already connected to n, perform a new preferential attachment instead). Otherwise, with a probability $1 - q$, perform another preferential attachment.*

The model has one control parameter more than the BA model, namely the probability q of creating triangles. Since the average number of trials to form triangles per added node is equal to $(m - 1)q$, the value of the clustering coefficient in the HK model can be tuned by changing q. When $q = 0$, the model reduces to the BA model, while for $q \neq 0$ it is possible to show that it produces graphs with a finite clustering coefficient C and, at the same time, with a power-law degree distribution $p_k \sim k^{-3}$.

We have performed a series of numerical simulations of the model and constructed graphs of different order N for $m = 3$ and various values of q, namely $q = 0, 0.2, 0.4, 0.6, 0.8$ and 1. Results are reported in Figure 6.6. Panel (a) shows that, for each q, the clustering coefficient quickly converges to a constant value as the order N of the graph increases. Values of C which do not depend on N are already obtained at $N \sim 5000$ for each value of q considered. Panel (b) indicates that the clustering coefficient of the networks produced by the model increases monotonically with the parameter q, up to $C \approx 0.6$. It is interesting to notice that the probability q of creating a triangle does not affect the shape of the degree distribution of the network. In fact, as shown in panel (c), the degree distributions remain, for any value of q, indistinguishable from the power law with exponent $\gamma = 3$ generated by the linear preferential attachment in the BA model. Finally, the model also reproduces another interesting feature of the clustering coefficient of real-world networks that we have explored in Example 4.4 of Chapter 4, namely the dependence of the

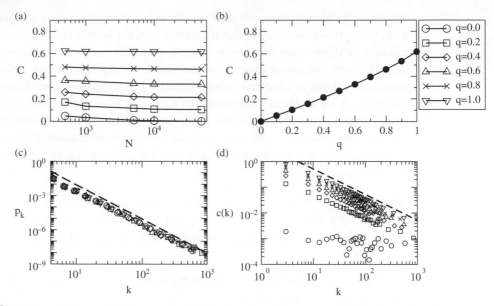

Fig. 6.6 The HK model allows scale-free networks with finite clustering to be constructed. (a) The clustering coefficient is reported as a function of the order N of the graph and (b) for different values of q. (c) Degree distribution and (d) average clustering coefficient of nodes of degree k for networks with $N = 50,000$ nodes.

node clustering coefficient C on the node degree k. For each value of q we have computed the average clustering coefficient $c(k)$ of nodes with degree k as defined in Eq. (4.5). The quantity $c(k)$ is reported as a function of k in panel (d). The results indicate that, for every value $q \neq 0$, the node clustering coefficient $c(k)$ has a power-law dependence on k, which converges to $c(k) \sim k^{-1}$ when q increases, thus reproducing the same exponent found for the movie actor collaboration network in Example 4.4. Conversely, when $q = 0$ we find that the value of the node clustering coefficient is very small and does not depend on the node degree k. In this case we expect $c(k) \to 0$ for each value of k when $N \to \infty$. This is perfectly consistent with the behaviour we would observe in a random graph. In fact, the clustering coefficient of a node i is equal to the density of the subgraph induced by the first neighbours of i. Since in a random graph each pair of nodes has the same probability of being connected by a link, we expect that the density of the subgraph induced by the first neighbours of a node does not depend on the node degree. Therefore in finite random graphs the clustering coefficient will be small, but different from zero, and such that the average clustering coefficient $c(k)$ of nodes of degree k will not depend on the value of k.

6.3 The Importance of Being Preferential and Linear

The BA model has many analogies with a model of citation networks published more than twenty years before [263]. The author of this model is the same Derek de Solla Price of the empirical studies on citation networks discussed in Section 6.1. In order to reproduce the power laws observed in the degree distribution of citation networks, de Solla Price refor-mulated, in terms of graph growth, the so-called *Yule process*, a well-known mechanism to explain the occurrence of power laws in a variety of contexts having nothing to do with networks and graphs. The process was originally introduced by George Udny Yule in the context of taxonomy to explain why the distribution of the number of biological species in a genus is a power law [319], and then applied by Herbert Simon to the distribution of words in prose samples by their frequency of occurrence, and to the distributions of cities by population [285]. The Yule process, is indeed a *rich gets richer* mechanism in which the most populous genera get more species in proportion to the number they already have (see Problem 6.3a). Yule and later Simon showed mathematically that this mechanism produces what is now called the Yule distribution, which is a distribution following a power law in its tail. What de Solla Price did was to propose a model for the growth of citation networks based on the assumption that the rate at which a paper gets new citations is proportional to the number of citations it already has.

Definition 6.3 (The de Solla Price model)   *Let N, n_0 be two positive integer numbers, and m be a real positive number (with $m < n_0 \ll N$). At time $t = 0$, one starts with n_0 unconnected nodes. The graph grows by iteratively repeating at time $t = 1, 2, 3, \ldots, N - n_0$ the following two steps:*

1 *A new node, labelled by the index n, being $n = n_0 + t$, is added to the graph. The node arrives together with k_n^{out} out-going links. The value of k_n^{out} can vary from one node to another, and is on average equal to m.*
2 *The k_n^{out} arcs connect the new node to k_n^{out} different nodes already present in the system. The probability $\Pi_{n \to i}$ that a new arc connects node n to node i (with $i = 1, 2, \cdots, n - 1$) at time t is:*

$$\Pi_{n \to i} = \frac{k_{i,t-1}^{in} + 1}{\sum_{l=1}^{n-1} (k_{l,t-1}^{in} + 1)}. \tag{6.14}$$

The two rules above are very close to those of the BA model, with the main difference being that the *de Solla Price model* produces directed graphs. The preferential attachment is a linear function of the in-degree k_i^{in} of node i, i.e. of the number of citations received by paper i. Assuming that the probability is proportional to $k^{in} + 1$ solves the problem that each node starts with zero in-degree, and hence would forever have zero probability of gaining new arcs. The offset term $+1$ can be justified by saying that one can consider the initial publication of a paper to be its first citation. Notice that m is the average number of papers cited by a new paper. This value is supposed to be constant over time and, differently from the BA model, it can take non-integer values, including values less than 1. It can be proven that, for large N, the de Solla Price model produces graphs with in-degree distribution that, for large values of in-degree k, goes as a power law:

$$p_k^{in} \sim k^{-\gamma^{in}} \qquad \gamma_{in} = 2 + 1/m. \tag{6.15}$$

Thus, for $m \geq 1$, the exponent γ takes values between 2 and 3. Interestingly enough, the value of the exponent does not depend on the value of the offset term. In fact, by using an attachment probability $\Pi_{n \to i} \sim k_{i,t-1}^{\text{in}} + a$ with offset parameter $a \neq 1$, instead of the attachment probability $\Pi_{n \to i} \sim k_{i,t-1}^{\text{in}} + 1$, we get again the same exponent as in Eq. (6.15) (see Problem 6.3b).

Example 6.5 *(Solving the model)* The power law distribution in Eq. (6.15) can be derived by writing down and solving the *rate equations* for the de Solla Price model [235]. For simplicity we shall denote the in-degree by k instead of k^{in}. The average probability (over the ensemble of different realisations) $\overline{T}_k(t)$ that a new arc arriving at time t points to *any* existing node with in-degree k is thus:

$$\overline{T}_k(t) = \frac{(k+1)\overline{n}_{k,t-1}}{\sum_k (k+1)\overline{n}_{k,t-1}} = \frac{(k+1)\overline{n}_{k,t-1}}{(m+1)n_{t-1}}. \tag{6.16}$$

where $\overline{n}_{k,t-1}$ is the average number of vertices with in-degree k at time $t-1$, n_{t-1} is the total number of nodes and, in the last equality, we have used the fact that m is also the mean in-degree of the network: $\sum_k k\overline{n}_{k,t-1} = mn_{t-1}$. The rate equations for the model read:

$$\overline{n}_{k,t} - \overline{n}_{k,t-1} = \frac{m}{(m+1)n_{t-1}}\left[k\overline{n}_{k-1,t-1} - (k+1)\overline{n}_{k,t-1}\right] + \delta_{k,0} \quad k \geq 0.$$

Expressing the equations in terms of in-degree probability distributions at time t and at time $t-1$, respectively $p_{k,t}^{\text{in}} = \overline{n}_{k,t}/n_t$ and $p_{k,t-1}^{\text{in}} = \overline{n}_{k,t-1}/n_{t-1}$, and looking for stationary solutions, we get the following recurrence relations in k:

$$p_k^{\text{in}} = \frac{m}{m+1}\left[kp_{k-1}^{\text{in}} - (k+1)p_k^{\text{in}}\right] + \delta_{k,0} \quad k \geq 0$$

which are analogous to those obtained in Eq. (6.9) for the BA model. Hence, in order to get p_k^{in}, we need to solve the iterative equations:

$$p_k^{\text{in}} = \begin{cases} \frac{k}{k+2+1/m}p_{k-1}^{\text{in}} & \text{for } k \geq 1, \\ \frac{m+1}{2m+1} & \text{for } k = 0. \end{cases} \tag{6.17}$$

Rearranging, we find:

$$p_k^{\text{in}} = \frac{k}{k+2+1/m}\frac{k-1}{k+1+1/m}\frac{k-2}{k+1/m} \cdots \frac{2}{4+1/m}\frac{1}{3+1/m}p_0^{\text{in}}$$

$$= \frac{k!\,(1+1/m)!\,(2+1/m)}{(k+2+1/m)!}\frac{m+1}{2m+1} = \frac{k!\,(1+1/m)!}{(k+2+1/m)!}(1+1/m).$$

The latter two expressions have meaning only if $1/m$ is an integer, so that the arguments of the factorials are integer numbers. In the general case in which $1/m$ is not integer, one has to use the natural extension of the factorial to real arguments, which is the Γ function, defined as [75]:

$$\Gamma(x) = \int_0^\infty t^{x-1}e^{-t}\,dt, \quad x > 0$$

where, if x is integer number n, we have $\Gamma(n) = (n-1)!$. Hence we have:

$$p_k^{\text{in}} = \left(1 + \frac{1}{m}\right)\frac{\Gamma(k+1)\Gamma(2+1/m)}{\Gamma(k+3+1/m)} = \left(1 + \frac{1}{m}\right)B(k+1, 2+1/m) \tag{6.18}$$

where we have made use of the Legendre's beta function. The beta function, also known as Euler's first integral, is defined as $B(x,y) \equiv \int_0^1 t^{x-1}(1-t)^{y-1}dt$, and can be expressed in terms of Γ functions as $B(x,y) = \frac{\Gamma(x)\Gamma(y)}{\Gamma(x+y)}$. Notice that, for large x and fixed values of y, we can use Stirling's approximation on Γ, which leads to the following asymptotics (see Problem 6.3c)

$$B(x,y) \sim \Gamma(y)x^{-y}.$$

Hence, for fixed values of m, Eq. (6.18) implies that, for large in-degrees k, the degree distribution goes as $p_k^{in} \sim k^{-2-1/m}$, as in Eq. (6.15).

Notwithstanding the differences, both the BA model and the de Solla Price model produce power-law degree distributions, and they both use a linear preferential attachment rule in the network growth. Now, a question comes natural. Is the linear preferential attachment necessary for the emergence of a power-law degree distribution? To address this question, we consider a variation of the BA model in which the new links connect with uniform probability to the nodes of the graphs [20].

Definition 6.4 (The uniform-attachment model)   *Given three positive integer numbers N, n_0 and m (with $m \leq n_0 \ll N$), the graph grows, starting at time $t = 0$ with a complete graph with n_0 nodes, and by iteratively repeating at time $t = 1, 2, 3, \ldots, N - n_0$, the two steps:*

1 A new node, labelled by the index n, being $n = n_0 + t$, is added to the graph. The node arrives together with m edges.
2 The m edges link the new node to m different nodes already present in the system with equal probability, i.e. the probability $\Pi_{n \to i}$ that a new link connects the new node n to node i (with $i = 1, 2, \cdots, n - 1$) is:

$$\Pi_{n \to i} = \frac{1}{n_{t-1}}$$

where $n_t = n_0 + t$ is the number of nodes in the graph at time t.

It is easy to show that the uniform attachment rule produces, for large times, networks whose degree distribution decays exponentially as:

$$p_k = ce^{-\beta k} \tag{6.19}$$

where c and β are two constants only dependent on the model parameter m.

Example 6.6   *(Solving the model)*   Again, we can write down and solve the rate equations which, in this case, read:

$$\bar{n}_{k,t} - \bar{n}_{k,t-1} = m\frac{\bar{n}_{k-1,t-1} - \bar{n}_{k,t-1}}{n_{t-1}} + \delta_{k,m} \quad k \geq m.$$

Expressing the equations in terms of the degree distributions $p_{k,t} = \bar{n}_{k,t}/n_t$ and $p_{k,t-1} = \bar{n}_{k,t-1}/n_{t-1}$, and looking for stationary solutions, we get the following recurrence relation in k:

$$p_k = m\left[p_{k-1} - p_k\right] + \delta_{k,m} \quad k \geq m.$$

In order to find p_k we need then to solve the following iterative equations:

$$p_k = \begin{cases} \frac{m}{m+1}p_{k-1} & \text{for} \quad k > m \\ \frac{1}{(m+1)} & \text{for} \quad k = m \end{cases}$$

whose solution:

$$p_k = \left(\frac{m}{m+1}\right)^{k-m} \frac{1}{m+1}$$

can be rewritten, making use of the transformation $y = a^x \Leftrightarrow y = e^{x \ln a}$, as:

$$p_k = \frac{1}{m+1} \, e^{m \ln(1+\frac{1}{m})} \, e^{-k \ln(1+\frac{1}{m})}.$$

This expression is equal to Eq. (6.19) with $c = \frac{1}{m+1} e^{m \ln(1+1/m)}$ and $\beta = \ln(1 + 1/m)$, and for $m \gg 1$ reduces to:

$$p_k = \frac{e}{m} e^{-\frac{k}{m}}$$

since we can approximate $\ln(1 + 1/m) \simeq 1/m$.

The uniform-attachment model does not generate scale-free distributions. Thus, preferential attachment is a necessary ingredient in a model of graph growth if we want to reproduce the degree distributions observed in real systems. In addition to this, Pavel Krapivsky, Sidney Redner and Francois Leyvraz have shown that the preferential attachment has indeed to be strictly *linear*. In Ref. [187] they have considered a generalisation of the BA model in which the attachment probability is non-linear, and goes as a power of node degree k^α, with a tunable exponent $\alpha \neq 1$. This model can be solved similarly to the models above (see Problem 6.3d), and it is possible to prove that it gives rise to two different behaviours according to the value of α. For $\alpha < 1$, the degree distribution is a power law times a stretched exponential whose exponent is a complicated function of α. For $\alpha > 1$ a single vertex acquires a finite fraction of all the connections in the network, and for $\alpha > 2$ there is a non-zero probability that this node will be connected to every other node on the graph. The remainder of the nodes have an exponentially decaying degree distribution.

In conclusion, it turns out that an attachment probability linearly proportional to the node degree is the "magic ingredient" to grow scale-free networks. When preferential attachment is not perfectly linear, then we cannot have a perfect power law. How is it then possible for a real system to select a preferential attachment that is exactly linear? This question comes together with a drawback of the BA model. The preferential attachment rule requires in fact a *global knowledge* on the system: a new node adding to the system needs to have full information on the degree sequence of all the previous nodes. This requirement is certainly unrealistic. Processes shaping networks in a real world are usually *local*, with a node having only a minimal knowledge of the existing network, often limited to a few other nodes and their neighbours [139]. For instance, in citation networks, when a scientist enters a new field, he is often familiar with only a few of the published papers. Then, little by little, he discovers other papers on the subject, mainly by following the references included in the papers he knows. When, finally, the scientist publishes a new paper, the paper includes, in its list of references, papers discovered in this way. We can reproduce in a simplified way this process by considering the following model in which the new nodes attach to first neighbours of randomly chosen nodes [41].

Definition 6.5 (The first-neighbour model) *Given two positive integer numbers: N, and m (with $m \ll N$), starting at time $t = 0$ with a complete graph with $n_0 = m + 1$ nodes, the graph grows by iteratively repeating at time $t = 1, 2, 3, \ldots, N - n_0$, the two steps:*

1 A new node, labelled by the index n, being $n = n_0 + t$, is added to the graph. The node arrives together with m edges.
2 The m edges link the new node to m different nodes already present in the system. A node j is randomly selected with uniform probability among the nodes in the graph. The m edges link the new node n to m randomly chosen neighbours of j.

The model does not include an explicit linear preferential attachment rule as the BA model. However, the first-neighbour rule automatically induces a preferential attachment which is *exactly linear*. We know, in fact, from Eq. (3.38) that the first neighbour of a randomly chosen node has a degree k with a probability proportional to k. Hence the rate of increase of a node degree will be proportional to its current degree. It is straightforward to see that this mechanism gives rise to power-law distributions. In fact the model turns out to be described by the same rate equations we have written for the BA model in Eq. (6.6). More elaborated and realistic models based on *random walkers on the graph* have been proposed for citation networks. For instance, Alexei Vázquez in Ref. [304], and Tim Evans and Jari Saramäki in Ref. [107] have considered various cases of walks of length larger than one, in which new nodes can link also to higher-order neighbours of randomly chosen nodes.

Walking on a graph is not the only local mechanism to induce a linear preferential attachment [305]. An alternative possibility is that of *node copying*, in which a newly arriving node copies, at least in part, the first neighbours of a randomly chosen node of the graph. This mechanism is quite common in biological systems and, for instance, it has been used to develop various models of protein interaction networks [288, 307]. In fact, genes that code for proteins can duplicate in the course of the evolution. Since the proteins coded for by each copy are the same, their interactions are also the same, that is the new gene copies its edges in the interaction network from the old gene. Subsequently, the two genes may develop differences because of evolutionary drift or selection. In one of the most common models, each node in the network represents a protein that is expressed by a gene, and the network grows at each time, following a step of duplication and a step of divergence [307].

Definition 6.6 (Protein duplication–divergence model) *Given two positive integer numbers, N and n_0 (with $n_0 \ll N$), and two real numbers p and q, such that $0 \leq p < 1$ and $0 \leq q < 1$, the graph grows, starting with n_0 isolated nodes, and iteratively repeating at time $t = 1, 2, 3, \ldots, N - n_0$ the two following steps:*

1 (Duplication). A new node labelled by the index n, being n = n$_0$ + t, is added to the graph. A node i is selected at random, and n is connected to all the neighbours j of i. With probability p also a link between n and i is established.

2 (Divergence). For each of the nodes j linked to i and n, we choose randomly one of the two links (i, j) or (n, j) and remove it with probability q.

The model has two control parameters, p and q. The first mechanism, the duplication, mimics the fact that the new protein, the result of duplication, is identical to the old protein, hence, it interacts with other proteins in the very same way. In the duplication mechanism, p is a parameter that control the existence of an interaction between the duplicate and the original protein. The second mechanism accounts for divergence, and q represents the loss of interactions between the duplicates and their neighbours due to the divergence of the duplicates. The interesting thing is that this simple model is able to reproduce real systems. For instance, it has been proven that, for $p = 0.1$ and $q = 0.7$, the model produces a reasonable agreement with the properties of the real *S. cerevisiae* protein interaction network [307].

6.4 Variations to the Theme

The BA model produces graphs with power-law degree distributions with $\gamma = 3$, an exponent which is close to that found in citation networks. There are, however, other real growing networks, such as the World Wide Web or various scientific collaboration networks, whose degree distributions are characterised by different values of the exponent. We have already come across a model, the de Solla Price's model of Definition 6.3 which can produce values of γ different from 3. The first model we discuss here is a much simpler way to grow scale-free networks with a tunable exponent. The model is known as the *Dorogovtsev–Mendes–Samukhin (DMS) model*, and is based on a variation of the BA rules which considers a linear preferential attachment with an offset constant that can be tuned [97].

Definition 6.7 (The DMS model)   *Given three positive integer numbers N, n$_0$ and m (with m < n$_0$ ≪ N), and a real number a (with −m ≤ a) known as the offset constant, the graph grows, starting at time t = 0 with a complete graph with n$_0$ nodes, and iteratively repeating at time t = 1, 2, 3, . . . , N − n$_0$, the following two steps:*

1 A new node, labelled by the index n, being n = n$_0$ + t, is added to the graph. The node arrives together with m edges.

2 The m edges link the new node to m different nodes already present in the system. The probability $\Pi_{n \to i}$ that n is connected to node i (with i = 1, 2, · · · , n − 1) at time t is:

$$\Pi_{n \to i} = \frac{k_{i,t-1} + a}{\sum_{l=1}^{n-1}(k_{l,t-1} + a)}. \qquad (6.20)$$

The first step of each iteration is the same as in the BA model, while the attachment probability of the DMS model depends on the offset constant a, and reduces to Eq. (6.1) for $a = 0$. The value of a is set equal for all nodes, and does not change in time. There are some natural constraints to the range in which a can vary. Since m is the smallest degree of a node, then a cannot be smaller than $-m$ if we want a non-negative probability for every node to acquire new links. We will now show that the model produces a scale-free graph with an exponent of the degree distribution equal to [97, 186]:

$$\gamma = 3 + a/m. \tag{6.21}$$

Hence, as the offset constant a grows from $-m$ to ∞, γ increases from 2 to ∞. In particular, for $a = 0$ we get back to the BA model and, consequently, to the same exponent $\gamma = 3$ we derived in Section 6.2. Negative values of a are needed to get values of γ smaller than 3, as those observed in some real-world networks. In Appendix A.13.2 we discuss an algorithm to sample graphs from the DMS model.

Example 6.7 *(Degree distribution of the DMS model)* The value of the exponent γ in Eq. (6.21) can be derived, as we did in Section 6.2 for the BA model, by writing down and solving the rate equations of the model:

$$\bar{n}_{k,t} - \bar{n}_{k,t-1} = m\frac{(k+a-1)\bar{n}_{k-1,t-1} - (k+a)\bar{n}_{k,t-1}}{2l_{t-1} + an_{t-1}} + \delta_{k,m} \quad k \geq m$$

where the denominator is obtained by summing $\sum_k (k+a)\bar{n}_{k,t-1}$. In the stationary state, the degree distribution of the DMS model obeys the following equations:

$$p_k = \frac{m}{2m+a}\left[(k+a-1)p_{k-1} - (k+a)p_k\right] + \delta_{k,m} \quad k \geq m.$$

Hence:

$$p_k = \begin{cases} \dfrac{k+a-1}{k+a+2+a/m}p_{k-1} & \text{for } k > m, \\[3mm] \dfrac{2+a/m}{m+a+2+a/m} & \text{for } k = m. \end{cases}$$

We therefore have:

$$p_k = \frac{k+a-1}{k+a+2+a/m}\frac{k+a-2}{k-1+a+2+a/m} \cdots \frac{m+a}{m+1+a+2+a/m}p_m$$

from which, by following the same steps adopted in the derivation of Eq. (6.18), we finally obtain:

$$p_k = \frac{B(k+a, 3+a/m)}{B(m+a, 2+a/m)},$$

where we have made use again of the Legendre beta function. For large k the expression above reduces to a power law with exponent $\gamma = 3 + a/m$. Notice that the exponent γ in this case depends both on the values of a and m. This is different from the case of the model by de Solla Price where γ only depends on m (see Problem 6.3b).

The BA model and the DMS model lack a number of important processes that can shape the growth of a real network, such as the rewiring of links, or the removal of nodes and links. Right after publishing the BA model, the same two authors proposed a new model, known as *Albert and Barabási (AB) model*, which also includes the appearance of new edges between existing nodes, and the movement of existing edges [4]. Interestingly enough, the addition of the two new rules to the basic preferential attachment rule of the BA model also allows the exponent of the degree-distributions to be tuned.

Definition 6.8 (The AB model)   *Given three positive integer numbers, N, n_0 and m (with $m \leq n_0 \ll N$), and two real numbers p and q, such that $0 \leq p < 1$ and $0 \leq q < 1 - p$, the graph grows, starting with n_0 isolated nodes, and performing at each time step $t = 1$, $2, 3, \ldots, N - n_0$ one of the following three operations.*

(a) *With probability p, m new links are added between nodes already present in the graph. For each new link, a node j of the graph is randomly selected as one end of the link, while the other end i is chosen with a probability:*

$$\Pi_{j \to i} = \frac{k_{i,t-1} + 1}{\sum_l (k_{l,t-1} + 1)}. \tag{6.22}$$

This operation is repeated m times.

(b) *With probability q, m links are rewired. First, a node j and a link (j, l) connected to it are randomly selected. Then, the link (j, l) is removed and replaced with a new link (j, i) that connects node j with node i chosen with probability as in Eq. (6.22). This operation is repeated m times.*

(c) *With probability $1 - p - q$ a new node j is added to the graph. The new node arrives with m new links. Each of the new links is connected to a node i already present in the system, with probability as in Eq. (6.22).*

The graph growth for $m = 2$ is illustrated in Figure 6.7. Suppose at time $t = 2$ we have the graph with five nodes and four links shown in the upper panel. At time step $t = 3$, one of the three possible events illustrated in the lower panels can take place. The first event is selected with a probability p, and consists in the addition of $m = 2$ new links to the graph. One end of each new link is selected randomly, while the other is selected by preferential attachment as in Eq. (6.22). The new added links are shown as dashed lines in panel (a). The second event is selected with a probability q and consists in the rewiring of $m = 2$ links. As an example, in panel (b), link AD is disconnected from its A end and connected preferentially to the highly connected node E. Analogously, link BE is disconnected from its E end and connected preferentially to the highly connected node C. The third possible process, selected with a probability $1 - p - q$, consists in the addition of a new node with its $m = 2$ links. In panel (c) the new node F is connected to two nodes of the system, namely E and C, selected with a preferential attachment probability.

Notice that the attachment probability in Eq. (6.22) is proportional to $k_{i,t-1} + 1$. This is a particular case $(a = 1)$ of the attachment probability of the DMS model. In fact in the

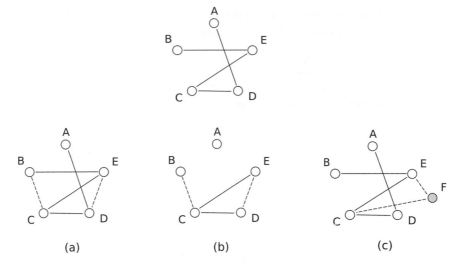

Fig. 6.7 Growth process of the AB model for $n_0 = 3$ and $m = 2$. At time $t = 3$, the graph shown in the upper panel can grow by one of the three mechanisms sketched in the lower panels: (a) addition of links BC and ED with probability p, (b) rewiring of links BE and AD respectively into BC and DE with probability q, (c) addition of node F and links CF and EF with probability $1 - p - q$.

$p = q = 0$ limit, the model reduces exactly to the DMS model with $a = 1$. At a given time step t, the growth process of the AB model results in a graph with:

$$\bar{n}_t = n_0 + (1 - p - q)t \quad \text{average number of nodes}$$
$$\bar{l}_t = (1 - q)mt \quad \text{average number of edges.} \tag{6.23}$$

Notice that, at variance with the BA model, in the AB model the number of nodes and the number of links are themselves random variables. For large times, this corresponds to a graph with an average degree $\langle k \rangle = (1 - q)2m/(1 - p - q)$, because for large t we can neglect the constant n_0 compared to the term linearly increasing with time. The average degree is a well-defined positive number, since we assumed $p + q < 1$ by definition.

Example 6.8 *(Time evolution of node degrees in the mean-field approximation)*   We can extract information on how the degree of a node grows in time by means of the same *mean-field* arguments used in the Example 6.4 for the BA model [4]. Again we approximate the degree of node i at time t, $k_{i,t}$, as a continuous real variable $\bar{k}_i(t)$ that represents the ensemble average and changes continuously in time. The differential equation ruling the time evolution of $\bar{k}_i(t)$ reads (see Problem 6.4):

$$\frac{d\bar{k}_i}{dt} = (p - q)m\frac{1}{\bar{n}(t)} + m\frac{\bar{k}_i + 1}{\sum_j(\bar{k}_j + 1)} \tag{6.24}$$

where $\bar{n}(t)$ is the average number of nodes in the network at time t. By making use of Eqs. (6.23), we can now rewrite the denominator in the second term of the right-hand side as:

$$\sum_j (\bar{k}_j + 1) = 2\bar{l}(t) + \bar{n}(t) = n_0 + mt \frac{2m(1-q)+1-p-q}{m}$$

For large time t, we finally get:

$$\frac{d\bar{k}_i}{dt} = \frac{1}{Bt} \left(\bar{k}_i + A + 1 \right)$$

where A and B are two numbers depending on the three control parameters of the model: p, q and m:

$$A(p,q,m) = (p-q) \left[\frac{2m(1-q)}{1-p-q} + 1 \right],$$

$$B(p,q,m) = \frac{2m(1-q)+1-p-q}{m}.$$

This differential equation can be easily solved with the initial condition that node i has been introduced at time $t = t_i$, such that $\bar{k}_i(t = t_i) = m$. We get:

$$\bar{k}_i(t) = [A + m + 1] \left(\frac{t}{t_i} \right)^{\frac{1}{B}} - A - 1, \tag{6.25}$$

telling us that, as in the BA model, the degree of any node increases as a power law of the time. However, in the AB model the value of the growth exponent is $1/B$, and can therefore be modulated as a function of the three parameters q, p and m. For instance, in the limiting case $q = 0$, $p = 0$, we get that $\bar{k}_i(t)$ grows as $(t/t_i)^{(m/(2m+1))}$. Notice that the AB model does not exactly reduces to the BA model for $q = 0$, $p = 0$, because in the AB model the attachment rule is proportional to $k + 1$ and not to k. This explains why, in the limit $q = 0$, $p = 0$, we do not get the exponent $1/2$.

The degree distribution p_k of the AB model can be obtained by writing down and solving the rate equations. Analogously, it can also be derived in the mean-field approximation from the expression of $\bar{k}_i(t)$ found in the Example 6.8 [4]. The model has indeed a very rich behaviour as a function of the two parameters p and q, as shown in the phase diagram reported in Figure 6.8. We notice, in particular, the existence of two separate regions in the parameter space (p, q), limited by the boundary curve $q = q_{max}(p)$. In the first region, corresponding to $q < q_{max}$ (dark-grey area in the figure), the degree distribution p_k is scale-free. More precisely, we get:

$$p_k \sim (k + A + 1)^{-(B+1)} \tag{6.26}$$

where $A(p,q,m)$ and $B(p,q,m)$ are the functions of the three parameters p, q, and m defined in Example 6.8. Such expression for p_k has a power-law tail for large k, with an exponent ranging between 2 and ∞ according to the values of p, q and m. In the second region, corresponding to $q > q_{max}$ (light-grey area), the degree distribution p_k is instead an exponential

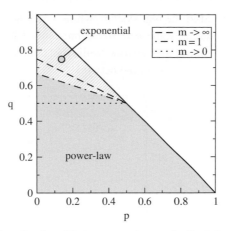

Fig. 6.8 Phase diagram of the AB model as a function of the two parameters p and q. The dark-grey and the light-grey areas correspond respectively to the scale-free regime and to the exponential regime of the model in the case $m = 1$. The boundary between the two phases, shown as dash-dotted line for $m = 1$, converges to the dotted line when $m \to 0$ and to the dashed line when $m \to \infty$.

function. As shown in Figure 6.8, the form of the boundary $q_{max}(p)$ separating the scale-free from the exponential region depends on the values of the parameter m. The dark-grey area corresponds to the scale free regime in the case $m = 1$, while the boundaries in the two limits $m \to 0$ and $m \to \infty$ are shown in the figure respectively as dotted and dashed lines. In particular, we always have $q_{max}(p) = 1 - p$ for $p \geq 0.5$. For $p < 0.5$ instead the phase boundary is a line with slope $-m/(1 + 2m)$ which depends on m. This means that the boundary approaches a horizontal line $q_{max}(p) = 0.5$ in the limit $m \to 0$ and the dashed line with slope -1/2 in the limit $m \to \infty$.

Example 6.9 *(The degree distribution of movie actors)*   The AB model can be used to model the movie actor collaboration network introduced in Section 2.5. The main purpose is to reproduce the degree distribution p_k of the real network, reported as circles in the figure below, that is a power law for large values of k, while it shows a saturation at small k [4]. The movie actor collaboration graph is indeed a perfect system to be described by the AB model. In fact, it is a system which grows by the addition of new actors. New links can arrive with new nodes, but the addition of new links among actors already in the network is also possible. Furthermore, existing links cannot be eliminated in the movie actor network, since a collaboration, i.e. the participation of two actors in the same movie, cannot be modified after the movie is out. Hence we set $q = 0$ in the AB model and we are left with two free parameters only, namely p and m. This means that, with probability p, m new internal links among existing actors are generated by new movies, while with probability $1 - p$ a new actor is introduced and linked to m other actors already in the network. It is possible to determine the values of m and p by fitting Eq. (6.26) to the observed distribution, and setting $q = 0$. The resulting curve obtained for $m = 1$ and $p = 0.972$ is shown in the figure as a solid line. Notice that the model reproduces the saturation at intermediate values of k observed in the real system, while is still not able to capture the bending for degrees smaller than 5. A value $p = 0.972$ indicates that only 2.8 per cent of new links is due to new actors joining the network, while most of the new links, namely 97.2 per cent of them, connect actors already in the network.

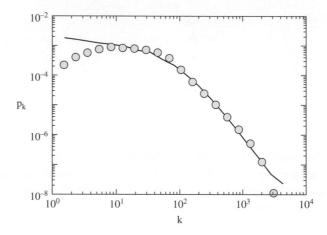

Furthermore, a value $m = 1$ means that only one new link is formed on average with every new movie, indicating that, although many actors collaborate in a movie, most of them are already linked because they have already played together in previous movies.

Apart from the de Solla Price model for citation networks, all the other models considered so far, such as the BA, the DMS, and also the AB model, are methods to construct undirected graphs. This turns out to be a problem when we want to reproduce the growth of a real directed network such as the World Wide Web. The Web has indeed a slightly more complicated structure than that of citation networks. For instance, it contains directed cycles, while the de Solla Price model produces networks with no directed cycles. These are a good representation of real citation networks, which are indeed directed and acyclic graph, but a poor representation of the World Wide Web. In addition to this, not only the in-degree distribution, but also the out-degree distribution of the World Wide Web follows a power law, whereas only the average value of the out-degree can be controlled in the de Solla Price model.

A number of authors have proposed new growth models specifically devised for the World Wide Web, and addressing the problems mentioned above. In particular, Pavel Krapivsky, Geoff Rodgers and Sidney Redner have studied a directed graph model of the World Wide Web in which nodes, but also directed arcs between already existing nodes, are added to the network at a constant rate [188, 185]. In this model, newly arriving nodes point to pre-existing nodes with a probability proportional to their in-degree. Concerning new arcs, the out-going end of each new arc is attached to nodes in proportion to their out-degree, and the in-going end in proportion to in-degree, plus appropriate constant offsets. This model, known in the literature as the *Web Graph (WG) model*, or as the *KRR model* by the name of the authors, appears to be quite a reasonable model for the growth of the Web. It produces a directed graph, it allows arcs to be added after the creation of a vertex, it gives power laws in both the in- and out-degree distributions, just as observed in the real Web. By varying the offset parameters for the in- and out-degree attachment mechanisms, one can even tune the exponents of the two degree distributions to agree with those empirically observed.

Definition 6.9 (The KRR model) *Given one positive integer number T, a probability p, $0 < p < 1$, and two real numbers $a > 0$ and $b > -1$, the model generates directed graphs with $N = pT + 2$ nodes and $K = T + 1$ arcs. At time $t = 0$, one sets $n = 2$ and starts with a graph having two nodes, labelled as node 1 and node 2, and one arc pointing from node 1 to node 2. The graph grows by performing at each time step $t = 1, 2, 3, \ldots, T$ one of the following two operations:*

(a) *With probability p, set $n = n + 1$ and add a new node, labelled by the index n, to the graph. The node arrives together with one arc pointing to one of the existing nodes in the network. The attachment probability depends only on the in-degree of the target node. The probability $\Pi_{n \to i}$ that node n is connected to node i (with $i = 1, 2, \ldots, n-1$) at time t is:*

$$\Pi_{n \to i} = \frac{k^{in}_{i,t-1} + a}{\sum_{l=1}^{n-1}(k^{in}_{l,t-1} + a)}. \tag{6.27}$$

(b) *With probability $q = 1 - p$, a new arc is created from an existing node j to an existing node i, $(j, i = 1, 2, \ldots, n)$. The choices of the originating and target nodes depend on the out-degree of the originating node j and on the in-degree of the target node i:*

$$\Pi_{j \to i} = \frac{(k^{in}_{i,t-1} + a)(k^{out}_{j,t-1} + b)}{\sum_{l=1}^{n} \sum_{l'=1}^{n}(k^{in}_{l,t-1} + a)(k^{out}_{l',t-1} + b)}. \tag{6.28}$$

The graph growth is illustrated in Figure 6.9. In the model, both attachment and creation rates have the same linear dependence on the popularity of the target node. Such preferential attachment based on the in-degree of a node can be easily justified by considering that websites, like scientific papers, are indeed easier to find if there are more links pointing to them. In addition to this, the model also assumes that webpages with a large number of

(a)   (b)

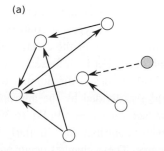
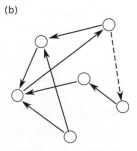

Fig. 6.9   Illustration of the growth processes in the KRR model. At a given time step, either a new node (with a new edge) is added with probability p (process (a)), or a new edge between two already existing nodes is created with probability $1 - p$ (process (b)). In (a) the newly arrived node is shaded in grey, while in both (a) and (b) the new link is represented as a dashed line.

out-going links are more active, and are therefore preferentially selected as origins of the new links. Notice that the constraints imposed on the values of the two offset parameters of the model, namely $a > 0$ and $b > -1$, ensure that the probabilities (6.27) and (6.28) are always positive for all the possible values of in- and out-degrees, $k^{\text{in}} \geq 0$ and $k^{\text{out}} \geq 1$. Let $\bar{n}_t$ be the average number of nodes in the graph at time t where, as usual, the average is intended over several realisations of the growth process, and let l_t^{in} and l_t^{out} be respectively the sum of the node in-degrees and out-degrees. At a given time step t the graph has:

$$\bar{n}_t = 2 + pt \quad \text{nodes}$$
$$l_t^{\text{in}} = l_t^{\text{out}} = 1 + t \quad \text{arcs.} \tag{6.29}$$

From this, we can immediately conclude that the average node in- and out-degrees, $\langle k^{\text{in}} \rangle = l_t^{\text{in}}/\bar{n}_t$ and $\langle k^{\text{out}} \rangle = l_t^{\text{out}}/\bar{n}_t$, are both time independent and equal to $1/p$ for large time t. To find the in- and out- degree distributions of the model, we can write down and solve the *rate equations* governing the time evolution for the average number of nodes with a certain degree. Since this time both in- and out-degree play a role, we will consider the function $\bar{n}_{k,k',t}$ defined as the average number of nodes with in-degree k and out-degree k', at time t. The quantities $\bar{n}_{k,k',t}$ obey the normalisation condition $\sum_k \sum_{k'} \bar{n}_{k,k',t} = \bar{n}_t \ \forall t$. Again, as for the BA model, the averages are to be intended as being performed over different realisations of the growth process. Considering average quantities allows us to treat the variables $\bar{n}_{k,k',t}$ as real numbers, and to simplify the problem. The rate equations for $\bar{n}_{k,k',t}$ read [188, 185]:

$$\bar{n}_{k,k',t} - \bar{n}_{k,k',t-1} = (p+q) \left[\frac{(k-1+a)\bar{n}_{k-1,k',t-1} - (k+a)\bar{n}_{k,k',t-1}}{l_{t-1}^{\text{in}} + a\bar{n}_{t-1}} \right]$$
$$+ q \left[\frac{(k'-1+b)\bar{n}_{k,k'-1,t-1} - (k'+b)\bar{n}_{k,k',t-1}}{l_{t-1}^{\text{out}} + b\bar{n}_{t-1}} \right]$$
$$+ p\, \delta_{k,0}\delta_{k',1} \quad k \geq 0, k' \geq 1. \tag{6.30}$$

Example 6.10 *(Derivation of the rate equations)* There are three groups of terms in the right-hand side of Eq. (6.30). The first group accounts for changes in the in-degree of target nodes. Such changes are due to two different types of contributions: the arrival of an arc together with a new node, with probability p, or the creation of a new arc between two already existing nodes, with probability q. For both these processes, the appearance of an arc to a node with in-degree k leads to a loss in the number of such nodes. This occurs at a rate equal to $(p + q)(k + a)\bar{n}_{k,k',t}$, divided by the appropriate normalisation factor $\sum_{k,k'}(k + a)\bar{n}_{k,k',t} = l_t^{\text{in}} + a\bar{n}_t$. The factor $p + q = 1$ in Eq. (6.30) has been written only to make explicit the two different types of contributions. Similarly, the second group of terms accounts for changes in the out-degree. These changes occur with a probability q because they are only due to the creation of new links between already existing nodes. Finally, the last term, $p\, \delta_{k,0}\delta_{k',1}$, accounts for the introduction of new nodes with no incoming links and one outgoing link.

As a useful self-consistency check of Eq. (6.30), we can easily verify that the total number of nodes, $\bar{n}_t = \sum_{k,k'} \bar{n}_{k,k',t}$, obeys $\bar{n}_{t+1} - \bar{n}_t = p$, which is in perfect agreement with the first of Eqs. (6.29). In the same spirit, the sum of all the in-degrees $l_t^{\text{in}} = \sum_{k,k'} k \bar{n}_{k,k',t}$ and the sum of all the out-degrees $l_t^{\text{out}} = \sum_{k,k'} k' \bar{n}_{k,k',t}$, obey $l_{t+1}^{\text{in}} - l_t^{\text{in}} = l_{t+1}^{\text{out}} - l_t^{\text{out}} = 1$, in agreement with the second of Eqs. (6.29). We can define a probability distribution in terms of the quantities $\bar{n}_{k,k',t}$ as:

$$p_{k,k',t} = \frac{\bar{n}_{k,k',t}}{\bar{n}_t}.$$

The in-degree and the out-degree distributions produced by the model at a given time t, can be expressed in terms of $p_{k,k',t}$ as $p_{k,t}^{\text{in}} = \sum_{k'} p_{k,k',t}$ and $p_{k',t}^{\text{out}} = \sum_k p_{k,k',t}$. Hence, the asymptotic behaviour of the degree distributions can be obtained by solving the rate equations in the long time limit. One finds that both the in- and out-degree distributions have power law forms for large degrees, namely:

$$p_k^{\text{in}} \sim k^{-\gamma_{\text{in}}}, \qquad \gamma_{\text{in}} = 2 + pa, \qquad\qquad (6.31)$$

$$p_k^{\text{out}} \sim k^{-\gamma_{\text{out}}}, \qquad \gamma_{\text{out}} = 1 + q^{-1} + bpq^{-1}. \qquad (6.32)$$

Notice that the two exponents of the degree distributions are functions of the parameters of the model. In particular γ_{in} depends on the in-degree offset a, and on p, while γ_{out} depends on the out-degree offset b, and on p and q. Notice also that both the exponents are greater than 2.

Example 6.11 *(Degree distributions of the KRR model)* Expressing Eqs. (6.30) in terms of the probability distributions $p_{k,k',t} = \bar{n}_{k,k',t}/\bar{n}_t$ and $p_{k,k',t-1} = \bar{n}_{k,k',t-1}/\bar{n}_{t-1}$, and making use of the expressions $\bar{n}_t = pt$ and $l_t^{\text{in}} = l_t^{\text{out}} = t$, valid for large times, we get:

$$\frac{\bar{n}_{t-1}}{p} \left[p_{k,k',t} - p_{k,k',t-1} \right] + p_{k,k',t} = \frac{(k-1+a)p_{k-1,k',t-1} - (k+a)p_{k,k',t-1}}{1+ap}$$

$$+ q \frac{(k'-1+b)p_{k,k'-1,t-1} - (k'+b)p_{k,k',t-1}}{1+bp}$$

$$+ \delta_{k,0} \, \delta_{k',1} \qquad k \geq 0, k' \geq 1.$$

Now, assuming that $p_{k,k',t}$ tends to a stationary solution $p_{k,k'}$ as $t \to \infty$, we obtain the following recurrence relations in k and k':

$$[k + c(k'+b) + d]p_{k,k'} = (k-1+a)p_{k-1,k'} + c(k'-1+b)p_{k,k'-1}$$

$$+ (1+pa) \, \delta_{k,0} \, \delta_{k',1} \qquad k \geq 0, k' \geq 1 \qquad (6.33)$$

where we have introduced the two constants c and d, set as:

$$c = q \frac{1+pa}{1+pb} \quad \text{and} \quad d = 1 + (1+p)a.$$

We can now obtain the recurrence relations for the in-degree distribution, $p_k^{\text{in}} = \sum_{k'} p_{k,k'}$, and the out-degree distribution, $p_{k'}^{\text{out}} = \sum_k p_{k,k'}$, by summing Eqs. (6.33) over k', or over k, respectively. For instance, by summing over k', we get:

$$(k + cb + d)p_k^{\text{in}} + c \sum_{k'} k' p_{k,k'} = (k - 1 + a)p_{k-1}^{\text{in}} + c \sum_{k'} (k' - 1)p_{k,k'-1} + cbp_k^{\text{in}}$$

$$+ (1 + pa)\,\delta_{k,0} \qquad k \geq 0$$

which gives:

$$(k + d)p_k^{\text{in}} = (k - 1 + a)p_{k-1}^{\text{in}} + (1 + pa)\delta_{k,0} \qquad k \geq 0.$$

The solution to these equations can be expressed in terms of ratios of gamma functions. For instance, from the first set of relations we get the following iterative equations:

$$p_k^{\text{in}} = \begin{cases} \frac{k-1+a}{k+d} p_{k-1}^{\text{in}} & \text{for} \quad k \geq 1 \\[2mm] \frac{1+pa}{d} & \text{for} \quad k = 0 \end{cases}$$

which give:

$$p_k^{\text{in}} = \frac{k - 1 + a}{k + d} \frac{k - 2 + a}{k - 1 + d} \frac{k - 3 + a}{k - 2 + d} \cdots \frac{2 + a}{3 + d} \frac{1 + a}{2 + d} \frac{a}{1 + d} p_0^{\text{in}}$$

$$= \frac{(k - 1 + a)!}{(a - 1)!} \frac{d!}{(k + d)!} p_0^{\text{in}}.$$

The latter expression have meaning only if a and d are integers, so that the arguments of the factorials are integer numbers. In the general case in which a and d are not integers we can make use of the Γ functions. We can finally write the in-degree distribution as:

$$p_k^{\text{in}} = \frac{1 + pa}{d} \frac{\Gamma(k + a)\,\Gamma(d + 1)}{\Gamma(a)\,\Gamma(k + d + 1)}. \tag{6.34}$$

Analogously, by summing Equations (6.33) over k, we get:

$$\left(k' + \frac{1}{q} + \frac{b}{q}\right) p_{k'}^{\text{out}} = (k' - 1 + b)p_{k'-1}^{\text{out}} + \frac{1 + pb}{q}\delta_{k',1} \qquad k' \geq 1$$

which give the out-degree distribution:

$$p_{k'}^{\text{out}} = \frac{1 + pb}{1 + q + b} \frac{\Gamma(k' + b)\,\Gamma(2 + q^{-1} + bq^{-1})}{\Gamma(k' + 1 + q^{-1} + bq^{-1})\,\Gamma(1 + b)}. \tag{6.35}$$

From Eqs. (6.34) and (6.35), and from the asymptotics of the gamma functions, one derives the asymptotic behaviour of the in- and out-degree distributions reported in Eqs. (6.31) and (6.32) [188, 185].

The predictions of the KRR model can be compared to the empirical observations on the in- and out-degree distributions of the World Wide Web discussed in Chapter 5. For instance, the values of the two degree exponents extracted from the AltaVista WWW network, the largest sample of the World Wide Web reported in Table 5.1, are $\gamma_{\text{in}} \approx 2.1$ and $\gamma_{\text{out}} \approx 2.7$. In addition to this we know that the AltaVista WWW network has $\langle k^{\text{in}} \rangle = \langle k^{\text{out}} \rangle = 7.5$, so we set $p^{-1} = 7.5$ in the KRR model. Now, the expressions of the exponents in Eq. (6.31) and in Eq. (6.32) depend on the two parameters a and b, and by fixing them to the values $a = 0.75$ and $b = 3.55$ the model matches the observed values of the in- and out-degree exponents. However, it is important to notice that all nodes belong to a single weakly connected component in the KRR model. Conversely, the real World Wide Web consists of various weakly and strongly connected components as sketched in Figure 5.2, and this can be taken into account in more refined and complicate growth models [66].

We now come back to the discussion of Section 5.2 on the presence of a natural cut-off in the degree distribution of scale-free networks. There are several reasons why the degree of the nodes cannot grow forever in a network. For instance, in collaborations among movie actors there is an effect of *ageing*: even very highly connected actors will eventually stop receiving new links after a certain time. The same thing can happen, even if more gradually, in citation networks, where we expect on average a reduction in the probability for a paper to acquire new links as the years from its publication go by. Also, there can be *cost constraints*. For instance, the connectivity of the nodes of a transportation network can be limited for economic reasons. The ageing of the vertices and the cost of links certainly limit the preferential attachment mechanism preventing a perfect scale-free distribution of connectivities at all scales, and there are models to take this into account as shown in the following example.

Example 6.12 *(Ageing and cost constraints)* In Ref. [9] Luís A. Nunes Amaral, Antonio Scala, Marc Barthèlemy, and H. Eugene Stanley have proposed two variations to the BA model in which the nodes have a finite lifetime to acquire new links. In practice, the two models work exactly as the BA, with the only difference that the nodes are into one of two possible states, active or inactive. All new vertices are created active, but over time they may become inactive, and once inactive they cannot receive links any more. The two models differ in the way nodes get inactive. In the first variation to the BA model, a node can become inactive at each time step with a constant probability This implies that the time a vertex remains active decays exponentially. In the second variation a node becomes inactive when it reaches a maximum number of links, k_{max}. Both models lead to cut-offs on the power law decay of the tail of the power-law degree distributions.

We close this section by mentioning a very useful model proposed by Jesús Gómez-Gardeñes, and Yamir Moreno from the University of Zaragoza [138]. The model is based on a combination of uniform and preferential attachment mechanisms that can be tuned by the value of a control parameter $0 \leq p \leq 1$. The nice thing about this model is that for $p = 0$ it produces graphs with scale-free degree distributions as the BA model, while in the limit $p = 1$ it coincides with the ER random graph model. It is therefore possible for intermediate values of the control parameter p, to directly interpolate between the two main models considered in this book.

6.5 Can Latecomers Make It? The Fitness Model

All the growth models considered so far implicitly assume that all nodes are equivalent. This means that, at a given time, two nodes can only differ because they have a different degree. Since the growth mechanisms solely depend on the node degrees, when a newly arrived node has to choose between two nodes with the same degree, it will link with either of them with the same probability. As a consequence of this, in the BA model and in all its variations the oldest nodes in the network are always those with the highest number of links simply because they had the longest time to acquire new edges.

There are instead various indications that in real systems the degree of a node and its growth rate do not depend on the node age alone. For instance, in the APS citation network the top-ten publications include classical papers published in the '50s and '60s, but also papers which made to the top in just one or two decades. By looking at the left panel of Figure 6.10 (see also the discussion in Example 6.1) it is clear that newer papers can sometimes get more citations than older ones. This has happened to the 1965 paper by Kohn and Sham (KS), that in late '80s has overcome the classical work on superconductivity by Bardeen–Cooper–Schrieffer (BCS). The BCS paper has also been overtaken in 2006 by the more recent 1996 paper by Perdew–Burke–Ernzerhof (PBE), which will probably also replace soon the KS paper at the top of the ranking. Similarly, age is rarely the only factor determining the popularity of a website in the WWW. As shown in the table reported in Figure 6.10, none of the five most linked websites in January 2014 was created before 1998, and six of the top-twenty sites were less than ten years old at the time in which the ranking was compiled. It is true that some of the early birds of the WWW have managed to acquire a relatively large number of incoming links, like `imdb.com`, which appeared in 1993 and ranked 48th in January 2014 with more than 400,000 incoming hyperlinks, or `theguardian.com`, created in 1995 and ranked 66th with around 350,000 incoming links. However, many historical websites which contributed to the early development of

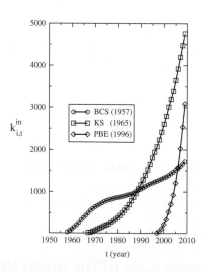

Rank	Domain	Year	$k^{in} \times 10^6$
1	facebook.com	2004	9.6
2	twitter.com	2006	6.5
3	google.com	1998	5.9
4	youtube.com	2005	5.4
5	wordpress.org	2003	4.1
6	adobe.com	1996	3.5
7	blogspot.com	2000	3.1
8	wikipedia.org	2001	2.9
9	linkedin.com	2002	2.4
10	wordpress.com	2000	2.3
11	yahoo.com	1994	1.9
12	amazon.com	1995	1.5
13	flickr.com	2003	1.4
14	pinterest.com	2009	1.4
15	tumblr.com	2006	1.2
16	w3.org	1994	1.2
17	apple.com	1996	1.1
18	myspace.com	1996	1.0
19	vimeo.com	2004	1.0
20	microsoft.com	1994	1.0

Fig. 6.10   Can latecomers make it to the top? The total number $k_{i,t}^{in}$ of citations received over time by three APS papers in the top-ten list of Table 6.1 (figure on the left), and the top-twenty Web domains by number of incoming hyperlinks (in millions) in January 2014 according to moz.com (table on the right) suggest that in real systems, overtaking of older nodes by latecomers is not only possible but quite common.

the WWW, like those of Berkeley (`berkeley.edu`), Cambridge (`cam.ac.uk`), and the University of North Carolina at Chapel Hill (`unc.edu`), are relatively low in the ranking (respectively 181th, 492nd and 493rd). And `cern.ch`, the first website ever created, is not even among the top 500 most linked websites in January 2014.

This means that reality can be more complicated but also more interesting than simple models. Intuitively, the success of products, papers and websites, and their ability to acquire new buyers, citations or hyperlinks, is not simply determined by their date of appearance but also depends on some kind of intrinsic quality which allows them to attract attention and eventually overtake older competitors. In 2001 Ginestra Bianconi and Albert Lás-zló Barabási introduced a generalisation of the BA model to take into account a possible intrinsic difference between nodes. In their model they first assigned to the network nodes an intrinsic *node attractiveness* or *node fitness* characterising the node ability to compete for links. Then, they considered an attachment rate depending both on the degree and on the attractiveness of a node. Notice that the attractiveness of a node plays here a completely different role from that of the offset constant in the DMS model. In fact, the attractiveness changes from node to node, while the offset constant is the same for all nodes. Furthermore, the attachment probability here depends on the product of degree and attractiveness, and not on the sum of the two quantities. This is today known as the *fitness model* [35, 34].

Definition 6.10 (The fitness model)   *Let N, n_0 and m be three positive integer numbers (with $m \leq n_0 \ll N$), and $\rho(a)$ a probability function defined in $(0, \infty)$. For $n = 1, \ldots, N$ we associate with the node n an attractiveness $a_n = a$, where a is chosen from the distribution $\rho(a)$. At time $t = 0$, one starts with a complete graph with n_0 nodes, labelled as $1, 2, \ldots, n_0$, and $l_0 = \binom{n_0}{2}$ edges. The graph grows by iteratively repeating at time $t = 1, 2, 3, \ldots, N - n_0$, the following two steps:*

1 A new node, labelled by the index n, being $n = n_0 + t$, is added to the graph. The node arrives together with m edges.
2 The m edges link the new node to m different nodes already present in the system. The probability $\Pi_{n \to i}$ that n is connected to node i (with $i = 1, 2, \cdots, n - 1$) at time t is:

$$\Pi_{n \to i} = \frac{a_i k_{i,t-1}}{\sum_{l=1}^{n-1} a_l k_{l,t-1}} \tag{6.36}$$

where $k_{i,t}$ is the degree of node i at time t, and $a_i > 0$ is the attractiveness of node i.

To find out the degree distribution of the graphs produced by the model we can use the usual approach based on rate equations. Nodes need now to be characterised both by their degree and their attractiveness. Hence, we define the functions $\bar{n}_{k,t}(a)$ with k being a discrete number, and a a positive real number. The quantity $\bar{n}_{k,t}(a)$ is the average number of nodes with degree k and attractiveness a at a given time t. It obeys the normalisation conditions $\int_0^\infty da \sum_k \bar{n}_{k,t}(a) = n_t$ $\forall t$, where n_t is the total number of nodes at time t. The rate equations of the model read [185]:

$$\bar{n}_{k,t}(a) - \bar{n}_{k,t-1}(a) = m \frac{a(k-1)\bar{n}_{k-1,t-1}(a) - ak\bar{n}_{k,t-1}(a)}{\int_0^\infty da \sum_k a k \bar{n}_{k,t-1}(a)} + \rho(a)\delta_{k,m} \quad k \geq m \tag{6.37}$$

As in the BA model, we have a loss and a gain term. The loss term describes new edges connecting to nodes with k edges and turning them into nodes with $k + 1$ edges. The gain term comprises two parts: the first accounts for new edges connecting to nodes with $k - 1$ edges, thus increasing their degree to k, while the second part describes the addition of a new node with m edges and fitness a chosen at random from distribution $\rho(a)$. The degree distributions produced in the long time limit can be derived analytically by finding the stationary solution of Eqs. (6.37), as we did for the BA model [185]. The final result is intriguing. Given an arbitrary attractiveness distribution function $\rho(a)$, the model produces graphs with a power-law degree distribution for any fixed value of the attractiveness a:

$$p_k(a) \sim k^{-\gamma(a)} \qquad \gamma(a) = 1 + c/a \tag{6.38}$$

with an exponent γ that depends on the value of a, and where c satisfies the following equation:

$$1 = \int_0^\infty da\, \rho(a) \left(\frac{c}{a} - 1\right)^{-1}. \tag{6.39}$$

Of course, we have to assume that $\rho(a)$ is defined on a bounded support, so that the integral above does not diverge and gives a finite value of c. In conclusion, in the fitness model we observe a completely novel phenomenon known as *multiscaling*. That is for each value of attractiveness we get a degree distribution that scales, as a function of k, as usually as a power law. However, the value of scaling exponent γ is different for each class of node attractiveness.

Example 6.13   *(Stationary solution of the rate equations)*   The rate equations (6.37) can be solved by following the same approach we have used to analyse previous models. We first define the distribution $p_{k,t}(a) = \bar{n}_{k,t}(a)/\int da \sum_k \bar{n}_{k,t}(a) = \bar{n}_{k,t}(a)/n_t$, which for large times can be approximated as $p_{k,t}(a) = \bar{n}_{k,t}(a)/t$. Rewriting Eq. (6.37) in terms of $p_{k,t}(a)$ and $p_{k,t-1}(a)$ we obtain:

$$(t-1)\left[p_{k,t}(a) - p_{k,t}(a)\right] + p_{k,t}(a) = m\frac{a(k-1)p_{k-1,t-1}(a) - akp_{k,t-1}(a)}{\int_0^\infty da \sum_k kap_{k,t-1}(a)} + \rho(a)\delta_{k,m} \quad k \geq m.$$

Assuming now that $p_{k,t}(a)$ tends to a stationary solution $p_k(a)$ as $t \to \infty$, and that the first term in the left-hand side goes to zero, it follows that the asymptotic distribution $p_k(a)$ must satisfy the recurrence relation in k:

$$p_k(a) = \frac{a\left[(k-1)p_{k-1}(a) - kp_k(a)\right]}{1/m\int_0^\infty da \sum_k kap_k(a)} + \rho(a)\delta_{k,m} \quad k \geq m.$$

By defining a new constant c as:

$$c = \frac{1}{m}\int_0^\infty da \sum_k kap_k(a) \tag{6.40}$$

we get following iterative equations:

$$p_k(a) = \begin{cases} \frac{k-1}{k}\left(1+\frac{c}{ak}\right)^{-1} p_{k-1}(a) & \text{for} \quad k > m \\[2mm] \frac{c\rho(a)}{ak}\left(1+\frac{c}{ak}\right)^{-1} & \text{for} \quad k = m \end{cases}$$

which give (see problem 6.5a):

$$p_k(a) = \frac{c\rho(a)}{k\,a} \prod_{j=m}^{k} \left(1+\frac{c}{j\,a}\right)^{-1}. \tag{6.41}$$

Finally, the distribution $p_k(a)$, i.e. the probability that a node has attractiveness a and degree equal to k, can be written in terms of gamma functions as [185]:

$$p_k(a) = \frac{c\,\rho(a)}{a} \cdot \frac{1}{k} \cdot \frac{k(k-1)\ldots m}{(k+\frac{c}{a})(k-1+\frac{c}{a})\ldots(m+\frac{c}{a})}$$

$$= \frac{c\,\rho(a)}{a} \cdot \frac{\Gamma\left(m+\frac{c}{a}\right)}{\Gamma(m)} \cdot \frac{\Gamma(k)}{\Gamma\left(k+1+\frac{c}{a}\right)}.$$

Notice, however, that such an expression depends self-consistently on the constant c. The value of c can be determined by substituting the expression of $p_k(a)$ just found into the definition of c of Eq. (6.40). We get:

$$m\Gamma(m) = \int_0^\infty da\,\rho(a)\,\Gamma\left(m+\frac{c}{a}\right) \sum_{k=m}^\infty \frac{\Gamma(k+1)}{\Gamma\left(k+1+\frac{c}{a}\right)}$$

where we have made use of the main property of Gamma functions, namely $n\,\Gamma(n) = \Gamma(n+1)$. Now, by using the identity: $\sum_{k=1}^\infty \frac{\Gamma(k+u)}{\Gamma(k+v)} = \frac{\Gamma(u+1)}{(v-u-1)\,\Gamma(v)}$ to simplify the summation over the degree k [185], we obtain the implicit relation for c reported in Eq. (6.39).

This condition on c leads to two alternative possibilities. If the support of a is unbounded, then the integral diverges and there is no solution for c. In this limit, the most attractive node is connected to a finite fraction of all links. Conversely, if the support of a is bounded by an upper limit attractiveness cut-off $a_{\max}$, then the integral can be solved, and gives a finite value of c. In this case, for each value of a, $p_k(a)$ has a power law dependence on k, with an exponent that is the function of a reported in Eq. (6.38) (see Problem 6.5c).

The fact that $p_k(a)$ is a power law for each value of a, does not mean that the model produces graphs with a power-law degree distribution p_k. The distribution p_k is in fact given by a weighted sum of power laws with different exponents:

$$p_k = \int_0^\infty da\,\rho(a)\,p_k(a). \tag{6.42}$$

Let us consider the simplest possible case that produces multiscaling, that of a *uniform fitness distribution*. In particular, let us assume $\rho(a) = 1$ when $a \in [0, 1]$, and $\rho(a) = 0$ otherwise [34]. The constant c can be determined from Eq. (6.39), which gives $c = 1.255\ldots$ (see problem 6.5b). Thus, according to Eq. (6.38) and Eq. (6.42), the degree distribution is given by:

$$p_k \sim \frac{k^{-2.255}}{\ln k}, \tag{6.43}$$

that is, a generalised power law with an inverse logarithmic correction.

Example 6.14 *(A well-known limit of the fitness model)*   When all nodes have the same attractiveness, the attachment probability of the fitness model is proportional to the node degree only, and we expect that the model reduces to the BA model. To verify this, let us assume $\rho(a) = \delta(a - 1) \; \forall a$. In such a case Eq. (6.39), defining the value of c, reads $1 = 1/(c - 1)$, which gives $c = 2$. Inserting this value of c in Eq. (6.38), and making use of Eq. (6.42), we get $p_k \sim k^{-3}$, the well-known scaling behaviour of the BA scale-free model.

A very interesting point is to determine which are the nodes that finally have the highest values of degree in the fitness model. In particular, we expect that, even relatively young nodes can acquire links at a higher rate than nodes with a larger number of links, provided they have a large enough fitness. In Figure 6.11 we consider the fitness model with a uniform fitness distribution in $[0, 1]$, and we plot the degree of four representative nodes as a function of time in a network of $N = 8 \times 10^4$ nodes. The first node has been introduced at time $t = 4$ and has attractiveness $a = 0.2$. The second one arrives at time $t = 5$ with attractiveness $a = 0.4$, the third one arrives at $t = 93$ with $a = 0.7$ while the last one arrives only at time $t = 559$, but has the highest value of attractiveness $a = 0.99$. The numerical simulations show that nodes with higher attractiveness acquire links at a higher rate (higher slopes of the straight lines in a double logarithmic scale). It is therefore just a matter of time for the latest (yet fittest) node to overcome the other three nodes, even if they joined the network much earlier. In the example shown in Figure 6.11, the degree of node 559 becomes larger than those of the other three nodes for $t \simeq 20,000$.

It is possible to prove that in the fitness model the time evolution of the degree of a node i follows a power law:

$$k_{i,t}(a_i) = m \left(\frac{t}{t_i} \right)^{a_i/c} \qquad t \geq t_i \tag{6.44}$$

as in the BA model, where t_i is the time at which node i has been introduced. However, in this model the exponent is a function of the attractiveness a_i of the node.

This result indicates that nodes with higher fitness acquire links at a higher rate than less fit nodes. This provides an explanation to why in the World Wide Web more recent websites, like `facebook.com` and `twitter.com`, have been able to rapidly overcome other websites that have been in the system since from the beginning: simply, they might have been associated with higher values of fitness, which allowed them to quickly fill the gap due to their late arrival. This is today known as *fit gets richer mechanism*.

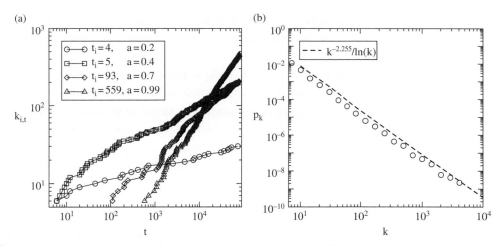

Fig. 6.11 In the fitness model, nodes with a higher value of fitness *a* eventually overtake previously arrived nodes with lower fitness, since their degree $k_{i,t}$ increases at a faster rate. (a) In the example shown, nodes 5 ($a_5 = 0.4$), 93 ($a_{93} = 0.7$) and 559 ($a_{559} = 0.99$) easily overtake node 4 ($a_4 = 0.2$) due to their larger attractiveness. Notice that, despite arriving much later, node 93 reaches at the end node 5, while node 559 overtakes all the other three for *t* slightly larger than 10,000, and eventually reaches the 44th place in the overall degree ranking. (b) The resulting degree distribution is a power law with a logarithmic correction $p_k \sim k^{-\gamma} / \log(k)$, where $\gamma = 2.255$ (dashed line).

Example 6.15 (*Time evolution of node degrees in the mean-field approximation*) As in Example 6.4, let us work in the mean-field approximation. This means considering the evolution of the average degree $\bar{k}_{i,t}(a_i)$ of a node i with a given attractiveness $a = a_i$ over different realisations of the growth process. By considering time as a continuous variable, we get the following differential equation describing the evolution of $\bar{k}_i(t, a_i)$ in the *mean-field* approximation:

$$\frac{d\bar{k}_i}{dt} = m\Pi(\bar{k}_i, a_i) = m\frac{a_i\bar{k}_i}{\sum_j a_j\bar{k}_j}$$

where a_i is the fitness of node i. The sum in the denominator goes over all nodes in the system at time t, and can be approximated as done before in Example 6.13 as:

$$\sum_j a_j\bar{k}_j = \int_0^\infty da \sum_k a\,k\,\bar{n}_{k,t}(a) = \int_0^\infty da \sum_k a\,k\,n_t p_{k,t}(a) = cmt$$

where the last equality is valid for large t, and we have used Eq. (6.40). We finally have:

$$\frac{d\bar{k}_i}{dt} = \frac{a_i\bar{k}_i}{ct}. \tag{6.45}$$

This differential equation can be easily solved with the initial condition that node i, with attractiveness a_i has been introduced at time $t = t_i$, such that $\bar{k}_i(t = t_i, a_i) = m$. The solution is that reported in Eq. (6.44). From this expression we notice that the growth exponent depends on the attractiveness of the node, and also that the value a_i/c needs to be bounded, namely $0 < a_i/c < 1 \ \forall i$, because a node always increases the number of links in time, and since the total number of links is increasing linearly with time, the degree of a node i cannot grow faster than t.

Finally, notice that an alternative way to find the value of c is by making use of Eq. (6.44). In fact, we can evaluate $\sum_j a_j k_j$ by considering that at a given time t we have all the nodes that have been introduced at any time $t' \leq t$. Hence, if we indicate as $k(t, t', a) = m(t/t')^{a/c}$ the degree, at time t, of a node introduced at time t' with attractiveness a, we can write:

$$\sum_j a_j \bar{k}_j = \int_0^\infty da \ \rho(a) \int_1^t a \ k(t, t', a) \ dt' = \int_0^\infty da \ \rho(a) \int_1^t a \ m \left(\frac{t}{t'} \right)^{a/c} dt'$$

$$= m \int_0^\infty da \ \rho(a) \frac{a}{1 - \frac{a}{c}} \left[t - t^{a/c} \right].$$

Since we have assumed $a/c < 1$, we finally have:

$$c = \lim_{t \to \infty} \frac{1}{mt} \sum_j a_j \bar{k}_j = \int da \ \rho(a) \frac{a}{1 - \frac{a}{c}} \tag{6.46}$$

which is exactly the same relation as that in Eq. (6.39).

6.6 Optimisation Models

We conclude this chapter by showing that scale-free graphs can also be produced by optimisation principles. We will focus on an extremely simple one-dimensional model to grow trees with scale-free degree distributions, known as the *Border Toll Optimisation process*, in which each new node chooses the node to link based on the minimisation of a cost function. The model, proposed by Raissa D'Souza et al. [98], has only one control parameter and can be seen as a simplified version of more realistic models for the growth of the Internet [109].

Definition 6.11 (The Border Toll Optimisation (BTO) model) *Let N be a positive integer number, and α a positive real number. One starts with a single root node, labelled $n = 0$, at position $x_0 = 0$. The nodes of the graphs, labelled $n = 1, 2, \ldots, N$, arrive one at a time, at random positions $x_1, x_2 \ldots, x_N$ in the unit interval $[0,1]$. The newly arrived node n links to the node i $(i = 0, 1, 2, \ldots, n - 1)$ which minimises the cost function:*

$$c_{n \to i} = \alpha \ n_{ni} + d_{i0} \tag{6.47}$$

with the constraint $x_i < x_n$ and $\alpha > 0$. The quantity n_{ni} is the number of nodes in the interval between node n and node i, while d_{i0} is the distance on the graph from node i to the root node. Furthermore, if there are several nodes which satisfy the constraint and minimise the cost function, we choose the one whose position x_i is closest to x_n.

The dynamics of the model is controlled by the value of the parameter α. Notice that the nodes are distributed at random in the interval $[0, 1]$ with a uniform probability, but once the positions are fixed, the construction of the graph is deterministic. This is because the choice of the nodes to link is univocally determined by the cost function in Eq. (6.47), which does not include any stochastic term. The cost function of the BTO model consists of two terms, so that the attachment of new nodes to old ones results from a trade-off between two competing effects. When a new node n arrives at a position x_n on the line, it tries to connect directly to the origin or to a node i which is close (on the graph) to the origin. In fact, on the one hand node n has to pay a cost equal to the graph distance d_{i0}, i.e. a unit cost for each hop from node i to the origin. On the other hand, the new node has also to pay a border toll α for each node on the line between x_n and x_i. Hence, it tries to connect to a node i such that there are not too many nodes n_{ni} in the interval (x_i, x_n). In the end, node n chooses the node i that minimises the cost in Eq. (6.47). The value of the control parameter $\alpha > 0$ determines the relative weighting between the two terms. We will see that the most interesting cases are when α is smaller than 1.

An example of the growth of the BTO model for $\alpha = 1/3$ is shown in Figure 6.12. At the time of their arrival both node 1 and node 2 have no other choice than that of connecting directly to node 0. The situation is different for node 3, which can choose between connecting directly to node 0 or instead to node 2. The costs to pay in the two cases are respectively equal to $c_{3 \to 0} = 1/3\, n_{30} + d_{00} = 1/3 + 0$ and to $c_{3 \to 2} = 1/3\, n_{32} + d_{20} = 0 + 1$. Since $c_{3 \to 0} < c_{3 \to 2}$, node 3 decides to connect directly to the root node. When node 4 arrives at time $n = 4$ it has four different possibilities to choose from, one for each of the four nodes on its left-hand side, respectively with costs $c_{4 \to 0} = 1$, $c_{4 \to 2} = 5/3$, $c_{4 \to 3} = 4/3$, $c_{4 \to 1} = 1$. Since there are two nodes of minimal cost, namely node 0 and node 1, node 4 chooses to connect to the one which is closest to its position, that is node 1. Of course, even with the same distribution of nodes in the interval $[0, 1]$ as in Figure 6.12, the growth of the network for $\alpha \neq 1/3$ could be different from the one reported. To show this, let us now consider the growth for a generic value $\alpha > 0$. In this case, the cost for each of the two possible choices of node 3 would respectively be $c_{3 \to 0} = \alpha$ and $c_{3 \to 2} = 1$. This means that node 3 will connect to node 2 if $\alpha \geq 1$, while it will again connect to node 0 as in Figure 6.12 for any $0 < \alpha < 1$. The costs for node 4 would then be: $c_{4 \to 0} = 3\alpha$, $c_{4 \to 2} = 2\alpha + 1$, $c_{4 \to 3} = \alpha + 1$, $c_{4 \to 1} = 1$. Hence for $\alpha \geq 1/3$ node 4 will connect again to node 1 as in Figure 6.12. Conversely, for any $\alpha < 1/3$, node 4 will find more convenient to attach directly to the root node.

The example in Figure 6.12 illustrates a general and important property of the model. Given the cost function in Eq. (6.47), it is possible to prove that any new node n has only two possibilities: it will either connect to the node i^* directly to its left, or to the so-called parent p_{i^*} of node i^*, that is the first left neighbour of i^* on the graph (see Problem 6.6a). This implies that the BTO is a local model, since it requires only a local knowledge of the system. In fact, a new node needs only to have information about the closest node to its left on the line, and about the distance on the graph between the parent of that node and the root node 0. The two prices to pay for node n, respectively to attach to i^* or to attach to p_{i^*} are:

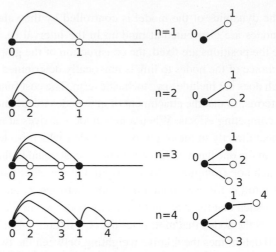

Fig. 6.12   In the BTO model the nodes arrive at random positions in the interval [0, 1] and choose to attach to the node on their left which minimises the cost function in Eq. (6.47). In practice, a new node n attaches either to the node i immediately to its left or to the parent p_i of the latter. Here we show the first four iterations of the model for $\alpha = 1/3$. Fertile nodes are indicated by filled circles, while empty circles denote infertile ones.

$$c_{n \to i^*} = d_{i^*0}$$
$$c_{n \to p_{i^*}} = \alpha n_{np_{i^*}} + d_{p_{i^*}0} = \alpha n_{np_{i^*}} + d_{i^*0} - 1$$

because we have $n_{ni^*} = 0$ and $d_{p_{i^*}0} = d_{i^*0} - 1$. Comparing $c_{n \to i^*}$ to $c_{n \to p_{i^*}}$ we can immediately realise that the model has two different regimes according to whether $\alpha < 1$ or $\alpha \geq 1$. The case $\alpha \geq 1$ is trivial, since $c_{n \to i^*} < c_{n \to p_{i^*}}$ and the new node n will always connect to the node i^* directly to its left. The case $\alpha < 1$ is instead more interesting. In fact, node n will link to p_{i^*} when $n_{np_{i^*}} < 1/\alpha$, that is if node p_{i^*} has less than $1/\alpha$ children which are closer to it than n. Otherwise node n will link to i^*.

In general, if we set $a = \lceil 1/\alpha \rceil$, i.e. we call a the smallest integer greater than or equal to $1/\alpha$, and denote as k_p the number of children of a node p, we have that when $k_p < a$, any new node that arrives immediately to the right of p on the interval [0, 1] or immediately to the right of one of its k_p children, will end up linking directly to p. This is the case for the instance of node 0 in Figure 6.12 that, at time $n = 2$, has a degree $k = 2 < \lceil 1/\alpha \rceil = 3$. Therefore, any node that at time $n = 3$ arrives adjacent to node 0 on the line, or to one of its two children, will be linked to 0. This is what happens indeed with node 3. Notice that when $k_p < a$ none of the children of p can get new links. Such nodes are said "infertile" and are shown as white circles in the figure. Notice also that node p will attract a number of nodes that is proportional to $k_p + 1$. In fact, if we assume that the points are all at the same distance on the interval [0, 1], which is a reasonable assumption for sufficiently large values of t, then node p will attract all the points which fall within the interval to the right of p itself, or within the intervals to the right of its k_p children of p, i.e. it will get new links with a probability proportional to $k_p + 1$. Instead when $k_p \geq a$, a node that arrives to the right of p, or to the right of one of the closest $a - 1$ children of p, will connect to p.

Conversely, if the newly arrived node lands to the right of a farther away child of p, that child will acquire the new edge. This is, for instance, the case of node 4 at time $n = 4$ which arrives close to node 1 and therefore is linked to node 1 and not to node 0, since node 0 already has $k_p = 3 \geq a$ edges. In practice, this means that the attractiveness of node p saturates at a, and all but the closest $a - 1$ children of p become "fertile", i.e. they can attract links from new nodes. Fertile nodes are represented as filled circles in Figure 6.12. Since in the example reported we have $\alpha = 1/3$, we see that node 1 becomes fertile at time $n = 3$, when its parent 0 has acquired 3 links. In particular, the probability for a fertile node p to acquire the next edge from the newly arrived node n is equal to $(k_p + 1)/n$ if $k_p < a$, while it is equal to a/n if $k_p \geq a$, where n coincides with the total number of intervals present at the arrival of node n [98].

The saturation of the probability to acquire new edges for an existing node is evident in Figure 6.13(b) where we show, for different values of α, the time evolution of the degree of the largest hub, namely node 0, of a graph with $N = 10^5$ nodes constructed by the BTO model. We notice that $k_{i,t}$ grows up linearly, until a critical value $k_c = \lceil 1/\alpha \rceil$ is reached, and then the acquisition of new edges slows down considerably. This effect is common to all fertile nodes.

The authors of Ref. [32] have proven that the degree distribution of a graph obtained through the BTO model with a parameter α is indeed a power-law $p_k \sim k^{-\gamma}$ with exponent $\gamma = 2$ up to the threshold a, and exhibits an exponentially decaying tail for degrees larger than a. More explicitly:

$$p_k \sim \begin{cases} k^{-2} & \text{for } k \leq a \\ \exp(-k/\kappa) & \text{for } k > a, \end{cases}$$

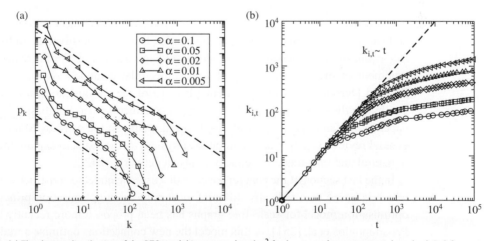

Fig. 6.13 (a) The degree distribution of the BTO model is a power law k^{-2} for $k \leq a$ and an exponential $\exp(-k/\kappa)$ for $k > a$. The curves are vertically displaced to enhance readability, the dashed lines correspond to k^{-2} and the dotted arrows indicate the position of the cut-off degree $1/\alpha$. (b) The degree $k_{i,t}$ of a node i increases linearly in time until $k_{i,t} < a$, and then saturates to a sub-linear growth.

with $\kappa = [\log(1 + 1/a)]^{-1}$. Figure 6.13(a) reports the degree distributions obtained for various values of α ($N = 10^5$) in a double logarithmic scale We notice that the central part of each distribution follows a power law $p_k \sim k^{-\gamma}$ with $\gamma \approx 2$. Then p_k starts decreasing exponentially for values of degree larger than a cut-off equal to a.

The growth rules of the BTO allow a simple numerical implementation of the model. For instance, going again to the example in Figure 6.12, when node 4 arrives we do not need to consider all the previous nodes, but only the one immediately to its left (in this case node 1), and its parent, node 0. Since $n_{40} = 3$ which is equal to $1/\alpha$, then node 4 links to node 1.

One drawback of the original formulation of the BTO model is that the value of the power-law exponent is fixed, and the parameter α determines only the value of k after which p_k turns into an exponential. In order to better describe real systems, the same authors of Ref. [32] have also considered a generalisation of the BTO model with two independent parameters, which allow both the exponent γ of the power-law of the degree distribution and the value of the cut-off degree to be tuned.

6.7 What We Have Learned and Further Readings

Real-world networks are very often the result of a dynamic process where the number of nodes and also the number of links change (generally increase) over time. In this chapter we have introduced a new type of real-world networks, namely *networks of citations* among scientific papers, and we have studied the basic mechanisms that rule their growth. We have then shown that power-law degree distributions can indeed be obtained by modelling the growth of a network and adopting a so-called *linear preferential attachment* mechanism in which newly arriving nodes select and link existing nodes with a probability linearly proportional to their degree. We have therefore introduced the *Barabási–Albert model* and various other generalisations of this model, and we have solved them by means of the *rate equations* approach and in the *mean-field* approximation. For the exact mathematical treatment of the BA model and of some other related models we suggest the book by Richard Durrett [99], while for a complete list of the various models to grow scale-free graphs see the review paper by Réka Albert and Albert László Barabási [3] and the review paper by Stefano Boccaletti et al. [43]. We also suggest the work by Mark Newman for a general review on power laws and on the physical mechanisms to explain their occurrence in natural and man-made systems, not only in networks [242].

In the last section of the chapter we have discussed a model to construct scale-free graphs which is based on a radically different idea, namely on optimisation principles. Another optimisation model for scale-free graphs has been proposed more recently by Fragkiskos Papadopoulos et al. [251]. In this model the new connections optimise a trade-off between popularity, which means linking to the oldest nodes in the network, and *homophily*, that is linking to similar nodes in an appropriate space of node features. Very interestingly the model shows that linear preferential attachment can also emerge from optimisation principles. In this way, optimisation provides an alternative explanation to the commonly accepted mechanism (namely, attaching to first neighbours of randomly selected nodes as in the model in Definition 6.5) to produce an exactly linear preferential attachment, and fuels an old debate between luck and reason [17].

Problems

6.1   **Citation Networks and the Linear Preferential Attachment**

Compare the productivity of three authors, A, B and C. Suppose author A has published 5 papers, respectively with 25, 4, 7, 5 and 2 citations, author B has published 7 papers, respectively with 6, 1, 20, 5, 2, 3 and 1 citations, and author C has published 6 papers, respectively with 6, 10, 2, 14, 8 and 6 citations. Which of the three authors has got the largest h-index?

6.2   **The Barabási–Albert (BA) Model**

Derive the degree distribution of the BA model in the *mean-field* approximation. Suppose to be at time t in the growth of the BA model, and consider the generic node i introduced at time t_i (with $t_i = 1, 2, \ldots, t$). By making use of the expression (6.11) which gives the time evolution of the degree $k_{i,t}$ as a function of t and t_i, evaluate the cumulative distribution $P(k_{i,t} \geq k)$ that node i has, at time t, a degree $k_{i,t}$ larger than or equal to k. Then, find the degree distribution $p(k)$ by differentiating:

$$p(k) = \frac{dP(k_{i,t} < k)}{dk}$$

and compare the result with the expression in Eq. (6.10) obtained by means of the rate equations approach.

6.3   **The Importance of Being Preferential and Linear**

(a) The *Yule process* is a well-known mechanism to generate power laws, originally introduced to explain why the distribution of the number of biological species in a genus is a power law. A genus is a principal taxonomic category that ranks above species and below families. The Yule process starts with the coexistence of different genera, each one consisting of a certain number k of species, and considers the addition of new species, by the mechanism of *speciation*, that is by the splitting of one existing species into two. Usually, the new species is added to one of the already existing genera, while occasionally it is so different from the others as to be considered the founder member of an entire new genus. This is mathematically modelled assuming that, at each time step, a new species initiates a new genus, while $m > 1$ other species are instead added to other preexisting genera, which are selected in proportion to the number of species k they already have. This is plausible since in a genus with k species each of the k species has the same chance per unit time of dividing in two. Denote by $n_t = t$ the number of genera at time t, and by $p_{k,t}$ the fraction of genera that have k species at time t. Write down the rate equations for the number of genera with k species, and solve them in the limit of large times.

(b) Consider a modification of the de Solla Price model in which Eq. (6.14) is replaced by:

$$\Pi_{n \to i} = \frac{k_{i,t-1}^{\text{in}} + a}{\sum_{l=1}^{n-1}(k_{l,t-1}^{\text{in}} + a)}$$

where $a > 0$. Following the same arguments as in the Example 6.5, prove that the in-degree distribution of the network is:

$$p_k = \left(\frac{1 + 1/m}{a + 1 + 1/m}\right) \frac{B(k + a, 2 + 1/m)}{B(a, 2 + 1/m)}.$$

This expression, for large values of k, has a power-law behaviour with exponent $\gamma = 2 + 1/m$, as in the original model with $a = 1$.

(c) Stirling formula for the factorial can be used to approximate the Γ function for large values of the arguments:

$$\Gamma(x + 1) \sim \exp(x(\ln x - 1)).$$

Use such approximation to prove that, for fixed y and large values of x one has $B(x, y) \sim \Gamma(y) x^{-y}$.

(d) Consider a generalisation of the BA model in which the added nodes are linked to previously existing nodes with a probability that is a nonlinear function of the node degree, with an exponent $\alpha \geq 0$. Namely, the probability $\Pi_{n \to i}$ that the new node n is connected to node i (with $i = 1, 2, \cdots, n - 1$) of Eq. (6.1) is now replaced by:

$$\Pi_{n \to i} = \frac{k_{i,t-1}^{\alpha}}{\sum_{l=1}^{n-1} k_{l,t-1}^{\alpha}}.$$

Write down the rate equations of this model, that is the analogous of Eqs. (6.6) we have derived for the BA model.

6.4 Variations to the Theme

Derive the differential equation (6.24) controlling the time evolution of $\bar{k}_i(t)$ in the AB model in the so-called *mean-field* approximation, as we did in the Example 6.4 for the BA model. Such differential equation can be obtained by adding the contributions of the three mechanisms of the model, denoted respectively as (a), (b) and (c) in Definition 6.8. For instance, process (a), namely the addition of m new links with a probability p, leads to the contribution [4]:

$$\left(\frac{d\bar{k}_i}{dt}\right)_{(a)} = pm\frac{1}{\bar{n}(t)} + pm\frac{\bar{k}_i + 1}{\sum_j(\bar{k}_j + 1)}$$

where $\bar{n}(t)$ is the average number of nodes at time t. The first term on the right-hand side corresponds to the random selection of one end of the new link, while the second term represents the preferential attachment used to select the other end of the link. Derive now $\left(\frac{d\bar{k}_i}{dt}\right)_{(b)}$ and $\left(\frac{d\bar{k}_i}{dt}\right)_{(c)}$ and add up the three contributions to get Eq. (6.24).

6.5 Can Latecomers Make It? The Fitness Model

(a) Derive the recursion relation Eq. (6.41) for the degree distribution $p_k(a)$ of the fitness model.

(b) Consider the fitness model with a uniform fitness distribution, namely $\rho(a) = 1$ when $a \in [0, 1]$, and $\rho(a) = 0$ otherwise. Prove that, in this case, the constant c is equal to 1.255.

(c) Start with the expression for the degree distribution $p_k(a)$ of nodes with attractiveness a found in Example 6.13 and prove that, for large values of k, it reduces to a power law $p_k(a) \sim k^{-\gamma(a)}$ with an exponent $\gamma(a) = 1 + c/a$ as in Eq. (6.38). HINT: express $p_k(a)$ in terms of the beta function $B(x, y) = \frac{\Gamma(x)\Gamma(y)}{\Gamma(x+y)}$, and make use of the asymptotics $B(x, y) \sim \Gamma(y)x^{-y}$ valid for large x and fixed values of y.

6.6 Optimisation Models

(a) Prove that a new node n in the BTO model will connect either to the node i^* directly to its left, or to the parent p_{i^*} of node i^*. Consider the example shown below, corresponding to a growth with $\alpha = 1/4$ [98]. First show that, for any node i standing on the line between node p_{i^*} and node i^*, we have $c_{n \to i} > c_{n \to i^*}$. Then, show that the new node n cannot connect any node to the left of p_{i^*}.

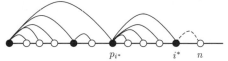

(b) Optimisation principles have also been used to produce scale-free graphs by non-growing models. Ramon Ferrer i Cancho and Ricard Solé have proposed a method to produce graphs with a fixed number of nodes N and having, at the same time, few links and a small typical distance between nodes. The idea is to construct a graph G by changing the number and the arrangement of the links in order to minimise the following cost function [67]:

$$c(G, \alpha) = \alpha \frac{L}{(N+1)/3} + (1 - \alpha) \frac{\sum_{i<j} a_{ij}}{\binom{N}{2}}$$

where $A = \{a_{ij}\}$ is the adjacency matrix of the graph and $0 \leq \alpha \leq 1$ is a parameter to tune the weight of the two different contributions. The first term in $c(G, \alpha)$ is the average distance L between two nodes in the graph (see definition in Eq. (3.32)), normalised by the maximum value L can take in a connected graph with N nodes, that is equal to $L^{\text{chain}} = (N+1)/3$, as we have seen in Example 3.12. The second term is the graph density, i.e. the ratio between the number of links in the graph an the maximum possible value of such number. Write a computer program to construct graphs that minimise the cost function for different values of α, and analyse the degree distributions you find. Also plot K, L and the cost as a function of α. HINT: we suggest you implement the following algorithm. Start with a connected random graph. At each time step, switch a few elements of the adjacency matrix of graph G, from 0 to 1 or from 1 to 0, ensuring you maintain the connectedness of the graph. Call the new graph G'. Then accept the new adjacency matrix if $c(G', \alpha) < c(G, \alpha)$. Otherwise, try again with a different set of changes.

7

Degree Correlations

In the previous two chapters we have concentrated on the study of the degree distribution of a network. Evaluating the degree distribution p_k of a graph requires the number of links of each node to be counted. However, the function p_k alone is sufficient to describe how the degree is distributed among the graph nodes only when there are no degree–degree correlations, that is, when the probability for an edge of a node of degree k to be connected to a node of degree k' depends on k' only, and not on the degree k. In the previous chapters we have implicitly assumed that all networks were uncorrelated. In most cases, however, this assumption does not hold. *Degree–degree correlations* are indeed present in real-world networks such as the Internet and the movie actors collaboration network. For instance, in the *Internet* we find that there are more links between high-degree nodes and low-degree nodes than expected if the network was truly random. Conversely, in the movie actor network, as well as in many other social networks, hubs tend to be more connected to hubs. In this chapter we will learn the mathematics necessary to describe degree–degree correlations, focusing in particular on the methods to quantify the *assortativity* or the *disassortativity* of a real-world network, or to detect the presence of the so-called *rich club*. We will also study how to construct graphs with positive or negative degree–degree correlations.

7.1 The Internet and Other Correlated Networks

As a first example, let us consider the collaboration network of movie actors we introduced in Chapter 2. It is certainly an interesting question to ask whether gregarious actors preferentially play movies with other gregarious actors or with the famous ones. To answer this question, we have to check whether in the graph introduced in DATA SET 2, the nodes with a small degree k preferentially associate with other low-degree nodes, or to high-degree ones. A simple way to do this check is to compare the typical degree of the first neighbours of low-degree nodes, to the typical degree of the first neighbours of high-degree nodes. Let us consider, for instance, two classes of nodes: those with a degree k ranging from 1 to 10, and those with k ranging from 1000 to 1010. If we compute the average degree k' of the neighbours of nodes in the first class, we find a value equal to 143.6. This is reported in Figure 7.1(a). Now, do not be misled by the fact that this number is much larger than 10. This is due to the long-tailed degree distribution of the network. We know, in fact, from Section 3.6 that, in an uncorrelated graph, the average degree of the neighbours of a randomly chosen node, $\langle k_{\mathrm{nn}} \rangle$, is equal to $\langle k^2 \rangle / \langle k \rangle$. In a graph with a scale-free distribution the

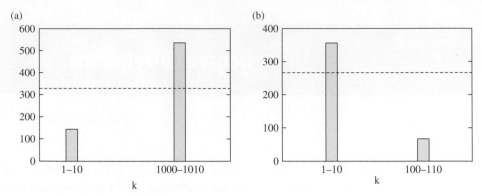

Fig. 7.1 Average degree k' of the first neighbours of two classes of nodes. Two real networks are shown, namely (a) the movie actors collaboration and (b) the Internet at the AS level, as it appears in May 2001. The dashed lines represent the average degree of first neighbours of randomly chosen nodes in the randomised version of the two networks.

quantity $\langle k^2 \rangle / \langle k \rangle$ can indeed be a number much larger than $\langle k \rangle$ because, as discussed in Chapter 5, a node with high degree has higher chances to appear as a first neighbour of a randomly chosen node than as a randomly chosen node itself. For instance, in the case of the movie actors we get $\langle k^2 \rangle / \langle k \rangle = 324.8$, which is a number even larger than 143.6. This value is reported as a dashed line in Figure 7.1(a). If we focus now on the second class of nodes, i.e. on nodes with degree ranging from 1000 to 1010, the average degree k' of their neighbours is equal to 534.8. Therefore, as shown in Figure 7.1(a), the value of k' depends on that of k, and is positively correlated to k. In practice this indicates that movie actors work preferentially with other actors of about the same importance. This is not surprising if we think that stars appear often together in high-budget movies, while the cast in low-cost movies is composed almost exclusively by less famous actors.

Are degree–degree correlations typical of social systems only, or we can also find them in other networks? As an example of a man-made network, we will study here the Internet, a world-wide communication infrastructure which allows the interchange of data and information between computers. We need to stress here that the Internet is different from the World Wide Web studied in Section 5.1. In fact, the former is a system made of electronic devices interconnected through cables and routing protocols, and in this it is more similar to other media infrastructures like the telephone or the radio. The latter is a network of text pages accessible over the Internet and connected through hyperlinks, and in this sense is more similar to other information systems like newspapers, research articles, etc. From a technical point of view, the Internet is traditionally defined as the collection of computer networks able to communicate through the TCP/IP protocols (Transmission Control Protocol/Internet Protocol). Despite being historically correct and sufficient for the aims of this book, we have to say that this definition of the Internet is not accurate enough, since the actual exchange of information between computers in the Internet is indeed made possible by dozens of other protocols, responsible for different aspects of the communication. Each of the networks of the Internet is composed by *hosts* (computers, workstations, laptops, smartphones, etc.), *servers* (powerful hosts which make available resources and information to hosts and other servers) and *routers* (special computers in

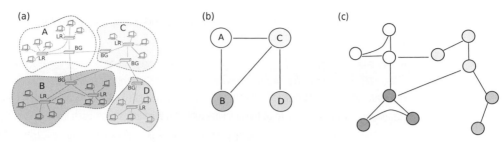

Fig. 7.2 The Internet is a communication infrastructure which connects computers across the world (a). The traffic over the Internet is managed by local routers (LR) and border gateways (BG). The Internet can be mapped into a network at two different scales: the AS network (b) and the router network (c).

charge of managing the packets travelling between hosts), which might be interconnected through different physical media, e.g. copper wires, fibre cables, radio channels or even satellite connections. A collection of networks controlled by the same entity or network operator is called *autonomous system* (*AS*). Each AS has its own internal rules and routing algorithms, and can exercise traffic preferences and restrictions to all the data packets exchanged within the AS, according to the old saying "What happens in the AS, remains in the AS". A schematic diagram of the Internet is shown in the left panel of Figure 7.2, where the computer icons represent hosts and servers, and each of them is connected to a *local router* (LR). The four large shaded areas represent different autonomous systems. It is important to notice that all the traffic between hosts/servers belonging to the same AS is managed by the local routers of the AS, which make use of the so-called *interior routing protocols*. Conversely, the communication between hosts/servers belonging to different autonomous systems is mediated by a chain of local routers and *border gateways* (*BG*). The latter are in charge of forwarding data packets from their AS to the destination AS, and vice versa. Each border gateway continuously exchanges information with other border gateways about the actual topology of the Internet, i.e. about the path each packet has to traverse in order to get to its destination, by means of efficient distributed algorithms called *exterior routing protocols*.

The difference in the roles played by local routers and border gateways is the secret behind the success of the Internet, and has played a fundamental role in the expansion of the initial network, which included just a few dozens of hosts grouped in autonomous systems controlled by universities and military departments in the USA, to a worldwide communication infrastructure counting thousands of independent autonomous systems and hundreds of millions of hosts. In fact, the only requirement for a new AS willing to join the Internet is to connect to some of the existing autonomous systems by means of border gateways which support the current border gateway protocols, independently of the actual structure of the interconnections of hosts, servers and routers within the new AS. For this reason, the Internet has effectively grown in a self-organised way, without any centralised authority to control its structure, and it is therefore very interesting to study its network [253]. There are several topological differences between the network of the Internet at the AS level and the router network, especially in properties such as the degree–degree correlations in which we are interested in this chapter.

Here, we will be focusing on the fifteen network samples of the *Internet at the AS level* described in DATA SET Box 7.1.

Box 7.1   **DATA SET 7: The Network of the Internet**

The structure of the *Internet* can be mapped into a network at two different scales, namely at the level of *autonomous systems* (*AS*), and at the level of *routers*. In the former case, each node of the graph represents an AS, and an edge indicates a connection between two ASs, as depicted in Figure 7.2(b). In the latter case, the nodes of the graph are local routers and border gateways, and edges indicate the existence of a physical connection between them, as shown in Figure 7.2(c). As an example of the Internet at the AS level, we show in the figure a subgraph of the National Research Networks of European countries. Here, each node is an independent AS corresponding to a national research network infrastructure, and links represent peering agreements between ASs [158].

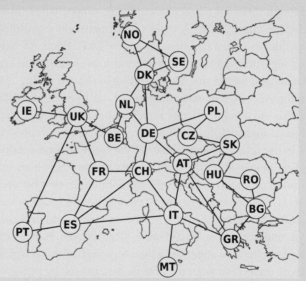

The networks provided in this data set are fifteen different snapshots of the Internet at the AS level in the period 1997–2001, and have been constructed from data made available by the Center for Applied Internet Data Analysis (CAIDA) [160]. The order of the networks ranges from $N = 3,015$ nodes in November 1997 to $N = 11,174$ in May 2001.

Since DATA SET 7 contains samples of the same network at different times, it gives us the opportunity to explore the growth of the Internet. Following the work of M. Ángeles Serrano, Marián Boguñá and Albert Díaz-Guilera [282] we plot, in Figure 7.3, the temporal evolution of the number of nodes, $N(t)$, and of the number of links, $K(t)$, in a linear–logarithmic scale. The linear behaviours found indicate that the Internet has grown in an exponential way. The two lines are the best fit estimates with exponential growth $N(t) \sim N_0 e^{at}$ and $K(t) \sim K_0 e^{bt}$, and the respective exponents found are $a = 0.030$ and $b = 0.033$. In Table 7.1 we report the precise values of N, K, average degree $\langle k \rangle$, clustering

Table 7.1 Basic properties of five of the fifteen networks of the Internet at the AS level in DATA SET 7.

Year	Nov 1997	Oct 1998	Oct 1999	Oct 2000	May 2001
N	3015	4,180	5,861	8,836	11,174
K	5156	7,768	11,313	17,823	23,409
$\langle k \rangle$	3.4	3.7	3.9	4.0	4.2
C	0.18	0.23	0.25	0.29	0.30
L	3.8	3.8	3.7	3.6	3.6

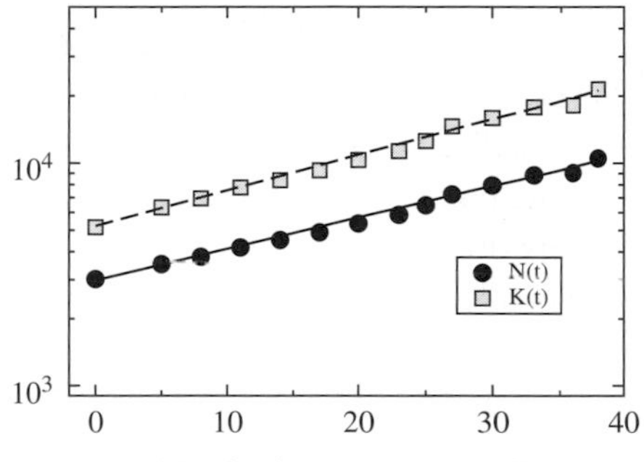

Fig. 7.3 Growth of the number of nodes N and links K in the network of the Internet at the AS level in the period from November 1997 to May 2001.

coefficient C and characteristic path length L for five of the fifteen samples. We notice that, while the number of nodes in May 2001 is 3.7 times larger than that observed in 1997, the average degree has increased only by less than one. In the same time interval, the clustering coefficient has almost doubled, while the characteristic path length has remained almost constant. Same thing for the degree distribution that is a power law with a nearly constant exponent 2.2 [306].

We are now ready to investigate degree correlations in the Internet at the AS level. In order to do so, we report in Figure 7.1(b) the average degree k' of the first neighbours of two classes of nodes in the network of the Internet as it appears in May 2001. As in the case of the movie actors, the first class of nodes has a degree k ranging from 1 to 10. Since the average degree of the Internet is smaller than that of the movie actors, as a second class of nodes we consider here those with k ranging from 100 to 110. Similarly to the case of the movie actors, we find that the network of the Internet has degree–degree correlations, However, in the Internet we observe a decreasing trend: the average degree of first neighbours of nodes in the first class is equal to 356.1, while the average degree of neighbours of nodes with k ranging from 100 to 110 is 67.0. The dashed line, as usual, corresponds to the expected average degree of neighbours of randomly chosen nodes $\langle k^2 \rangle / \langle k \rangle = 265.6$ if the network was uncorrelated. From Figure 7.1 it is therefore clear that the Internet is a correlated network, but its degree–degree correlations are of a different type of those found in the network of movie actors. In the next section we will learn how to describe and quantify correlations in a network.

7.2 Dealing with Correlated Networks

Historically, the Internet was indeed the first complex network where degree–degree correlations were explored and found. This result was published in *Physical Review Letters* in 2001, in a seminal work by Romualdo Pastor-Satorras, Alexei Vázquez and Alessandro Vespignani [254]. In this paper, and in a series of following ones, the authors also proposed a way to detect degree–degree correlations in complex networks [306, 255]. In this and in the next two sections we will introduce the mathematics to treat degree–degree correlations. We will deal with undirected graphs, and we will learn, step by step, the main methods which are commonly used today to detect and quantify degree–degree correlations in real-world networks.

The most general way to describe the connectivity properties of a correlated graph is to consider, together with the degree distribution p_k, the so-called *conditional probability* $p_{k'|k}$, or alternatively the *joint probability* $p_{kk'}$, defined as follows [254].

Definition 7.1 (Conditional probabilities)   *The* conditional probability $p_{k'|k}$ *is the probability that a link from a node of degree k is connected to a node of degree k'. Conditional probabilities satisfy the normalisation condition:*

$$\sum_{k'} p_{k'|k} = 1 \quad \forall k. \tag{7.1}$$

Definition 7.2 (Joint probabilities)   *The* joint probability $p_{kk'}$ *is the probability that a randomly chosen link goes from a node of degree k to a node of degree k'. Joint probabilities satisfy the normalisation condition:*

$$\sum_{k,k'} p_{kk'} = 1. \tag{7.2}$$

Joint and conditional probabilities characterise completely the type of degree–degree correlations in a network. Although both probabilities evaluate the likelihood that a link ends in a node of degree k', they are different objects and have different normalisations as you can immediately notice by comparing Eq. (7.1) to Eq. (7.2). In fact, in the conditional probability $p_{k'|k}$ one considers only links from nodes of degree k, and therefore the normalisation is obtained by summing over k' only. Eq. (7.1) means that a link from a node of degree k has to be connected either to a node of degree 1, or to a node of degree 2, and so on. Conversely, the joint probability $p_{kk'}$ considers links from nodes of any degree k, and it has to be summed over k and k' to give 1.

We observe that the probability $p_{kk'}$ of going from a node of degree k to a node of degree $k' \neq k$ is the same as the probability $p_{k'k}$ of going from a node of degree k' to a node of degree k, because we are considering undirected graphs, therefore matrix $\{p_{kk'}\}$ is symmetric. Consequently, the probability that a randomly selected edge of the graph connects nodes of degrees k and $k' \neq k$ is $p_{kk'} + p_{k'k} = 2p_{kk'}$. Of course, the probability that a randomly chosen edge joins a node of degree k to another node of degree k is simply p_{kk}.

In practice, the joint and conditional probabilities are obtained by counting the number of edges between pairs of nodes of degree k and k'. Let us first define a matrix $E \equiv \{e_{kk'}\}$ in the following way. The entry $e_{kk'}$ is equal to the number of edges between nodes of degree k and nodes of degree k' for $k \neq k'$, while e_{kk} is *twice* the number of links connecting two nodes having both degree k. With such a definition, matrix E is symmetric and satisfies the conditions:

$$\sum_{k'} e_{kk'} = kN_k \tag{7.3}$$

$$\sum_{k} \sum_{k'} e_{kk'} = \langle k \rangle N = 2K. \tag{7.4}$$

The first relation states that the number of edges, more precisely the number of stubs, emanating from all the nodes of degree k is kN_k, where N_k is the number of nodes of degree k. The second relation tells us that the number of stubs emanating from nodes of any degree is twice the total number of links in the graph.

Example 7.1 *(Counting edges)* Consider the graph with $N = 6$ and $K = 8$ shown in the figure. The graph has one node with $k = 1$, two nodes with $k = 2$, one node with $k = 3$ and two nodes with $k = 4$.

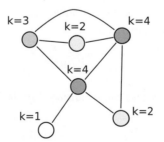

To construct matrix E we need to count the number of edges between nodes of degree k and nodes of degree k'. To perform such a counting we can label the eight edges and scan them one by one. Suppose the first edge we scan is that connecting a node with degree 3 to a node of degree 2. Then, we increase both entry e_{23} and e_{32} by 1. In general, if the edge connects a node of degree k and a node of degree k', we set $e_{kk'} = e_{kk'} + 1$ and $e_{k'k} = e_{k'k} + 1$. Notice that, if the edge connect two nodes of the same degree k, then this procedure corresponds to increase the diagonal term e_{kk} by 2: $e_{kk} = e_{kk} + 2$. In our graph there is only one edge connecting two nodes with the same degree, namely $k = 4$. We therefore have set $e_{44} = 2$. This is better understood if we count the edges in a different way, that is by scanning one by one all the nodes of the graph. In this way each edge is counted twice. When we scan the node with degree 1, we find that its edge goes to a node of degree $k = 4$ and we set $e_{14} = e_{14} + 1$. Only when the same edge is scanned starting from the node with degree $k = 4$, we will set $e_{41} = e_{41} + 1$. Also the edge connecting the two nodes with $k = 4$ will be scanned twice, and each time we will set $e_{44} = e_{44} + 1$. Matrix E finally reads:

$$E = \begin{pmatrix} 0 & 0 & 0 & 1 \\ 0 & 0 & 1 & 3 \\ 0 & 1 & 0 & 2 \\ 1 & 3 & 2 & 2 \end{pmatrix}, \tag{7.5}$$

Notice that the sum $\sum_{k'} e_{1k'}$ of the first row of E is equal to the number of links connected to nodes of degree 1, namely $1 \cdot N_1 = 1$. The sum $\sum_{k'} e_{2k'}$ of the second row is equal to the number of links connected to nodes of degree 2, that is $2 \cdot N_2 = 4$, and so on. In general we have $\sum_{k'=1}^{4} e_{kk'} = kN_k$, and $\sum_{k=1}^{4} \sum_{k'=1}^{4} e_{kk'} = 2K = 16$.

The conditional probability $p_{k'|k}$ can be expressed in terms of $e_{kk'}$ as:

$$p_{k'|k} = \frac{e_{kk'}}{\sum_{k'} e_{kk'}} = \frac{e_{kk'}}{kN_k} \tag{7.6}$$

We can also express the *joint probability distribution* $p_{kk'}$ in terms of $e_{kk'}$. Since $p_{kk'}$ is the probability that a randomly chosen link goes from a node of degree k to a node of degree k', this time we have:

$$p_{kk'} = \frac{e_{kk'}}{\sum_{k,k'} e_{kk'}} = \frac{e_{kk'}}{\langle k \rangle N} = \frac{e_{kk'}}{2K}. \tag{7.7}$$

Joint and conditional probabilities contain the same information, although there are important differences. For instance $p_{kk'}$ is symmetric, while $p_{k'|k}$ is not. From Eq. (7.6) and Eq. (7.7) we can write:

$$p_{kk'} = \frac{kN_k}{2K} p_{k'|k} = \frac{kp_k}{\langle k \rangle} p_{k'|k}.$$

Hence the *joint* and the *conditional probability distributions* $p_{kk'}$ and $p_{k'|k}$ are related through:

$$p_{kk'} = q_k p_{k'|k} \tag{7.8}$$

where q_k is the degree distribution of the nodes to which a randomly chosen edge leads, defined in Eq. (3.38).

This expression is in agreement with the standard definitions of joint and conditional probabilities. In fact, given two events a and b, the conditional probability $p(b|a)$ of b given a is the probability of having b if a is known to occur. The joint probability $p(a, b)$ of occurrence of both event a and event b is equal to the probability $p(a)$ of having a, times the conditional probability of having b given a: $p(a, b) = p(a)p(b|a)$. Probability $p(a)$ is the so-called *marginal probability* associated with the joint probability $p(a, b)$. The marginal probability is the probability of one variable taking a specific value irrespective of the values taken by the other variable, and can in fact be written as: $p(a) = \sum_b p(a,b)$.

In the case of Eq. (7.8), the joint probability $p_{kk'}$ is equal to the probability q_k of finding a node of degree k by following a link in the graph, times the probability $p_{k'|k}$ that a link from a node of degree k is connected to a node of degree k'. Notice that summing Eq. (7.8) over k', and using Eq. (7.1), gives:

$$\sum_{k'} p_{kk'} = q_k \tag{7.9}$$

which means that q_k is the marginal probability associated with the joint probability $p_{kk'}$.

Another important property relating the degree distribution and the conditional probability is the following:

$$k\, p_{k'|k}\, p_k = k'\, p_{k|k'}\, p_{k'}. \tag{7.10}$$

This equality, known as *detailed balance condition*, states the conservation of edges among nodes. In fact, multiplying both sides by N we get: $kN_k\, p_{k'|k} = k'N_{k'}p_{k|k'}$ In the left-hand side, kN_k is the total number of edges departing from nodes of degree k, and $p_{k'|k}$ is the fraction of them ending into nodes of degree k'. Therefore the product is the number of edges from nodes of degree k to nodes of degree k'. Finally, relation (7.10) indicates that the number of edges pointing from vertices of degree k to vertices of degree k' is equal to the number of edges pointing from vertices of degree k' to vertices of degree k. The detailed balance condition (7.10) emerges naturally from expression (7.8) or (7.6) and from the fact that $e_{kk'}$ and $p_{kk'}$ are symmetric.

We are now ready to give a precise definition of graph with or without degree–degree correlations in terms of joint and conditional probabilities.

Definition 7.3 (Correlated and uncorrelated graphs)   *A graph has no degree–degree correlations and is said to be* uncorrelated *if, for any possible value of k, the conditional probability $p_{k'|k}$ does not depend on the degree k. Otherwise, the graph is said to be* correlated.

We can easily find the precise expression for the conditional probability in uncorrelated graphs. In fact, by summing over k both sides of the detailed balance condition in Eq. (7.10), and using the normalisation condition $\sum_k p_{k|k'} = 1$ we get:

$$\sum_k k\, p_{k'|k}\, p_k = k'\, p_{k'}.$$

If the graph is uncorrelated, the quantity $p_{k'|k}$ does not depend on k and can be moved out of the summation, leading to the result below.

In uncorrelated networks the conditional probability reduces to the following expression:

$$p_{k'|k}^{\mathrm{nc}} = \frac{k' p_{k'}}{\langle k \rangle} = q_{k'}. \tag{7.11}$$

In the formula above we have used the superscript "nc" to refer to networks with no degree–degree correlations. If a graph has no degree–degree correlations, the probability of ending up in a node of degree k', starting from a node of degree k, reduces to $q_{k'}$, that is the probability of ending up in a node of degree k' by choosing an edge at random with uniform probability. This result better explains some of the properties of uncorrelated graphs used in Section 3.6 for ER random graphs and in Section 5.4 for random graphs with assigned p_k. The same result of Eq. (7.11) can be expressed in terms of joint probabilities as below.

> In uncorrelated networks the joint probability factorises as:
>
> $$p_{kk'}^{\text{nc}} = q_k q_{k'}. \tag{7.12}$$

Therefore, in a network with no degree–degree correlations, the probability that a randomly chosen link goes from a node of degree k to a node of degree k' is equal to the product of the two marginal probabilities q_k and $q_{k'}$

Example 7.2 *(The average degree of the first neighbours of a node)* We have seen in Section 3.6 and in Section 5.4 that, in uncorrelated graphs, the average degree $\langle k_{\text{nn}} \rangle$ of the first neighbours of a node chosen at random is equal to $\langle k^2 \rangle / \langle k \rangle$. We want now to evaluate $\langle k_{\text{nn}} \rangle$ for a graph with degree–degree correlations and an assigned conditional probability $p_{k'|k}$. We need to start from the original definition of $\langle k_{\text{nn}} \rangle$ given in Eq. (3.37) and express the summation in terms of degree classes instead of graph nodes. When we select the first node at random, we find a node with degree k with a probability p_k. From this node we then end up into a node of degree k' with a probability equal to $p_{k'|k}$, that is in general different from $q_{k'}$. Hence, the average nearest neighbours degree in correlated networks can be written as:

$$\langle k_{\text{nn}} \rangle = \sum_k p_k \sum_{k'} k' p_{k'|k}. \tag{7.13}$$

This expression is of general validity and extends Eq. (3.40), which instead works only for uncorrelated graphs (see also Problem 7.2(d) for the several different ways of averaging the degrees of the nodes of a graph). In particular, it is also valid for uncorrelated graphs and can be used to get Eq. (3.40). In fact, if a graph has no correlations we can write $p_{k'|k}^{\text{nc}} = q_{k'}$ and Eq. (7.13) reduces as expected to $\langle k_{\text{nn}} \rangle = \sum_{k'} k' q_{k'} = \langle k^2 \rangle / \langle k \rangle$.

The Structural Cut-Off in Correlated Networks

We have already seen in Chapter 5 that, together with a *natural cut-off* in the largest possible degree due to the finite order of a network, there is also a *structural cut-off* which appears when we do not allow multiple edges in a network. We have also found that the structural cut-off grows with the number of nodes as in Eq. (5.48), if the network is uncorrelated. However, the structural cut-off can also be present in networks with degree–degree correlations. Following again the work by Marián Boguñá, Romualdo Pastor-Satorras and

Alessandro Vespignani [45], we will derive here a general expression for the structural cut-off as a function of the joint degree distribution $p_{kk'}$ of a network. Let us start by considering the ratio $R_{kk'}$ between the number of edges $e_{kk'}$ connecting nodes of degree k and nodes of degree k', and the maximum possible value for this quantity, $e_{kk'}^{\max}$. We can evaluate the maximum number of edges between nodes of degree k and nodes of degree k' as the smallest of the three following quantities: the number of edges from nodes of degree k, the number of edges from nodes of degree k', and the product $N_k N_{k'}$. In fact, the maximum number of edges $e_{kk'}^{\max}$ cannot be larger than the total number of edges from nodes of degree k, which is equal to kN_k, or than the number of edges from nodes of degree k', which is equal to $k'N_{k'}$. But, at the same time, there is a third constraint that comes from the fact that we do not allow multiple edges between the same pair of nodes: since there can be at most one link between two nodes, then $e_{kk'}^{\max}$ cannot be larger than $N_k N_{k'}$.

Example 7.3   Suppose we want to compute $e_{34}^{\max}$ in the network shown in the figure, where we have three nodes of degree 3 (white circles), and two nodes of degree 4 (grey circles).

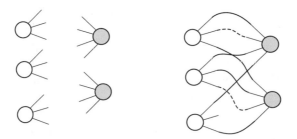

On the one hand, the number $e_{34}^{\max}$ of links between nodes of degree 3 and nodes of degree 4 cannot be larger than the total number of links from nodes of degree 3 or from nodes of degree 4. Since we have $3N_3 = 9$ links from nodes of degree 3 and $4N_4 = 8$ links from nodes of degree 4, as shown in the left panel, the number $e_{34}^{\max}$ can be at most equal to 8. However, there is still another constraint to consider if we do not allow multiple links. In fact, since $N_3 = 3$ and $N_4 = 2$, the maximum number of pairs of nodes, respectively of degree 3 and 4, we can connect is equal to 6 (the corresponding edges are reported as solid lines in the right panel of the figure). Therefore, in the example considered, $e_{34}^{\max} = 6$. The situation would be different if multiple edges would be allowed. In that case we would have $e_{34}^{\max} = 8$, considering also the dashed-line links shown in the right panel.

Summing up, we have:

$$R_{kk'} = \frac{e_{kk'}}{e_{kk'}^{\max}} = \frac{\langle k \rangle N p_{kk'}}{\min[kN_k, k'N_{k'}, N_k N_{k'}]}$$

where, at the numerator, we have used Eq. (7.7) to express $e_{kk'}$ as a function of the joint probability distribution $p_{kk'}$. Suppose now that $k > N_{k'}$ and $k' > N_k$, so that the third term in the denominator is the smallest one. In this case the value of $e_{kk'}^{\max}$ is determined by the constraint of forbidding multiple links. The expression above reduces to:

$$R_{kk'} = \frac{\langle k \rangle N p_{kk'}}{N_k N_{k'}} = \frac{\langle k \rangle p_{kk'}}{N p_k p_{k'}}.$$

By definition the ratio $R_{kk'}$ must be smaller or equal to 1 for any value of k and k'. We can now define a structural cut-off $k_{\max}^{\text{struct}}$ as the largest value of the degree k such that [45]:

$$R_{kk'} \leq 1 \quad \forall k'. \tag{7.14}$$

For correlated networks the position of the cut-off will depend on the nature of the correlations through the specific form of $p_{kk'}$. In the case of uncorrelated networks, Eq. (7.14) leads to Eq. (5.48), as shown in the following example.

Example 7.4 *(A simple expression for uncorrelated networks)* For uncorrelated networks the joint distribution factorises, namely $p_{kk'} = q_k q_{k'}$, and we get the following simple expression for $R_{kk'}$:

$$R_{kk'} = \frac{\langle k \rangle q_k q_{k'}}{N p_k p_{k'}} = \frac{kk'}{\langle k \rangle N}.$$

Now, if $k_{\max}^{\text{struct}}$ is the largest degree in the network, since the highest value of the ratio $R_{kk'}$ is obtained when $k = k' = k_{\max}^{\text{struct}}$, we can derive $k_{\max}^{\text{struct}}$ from the condition:

$$\frac{(k_{\max}^{\text{struct}})^2}{\langle k \rangle N} = 1. \tag{7.15}$$

We therefore get the structural cut-off for uncorrelated networks given in Eq. (5.48).

7.3 Assortative and Disassortative Networks

In order to quantify degree–degree correlations in a real network we can compare the joint probability $p_{kk'}$ of the network with the joint probability in the uncorrelated case. However, a direct plot of $p_{kk'}$ is unpractical in most cases because the quantities $e_{kk'}$ are small integer numbers, and are strongly affected by statistical fluctuations, especially in networks with power-law degree distributions. A compact indicator is obtained by considering the difference between $p_{kk'}$ and $p_{kk'}^{\text{nc}}$, and taking a norm of such a quantity, for example:

$$d = \frac{\sum_{k,k'} |p_{kk'} - q_k q_{k'}|}{\sum_{k,k'} |p_{kk'}|}. \tag{7.16}$$

Notice that the denominator in the expression of d can be omitted, since the probabilities $p_{kk'}$ are non-negative and sum up to 1. When we compute the quantity in Eq. (7.16) for Internet AS we obtain $d = 0.70$, while for the network movie actors we get $d = 0.59$. Both values significantly differ from zero, clearly indicating the presence of degree correlations.

However, we know from Figure 7.1 that the degree correlation pattern of Internet AS is quite different from that of the movie actor network, but the distance d does not seem to be able to distinguish between these two cases.

An alternative method to test numerically the presence and sign of degree–degree correlations is to generalise the procedure used in Figure 7.1. Namely, to evaluate and plot as a function of k, the average degree of the nearest neighbours of nodes of degree k. This quantity is usually known as the *average nearest neighbours' degree function*, and is denoted as $\langle k_{nn}\rangle(k)$, where the subscript "nn" stands for nearest neighbours, as in Section 3.6. Notice that $\langle k_{nn}\rangle(k)$ is a function of k, and not a single number as the quantity defined in Eq. (3.37) of Section 3.6 and in Eq. (7.13). This is because now we do not consider the degree of the nearest neighbours of all the nodes in the graph, but only of those of degree k. The function $\langle k_{nn}\rangle(k)$ was proposed in Ref. [254] by Romualdo Pastor-Satorras, Alexei Vázquez, and Alessandro Vespignani as a way to quantify correlations in the Internet. It can be expressed in terms of the conditional probabilities as in the following definition.

> **Definition 7.4 (Average nearest neighbours' degree function)**   *The average degree of the nearest neighbours of nodes of degree k, known as the* average nearest neighbours' degree *function, is defined as:*
>
> $$\langle k_{nn}\rangle(k) = \sum_{k'} k' p_{k'|k} \qquad (7.17)$$
>
> *where $p_{k'|k}$ is the conditional probability distribution of the network.*

The expression of the average degree $\langle k_{nn}\rangle(k)$ of the nearest neighbours of nodes of degree k given in Eq. (7.17) clearly requires the knowledge of $p_{k'|k}$, which is based on the classification of the nodes in degree classes. For practical purposes, it is more convenient to express and compute $\langle k_{nn}\rangle(k)$ directly from the adjacency matrix of the network. We can therefore give the following alternative definition based on the expression of the average degree $k_{nn,i}$ of the nearest neighbours of a specific node i, with $i = 1, 2, \ldots, N$, introduced in Eq. (3.36).

> **Definition 7.5 (Average nearest neighbours' degree function)**   *The average degree $\langle k_{nn}\rangle(k)$ of the nearest neighbours of nodes of degree k can be also obtained by considering the average degree $k_{nn,i}$ of the nearest neighbours of node i as in Eq. (3.36), and by averaging this quantity over all nodes with k links:*
>
> $$\langle k_{nn}\rangle(k) = \frac{1}{N_k} \sum_{i=1}^{N} k_{nn,i}\, \delta_{k_i k} \qquad (7.18)$$
>
> *where N_k is the number of nodes of degree k.*

The main difference between the two definitions given above is that summations in Eq. (3.36) and Eq. (7.18) are taken over node labels, rather than over degree classes as in Eq. (7.17).

Example 7.5 *(Relations between $\langle k_{nn} \rangle$ and $\langle k_{nn} \rangle(k)$)* The average degree $\langle k_{nn} \rangle$ of the nearest neighbours of a node introduced in Definition 3.8 can be written in terms of the average nearest neighbours' degree function $\langle k_{nn} \rangle(k)$ of a graph by further averaging this quantity over all the degree classes:

$$\langle k_{nn} \rangle = \sum_k p_k \cdot \langle k_{nn} \rangle(k). \tag{7.19}$$

Now, if we plug in the right-hand side of the above equation the expression of $\langle k_{nn} \rangle(k)$ given in Eq. (7.17) we get:

$$\langle k_{nn} \rangle = \sum_k p_k \sum_{k'} k' p_{k'|k}$$

which coincides with Eq. (7.13).

Analogously, if we plug in the right-hand side of Eq. (7.19) the expression of $\langle k_{nn} \rangle(k)$ given in Eq. (7.18), and we use Eq. (3.36), we get:

$$\langle k_{nn} \rangle = \sum_k p_k \frac{1}{N_k} \sum_{i=1}^N k_{nn,i} \, \delta_{k_i k} = \sum_{i=1}^N \frac{p_{k_i}}{N_{k_i}} k_{nn,i} = \frac{1}{N} \sum_{i=1}^N k_{nn,i} = \frac{1}{N} \sum_{i=1}^N \frac{1}{k_i} \sum_{j=1}^N a_{ij} k_j$$

which coincides with Eq. (3.37).

Being $\langle k_{nn} \rangle(k)$ an average over all nodes of degree k, we expect that this quantity is less affected by statistical fluctuations. As we will see, plotting $\langle k_{nn} \rangle(k)$ as a function of k turns out indeed to be very useful for studying correlations in real networks. This is because the following result holds.

> In the case of uncorrelated networks the average degree of the nearest neighbours of nodes of degree k does not depend on the value of k and is equal to:
>
> $$\langle k_{nn}^{nc} \rangle(k) = \frac{\langle k^2 \rangle}{\langle k \rangle}. \tag{7.20}$$

In fact, in a network with no correlations the conditional probability reduces to the one in Eq. (7.11). By plugging this expression into (7.17) we obtain Eq. (7.20), which shows that $\langle k_{nn}^{nc} \rangle(k)$ is independent of k. If, instead, the plot of $\langle k_{nn} \rangle(k)$ as a function of k is not flat, this implies that the network is correlated. We can give the following definition.

> **Definition 7.6 (Assortative and disassortative networks)** *When the average nearest neighbours degree function $\langle k_{nn} \rangle(k)$ increases with k, that is if nodes with many links tend to be connected to other nodes with many links, we say that the network is assortative. Conversely, when $\langle k_{nn} \rangle(k)$ is a decreasing function of k, that is if nodes with many links tend to be connected to nodes with low degree, we say that the network is disassortative.*

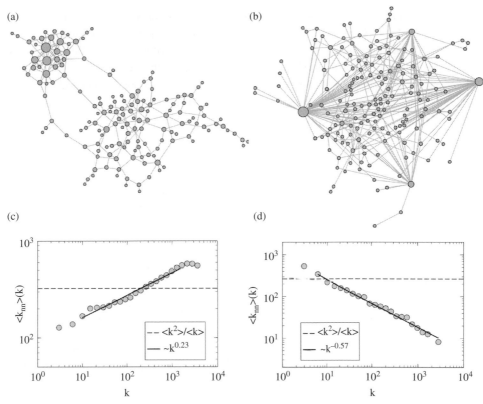

Fig. 7.4 An assortative (a) and a disassortative (b) network with the same number of nodes and links, $N = 170$ and $K = 275$. The size of each node is proportional to its degree. In the bottom panels we report the average nearest neighbours degree function $\langle k_{nn} \rangle (k)$ respectively for an assortative and a disassortative network, namely (c) the movie actor collaboration network and (d) the May 2001 snapshot of the Internet at the AS level. In both cases $\langle k_{nn} \rangle (k)$ is a power-law k^{ν}, with $\nu = 0.23$ and $\nu = -0.57$, respectively.

In Figure 7.4 we show how an assortative and a disassortative network look. In the assortative graph reported in panel (a) we observe that high-degree nodes are preferentially connected to each other, while low-degree nodes are usually connected to other low-degree nodes. Conversely, in the disassortative graph reported in panel (b) of the same figure high-degree nodes are preferentially connected to low-degree nodes.

We can now come back to study the network of movie actors and that of the Internet map at the AS level. In Figure 7.4(c)–(d) we plot the average degree of the nearest-neighbours of a node as a function of the node degree k, in the two cases. As a term of comparison, we also report as dashed lines the corresponding values of the average nearest neighbours degree function $\langle k_{nn}^{nc} \rangle (k)$ as given by Eq. (7.20) if the two networks were uncorrelated. The increase of $\langle k_{nn} \rangle (k)$ as a function of k found in the left panel means that in the movie collaboration network, actors with many links tend to be connected to other actors with many links, indicating that the network is *assortative*.

Conversely, in the network of Internet at AS level we observe that the average degree of the nearest neighbours of nodes of degree k is a decreasing function of k. The plot clearly indicates the existence of nontrivial degree–degree correlations of *disassortative* type. That the Internet is disassortative is quite reasonable, since the high-degree vertices of the network mostly represent connectivity providers, such as telephone companies and other communications carriers, which typically have a large number of connections to clients who themselves have only a single or just a few connections. Interestingly enough, in both cases shown in Figure 7.4(c) and in Figure 7.4(d) the quantity $\langle k_{nn} \rangle(k)$ can be nicely fitted as a power-law function:

$$\langle k_{nn} \rangle(k) \sim k^{\nu} \tag{7.21}$$

We get a positive exponent $\nu = 0.23$ for the network of movie actors and a negative exponent $\nu = -0.57$ for the Internet. Together with the scale-free behaviour found in the degree distributions of such two networks, this result emphasises the presence of a certain hierarchy also in the degree correlations.

The Rich Club

A network property that is also an indication of the presence of a certain type of correlations is the so-called *rich-club behaviour*. This is the tendency in a network of *all* the nodes with high degree to be linked together, forming what is known as the *rich club* of the network. The rich-club behaviour of a network can be quantified by defining the *rich-club coefficient*, as proposed by Vittoria Colizza, Alessandro Flammini, M. Ángeles Serrano and Alessandro Vespignani in Ref. [83].

> **Definition 7.7 (The rich-club coefficient)**   *The rich-club coefficient $\phi(k)$ is defined as:*
>
> $$\phi(k) = \frac{2e_{>k>k}}{N_{>k}(N_{>k} - 1)} \tag{7.22}$$
>
> *where $e_{>k>k}$ indicates the number of links among the $N_{>k}$ nodes having degree higher than k.*

The coefficient $\phi(k)$ denotes the fraction of links connecting pairs of nodes with degree higher than k [323]. Notice, in fact, that $e_{>k>k}$ is the number of links between the $N_{>k}$ nodes with degree higher than k, while $N_{>k}(N_{>k} - 1)/2$ is equal to the maximum possible number of links among such nodes. However, to take into account the fact that nodes with high degrees have by definition a larger probability of sharing edges than low-degree nodes, we have to compare the quantity in Eq. (7.22) with the same quantity evaluated in a randomised version of the network that preserves the degree distribution. Finally, a properly normalised quantity we can plot as a function of k to investigate the rich-club behaviour is the ratio [83]:

$$\rho(k) = \frac{\phi(k)}{\phi^{nc}(k)} \tag{7.23}$$

where $\phi^{nc}(k)$ is the rich-club coefficient for the randomised network. As usual, the superscript "nc" stands for no degree–degree correlations.

Finding values of the normalised rich-club coefficient $\rho(k)$ larger than 1 for large values of k, and increasing with the degree k indicates the presence of a *rich club* in a network.

In Figure 7.5 we show a study of the rich-club behaviour for the movie actor network and for the Internet. Networks with assortative degree–degree correlations are good candidates to exhibit a rich club, since in those networks nodes with high degree tend to be preferentially connected with nodes having similar degree. We find indeed a strong signature of a rich-club structure in the case of the actor collaboration network. The normalised rich-club coefficient ρ starts already to deviate from the value of 1 for degrees larger than $k \simeq 30$ (which is about one half of the average degree of the network), is equal to 2 for nodes with about 500 edges and peaks at around 3 for degrees between 1000 and 2000. This means that in the movie actor network the number of links connecting pairs of nodes with degree higher than 500 is twice as larger as in a randomised version of the network with the same degree distribution. Similar results can be found for other assortative networks, such as scientific collaboration networks, where the presence of a rich-club effect indicates that scientists with a high number of publications tend to collaborate among each others, forming scientific elites in the various domains of science. Notice, however, that the rich-club phenomenon is not necessarily associated with assortative mixing. Cases of disassortative networks showing a rich-club phenomenon can in fact be found (see Problem 7.3b). For the disassortative network of the Internet introduced in this chapter we observe instead small variations from $\rho = 1$ [83]. In particular, for intermediate values of k the value of the normalised rich-club coefficient ρ is smaller than 1, indicating a small tendency to an anti rich-club behaviour.

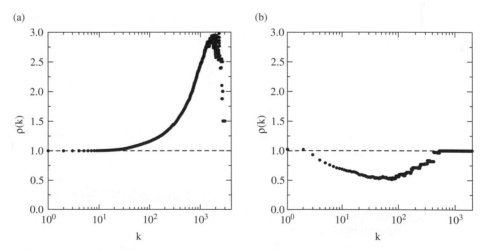

Fig. 7.5 The normalised rich-club coefficient $\rho(k)$ is plotted as a function of k for (a) the movie actor collaboration network and for (b) the Internet at the AS level. The movie actor network shows a clear signature of rich-club behaviour, since the probability for two nodes of large degree to be connected is as much as three times larger than in the randomised network.

Let us now explore whether there is a simple relation between rich-club behaviour and the presence of degree–degree correlations as quantified by the calculation of $\langle k_{nn}\rangle(k)$. Notice that the rich-club coefficient $\phi(k)$ can be expressed in terms of the degree distribution p_k and of the joint probability $p_{kk'}$. We can in fact write the number of links between the $N_{>k}$ nodes with degree higher than k as $e_{>k>k} = (\sum_{k'=k}\sum_{k''=k} e_{k'k''})/2$. By using the definition of the joint probability distribution $p_{kk'}$ in Eq. (7.7) we have:

$$\phi(k) = \frac{N\langle k\rangle \sum_{k'=k}\sum_{k''=k} p_{k'k''}}{\left[N\sum_{k'=k} p_{k'}\right]\left[N\sum_{k'=k} p_{k'} - 1\right]}. \tag{7.24}$$

Such an expression indicates that $\phi(k)$ is also a measure of correlations in a network. In fact, similarly to $\langle k_{nn}\rangle(k)$, the rich-club coefficient depends on the form of the joint probability distribution $p_{k'k''}$. However, $\phi(k)$ and $\langle k_{nn}\rangle(k)$ represent two different projections of $p_{k'k''}$, so that the rich club is a property that is not trivially related to the assortativity/disassortativity of a network. While $\langle k_{nn}\rangle(k)$ quantifies two-body correlations and so accounts for local properties of the nodes with degree k, the rich club is a global property of the subgraph induced by the subset of nodes with degree larger than k. The double summation in Eq. (7.24) is indeed a convolution of the joint probability distribution $p_{kk'}$ so that the same rich-club features can be obtained in networks with different types of degree–degree correlations [83].

The expression of $\phi(k)$ in terms of $p_{kk'}$ given in Eq. (7.24) also allows a useful analytical formula to be derived for the rich-club coefficient in uncorrelated networks, as shown in the following example.

Example 7.6 *(The rich-club coefficient in uncorrelated networks)* From Eq. (7.24) we can get an approximate formula for $\phi^{nc}(k)$, the rich-club coefficient in uncorrelated networks. Making use of the factorisation of the joint probability for uncorrelated networks, $p_{kk'}^{nc} = q_k q_{k'}$, and substituting into Eq. (7.24), we get:

$$\phi^{nc}(k) = \frac{\sum_{k'=k} k' p_{k'} \sum_{k''=k} k'' p_{k''}}{\langle k\rangle \left[\sum_{k'=k} p_{k'}\right]\left[N\sum_{k'=k} p_{k'} - 1\right]} \simeq \frac{\left[\sum_{k'=k} k' p_{k'}\right]^2}{N\langle k\rangle \left[\sum_{k'=k} p_{k'}\right]^2}. \tag{7.25}$$

It is important to notice that this expression for $\phi^{nc}(k)$ is valid only for uncorrelated networks whose maximum degree k_{max} is smaller than the structural cut-off k_{max}^{struct}. In fact, by evaluating Eq. (7.25) for $k = k_{max}$ we obtain:

$$\phi^{nc}(k_{max}) = \frac{k_{max}^2 p_{k_{max}}^2}{N\langle k\rangle p_{k_{max}}^2} = \frac{k_{max}^2}{N\langle k\rangle}.$$

We also have:

$$\phi^{nc}(k_{max}) \leq 1 = \frac{(k_{max}^{struct})^2}{N\langle k\rangle}$$

where the inequality derives from the fact that $\phi(k)$ is, by definition, a fraction of edges, so its value cannot be larger than 1, while the equality comes from the definition of structural cut-off in Eq. (7.15). Comparing the two expressions of $\phi^{nc}(k_{max})$ we thus have $k_{max} \leq k_{max}^{struct}$.

By approximating the summations in Eq. (7.25) with integrals, and in the limit of large networks, we get:

$$\phi^{nc}(k) \simeq \frac{1}{N\langle k \rangle} \left[\frac{\int_k^\infty k' p_{k'}}{\int_k^\infty p_{k'}} \right]^2 .$$

Now, by considering a scale-free distribution $p_k \sim k^{-\gamma}$ with $2 < \gamma \leq 3$, and evaluating the integrals, we obtain:

$$\phi^{nc}(k) = \frac{\gamma - 1}{\gamma - 2} \frac{k^2}{N\langle k \rangle}. \tag{7.26}$$

Eq. (7.26) tells us that, in uncorrelated scale-free networks, the rich-club coefficient $\phi(k)$ is expected to grow as the square of the degree k. This expression is very useful because it can be directly plugged into Eq. (7.23) when we need to calculate and plot the normalised coefficient $\rho(k)$ of a real-world network as a function of k.

7.4 Newman's Correlation Coefficient

A compact measure of degree–degree correlations in terms of a single number can be obtained by the appropriate definition of a correlation coefficient. In multivariate statistics, the so-called *Pearson's correlation coefficient* is one of the most commonly adopted quantities to measure the strength of linear correlation between two variables (see Box 7.2). In this section we will show how to define a Pearson's correlation coefficient to quantify degree–degree correlations in a network, and how such a definition, at variance with the quantity d introduced in Eq. (7.16), allows us to disentangle assortative from disassortative networks.

Let us start from scratch. Instead of focusing directly on the degrees of the two nodes at the ends of the links of a network, let us consider the most general case of any two random variables X and Y. Assume that the two variables are described by the joint probability $p(x, y) = \text{Prob}(X = x, Y = y)$. The extent to which the two random variables vary together can be measured by their *covariance* $\text{cov}(X, Y)$, defined as:

$$\text{cov}(X, Y) = E[(X - E[X])(Y - E[Y])] = E[XY] - E[X]E[Y] \tag{7.27}$$

where, for any random variable, E denotes the population expectation value. For example, if X and Y take discrete values, we can express averages as summations, and we get:

$$E[X] = \sum_{x,y} xp(x,y) = \sum_{x} xp_X(x)$$

$$E[Y] = \sum_{x,y} yp(x,y) = \sum_{y} yp_Y(y)$$

$$E[XY] = \sum_{x,y} xyp(x,y)$$

where we have indicated as $p_X(x) = \sum_y p(x,y)$ and $p_Y(y) = \sum_x p(x,y)$ the marginal probabilities associated respectively with X and Y. In practice, in Eq. (7.27), for each pair sample (x, y), one takes their differences from their mean values and multiplies these differences together. This product is positive, if the values of X and Y vary together in the same direction from their means. Conversely, the product is negative if they vary in opposite directions. The larger the magnitude of the product, the stronger the strength of the relationship. Finally, the covariance $cov(X, Y)$ is defined as the mean value of this product, evaluated over the possible pairs of x and y. A positive value of $cov(X, Y)$ indicates that higher than average values of one variable tend to be paired with higher than average values of the other variable. A negative value of $cov(X, Y)$ indicates that higher than average values of one variable tend instead to be paired with lower than average values of the other variable. Regarding a zero value of the covariance, it is straightforward to prove that, if two random variables X and Y are *independent*, i.e. if $p(x, y) = p_X(x) \times p_Y(y)$, then $cov(X, Y) = 0$ (see Box 7.2). However, the inverse is not true: $cov(X, Y) = 0$ means only that $E[XY] = E[X]E[Y]$, while it does not necessarily imply that X and Y are independent. For instance, a nonlinear relationship between the two variables can exist that would still result in a covariance value of zero, as shown in the following example.

Example 7.7   Consider two random variables X and Y taking real values and described by the joint probability density:

$$p(x, y) = \sum_{i=1}^{n} w_i p_i(x) q_i(y)$$

where w_i, $i = 1, 2, \ldots, n$ are positive numbers summing to one, while $p_i(x)$ and $q_i(x)$, $i = 1, \ldots, n$, are the densities of random variables with the same mean, say zero for simplicity. Variables X and Y are not independent, since the density cannot in general be written as a product of a function of x times a function of y. However, we will now show that their covariance is zero. We can first write the covariance as:

$$cov(X, Y) = E[XY] - E[X]E[Y] = E[XY]$$

since we have $E[X] = 0$ and $E[Y] = 0$. Direct calculation shows also that:

$$E[XY] = \int_x dx \int_y dy \sum_i w_i \, xy \, p_i(x) \, q_i(y) = \sum_i w_i \int_x x p_i(x) \, dx \int_y y q_i(y) \, dy = 0$$

hence $cov(X, Y) = 0$.

Another extreme example of strongly dependent variables which have zero correlation is the following. Consider a random variable X with density $p(x)$, which we assume for simplicity to be in $[-1, 1]$, so that $p(x) = p(-x) > 0$, $x \in [-1, 1]$, and $\int_{-1}^{1} p(x)\, dx = 1$. Let us now consider variable $Y = f(X)$, with $f : [-1, 1] \to [-1, 1]$ an even function of X. Then the density function corresponding to the pair (X, Y) is given by:

$$p(x, y) = \frac{p(x)\delta(y - f(x))}{\int_{-1}^{1} p(z) f(z)\, dz}.$$

Direct calculation shows that:

$$\int_{-1}^{1} dx \int_{-1}^{1} dy\, x\, y\, p(x, y) = \int_{-1}^{1} x\, p(x) f(x)\, dx = 0,$$

therefore, even in this limit case of two dependent variables, we get a zero covariance.

The main problem with the covariance is that it depends on the actual values assumed by the two random variables. This problem can be solved by normalising the covariance by the product of the standard deviations of the two variables. In fact, by making use of the Schwarz inequality, it is possible to prove that $|\text{cov}(X, Y)| \leq \sigma_X \sigma_Y$. Hence, the so-called *Pearson correlation coefficient* r, defined as in Eq. (7.29) of Box 7.2, is a dimensionless measure of the strength of linear correlation between two variables X and Y, which takes values between -1 and $+1$.

Box 7.2   **Linear Dependence and the Pearson Correlation Coefficient**

The demand for a product and its price, the mass of different animal species and their metabolic rates, or the gross domestic product (GDP), the length of road infrastructures and the population of a city, are typical examples of statistically dependent quantities. In mathematical terms, two random variables X and Y are said to be *independent* if $\text{Prob}(X = x, Y = y) = p_X(x) \times p_Y(y)\, \forall x, y$, i.e. if the joint probability of obtaining the pair of values x and y can be factorised in terms of two corresponding marginal probabilities $p_X(x)$ and $p_Y(y)$. Otherwise, the two random variables are said to be *dependent*. The *Pearson correlation coefficient*, r, is the quantity typically used to measure linear correlations between two variables X and Y. It is defined as [143]:

$$r = \frac{\text{cov}(X, Y)}{\sigma_X \sigma_Y} = \frac{E[XY] - E[X]E[Y]}{\sqrt{E[X^2] - E^2[X]}\sqrt{E[Y^2] - E^2[Y]}} \tag{7.29}$$

where $\text{cov}(X, Y)$ is the covariance between the two random variables, and σ_X and σ_Y are the two corresponding standard deviations. The correlation coefficient r is equal to $+1$ in the case of a perfect linear correlation, -1 in the case of a perfect decreasing linear relationship (anticorrelation), and takes values in the range $(-1, 1)$ in all other cases, indicating the degree of linear dependence between the variables. Notice that the Pearson coefficient only measures the strength of linear dependence between the two variables. Thus, a value $r = 0$ indicates that there is no linear correlation between X and Y, but the two variables can still be dependent and connected by a nonlinear relationship.

Let us come back now to our original problem of measuring whether or not the degrees of the nodes at the end of the links of a graph are independent. The two random variables X and Y are, in this case, the degrees of pairs of nodes connected by a link, and their distribution can be described by the joint probability $p_{kk'} = \text{Prob}(X = k, Y = k')$ introduced in Section 7.2, where by the symbols k and k' we indicate the possible values of such two degrees. By definition, the function $p_{kk'}$ is symmetric and obeys to the following normalisations:

$$\sum_{kk'} p_{kk'} = 1, \qquad \sum_{k'} p_{kk'} = q_k, \qquad \sum_{k} p_{kk'} = q_{k'}, \tag{7.30}$$

where the last two equations indicate that q_k is the marginal probability of finding a node of degree k at the end of a link, no matter the degree of the node at the other end. Hence, consistently with Eq. (7.12), we can say that the degrees of pairs of nodes at the end of the links of a graph are independent if $p_{kk'}$ is equal to the product $q_k q_{k'}$ of the two marginal probabilities $\forall k, k'$. Conversely, in a correlated network, $p_{kk'}$ will differ from $q_k q_{k'}$. The amount of linear correlations can be quantified by the covariance between the two degrees which, using Eq. (7.27), reads:

$$c = \sum_{k} \sum_{k'} kk' p_{kk'} - \sum_{k} kq_k \sum_{k'} k' q_{k'}.$$

Notice that the first term in the expression of c is nothing other than the average of the quantity kk' over all the graph links. In fact, by denoting averages over the edges of the graph by the symbol $\langle \ldots \rangle_{l \in \mathcal{L}}$, we have:

$$\langle kk' \rangle_{l \in \mathcal{L}} = \frac{\sum_i \sum_j a_{ij} k_i k_j}{\sum_i \sum_j a_{ij}}$$

$$= \frac{\sum_k \sum_{k'} e_{kk'} kk'}{2K} = \sum_{k} \sum_{k'} p_{kk'} kk'$$

where we have transformed the summations over node labels into summations over degree classes. Analogously, the second term in the expression of the covariance c is the same quantity $\langle kk' \rangle_{l \in \mathcal{L}}$ evaluated, this time, in the case in which the two degrees at the end of a link are independent random variables.

For the purpose of comparing different networks, we normalise the covariance c by dividing it by its maximal value $c_{\max}$ which, in agreement with Eq. (7.29), is equal to the variance of the marginal distribution q_k, namely $\sigma_q^2 = \sum_k k^2 q_k - \left[\sum_k kq_k \right]^2$. It is easy to check that such maximal value is indeed achieved when links exist only between nodes with exactly the same degree, i.e. $p_{kk'} = q_k \delta_{kk'}$. Finally, we get an expression for the Pearson correlation coefficient of the degrees of pairs of nodes connected by an edge. Such quantity was proposed for the first time by Mark Newman as a measure of degree–degree correlations in networks, and is therefore known as the *Newman's correlation coefficient* [230, 232].

Definition 7.8 (Newman's correlation coefficient) *The degree–degree correlation coefficient r is defined as:*

$$r = \frac{\sum_k \sum_{k'} kk'(p_{kk'} - q_k q_{k'})}{\sum_k k^2 q_k - \left[\sum_k k q_k\right]^2} \qquad (7.31)$$

and is a quantity normalised to vary in $-1 \leq r \leq 1$*. A positive (negative) value of r indicates that the network is* assortative *(*disassortative*).*

If the degrees at the two ends of an edge are independent, then Newman's correlation coefficient r is, by definition, equal to zero. Therefore, finding a value of $r \neq 0$ is an indication that a network has degree–degree correlations.[1] Moreover, a positive value of r will mean that the correlations are positive, that is the network is assortative, while a negative value of r will tell us that the network is disassortative.

For the practical purpose of evaluating r for a given network, we can rewrite summations over degree classes in Eq. (7.31) as summations over nodes. We finally get an expression which can be directly calculated from the degree of each node, without the need to construct the joint probability $p_{kk'}$ (see Problem 7.4a):

$$r = \frac{2K \sum_i \sum_j a_{ij} k_i k_j - \left(\sum_i k_i^2\right)^2}{2K \sum_i k_i^3 - \left(\sum_i k_i^2\right)^2}. \qquad (7.32)$$

In some cases, it can be even more useful to express the Newman's correlation coefficient r in terms of summations over the links of a graph, instead of in terms of summations over the nodes. After some simple algebra (see Problem 7.4(b)), we can rewrite Eq. (7.32) as:

$$r = \frac{4K \sum_e k_e k'_e - \left[\sum_e (k_e + k'_e)\right]^2}{2K \sum_e (k_e^2 + k_e'^2) - \left[\sum_e (k_e + k'_e)\right]^2} \qquad (7.33)$$

where now $\sum_e$ denotes a summation over the edges of the graph, with $e = 1 \ldots K$, and k_e and k'_e are the degrees of the two nodes at the ends of the eth edge.

Example 7.8 *(Directed graphs)* In this chapter we have considered how to treat degree–degree correlations in undirected graphs. However, all the formalism we have introduced can be extended also to the case of directed networks. Each node in a directed network has in- and out-degree, respectively defined as the number of ingoing and outgoing links. We need then to consider the joint probability $p_{k^{\text{out}} k^{\text{in}}}$, i.e. the probability that a randomly chosen arc (directed link) leads from a node of out-degree k^{out} into a node of in-degree k^{in}. Differently from matrix $\{p_{kk'}\}$, matrix $\{p_{k^{\text{out}} k^{\text{in}}}\}$ can in general be asymmetric. We can finally define a *out-degree–in-degree correlation coefficient r* for a directed network as:

---

[1] Notice however that, if we find $r = 0$, this does not imply that the graph is uncorrelated, i.e. that the degree of connected nodes are *independent* variables and the joint probability distribution factorises into the product of the two marginal probabilities. Such condition of independence, used in Definition 7.3 to define an uncorrelated graph, implies $r = 0$ and $\langle k_{\text{nn}} \rangle$ being independent of k. Therefore it is a stronger condition than the latter two.

$$r = \frac{\sum_{k^{\text{out}}k^{\text{in}}} k^{\text{out}}k^{\text{in}}(p_{k^{\text{out}}k^{\text{in}}} - q_{k^{\text{out}}}^{\text{out}}q_{k^{\text{in}}}^{\text{in}})}{\sigma^{\text{out}}\sigma^{\text{in}}}, \tag{7.34}$$

where $q_{k^{\text{out}}}^{\text{out}} = \sum_{k^{\text{in}}} p_{k^{\text{out}}k^{\text{in}}}$ and $q_{k^{\text{in}}}^{\text{in}} = \sum_{k^{\text{out}}} p_{k^{\text{out}}k^{\text{in}}}$ are the two marginal probability distributions, and σ^{out} and σ^{in} are the associated standard deviations.

For the practical purpose of computing the correlation coefficient of a given network, it can be convenient to express r in terms of summations over the arcs of the corresponding graph. We get [232]:

$$r = \frac{\sum_e k_e^{\text{out}}k_e^{\text{in}} - K^{-1}\left(\sum_e k_e^{\text{out}}\right)\left(\sum_e k_e^{\text{in}}\right)}{\sqrt{\left[\sum_e(k_e^{\text{out}})^2 - K^{-1}\left(\sum_e k_e^{\text{out}}\right)^2\right]\left[\sum_e(k_e^{\text{in}})^2 - K^{-1}\left(\sum_e k_e^{\text{in}}\right)^2\right]}}, \tag{7.35}$$

where k_e^{out} and k_i^{in} are the out-degree and the in-degree of the nodes that the directed edge e leads out of and into respectively, and K is the number of arcs in the graph.

Finally, notice that, in order to completely describe correlations in directed networks, in principle we must also be able to characterise the correlations between the in- and the out-degree *at a given node*. This can be done by defining another joint probability distribution, $f_{k^{\text{in}}k^{\text{out}}}$, which represents the probability that a randomly chosen node has in-degree equal to k^{in} and out-degree equal to k^{out}.

In Table 7.2 we report the values of r obtained for a number of real-world networks and models. We have considered social, technological and biological networks, in most cases from the data sets introduced in this book. The table reveals an interesting feature: all the social collaboration networks examined are significantly assortative by degree. This is indeed the most common situation in social systems, where usually high-degree individuals tend to be connected to other high-degree individuals. Such a tendency can be indeed very strong in collaboration networks. For instance, we found a value $r = 0.24$ for the case of movie actors, and $r = 0.65$ for the coauthorship network of high-energy physicists, definitely a very assortative environment. Conversely, almost all the technological and biological networks are disassortative with high-degree vertices preferentially connected to low-degree ones. Notice, however, that the correlation coefficients found in these cases are, in absolute value, much smaller than those observed in social networks. And in some cases, such as for instance in the US power grid, the network of high-voltage transmission lines in the Western USA, the value of r is so small that degree–degree correlations can in practice be neglected. Notice that we have reported citation networks together with man-made networks because, in a sense, they look more similar to the World Wide Web than to social networks (they both describe directed links between documents), and in fact they display negative, even if small, values of r. The strongest disassortativity was found for the Internet at the AS level, the same network we studied in Figure 7.4 and in Figure 7.5.

In Table 7.2 we also report the correlation coefficients obtained for three different models, namely ER random graph model A, and the BA and DMS models. In all cases we have constructed networks with $N = 200,000$ nodes. Of course, for ER random graphs, since edges are placed at random without regard to the node degrees, it follows trivially that

Table 7.2 Newman's correlation coefficient for various real networks and models.

Network	N	r
Medline coauthorship	1,520,252	0.127
ArXiv coauthorship	52,909	0.363
astro-ph coauthorship	16,706	0.235
cond-mat coauthorship	16,726	0.185
hep-th coauthorship	8,361	0.294
SPIRES coauthorship	56,627	0.650
NCSTRL coauthorship	11,994	0.189
Movie actor collaboration	248,243	0.237
Internet AS	11,174	−0.195
Internet routers	190,914	0.025
Notre Dame WWW	329,729	−0.044
Stanford WWW	281,903	−0.096
Berkley-Stanford WWW	685,230	−0.119
Google WWW	875,713	−0.049
APS citations	450,084	−0.022
Scientometrics citations	1,655	−0.043
US power grid from Ref.[311]	4,941	0.003
C. elegans	279	−0.093
Protein interactions from Ref.[165]	1,870	−0.162
ER random graphs ($\langle k \rangle = 10$)	200,000	0.0008
BA model ($m = 4$)	200,000	−0.012
DMS model ($m = 4, a = -2$)	200,000	−0.046

$r = 0$ in the limit of large graph size because the graph has no degree–degree correlations. For instance, in the case reported in the table we find a value of 0.0008 in a network with $N = 200,000$ and average degree $\langle k \rangle = 10$. The other two models considered, BA and DMS, both produce scale-free networks. In particular, the DMS model allows the value of the exponent γ of the degree distribution to be tuned, and reduces to the BA model when $a = 0$. For instance, when $m = 4$ and $a = -2$ the DMS model produces, according to Eq. (6.21), scale-free networks with an exponent $\gamma = 2.5$. Interestingly enough, the value of the correlation coefficient r is also very small both in the case of the BA and the DMS model. However, it is important to remember that not all graphs with $r = 0$ are without degree–degree correlations. Newman's correlation coefficient is based on the assumption that the relationship between the two variables to investigate is linear, and getting a value of $r = 0$ simply implies that the mean degree correlation is zero when averaged over all degrees. Indeed, the BA and the DMS models provide examples of networks having degree–degree correlations, but with a correlation coefficient $r = 0$. We will now in fact prove that the joint probability $p_{kk'}$ does not factorise in the BA model, hence the networks produced are correlated. To show this, we will be considering for simplicity the case of the BA model with $m = 1$. In Chapter 6 we were interested in the degree distribution of the network. We therefore wrote down and solved the rate equations for the evolution of $n_{k,t}$, the number of nodes with k links at time t. Our focus is now on the number of edges connecting nodes of degree k and nodes of degree k'. We need then to write down and solve the rate equations governing the time evolution of the quantities $e_{kk'}$. By $e_{kk',t}$ we indicate, in this case, the average number of links between nodes of degree k and nodes of degree k' at time t, and as in Chapter 6, averages have to be intended as performed over infinite realisations of the growth process of the BA model with the same set of parameters. We will follow closely the work by Pavel Krapivsky and Sidney Redner [186]. At time t, a new node arrives and attaches to a node of the existing graph. The rate equations expressing the quantities $e_{kk',t}$ as a function of $e_{kk',t-1}$ read:

$$e_{kk',t} = e_{kk',t-1} + \frac{(k-1)e_{k-1,k',t-1} - ke_{kk',t-1}}{\sum_k kn_{k,t-1}} + \frac{(k'-1)e_{k,k'-1,t-1} - k'e_{kk',t-1}}{\sum_k kn_{k,t-1}}$$

$$+ \frac{(k'-1)n_{k'-1,t-1}\delta_{k1} + (k-1)n_{k-1,t-1}\delta_{k'1}}{\sum_k kn_{k,t-1}} \qquad k \geq 1, k' \geq 1 \quad (7.36)$$

where $n_{k,t}$ is the number of nodes with k links at time t. The first ratio on the right-hand side accounts for the change in $e_{kk'}$ due to the addition of a link onto a node of degree $k-1$ (gain), and that due to the addition of a link onto a node of degree k (loss). In fact, it is equal to the probability $(k-1)/\sum_k kn_k$ that the new link attaches to *one specific* node of degree $k-1$, times the number $e_{k-1,k',t-1}$ of links between nodes of degree $k-1$ and nodes of degree k', minus the probability $k/\sum_k kn_k$ that the new link attaches to *one specific* node of degree k, times the number $e_{kk',t-1}$ of links between nodes of degree k and nodes of degree k'. In other words, the two terms are respectively equal to the average number of links of type $(k-1,k')$, i.e. links between nodes of degree $k-1$ and nodes of degree k', that are transformed into links of type (k,k'), and the average number of links of type (k,k') that are transformed into $(k+1,k')$. The second ratio in the right-hand side accounts for the change in $e_{kk'}$ due to the addition of a link onto a node of degree $k'-1$ (gain), and that due to the addition of a link onto a node of degree k' (loss). Finally, the last two terms account respectively for the gain in $e_{1k'}$ and e_{k1} due to the addition of the new edge arriving at each time with the new node. Notice that the equations we have derived are, as expected, symmetric with respect to changing k and k'.

Example 7.9   To better understand the different contributions in the rate equations of the BA model let us consider a simple example. At time t, a new nodes n arrives and attaches

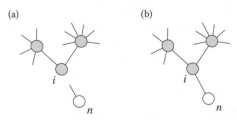

(a)   (b)

to a node i of the existing graph. Suppose now i has $k = 2$ links, one connecting node i to a node with 5 links and one connecting node i to a node with 7 links, as shown in panel (a) of the figure above. By acquiring a link from the new node n, node i will become a node of degree 3, and this will affect the properties of the edges of i as shown in panel(b) of the figure. In fact, both the numbers of links $e_{2,5} = e_{5,2}$ between nodes of degree 2 and degree 5, and the number of links $e_{2,6} = e_{6,2}$ between nodes of degree 2 and degree 6 will decrease by one unit. At the same time, the quantities $e_{3,5} = e_{5,3}$ and $e_{3,6} = e_{6,3}$ will increase by one because the node i has acquired a new link. Notice that there is another contribution to take into account. In fact, the new node n has changed the properties of the two links of i, but also n has arrived with a new edge. So, due to this new edge, we have to increase $e_{3,1} = e_{1,3}$ by one. In the general case the different contributions are all taken into account in the rate equations reported above.

Since $\sum_k k n_{k,t-1} = 2l_{t-1} = 2n_{t-1}$, we can write:

$$e_{kk',t} - e_{kk',t-1} = \left[(k-1)p_{k-1,k',t-1} - kp_{kk',t-1}\right] + \left(k'-1)p_{k,k'-1,t-1} - k'p_{kk',t-1}\right] +$$

$$+ \frac{k'-1}{2}p_{k'-1,t-1}\delta_{k1} + \frac{k-1}{2}p_{k-1,t-1}\delta_{k'1} \qquad k \geq 1, k' \geq 1 \qquad (7.37)$$

where in the right-hand side of Eq. (7.36) we have made use of the definitions of the joint probability distribution $p_{kk',t-1} = e_{kk',t-1}/2l_{t-1}$ as in Eq. (7.7), and of the probability distribution $p_{k,t-1} = n_{k,t-1}/n_{t-1}$. Analogously to what done in Section 6.2, and using the fact that $l_t = l_{t-1} + 1$, the left-hand side of Eq. (7.37) can be written as:

$$e_{kk',t} - e_{kk',t-1} = 2l_{t-1}[p_{kk',t} - p_{kk',t-1}] + 2p_{kk',t}$$

which for large times converges to the quantity $2p_{kk'}$. Of course, here we are assuming that, for $t \to \infty$, the joint probability distribution $p_{kk',t}$ converges to a stationary distribution faster than $1/t$. Finally, for large times, Eqs. (7.37) reduce to the time-independent recursion relations [186]:

$$(k + k' + 2)p_{kk'} = (k-1)p_{k-1,k'} + (k'-1)p_{k,k'-1} + \frac{k'-1}{2}p_{k'-1}\delta_{k1} + \frac{k-1}{2}p_{k-1}\delta_{k'1}$$

valid for $k \geq 1$ and $k' \geq 1$. The latter can be reduced to a set of constant-coefficient inhomogeneous recursion relations by the substitution [186]:

$$p_{kk'} = \frac{\Gamma(k)\Gamma(k')}{\Gamma(k+k'+3)}f_{kk'}$$

which yields:

$$f_{kk'} = f_{k-1,k'} + f_{k,k'-1} + 2(k'+2)\delta_{k1} + 2(k+2)\delta_{k'1} \qquad k \geq 1, k' \geq 1. \qquad (7.38)$$

The solution to this set of equations can be obtained by solving them for the first few values of k and k', and then grasping the pattern of dependence on k and k'. Alternatively, the equations can be solved in a more systematic way by using the generating function method [186]. We finally get (see Problem 7.4c):

$$f_{kk'} = 2\frac{\Gamma(k+k'+1)}{\Gamma(k+2)\Gamma(k')} + 4\frac{\Gamma(k+k')}{\Gamma(k+1)\Gamma(k')} + 4\frac{\Gamma(k+k')}{\Gamma(k)\Gamma(k'+1)} + 2\frac{\Gamma(k+k'+1)}{\Gamma(k)\Gamma(k'+2)} \qquad (7.39)$$

so that the joint probability distribution $p_{kk'}$ for the BA model with $m = 1$ reads:

$$p_{kk'} = \frac{2}{(k+k'+1)(k+k'+2)} \left[\frac{1}{k(k+1)} + \frac{2}{k(k+k')} + \frac{2}{k'(k+k')} + \frac{1}{k'(k'+1)} \right]$$
(7.40)

The most important result is that the expression for the joint distribution in Eq. (7.40) does not factorise, since we can easily check that the inequality

$$p_{kk'} \neq q_k q_{k'} = \frac{16}{\langle k \rangle^2} \frac{1}{(k+1)(k+2)} \frac{1}{(k'+1)(k'+2)}$$

holds. This means that the BA model produces scale-free networks with non-trivial degree–degree correlations. The correlations arise spontaneously in the growth process, and this makes the BA model different from other models, such us the configuration model, that produce instead uncorrelated scale-free networks.

Coming back to the value of r for the BA model, which we found very close to zero in Table 7.2, this is because the Newman's correlation coefficient measures correlations relative to a linear model. And the result $r \simeq 0$ for the BA model simply indicates that no linear correlations are present in this model. The lesson to learn here is that Newman's correlation coefficient r is an easy-to-calculate and compact measure of degree–degree correlations. However, a better way to detect correlations is certainly to plot, as a function of k, the average degree $\langle k_{nn} \rangle(k)$ of the nearest neighbours of nodes of degree k, as we did in Figure 7.4 for the movie actor collaboration network and for the Internet. In Figure 7.6 we show such a plot for two networks constructed, respectively, by the BA and the DMS models with the same parameters as those reported in Table 7.2. In the case of the BA model (Figure 7.6(a)) we notice that the quantity $\langle k_{nn} \rangle(k)$ measured on one realisation of the network (circles) exhibits deviations, even if small ones, from the value $\langle k^2 \rangle / \langle k \rangle = 30.6$ (dashed line) expected in the randomised version of the same network. Disassortative degree–degree correlations are more evident in the case of the DMS model, which is characterised by a power-law function $\langle k_{nn} \rangle(k) \sim k^\nu$ with a negative fitted exponent $\nu = -0.45$ (solid line in Figure 7.6(b)). In this case, the difference from the uncorrelated value $\langle k^2 \rangle / \langle k \rangle = 148.5$ (dashed line) is striking. Alain Barrat and Romualdo Pastor-Satorras were able to derive analytical expressions for the average degree $\langle k_{nn} \rangle(k)$ of the nearest neighbours of nodes of degree k in the DMS model by means of a rate equation approach similar to the one we used above [25]. The expression obtained for the case $-m \leq a < 0$ is:

$$\langle k_{nn} \rangle(k) \simeq N^{2\beta - 1} k^{-2 + 1/\beta}, \quad \beta = m/(2m + a)$$
(7.41)

where N is the number of nodes in the network, while for $a = 0$, i.e. when the DMS reduces to the BA model, we have:

$$\langle k_{nn} \rangle(k) \simeq \frac{m}{2} \ln N.$$

The latter expression predicts that the average degree $\langle k_{nn} \rangle(k)$ of the nearest neighbours of nodes of degree k does not depend on k in the BA model, and is equal to 24.4 (horizontal dotted line in Figure 7.6(a)). However, such a value is different from the value

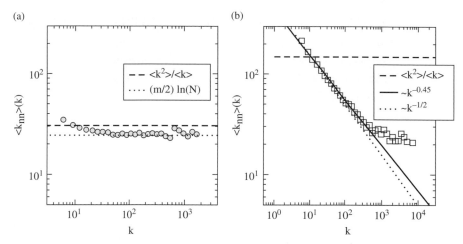

Fig. 7.6 The average degree $\langle k_{nn}\rangle(k)$ of the nearest neighbours of nodes of degree k is plotted as a function of k for the BA (circles, panel (a)) and the DMS models (squares, panel (b)) with the same parameters as in Table 7.2. The dashed lines in the two panels indicate the expected values $\langle k_{nn}\rangle(k) = \langle k^2\rangle/\langle k\rangle$ in uncorrelated graphs, while the dotted lines correspond to the predictions from Eq. (7.41).

$\langle k^2\rangle/\langle k\rangle = 30.6$ expected in the absence of correlations (dashed line in the same figure), and indeed reproduces for large k the results we have obtained through the direct numerical simulations of the model (circles). Conversely, in the case of the DMS model with $m = 4$ and $a = -2$ we have $\beta = 2/3$, and the expression found by Barrat and Pastor-Satorras predicts a power law dependence $\langle k_{nn}\rangle(k) \sim k^\nu$ with an exponent $\nu = -0.5$ (dotted line in Figure 7.6(b)), which is in perfect agreement with the exponent $\nu \simeq -0.45$ resulting from the fitting (solid line) of the numerical results.

7.5 Models of Networks with Degree–Degree Correlations

A general way to construct graphs with degree–degree correlations is to generate, on a computer, random graphs with a given matrix $\{p_{kk'}\}$. In particular, we will discuss here a class of models based on hidden variables proposed by Marián Boguñá and Romualdo Pastor-Satorras, which can be used to generate graphs with an assigned joint probability distribution $p_{kk'}$. In a hidden-variable model, nodes are assigned hidden variables, i.e. numbers which control the establishment of the edges between pairs of nodes. In the case of interest to us, the hidden variable associated with a node will be a positive integer number representing the desired degree of the node. A general formulation of a model with hidden variables is the following [46]:

Definition 7.9 (The hidden-variable model) *Consider a set of N isolated vertices, with each vertex i, $i = 1,\ldots,N$, being assigned a hidden variable h_i, that is a natural number* *independently drawn from a probability distribution $\rho(h)$. An undirected graph is generated by the following rule. For each (unordered) pair of distinct vertices i and j, with respective hidden variables h_i and h_j, an undirected edge is created with a probability $f(h_i, h_j)$, where the connection probability $f(h, h')$ is a symmetric function of h and h' with values in $[0, 1]$.*

Given the two functions $\rho(h)$ and $f(h, h')$, the model generates random graphs with neither loops nor multiple edges, and whose degree distribution and correlation properties depend on ρ and f. The authors of Ref. [46] have shown that, by an appropriate choice of $\rho(h)$ and $f(h, h')$, the hidden-variable model allows graphs with an assigned joint probability distribution $p_{kk'}$, i.e. with prescribed degree–degree correlations, to be produced. The main idea is to set the hidden variable h_i of each node i to be equal to the desired degree k_i^*. This can be obtained by sampling the values of h from the desired degree distribution, i.e. by setting the function $\rho(h)$ equal to p_h, where the function p_k is indeed the degree distribution associated with the assigned joint probability distribution $p_{kk'}$. We can then write:

$$\rho(h) \equiv p_h = \frac{\sum_{h'} p_{hh'}/h}{\sum_{h'} \sum_{h''} p_{h'h''}/h'}. \tag{7.42}$$

Notice in fact that by fixing $p_{kk'}$ we automatically fix also the degree distribution p_k of a graph. This is because Eq. (7.9) gives q_k in terms of $p_{kk'}$, so that we obtain an expression for p_k as a function of $p_{kk'}$ by inverting the relation $q_k = kp_k/\langle k \rangle$. In this way we get:

$$p_k = \frac{\langle k \rangle q_k}{k} = \frac{\langle k \rangle}{k} \sum_{k'} p_{kk'} \qquad k = 1, 2, \ldots$$

From the normalisation condition $\sum_k p_k = 1$, by summing over k, we obtain:

$$\langle k \rangle = \frac{1}{\sum_k \frac{1}{k} \sum_{k'} p_{kk'}}.$$

Now, by substituting such an expression of $\langle k \rangle$ in the form of p_k above, we finally get:

$$p_k = \frac{\sum_{k'} p_{kk'}/k}{\sum_{k'} \sum_{k''} p_{k'k''}/k'} \qquad k = 1, 2, \ldots \tag{7.43}$$

The authors of Ref. [46] have proven that, by setting the function $f(h, h')$ as:

$$f(h, h') \equiv \frac{\langle k \rangle p_{hh'}}{N \rho(h) \rho(h')} \tag{7.44}$$

with $\rho(h)$ given by Eq. (7.42), the hidden-variable model in Definition 7.9 generates an ensemble of random graphs with the assigned joint probability distribution $p_{kk'}$. In addition, it is possible to show that, with the choices of $\rho(h)$ and $f(h, h')$ as in Eq. (7.42) and Eq. (7.44), the expected degree $\overline{k}(h)$ of nodes of hidden variable h is:

$$\overline{k}(h) = h.$$

This means that, if the hidden variable of a node i is equal to the desired degree k_i^*, then the expected degree of i, i.e. the average degree over the ensemble of graphs generated by the model, will be exactly equal to k_i^*. Moreover, the fluctuations around k_i^* follow a Poisson distribution, since the conditional probability $g(k|h)$ for a node of hidden variable h to end up having k links is [46]:

$$g(k|h) = \frac{e^{-h} h^k}{k!}.$$

The hidden-variable model can be very handy when we need to produce ensembles of graphs with the same degree distribution and correlation properties of a given real network, i.e. when we are given a joint probability distribution $p_{kk'}$ derived from empirical observations. As an illustrative example, we show here how to model the Internet at the AS level. We have first constructed matrix E from the May 2001 network of the Internet AS in DATA SET 7, and the corresponding distribution $p_{kk'}$ from Eq. (7.7). In order to build synthetic networks of the Internet we have then used the hidden-variable model with the two functions $\rho(h)$ and $f(h, h')$ obtained by plugging in Eq. (7.42) and Eq. (7.44) the derived expression of $p_{kk'}$. The details of the algorithm used are discussed in Appendix A.14. In particular we have sampled 1000 graphs from the hidden-variable model. In Figure 7.7(a) we focus on the node degree of the obtained graphs. Namely, we consider nodes having respectively a value of hidden variable $h = 5, 20, 40$, and we plot the distribution of their actual degree $g(k_i = k|h_i = h)$. We notice that, for each value of h, the obtained distribution is indistinguishable from a Poisson with average equal to h. It is possible to show that the model reproduces very well the degree distribution of the Internet and its degree–degree correlation properties. For instance, in Figure 7.7(b) we plot the average nearest neighbours degree $\langle k_{nn} \rangle(k)$ as a function of k, obtained for the different realisations of hidden-variable model (black dots). The results of the model are in good agreement with the degree correlation patterns observed in the Internet (squares). This is also testified by the distribution of the exponent ν of the power-law fit of $\langle k_{nn} \rangle(k)$ in the sampled graphs, reported in the figure inset, which is peaked around the value $\nu = -0.57$ observed in the actual network.

Unfortunately, how to use the hidden-variable model to generate graphs with tunable degree–degree correlations is still an open problem. In fact, when we do not have a real network to extract the joint degree–degree distribution from, it turns out that setting $p_{kk'}$ in the model in order to obtain a desired function $\langle k_{nn} \rangle(k)$, or a prescribed value of the correlation coefficient r, can be very tricky. Mark Newman proposed considering a joint distribution $p_{kk'}$ with the following form [232]:

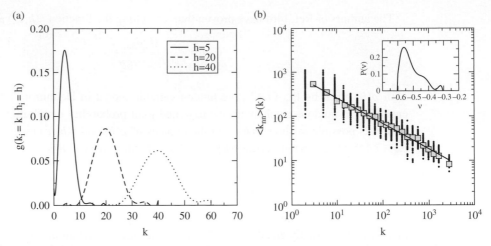

Fig. 7.7 (a) Conditional probability $g(k_i = k | h_i = h)$ for a node i with hidden variable $h_i = h$ to have a degree $k_i = k$. We have considered nodes with respectively $h = 5, 20$ and 40, and 1000 network realisations of the hidden-variable model with the same $p_{kk'}$ as the Internet. (b) Average nearest neighbours degree $\langle k_{nn} \rangle (k)$ as a function of k for the realisations of the hidden-variable model (black dots) compared with the one observed in the Internet AS network (grey squares). The solid line is a power-law fit $\langle k_{nn} \rangle (k) \sim k^\nu$ with $\nu = -0.57$. In the inset we report the distribution of the exponent ν over the different realisations of the model.

$$p_{kk'} = q_k q_{k'} + r \sigma_q^2 m_{kk'}, \tag{7.45}$$

where q_k is an assigned function representing the degree distribution of the nearest neighbours of a randomly chosen node, σ_q is the standard deviation of the distribution q_k, and $\{m_{kk'}\}$ is a symmetric matrix with all rows and columns summing to zero. With such constraints on the quantities $m_{kk'}$, the expression of $p_{kk'}$ given above satisfies all the required normalisations in Eqs. (7.30). Furthermore, if we impose:

$$\sum_{k,k'} kk' m_{kk'} = 1 \tag{7.46}$$

we can immediately notice from the definition in Eq. (7.31) that the quantity r in Eq. (7.45) coincides exactly with the Newman's correlation coefficient. Since we want to construct scale-free graphs, it is convenient to assume an exponentially truncated power-law degree distribution:

$$p_k \sim k^{-\gamma} e^{-k/\kappa} \qquad \text{for } k \geq 1$$

with exponent $2 < \gamma \leq 3$ and exponential cut-off parameter κ, which corresponds to:

$$q_k \sim k^{-\gamma+1} e^{-k/\kappa} \qquad \text{for } k \geq 1$$

It is now clear from the expression of q_k that the cut-off κ in the degree distribution is necessary to avoid the divergence of the average of the distribution q_k. Introducing a cut-off κ in the degree distribution also turns out to be useful because it provides us with an extra tunable parameter controlling the number of links in the network. For any choice of $m_{kk'}$ satisfying the appropriate constraints discussed above, the hidden-variable model with $p_{kk'}$ given in Eq. (7.45) allows ensembles of graphs parametrised by the correlation coefficient r to be produced. Therefore, the last thing we need to do is to find a way to construct matrix $\{m_{kk'}\}$. One possibility is to use the expression:

$$m_{kk'} = \frac{(q_k - x_k)(q_{k'} - x_{k'})}{\sum_{k,k'} kk'(q_k - x_k)(q_{k'} - x_{k'})} \tag{7.47}$$

which satisfies all the normalisation conditions if x_k is any distribution such that $x_k \geq 0 \, \forall k$, and $\sum_k x_k = 1$. To ensure that the quantities $p_{kk'}$ never become negative, we further need to impose that x_k is smaller than q_k for any value of k. We have therefore chosen a function x_k that decays faster than q_k. In particular, we have assumed that x_k has the same functional form as p_k but with a different cut-off parameter κ', where $\kappa' < \kappa$, namely $x_k \sim k^{-\gamma} e^{-k/\kappa'}$ for $k \geq 1$. However, this condition on x_k is necessary but not sufficient to guarantee that $p_{kk'}$ remains non-negative $\forall k, k' \geq 1$ and that $f(h, h')$ takes values in $[0, 1]$, especially when the absolute value of r is relatively large. In practice, this means that r is limited to take values in a sub-interval of $[-1, 1]$.

In order to test the model, we have chosen $p_{kk'}$ as in Eq. (7.45) with $m_{kk'}$ as in Eq. (7.47) and $\gamma = 2.5, \kappa = 100, \kappa' = 10$ and $k_{\max} = 100$. We have then sampled assortative (r=0.2), uncorrelated (r=0.0) and disassortative networks (r=−0.15) with $N = 10,000$ nodes. For each of the three values of r we have considered 1000 different realisations. In Figure 7.8(a) we report the plot of the typical degree distribution of the constructed graphs which, in all the three cases, is as expected in agreement with a power law $p_k \sim k^{-\gamma}$ with $\gamma = 2.5$ (solid line). As evident from Figure 7.8(b) the average degree $\langle k_{nn}\rangle(k)$ of the nearest neighbours

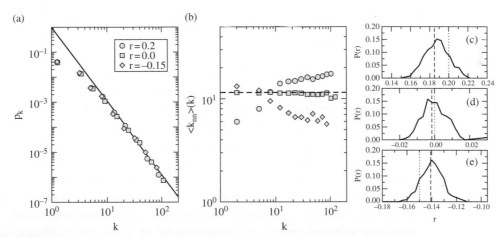

Fig. 7.8 The degree distribution (panel (a)) and the average degree of the nearest neighbours of nodes of degree k (panel (b)) for assortative ($r = 0.2$, circles), uncorrelated ($r = 0.0$, squares) and disassortative networks ($r = −0.15$) constructed by the hidden-variable model as explained in the text. The degree distributions in all the three cases are in agreement with a power law with exponent $\gamma = 2.5$. The dashed line in panel (b) indicates the value $\langle k^2 \rangle / \langle k \rangle$ expected in a network with no degree–degree correlations. In panels (c), (d) and (e) we report how the assortativity coefficient is distributed, over 1000 realisations of the three networks, around the value of r given as input to the hidden-variable model.

of nodes with degree k is an increasing function of k when we set $r > 0$ (circles) and a decreasing one when we set $r < 0$ (diamonds). For $r = 0$ (squares) we get an uncorrelated scale-free graph, as those discussed in Section 5.3, in which $\langle k_{nn} \rangle (k)$ is equal to $\frac{\langle k^2 \rangle}{\langle k \rangle}$ (dashed line). Notice also that the obtained value of the assortativity coefficient, measured for each realisation of the network can fluctuate around the desired value of r given as an input to the hidden-variable model. In Figure 7.8(c)-(e) we plot the distribution $P(r)$ of the assortativity coefficient r obtained over 1000 realisations of networks respectively with $r = 0.2, 0$ and -0.15. The dotted lines in each panel are the values of r given as input to the hidden-variable model to construct the three ensemble of networks, while the dashed line is the average over the obtained networks (the average of the distribution $P(r)$).

7.6 What We Have Learned and Further Readings

Reproducing the degree distribution or the degree sequence of a network is only the first step to model its structure. Real-world networks such as the movie actor collaboration network from Chapter 2 or the Internet, the network of physical wiring between clusters of computers introduced as the data set of this chapter, are in fact *correlated*. This means that in these networks the probability for an edge departing from a node of degree k to arrive at a node of degree k' depends not only on k' but also on the degree k. We therefore need something more than the degree distribution p_k to describe them. In this chapter we have introduced the mathematical framework to deal with degree–degree correlations. A complete description of a correlated network can be obtained in terms of the *joint probability* $p_{kk'}$ or of the *conditional probability* $p_{k'|k}$. Given a network, these two probabilities can be easily constructed by counting the number of edges between nodes of degree k and nodes of degree k'. However, it is difficult to extract directly from $p_{kk'}$ or $p_{k'|k}$ whether the network is correlated or not and the sign of correlations. In this chapter we have learned that a more compact information can be obtained by computing the *average nearest neighbours' degree* $\langle k_{nn} \rangle (k)$ of nodes of degree k, or by the *Newman's correlation coefficient* r. An increasing (decreasing) value of $\langle k_{nn} \rangle (k)$ as a function of k, or a positive (negative) value of r, indicate that the network is *assortative* (*disassortative*). We have also shown that networks with an assigned $p_{kk'}$ can be obtained by means of a *hidden-variable model*. Conversely, the problem of tuning the amount and sign of correlations in a network is yet to be solved in a satisfactory manner. In fact, there are several algorithms capable of generating networks with a certain level of assortativity, but most of them are characterised by caveats which end up limiting their applicability. Typical examples are procedures based on edge rewiring which often produce networks with very peculiar structures [232, 317], or models focusing on the tuning of the average degree of nearest neighbours function [312], which permit a fine-grained control on the degree correlation pattern of the resulting network, at the cost of a complicated algorithmic implementation.

The degree–degree correlations we have studied in this chapter are only a particular case of a more general characteristic of networks known as *mixing* by a node property. Typical examples of assortative mixing by a node property other than the degree can be easily found in social systems, where people usually prefer to link with others who are like them in terms of race, culture, language, or age. This kind of assortative linking, known also as *homophily*, has been widely studied in the social sciences [308, 278]. One example of disassortative mixing by node type is food webs, i.e. networks describing which species eat which in an ecosystem. Different types here represent plants, herbivores and carnivores, and while many edges link plants and herbivores, and herbivores and carnivores, there are only very few edges linking herbivores to other herbivores, or plants to plants. These other types of mixing can be characterised in a way in all similar to how we treated degree–degree correlations in this chapter [232].

The *Internet at the AS level* was the real-world network we studied in this chapter. For a complete and accessible review on the Internet and on the various data collection projects, their reliability and their limitations, we suggest the book by Romualdo Pastor-Satorras and Alessandro Vespignani [253].

Problems

7.1   The Internet and Other Correlated Networks

(a) Consider the movie actor collaboration network and the Internet at the AS level as in May 2001. Construct and plot the histograms of the differences in the degrees of the nodes at the two ends of an edge. Do these histograms support the same conclusions derived from the plots in Figure 7.1 that the two networks are respectively positively and negatively correlated? Can you think of any other graphical way to show that the two networks have degree–degree correlations?

(b) Repeat the same analysis as in Figure 7.1 for the snapshots of the Internet at different times given in DATA SET 7. Remember to normalise the results by the order of the networks.

7.2   Dealing with Correlated Networks

(a) Do you understand why the quantity e_{kk} in Eq. (7.7) and in Eq. (7.6) needs to be defined as *twice* the number of links connecting two nodes of degree k?

(b) Consider the graph described by the adjacency matrix

$$A = \begin{pmatrix} 0 & 1 & 0 & 1 & 1 \\ 1 & 0 & 1 & 1 & 1 \\ 0 & 1 & 0 & 0 & 0 \\ 1 & 1 & 0 & 0 & 0 \\ 1 & 1 & 0 & 0 & 0 \end{pmatrix}.$$

Write the joint probability distribution $p_{kk'}$ of the graph. Can you construct a different graph with the same $p_{kk'}$?

(c) Eq. (7.9) tells us that $\sum_{k'} p_{kk'} = q_k$, which means that q_k is the marginal probability associated with the joint probability $p_{kk'}$. Since $p_{kk'}$ is symmetric, we also have $\sum_k p_{kk'} = q_{k'}$. Are you able to prove this by using Eq. (7.8) and the detailed balance condition?

(d) There are several different ways to average the degrees of a graph with N nodes and K links. For instance, you can construct the average node degree $\langle k \rangle$ by selecting the N nodes with uniform probability. Alternatively, you can select the K links of the graph with uniform probability, and then evaluate the average degree of the nodes at the two ends of the links. This average from the can be computed as $\sum_k k q_k$ from the distribution q_k defined in Eq. (3.38). As a third quantity, you can construct the average degree $\langle k_{nn} \rangle$ of the first neighbours of a random node by using Eq. (7.2). This corresponds to summing the average degree of the first neighbours of all the nodes, and then dividing the obtained number by N. Consider the graph of Example 3.10 and construct these three different averages. Can you say from your results if the graph is correlated or uncorrelated?

7.3 Assortative and Disassortative Networks

(a) Extract and plot, as a function of time, the value of the exponent ν of the power-law $\langle k_{nn} \rangle(k) \sim k^{\nu}$ for the Internet at the AS level.

(b) Is it possible to have a rich-club behaviour in a network with disassortative degree–degree correlations? Can you sketch how a disassortative network with rich club behaviour might look?

7.4 Newman's Correlation Coefficient

(a) Are you able to derive the expression of the correlation coefficient r given in Eq. (7.32) from Eq. (7.31)?

(b) Derive Eq. (7.33) which expresses the Newman's correlation coefficient in terms of summations over the links of the network. HINT: Start from Eq. (7.32) and notice that $\sum_e (k_e + k'_e) = \sum_i k_i^2$, i.e. when you sum over the links of the graph the degrees of the two nodes at the ends of each link you get the sum of the squares of the degrees of all the nodes in the graph. This is because summing over the links corresponds to considering a node a number of times equal to its degree. If you are not able to prove this, verify it with a small graph as an example.

(c) Verify that the expression of $f_{kk'}$ given in Eq. (7.39) satisfies the recursion relations in Eq. (7.38). HINT: As a first step, write and solve the recursion relations for $k' = 1$.

7.5 Models of Networks with Degree–Degree Correlations

(a) Consider the following matrix:

$$E = \begin{pmatrix} 0 & 0 & 1 \\ 0 & 2 & 2 \\ 1 & 2 & 0 \end{pmatrix},$$

describing the number of edges of type (k, k') i.e. the number of edges between nodes of degree k and nodes of degree k'. This means, in practice, that there are two edges of type $(3, 2)$, one edge of type $(3, 1)$ and one edge of type $(2, 2)$.

Starting from these four edges, show how to associate the eight nodes at the ends of the four edges to construct a network with $N = 4$ and $K = 4$ and matrix E.

Does this mean that, for any assigned matrix $\{p_{kk'}\}$, it is possible to construct a network with corresponding degree–degree correlations by drawing K edges from the desired distribution $p_{kk'}$, i.e. by producing a list of K edges, each of them characterised by the degree k and k' of its two ends, and then grouping together and identifying the degree k ends randomly in groups of k to create the network?

(b) Consider the movie actor collaboration network in DATA SET 2. Construct the matrix E and use Eq. (7.7) to obtain the corresponding distribution $p_{kk'}$. Implement then a hidden-variable model selecting the functions $\rho(h)$ and $f(h, h')$ by plugging in Eq. (7.42) and Eq. (7.44) the derived expression of $p_{kk'}$. Sample 10 graphs from the hidden-variable model and plot the corresponding average nearest neighbour degree function $\langle k_{nn} \rangle(k)$ as a function of k, as done in Figure 7.7(b) for the Internet.

8

Cycles and Motifs

What are the building blocks of a complex network? We have seen in Chapter 4 that triangles are highly recurrent in social and biological networks, so that they can be considered as one of their elementary bricks. In this chapter we will discuss a general approach to define and detect the building blocks of a given network. The basic idea is to look not only at triangles but also at *cycles* of length larger than three, and at other small subgraphs, known as *motifs*, which occur in real networks more frequently than in their corresponding randomised counterparts. We will first derive a set of formulas to count the number of cycles in a graph directly from its adjacency matrix. As an application, we will use these formulas to find the number of cycles of different lengths in *urban street networks* and to compare various cities from all over the world. Notice that urban streets are a very special type of network. Their nodes have a position in Euclidean space and their links have a length, and as such they need to be described in terms of *spatial graphs*. The topology of spatial graphs is constrained by their spatial embedding, so that urban streets require special treatment. This will imply, in our case, the choice of appropriate spatial graphs to use as network null models when counting cycles in a city. In the second part of the chapter we will concentrate on other small subgraphs which are overabundant in some real networks and can therefore be very useful to characterise their microscopic properties. In particular, we will show that one specific motif, namely the so-called *feed-forward loop*, that emerges in the structure of the *transcription regulation network of* E. coli and of other biological networks, is there because it plays an important biological function. This relation between structure and function is the main reason why, by performing a so-called *motif analysis* and by looking at the profile of abundance of all possible subgraphs, it is possible to classify complex networks and to group them into different *network superfamilies*.

8.1 Counting Cycles

Enumerating the cycles in a graph, i.e. counting their number, can be of primary importance to study complex networks. In this section we show how to express the number of cycles in a graph in terms of the adjacency matrix of the graph. In particular, we will derive a set of formulas giving the number $n_G(\mathbb{C}_l)$ of cycles of length $l = 3, 4$ and 5 in an undirected graph G as functions of the different powers of the adjacency matrix A of the graph. In order to do so, we will closely follow Ref. [7].

The first important observation is that the number of closed walks of length l in a graph is equal to the trace of A^l. This is a direct consequence of a result we have already come across a few times in the previous chapters of this book, and it is now time to state it in the form of a theorem.

Theorem 8.1 (Number of Walks)   *Given an undirected or a directed graph G described by the adjacency matrix A, the number of walks of length l from node i to node j is equal to $(A^l)_{ij}$.*

Proof   The theorem can be proven by induction on l. For $l = 1$, the result follows from the very same definition of the adjacency matrix A, since a walk of length 1 is precisely a link. Suppose now that $(A^l)_{ij}$ is the number of walks of length l from i to j. We will prove that $(A^{l+1})_{ij}$ is equal to the number of walks of length $l + 1$ from i to j. If $(A^l)_{ij}$ is the number of walks of length l, then there are $(A^l)_{im}(A)_{mj}$ walks of length $l + 1$ from i to j in which node m is the penultimate node. Our result then follows because the total number of walks of length $l + 1$ from i to j is equal to: $\sum_m (A^l)_{im} A_{mj} = (A^{l+1})_{ij}$, Q.E.D.   □

The theorem above tells us that, in order to find the number of walks of a given length l, we simply need to evaluate the lth power of the adjacency matrix. As a particular case, the theorem states that $(A^l)_{ii}$ is equal to the number of closed walks of length l from a node i to itself. Hence:

$$\text{Trace}(A^l) = \sum_i (A^l)_{ii}$$

is equal to the total number of closed walks of length l in G. Having such simple formulas can be very useful. For instance, from the number of walks of different lengths in a graph it is possible to define other measures of centrality similar to those we have discussed in Section 2.3. We show two examples, respectively known as *Katz centrality* [174] and *subgraph centrality* [104].

Example 8.1   *(Katz centrality)*   In 1953 Leo Katz proposed measuring the centrality of the nodes of a connected graph by means of a weighted sum of all the powers of the adjacency matrix A [174]. The idea is simple and powerful. We have seen in Theorem 8.1 that powers of A give the number of walks of different lengths. Let us then construct the following matrix $S = \{s_{ij}\}$, where $0 < \alpha < 1$ is the so-called attenuation factor:

$$S = \alpha A + \alpha^2 A^2 + \alpha^3 A^3 + \ldots = \sum_{l=1}^{\infty} \alpha^l A^l \tag{8.1}$$

Entry s_{ij} measures the communication between two nodes i and j taking into account walks of any length. Since α is smaller than 1, considering higher powers of A with less weight means to attenuate the influence of longer walks. Now, the nice thing is that, as long as $|\alpha| < 1/\rho(A)$, where $\rho(A)$ is the *spectral radius* of the adjacency matrix A, i.e. the largest eigenvalue in absolute value, the infinite sum $\sum_{l=0}^{\infty} \alpha^l A^l$ converges to $(I - \alpha A)^{-1}$. We can therefore write:

$$S = \sum_{l=0}^{\infty} \alpha^l A^l - I = (I - \alpha A)^{-1} - I.$$

The *Katz centrality* c_i^K of a node i can then be measured as the sum of the elements of column i of matrix S [174]:

$$c_i^K = \sum_j s_{ji} = \sum_j (S^\top)_{ij}. \tag{8.2}$$

In this way a node i is more central if there are more ways to reach it from the other nodes with walks of various lengths, with of course longer walks contributing less than shorter walks. In vectorial form, the previous expression reads:

$$\mathbf{c}^K = S^\top \mathbf{1}$$

where $\mathbf{1}$ is the identity column vector. We finally have the Katz centrality vector:

$$\mathbf{c}^K = \left[(I - \alpha A^\top)^{-1} - I \right] \mathbf{1}. \tag{8.3}$$

Notice that the Katz centrality only differs by a constant vector from the α-centrality we studied in Section 2.3 (see Problem 8.1(a)), although the two measures were introduced based on different arguments.

Example 8.2 *(Subgraph centrality)* In 2005 Ernesto Estrada and Juan Rodríguez-Velázquez introduced a measure to quantify the importance of a node from its participation to network subgraphs (consisting of closed walks). This is based on the construction of matrix $G = \{g_{ij}\}$:

$$G = I + A + \frac{A^2}{2!} + \frac{A^3}{3!} + \ldots = \sum_{l=0}^{\infty} \frac{A^l}{l!} = e^A \tag{8.4}$$

that is again a linear combination of all powers of A, and differs from matrix S in the way the different terms are weighted. With such a choice of weights, matrix G coincides with the exponential of the adjacency matrix e^A. G is known as the graph *communicability matrix* since it has revealed useful to generalise the concept of node communication, usually only based on shortest paths [103, 102]. The *subgraph centrality* c_i^S of node i is then defined as:

$$c_i^S = g_{ii} = \sum_{l=1}^{\infty} \frac{(A^l)_{ii}}{l!} = (e^A)_{ii} \tag{8.5}$$

that is the weighted sum of closed walks of different lengths starting and ending at node i [104, 102]. In this way, nodes involved in many closed walks have higher values of subgraph centrality, with shorter closed walks contributing more to the centrality of a node than longer closed walks. The subgraph centrality can be calculated from the eigenvalues and the eigenvectors of the adjacency matrix of the graph (see Problem 8.2b).

After this short detour on centrality measures, let us come back to the main focus of this section, that of counting the number of cycles $n_G(\mathbb{C}_l)$ in a graph G. Theorem 8.1 tells us that $\text{Trace}(A^l)$ is equal to the number of closed walks of length l in a graph. Hence, in order to obtain the number of cycles, what we need to do is to compute the number of non-simple closed walks of length l. By non-simple closed walks we mean closed walks in which a generic node, excluding starting and ending nodes, is visited more than once, or the starting/ending node is repeated more than twice. In this way, by subtracting from $\text{Trace}(A^l)$ the number of non-simple closed walks of length l, we get the number of (simple) closed paths of length l in the graph. This number is just $2l$ times the number of cycles of length l, because a cycle of length l can be visited by starting from each one of the l nodes and going around clockwise or anticlockwise. We will now describe an easy way to count the number of non-simple closed walks of length l. However, before doing so, we need to go through a few definitions.

Let us denote by $\mathbb{C}_l$ a cycle of length l, with $l = 3, 4$ and so on. Following we introduce the definitions of *homomorphism*, *homomorphic image*, and that of *cyclic graphs*.

Definition 8.1 (Homomorphism)   *Let $G_1 = (\mathcal{N}_1, \mathcal{L}_1)$ and $G_2 = (\mathcal{N}_2, \mathcal{L}_2)$ be two graphs. A mapping $f : \mathcal{N}_1 \cup \mathcal{L}_1 \rightarrow \mathcal{N}_2 \cup \mathcal{L}_2$ is a* homomorphism *if for every $v \in \mathcal{N}_1$ we have $f(v) \in \mathcal{N}_2$, and for every link $l = (u, v) \in \mathcal{L}_1$ we have $f(l) = (f(u), f(v)) \in \mathcal{L}_2$.*

Definition 8.2 (Homomorphic image)   *Let f be a homomorphism from G_1 to G_2. If f is onto $\mathcal{N}_2 \cup \mathcal{L}_2$, we say that G_2 is a* homomorphic image *of G_1 according to the mapping f.*

Notice that in a homomorphism, graph G_2 can have a larger number of nodes and links than G_1, but also G_1 can have a larger number of nodes and links than G_2. However, if G_2 is a homomorphic image of G_1 the number of nodes and links of G_2 cannot be larger than the number of nodes and links of G_1, respectively. The two definitions above are better illustrated by means of the following examples.

Example 8.3   Suppose we are given the two graphs $G_1 = (\mathcal{N}_1, \mathcal{L}_1)$ and $G_2 = (\mathcal{N}_2, \mathcal{L}_2)$ shown in the following figure. Let us consider the following mapping from the nodes of G_1

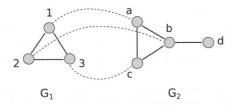

into the nodes of G_2: $f(1) = a, f(2) = b, f(3) = c$. The mapping f is a homomorphism since the edges of G_1 are also mapped into the edges of G_2. In fact, we have: $f(1, 2) = (a, b) \in \mathcal{L}_2, f(2, 3) = (b, c) \in \mathcal{L}_2, f(3, 1) = (c, a) \in \mathcal{L}_2$. However, G_2 is not the homomorphic image of G_1, because f is not onto $\mathcal{N}_2 \cup \mathcal{L}_2$. In fact, the mapping f does not involve all the nodes and edges in G_2. Namely, node d of G_2 is not associated with any node of G_1, and link (b, d) of G_2 is not associated with any of the links of G_1. Other possible

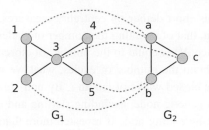

homomorphisms are the mappings $f(1) = a, f(2) = c, f(3) = b$; $f(1) = b, f(2) = a, f(3) = c$; $f(1) = b, f(2) = c, f(3) = a$; $f(1) = c, f(2) = a, f(3) = b$ and $f(1) = c, f(2) = b, f(3) = a$. For none of this functions f, G_2 is a homomorphic image of G_1.

Suppose now we are given the following two graphs G_1 and G_2. Consider the mapping from the nodes of G_1 into the nodes of G_2: $f(1) = a, f(2) = b, f(3) = c, f(4) = a, f(5) = c$. The mapping f is a homomorphism since the edges of G_1 are also mapped into the edges of G_2. In addition to this, G_2 is the homomorphic image of G_1 because for each node i_2 of G_2 there is a node i_1 of G_1 such that $f(i_1) = i_2$, and for each link l_2 of G_2 there is a link l_1 of G_1 such that $f(l_1) = l_2$.

As a final remark, note that if G_2 is a homomorphic image of G_1, and the two graphs have the same number of nodes and links, then G_1 and G_2 are isomorphic, i.e. the homomorphism f is actually an *isomorphism* (see Definition 1.2).

We are now ready to define *l*-cyclic graphs.

Definition 8.3 (*l*-cyclic graph) *A graph $H = (\mathcal{N}_H, \mathcal{L}_H)$ is said to be l-cyclic, for $l \geq 3$, if there exists a mapping f such that H is a homomorphic image of the cycle $\mathbb{C}_l$. The number of different homomorphisms from $\mathbb{C}_l$ onto H is denoted by $c_l(H)$. Clearly, H is l-cyclic if and only if $c_l(H) > 0$.*

From Example 8.3 it is clear that it is not possible to find a homomorphism from G_1 to G_2 that is *onto* $\mathcal{N}_2 \cup \mathcal{L}_2$ when the number of nodes in G_2 is larger than the number of nodes in G_1. This means that, when we are looking for the 3-cyclic graphs, we have only to consider graphs with at most three nodes, when we are looking for the 4-cyclic graphs, we have only to consider graphs with at most four nodes, and so on. It is simple to understand that the only 3-cyclic graph is $\mathbb{C}_3$ itself, and that we have $c_3(\mathbb{C}_3) = 6$. In fact, we have six different homomorphisms from $\mathbb{C}_3$ onto $\mathbb{C}_3$ as shown in the following example.

Example 8.4 *(The 3-cyclic graph)* When we look for 3-cyclic graphs, we have to consider only graphs with at most three nodes. These are the graphs H_1, H_2 and $\mathbb{C}_3$ shown in the right-hand side of the figure below. It can be easily shown that H_1 is not 3-cyclic, because it is not possible to find a mapping f such that H_1 is a homomorphic image of the triangle. To see this, let us name 1, 2, 3 the nodes of a triangle as shown in the left-hand side of the figure. If we consider for instance the mapping $f(1) = a$, $f(2) = b$, $f(3) = a$, then link $(1,2)$ of the first graph would be mapped into (a, a), which is not a link of H_1. The

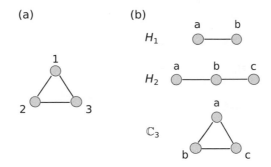

same thing happens for any other possible mapping f. Also H_2 is not 3-cyclic. In fact, if we consider the following mapping $f(1) = a, f(2) = b, f(3) = c$, then link $(1,2)$ of the first graph is mapped into link (a,b) of H_1, link $(2,3)$ is mapped into link (b,c), while the third link $(1,3)$ is mapped into (a,c) which is not a link of H_2. Conversely $\mathbb{C}_3$ is a 3-cyclic graph. We can in fact map node 1 of the first triangle into one of the three nodes of the second triangle. Once we fix this, let say $f(1) = a$, then there are only two possibilities for the remaining two nodes: either $f(2) = b, f(3) = c$ or $f(2) = c, f(3) = b$. Hence, there are a total of $3 \times 2 = 6$ different mappings from $\mathbb{C}_3$ onto $H = \mathbb{C}_3$.

The situation is more complex for l-cyclic graphs with $l \geq 4$. Figs. 8.1, 8.2, and 8.3 report the l-cyclic graphs for $l = 4, 5$ and 6, respectively. The l-cyclic graphs shown, which are not simple cycles as $\mathbb{C}_3, \mathbb{C}_4, \mathbb{C}_5$, are ordered according to the number of edges they contain and denoted by $H_1, H_2, \ldots$ as in Ref. [7]. We have, in particular, three 4-cyclic graphs, namely H_1, H_2 and $\mathbb{C}_4$, and three 5-cyclic graphs, $\mathbb{C}_3$, H_5 and $\mathbb{C}_5$. As shown in Example 8.5, there are two different homomorphisms from $\mathbb{C}_4$ onto H_1, four different homomorphisms from $\mathbb{C}_4$ onto H_2, and eight from $\mathbb{C}_4$ onto $\mathbb{C}_4$, while for the three 5-cyclic graphs we have $c_5(\mathbb{C}_3) = 30$, $c_5(H_5) = 10$ and $c_5(\mathbb{C}_5) = 10$. Finally, we have ten 6-cyclic graphs, from the smallest one with only two nodes, to the largest one with six nodes and six links.

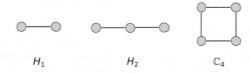

H_1 H_2 $\mathbb{C}_4$

Fig. 8.1   The three 4-cyclic graphs.

$\mathbb{C}_3$ H_5 $\mathbb{C}_5$

Fig. 8.2   The three 5-cyclic graphs.

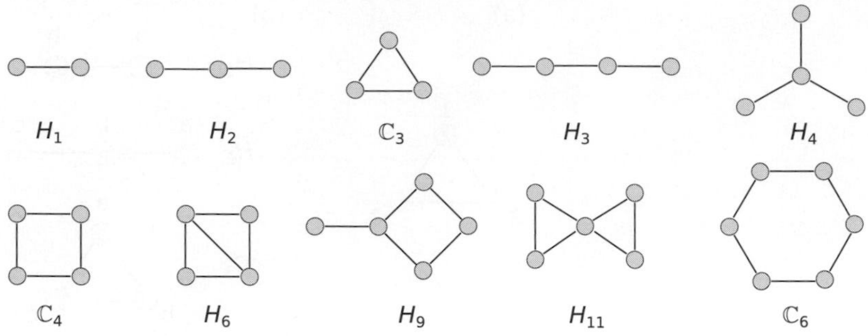

Fig. 8.3   The ten 6-cyclic graphs.

Example 8.5   *(4-cyclic and 5-cyclic graphs)*   When looking for the 4-cyclic graphs, we need to consider graphs with at most four nodes. It is then relatively easy to prove that there are only three 4-cyclic graphs, namely those shown in Figure 8.1. Here we show that:

$$c_4(H_1) = 2 \quad c_4(H_2) = 4 \quad c_4(\mathbb{C}_4) = 8 \tag{8.6}$$

The number of different homomorphisms from $\mathbb{C}_4$ onto H_1 is equal to 2 because $f(1)$ can be either a or b, and once we have fixed the first correspondence all the others are automatically determined. In the case of H_2, $f(1)$ can be either a, b or c. If $f(1) = a$, then necessarily $f(2) = b$, $f(3) = c$, $f(4) = b$. Likewise if $f(1) = c$, then necessarily $f(2) = b$, $f(3) = a$, $f(4) = b$. Instead, if $f(1) = b$ then there are two possibilities, either $f(2) = a$, $f(3) = b$, $f(4) = c$ or $f(2) = c$, $f(3) = b$, $f(4) = a$. Finally, $c_4(\mathbb{C}_4) = 8$ because node 1 of the first graph can be mapped into any of the four nodes of the second graph, and then we have two different directions to go around the graphs.

Analogously, it is not too difficult to show that:

$$c_5(\mathbb{C}_3) = 30 \quad c_5(H_5) = 10 \quad c_5(\mathbb{C}_5) = 10 \tag{8.7}$$

for the three 5-cyclic graphs shown in Figure 8.2.

We are finally ready to proceed to the main goal of this section, that of expressing the number $n_G(\mathbb{C}_l)$ of cycles of length l, with $l = 3, 4, \ldots$, in graph G, in terms of the adjacency matrix of G [7]. To calculate $n_G(\mathbb{C}_l)$ we need to find the number of subgraphs of G isomorphic to cycle $\mathbb{C}_l$ (for the isomorphism between two graphs see Definition 1.2 in Chapter 1). To do this, let us first denote as $n_G(H)$ the number of subgraphs of G isomorphic to a given graph H. Now, the total number of closed walks of length l in G can be written as a sum over all the l-cyclic graphs. Namely, we have:

$$\text{Trace}(A^l) = \sum_H c_l(H) n_G(H) \tag{8.8}$$

where we have indicated as H the generic l-cyclic graph, and as $c_l(H)$ the number of different homomorphisms from $\mathbb{C}_l$ onto H. In practice, this equation tells us that the number of closed walks of length l in a graph G is equal to the number of l-cyclic graphs in G times the number of different ways in which each l-cyclic graph can be visited in a walk of length l. For instance, if we are looking for the number of walks of length 4, Trace(A^4), the summation on the right-hand side of Eq. (8.8) is made of three terms, corresponding to the three 4-cyclic graphs in Figure 8.1, namely H_1, H_2 and $\mathbb{C}_4$.

Example 8.6 *(Number of walks of length 4)* Suppose you want to count the number of walks of length 4 in the graph G with $N = 5$ nodes and $K = 6$ links shown in the figure. In this case Eq. (8.8) reduces to:

$$\text{Trace}(A^4) = c_4(H_1)n_G(H_1) + c_4(H_2)n_G(H_2) + c_4(\mathbb{C}_4)n_G(\mathbb{C}_4).$$

To evaluate the right-hand side of this equation, the first 4-cyclic graph to consider is H_1. Counting the number of times graph H_1 appears in G is equivalent to count the number of

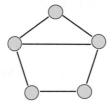

links in G, that is $K = 6$. Each link can be visited in $c_4(H_1) = 2$ different ways, so that the first of the three terms in the summation in the right-hand side of Eq. (8.8) is equal to $2 \times K = 12$.

The second term corresponds to the 4-cyclic graph H_2, that is a sequence of two links, i.e. two links with a common vertex. There are nine such 4-cyclic graphs in G, and since $c_4(H_2) = 4$ the second term in the summation is equal to 36. The last 4-cyclic graph to consider is $\mathbb{C}_4$. There is only one subgraph of G isomorphic to $\mathbb{C}_4$, and since $c_4(\mathbb{C}_4) = 8$ the last term in the summation is 8. Finally, the right-hand side of Eq. (8.8) is 12+ 36+ 8=56 and this number is equal to Trace(A^4), where A is the adjacency matrix of G.

The right-hand side of Eq. (8.8) can be divided into two different contributions. In fact, we have seen that the number of nodes N_H of a l-cyclic graph H cannot be larger than l. Hence $N_H \leq l$, and in fact $N_H < l$, unless $H = \mathbb{C}_l$. We thus can write:

$$\sum_H c_l(H)n_G(H) = \sum_{H:N_H<l} c_l(H)n_G(H) + c_l(\mathbb{C}_l)n_G(\mathbb{C}_l).$$

Since we have $c_l(\mathbb{C}_l) = 2l$ for every $l \geq 3$ (see Problem 8.1c), we therefore obtain the following formula [7].

The number of l-cycles in an undirected graph G can be written as:

$$n_G(\mathbb{C}_l) = \frac{1}{2l}\left[\text{Trace}(A^l) - \sum_{H:N_H<l} c_l(H)n_G(H)\right] \tag{8.9}$$

where A is the adjacency matrix of G, $n_G(H)$ is the number of subgraphs of G isomorphic to the l-cyclic graph H, $c_l(H)$ is the number of different homomorphisms from $\mathbb{C}_l$ onto H, and the summation is over all l-cyclic graphs with fewer than l nodes.

In the simplest possible case, $l = 3$, since the triangle is the only 3-cyclic graph, the formula above reduces to the following result.

The number of triangles in graph G can be written as:

$$n_G(\mathbb{C}_3) = \frac{1}{6}\text{Trace}(A^3) \tag{8.10}$$

where A is the adjacency matrix of G.

To use the formula for a generic $l > 3$ we need to evaluate $n_G(H)$ for all l-cyclic graphs H (see Problem 8.1(d)). For instance, for $l = 4$, we have:

$$n_G(\mathbb{C}_4) = \frac{1}{8}\left[\text{Trace}(A^4) - 4n_G(H_2) - 2n_G(H_1)\right].$$

Now $n_G(H_1)$ is equal to K, the number of links in G, while it is simple to prove that $n_G(H_2) = \sum_{i=1}^{N} k_i(k_i - 1)/2$, where N is the number of nodes in graph G. In fact, given a node i with a degree $k_i \geq 2$, there are $k_i(k_i - 1)/2$ different ways to construct graph H_2 centred at i, since we can choose the first edge in k_i ways and the second edge in $k_i - 1$ ways, and we need to divide by 2 to avoid double counting. This number should then be summed over the nodes of G. We finally get the following result:

The number of cycles of length $l = 4$ in graph G can be written as:

$$n_G(\mathbb{C}_4) = \frac{1}{8}\left[\text{Trace}(A^4) - 2\sum_{i=1}^{N} k_i(k_i - 1) - 2K\right] \tag{8.11}$$

where A is the adjacency matrix of G, and k_i is the degree of node i.

Analogously, the expression for the number of cycles of length five reads:

$$n_G(\mathbb{C}_5) = \frac{1}{10}\left[\text{Trace}(A^5) - 10n_G(H_5) - 30n_G(\mathbb{C}_3)\right]$$

where we need to evaluate $n_G(H_5)$, i.e. the number of subgraphs of G isomorphic to graph H_5. This number can be written in terms of the adjacency matrix as: $n_G(H_5) = 1/2\sum_i(A^3)_{ii}(k_i - 2)$, so that in definitive we have an expression of $n_G(\mathbb{C}_5)$ as a function of the adjacency matrix of the graph.

The number of cycles of length $l = 5$ in graph G can be written as:

$$n_G(\mathbb{C}_5) = \frac{1}{10}\left[\mathrm{Trace}(A^5) - 5\sum_{i=1}^{N}(A^3)_{ii}(k_i - 2) - 5\mathrm{Trace}(A^3) \right] \qquad (8.12)$$

where A is the adjacency matrix of G, and k_i is the degree of node i.

Other similar formulas for the number of cycles of higher order, up to length $l = 7$, can be found in Ref. [7].

8.2 Cycles in Scale-Free Networks

As an application of the formulas derived in the previous section, we will compute here the number of cycles in scale-free networks with different degree-exponents. Then, in the next section, we will introduce the first data set of this chapter, and we will focus on the search of cycles in a particular class of real-world networks, namely spatial networks.

Let us first notice that we can immediately get an expression for the number of cycles $n(\mathbb{C}_l)$ of length l in ER random graphs by using Eq. (3.21) of Chapter 3. We have:

$$\overline{n}(\mathbb{C}_l) = \frac{\langle k \rangle^l}{2l} \qquad (8.13)$$

as detailed in the following example.

Example 8.7 *(Cycles in ER random graphs)*   Eq. (3.21) gives the expected number of times, $\overline{n}(F_{n,l})$, a given subgraph $F_{n,l}$ with n nodes and l links is found in ensembles of ER random graphs. By setting $n = l$ and considering ensembles of graphs in which the probability p has an inverse linear dependence on N, $p(N) = \langle k \rangle N^{-1}$, i.e. graphs with a fixed average degree $\langle k \rangle$, we get:

$$\overline{n}(\mathbb{C}_l) = \frac{\langle k \rangle^l}{a_{\mathbb{C}_l}}.$$

Since the number of automorphisms, $a_{\mathbb{C}_l}$, of a cycle of length l is equal to $2l$ (see Definition 1.3 of automorphism, and the example in Figure 1.5), we obtain Eq. (8.13).

A remarkable aspect of Eq. (8.13) is that the average number of cycles of any given length in ER random graphs is constant and does not depend on N. In other words, the number of cycles does not increase with the order of the network. So, as already stated in Section 3.3, cycles in ER random graphs can be neglected. There is a simple way to understand this result for the case $l = 3$. A triangle is obtained when a third link is added to a triad (two adjacent edges). Now, from Eq. (3.21), we have that the average number of adjacent edges in a random graph with N nodes and $p(N) = \langle k \rangle N^{-1}$ is equal to $\langle k \rangle^2 N/2$. Hence, the number of triads increases linearly with the graph order N. To obtain the number of triangles in a random graph, we have to multiply the number of triads by the probability that a triad closes to form a triangle. This probability scales with the graph order as $1/N$, since a link emerging from a node at one extremity of the triad has one out of N possibilities to connect to the node at the other extremity. Therefore, the average number of triangles is equal to $\langle k \rangle^2 N/2N$ and does not depend on N. We will show in the following that random graphs with scale-free degree distributions exhibit a behaviour qualitatively different from that of ER random graphs. Namely, the number of cycles in scale-free graphs with exponent γ between 2 and 3 increases with the graph order N.

We start by studying cycles in scale-free networks with a fixed number of nodes. To this aim we have generated graphs with $N = 100$ and $\gamma = 2, 3$ and 5 from the configuration model, and we have counted cycles of different lengths. In Figure 8.4 we plot the average number of cycles of length l, $\bar{n}(\mathbb{C}_l)$, as a function of l, from $l = 3$ to $l = 12$. Up to $l = 5$, we have used Eq. (8.9), while for larger values of l we have implemented an algorithm for the enumeration of cycles due to Johnson [169] (see Appendix A.15). Although the number of cycles reported in the figure grows exponentially with l, it can be shown that Eq. (8.13) does not reproduce the exact values. A better agreement can be obtained by the expression:

$$n(\mathbb{C}_l) \simeq \lambda_1^l / 2l \tag{8.14}$$

where λ_1 is the largest eigenvalue of the adjacency matrix. This is an approximate and easy to compute expression, as detailed in the following example, which is valid for any graph independently from its degree distribution.

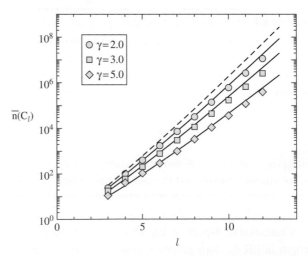

Number of cycles in scale-free networks with $N = 100$ and three values of γ. The direct counts (symbols) are well reproduced by the analytic predictions of Eq. (8.15) (solid lines). For $\gamma = 3$ we also report the approximation given by Eq. (8.14) (dashed line).

Example 8.8 *(A spectral formula to count cycles)*   The eigenvalues and associated eigenvectors of the adjacency matrix A of a graph G are intimately related to important topological features such as the diameter of the graph (for instance, the diameter $D(G)$ is smaller than the number of distinct eigenvalues of A), its connectivity properties, and the number of cycles in the graph (see Problem 8.2a). In particular, in an undirected graph, the number of closed walks of length l can be readily expressed in terms of the eigenvalues of A. The adjacency matrix A of an undirected graph is symmetric and all its eigenvalues are real. If we indicate as $\lambda_1 \geq \lambda_2 \geq \ldots \geq \lambda_N$ the eigenvalues of A, then $\text{Trace}(A^l) = \sum_i \lambda_i^l$. Now, if the largest eigenvalue is much larger than the others, $\lambda_1 \gg \lambda_2$ (in this case the graph is said to be *expanding*), the number of closed walks of length l is well approximated by λ_1^l. By using Eq. (8.9) and assuming that for small l the number of self-crossings in a walk is small, we obtain Eq. (8.14).

In Figure 8.4 we report as dashed lines the prediction of Eq. (8.14) for $\gamma = 3$. Unfortunately, there is some agreement with the actual number of cycles only for small values of l. Increasing the value of l, Eq. (8.14) deviates considerably since the number of walk crossings becomes non-negligible. It is also worth noticing that the number of cycles of a given length becomes smaller as γ increases. Two Italian physicists, Ginestra Bianconi and Matteo Marsili, have derived exact expressions for the number of cycles in uncorrelated random graphs with a given p_k [37].

In the case of small values of l, the number of cycles of length l reads:

$$\overline{n}(\mathbb{C}_l) \simeq \frac{1}{2l} \left(\frac{\langle k^2 \rangle - \langle k \rangle}{\langle k \rangle} \right)^l \tag{8.15}$$

where $\langle k \rangle$ and $\langle k^2 \rangle$ are the two first moments of the degree distribution.

The predictions of this formula are reported as solid lines in Figure 8.4, and are in perfect agreement with the actual cycle counts. We can derive Eq. (8.15) as a particular case of a more general expression for the number of short cycles in networks with degree–degree correlations, as illustrated in the following example.

Example 8.9 *(Short cycles in correlated networks)*   A simple argument allows us to derive the expression for the number of short cycles in a correlated network, and Eq. (8.15) as a particular case [36]. When dealing with short cycles we can neglect the number of self-crossings in a walk. We can therefore approximate the number of cycles with the number of walks, namely: $n(\mathbb{C}_l) \simeq \text{Trace}(A^l)/2l$. Let us now evaluate $\text{Trace}(A^l)$ for a graph described by degree distribution p_k, and conditional probability $p_{k'|k}$. The number of closed walks of length l can be obtained as a summation over the graph nodes, as in formula $\text{Trace}(A^l)$, or by grouping nodes in degree classes and then summing over k. In particular, on a closed walk, the probability that a node of degree k_1 is connected to a node of degree k_2 is equal to $(k_1 - 1)p_{k_2|k_1}$, since there are $k_1 - 1$ remaining links to follow not considering the one we came from, and $p_{k_2|k_1}$ is the probability that a link from a node of degree k_1 ends up in a node of degree k_2. The probability that the node of degree k_2 is followed on the walk by a node of degree k_3 is equal to $(k_2 - 1)p_{k_3|k_2}$, and so on until when we close the walk. Let us consider the simple case $l = 3$. The probability of having a walk starting and ending at a node of degree k_1 is then:

$$\sum_{k_2}\sum_{k_3}(k_1 - 1)p_{k_2|k_1}(k_2 - 1)p_{k_3|k_2}(k_3 - 1)p_{k_1|k_3}.$$

To evaluate the number of closed walks of length 3 we have also to sum this quantity over k_1. We finally get $\mathrm{Trace}(A^3) = \mathrm{Trace}(C^3)$, where $C_{k,k'} = (k - 1)p_{k'|k}$. In the most general case of cycles of length l we have:

$$n(\mathbb{C}_l) \simeq \frac{1}{2l}\mathrm{Trace}(C^l) \quad \text{where} \quad C_{k,k'} = (k - 1)p_{k'|k}. \tag{8.16}$$

In the particular case of uncorrelated graphs we have: $p_{k'|k} = q_{k'} = k'p_{k'}/\langle k \rangle$. Hence, $C_{k,k'} = (k - 1)k'p_{k'}/\langle k \rangle$ and Eq. (8.16) reduces to Eq. (8.15).

Formula (8.15) also shows that the difference between scale-free and ER random graphs is not only quantitative but qualitative. First of all, in the case of a Poisson degree distribution, we have $\langle k^2 \rangle = \langle k \rangle^2 + \langle k \rangle$ and Eq. (8.15) reduces to Eq. (8.13), which predicts that the number of cycles in ER random graphs does not depend on N. A similar result is obtained for scale-free graphs with $\gamma > 3$, since for $\gamma > 3$ the second moment of the degree distribution does not diverge with N. Conversely, in uncorrelated scale-free graphs with $2 < \gamma < 3$, the second moment of the degree distribution scales as $\langle k^2 \rangle \sim N^{(3-\gamma)/2}$. This is because $\langle k^2 \rangle \sim k_{max}^{3-\gamma}$ (see Eq. (5.27)) and the expression for the structural cut-off is $k_{max} \sim N^{1/2}$ (see Eq. (5.48)). We thus obtain [37]:

$$\overline{n}(\mathbb{C}_l) \sim N^{l(3-\gamma)/2}. \tag{8.17}$$

Hence, the number of short cycles for $\gamma = 2$ is larger than the number of short cycles for $\gamma = 3$ and $\gamma = 5$, as found in Figure 8.4. Moreover, in a scale-free graph with $2 < \gamma < 3$ the number of short cycles diverges with N, and this is a qualitative difference with respect to ER random graphs. In the following example we show that the number of cycles of length 3, 4 and 5 in a real scale-free network such as the Internet scales indeed as a power law of the number of nodes N.

Example 8.10 *(Short cycles in the Internet)*   We use the database of the Internet at the AS level introduced in Box 7.1. Following the work of Ginestra Bianconi, Guido Caldarelli and Andrea Capocci we study how the number of cycles in the network has changed over time [36]. In the figure below we report the number of cycles $n(\mathbb{C}_l)$ of length $l = 3, 4, 5$ (circles, squares and diamonds respectively) as a function of the order of the graph. Instead of

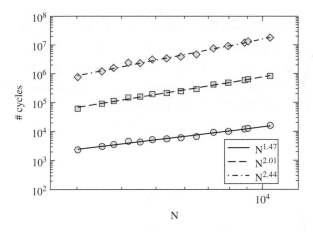

Eq. (8.10), Eq. (8.11) and Eq. (8.12), for the calculation we have used Johnson's algorithm to enumerate cycles described in Appendix A.15. From left to right the points correspond to the time evolution of the Internet. We observe a power law behaviour $n(\mathbb{C}_l) \sim N^{\beta_l}$ with exponents β_l depending on the cycle lengths. The three lines in the figure are power-law fits to the data, respectively with exponents: $\beta_3 = 1.47$, $\beta_4 = 2.01$, and $\beta_5 = 2.44$. In the case of the Internet, Eq. (8.15) is not enough to explain the empirical results. In fact, as shown in Chapter 7 the networks of the Internet introduced in Box 7.1 are all scale free, with the same exponent $\gamma \simeq 2.2$, so that, by using Eq. (8.17), we get the exponents: $\beta_3 = 1.2$, $\beta_4 = 1.6$, and $\beta_5 = 2.0$. The difference between such values and the exponents extracted numerically is due to the fact that the Internet is a correlated network. Therefore, a much better explanation of the scaling of the number of cycles in the Internet can be obtained by taking into account degree–degree correlations, i.e. by using Eq. (8.16) [36]. The numerical evaluation of such a formula produces the scaling exponents $\beta_3 = 1.41$, $\beta_4 = 1.89$, and $\beta_5 = 2.33$, which are in better agreement with those extracted empirically.

8.3 Spatial Networks of Urban Streets

Spatial networks are a special class of complex networks whose basic units are embedded in Euclidean space, and where the relations are not abstract things, such as friendships in networks of acquaintances or collaborations between individuals, but are real physical connections. Typical examples include neural networks, electronic circuits, power grids, ant galleries and transportation systems such as railroads, highways and urban street networks. Spatial networks are therefore better described in terms of *spatial graphs*, i.e. graphs whose nodes have a position in space and the links have an associated length. The topology of spatial networks is strongly constrained by their spatial embedding. For instance, long-range connections in a spatial network are expensive to construct. Also, the number of edges that can be connected to a single node is often limited by the availability of physical space, an

Box 8.1   **DATA SET 8: Urban Street Networks**

Allan Jacobs, in his book *Great Streets*, studied a wide array of urban spaces around the world [164]. The book contains a collection of 1-square-mile maps, each reproduced at the same scale to facilitate comparisons, of the street plans of representative cities around the world. In 2006 Paolo Crucitti, Vito Latora and Sergio Porta constructed the networks corresponding to each 1-square-mile map [89]. In practice, as shown

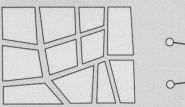 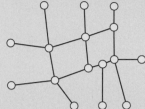

in the figure, the map of each city is transformed into a spatial network whose nodes represent street intersections, and the edges are streets. In this way, each node has a position in a two-dimensional square of side equal to 1 mile, and each edge has a real number associated that is equal to the length of the corresponding street. In Figure 8.5 we show four cities, namely Ahmedabad (India) (a), Venice (Italy) (b), Richmond (California) (c), and Walnut Creek (California) (d) as they appear in the original 1-square-mile maps, and the associated networks. The data set provided here consists of twenty 1-square-mile samples of cities from all over the world. Table 8.1 reports the basic properties of such networks.

this imposes some constraints on the degrees. In few words, spatial networks are different from other complex networks, and as such they need a special treatment.

In this section we will consider spatial networks derived from urban street patterns and we will show that, by counting the number of short cycles of different length, we can obtain important information on the structure of a city. The data set we study consists of twenty different *urban street networks*, and is illustrated in DATA SET Box 8.1. Differences between cities are clear from their maps and networks. In Figure 8.5 we show a comparison between Ahmedabad and Venice, two typical *self-organised patterns*, spontaneously emerged from a historical process outside of any central coordination, and Richmond and Walnut Creek, two examples of *planned patterns*, developed following one coordinating layout in a relatively short period of time. We immediately notice that Ahmedabad and Venice are densely interwoven, uninterrupted urban fabric. In particular, Venice is also shaped by the presence of the Grand Canal which separates the network into two parts only connected in two points, by the Rialto and the Accademia bridge. Richmond and Walnut Creek are instead more similar to a regular grid. Richmond is a typical example of a traditional grid-iron structure, with street block of different sizes, whereas Walnut Creek has a "lollipop" layout, i.e. a treelike structure with a lot of dead-end streets typical of postwar suburbs. In Table 8.1 we report the basic properties of all the twenty networks in DATA SET 8. The cities considered include large American cities, such as New York,

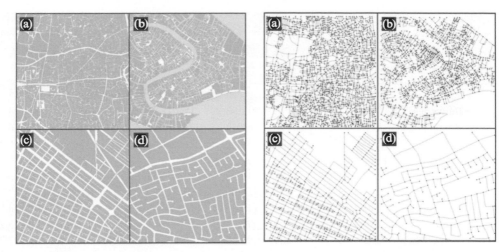

Fig. 8.5 Four 1-square-mile maps of urban streets and their associated urban street networks: Ahmedabad (India) (a), Venice (Italy) (b), Richmond (California) (c) and Walnut Creek (California) (d). Figure taken from Ref. [259].

Los Angeles and Brasilia, European cities such as London, Paris and Barcelona, but also cities in Africa and Asia. Among the twenty cities, Ahmedabad, Cairo and Venice are certainly the most representative examples of self-organised patterns, while Los Angeles, Richmond and San Francisco are typical examples of mostly planned patterns. In particular, two different parts of the city of Irvine (California), namely Irvine 1 and Irvine 2, have been selected because they represent two highly diverse types of urban fabrics. The first is a sample of an industrial area showing enormous blocks with few intersections, while the second is another typical residential early sixties "lollipop" low density suburb, which looks similar to the map of Walnut Creek shown in Figure 8.5. From the numbers reported in Table 8.1 we notice that the basic characteristics of the networks can widely differ, notwithstanding the fact we have considered the same amount of land of each city. For instance, Ahmedabad has 2870 nodes and a total of 121, 037 meters of streets in a 1-square-mile of land, while one of the two maps of Irvine has only 32 nodes and 11, 234 meters of streets. Also, Ahmedabad, Venice and Cairo are the cities with the smallest average street length, while planned cities, such as Irvine, San Francisco and Brasilia have the largest values of $\langle l \rangle$. Of course, what matters is not only the average street length $\langle l \rangle$, but also how the street length is distributed. As you can see from the following example, the distribution of street lengths can be very different from city to city and can be used to characterise city types.

Example 8.11 *(Street length distributions)* We have seen in Table 8.1 that the average length of a street, $\langle l \rangle$, can vary a lot from city to city. Here we look at the entire distribution of street lengths. Namely, we can define the probability $p(l)$ of observing a street of length l as the ratio $K(l)/K$, where $K(l)$ is the number of edges whose length is in the range ($l - 5$ meters; $l + 5$ meters), and K is the total number of edges in the network. In the figure

Table 8.1 Basic properties of the 20 urban street networks in DATA SET 8. N is the number of nodes, K is the number of edges, while W and $\langle l \rangle$ are respectively the total length of the edges and the average edge length, both expressed in metres.

	CITY	N	K	W	$\langle l \rangle$		CITY	N	K	W	$\langle l \rangle$
1	Ahmedabad	2870	4387	121037	27.59	11	New York	248	419	36172	86.33
2	Barcelona	210	323	36179	112.01	12	Paris	335	494	44109	89.29
3	Bologna	541	773	51219	66.26	13	Richmond	697	1086	62608	57.65
4	Brasilia	179	230	30910	134.39	14	Savannah	584	958	62050	64.77
5	Cairo	1496	2255	84395	37.47	15	Seoul	869	1307	68121	52.12
6	Irvine 1	32	36	11234	312.07	16	San Francisco	169	271	38187	140.91
7	Irvine 2	217	227	28473	128.26	17	Venice	1840	2407	75219	31.25
8	Los Angeles	240	340	38716	113.87	18	Vienna	467	692	49935	72.16
9	London	488	730	52800	72.33	19	Washington	192	303	36342	119.94
10	New Delhi	252	334	32281	96.56	20	Walnut Creek	169	197	25131	127.57

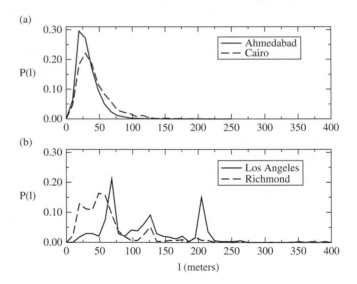

we report the edge distribution $p(l)$ for two self-organised cities, Ahmedabad and Cairo, and for two planned cities such as Los Angeles and Richmond [88]. We can immediately notice that cities of the first group show single peak distributions, while cities of the second group, due to their regular structure, exhibit multimodal distributions with two or three peaks corresponding to the typical street lengths of their grids. Other details on the basic structural properties of such networks can be found in Ref. [69], while a study of centrality can be found in Ref. [89].

In spatial graphs describing urban street patterns, the nodes have a position in a two-dimensional square. In general, the nodes of a spatial graph can have a position in three or higher dimensional space. The most general definition of *spatial graph* is the following.

> **Definition 8.4 (Spatial graphs)**   *In a spatial graph, there are D real numbers $x_i, y_i, \ldots$ associated with node i, $(i = 1, \ldots, N)$, representing the position of the node in a D-dimensional Euclidean space. In addition to this, a real positive number l_{ij}, representing the length of the edge, is associated with each edge (i, j).*

Here we are interested in counting the number of cycles, and possibly differentiating cities based on that. Since graphs corresponding to different cities have varying number of nodes, in order to compare two cities, we have to find a proper way to normalise the results. The graphs representing urban street patterns are, by construction, *plane graphs*. We will then need to compare their structural properties to those of other *planar graphs*.

Definition 8.5 (Planar graphs and plane graphs)   A planar graph *is a graph that can be embedded in the plane, i.e. it can be drawn in a plane so that no edges intersect. A drawing of a planar graph in the plane without edge intersections is called a* plane graph, *or* planar embedding *of the graph. We will refer to the regions defined by a plane graph as its* faces, *the unbounded region being called the* exterior face. *A nonplanar graph cannot be drawn in the plane without edge intersections.*
A maximal planar graph, *is a planar graph in which no link can be added without losing planarity.*

In practice, a graph is planar if it can be drawn on a plane with no edge intersections. A generalisation of planar graphs consists in graphs which can be drawn on a surface of a given genus. The genus is a topologically invariant property of a surface defined as the largest number of nonisotopic simple closed curves that can be drawn on the surface without separating it, i.e. the number of handles in the surface. Two closed curves on a surface are called isotopic if there is a smooth mapping from one to the other. If there is no such a mapping they are called nonisotopic. In this terminology, planar graphs have graph genus 0, since the plane (and the sphere) are surfaces of genus 0. Graphs of genus 1 are graphs that can be embedded on a surface of genus 1, i.e. on a torus, graphs of genus 2 are graphs that can be embedded on a surface of genus 2, i.e. on a double torus, and so on.

Example 8.12   *(Small planar graphs)*   All graphs with $N = 1, 2, 3, 4$ nodes are planar. Also, any tree of any order is planar. The number of planar graphs with $N = 1, 2, 3, 4, 5$ nodes is respectively equal to $1, 2, 4, 11, 33$, while the number of connected planar graphs is $1, 1, 2, 6, 20$. The six connected planar graphs with $N = 4$ nodes are shown in panel (a) of the figure. Note that, while graph planarity is an inherent property of a graph, it is still in general possible to draw nonplanar embeddings of planar graphs. For example, the two embeddings in the panel (b) both correspond to the planar tetrahedral graph (the four-node

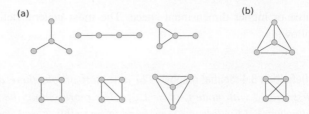

planar graph with the largest number of links, namely six), but while the top embedding is planar, the bottom embedding is not.

Example 8.13 *(Three houses and three wells)* This is a well-known puzzle having to do, at the same time, with human nature and graph theory. Three enemies live in three houses in the woods. They want to lay down tracks to the wells, so that each house has a direct path to each of the three existing wells in the wood. At the same time, in order to

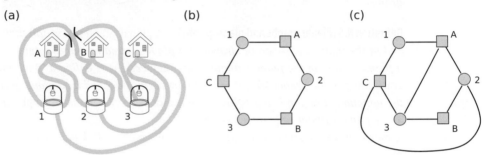

avoid conflicts, they do not want to meet: that is, they do not want any of the paths to cross. Can this be done? In terms of graphs, this is equivalent to asking whether the complete 3×3 bipartite graph $\mathbb{K}_{3,3}$ in panel (a) of the figure (the houses are labelled as A, B, C, and the wells as 1, 2, 3) is planar. We will give an intuitive proof of the impossibility of a planar drawing of $\mathbb{K}_{3,3}$. Our argument is based on the fact that every closed curve in the plane separates the plane into two regions, namely the region inside the closed curve, and the region outside. Let us start by connecting each house with two wells, in such a way to form a plane cycle as the hexagon illustrated in panel (b) of the figure. Now, in order to connect house A with well 3 without edge crossings we have two possibilities. Either we draw the link inside the hexagon or outside it. Let us assume the link (A,3) is inside the hexagon, then the only possibility to connect house C to well 2 is by drawing a link outside the hexagon as in the panel (c). At this point, there is no way to draw the link (B,1) avoiding intersections. Hence, graph $\mathbb{K}_{3,3}$ is nonplanar as shown by the intersection of the path (B,1) with the path (A,3).

Is there a general way to find which graph is planar? In the previous example we have shown that $\mathbb{K}_{3,3}$ is nonplanar. This graph, and another nonplanar graph, the complete graph with five nodes $\mathbb{K}_5$, are crucial graphs to answer this question. In fact, in 1930 Kasimir Kuratowski proved the following theorem [150]:

Theorem 8.2 (Kuratowski's)   *A graph is planar if and only if it has no subgraph homomorphic to $\mathbb{K}_5$ or to $\mathbb{K}_{3,3}$.*

In practice, it is difficult to use Kuratowski's criterion to quickly decide whether a given graph is planar. However, today there are in the literature alternative planarity tests to determine in time O(N) (linear time) whether a given graph with N nodes is planar or not [157]. In the following, we will see some simple theorems that, although they cannot be used as a general method to establish planarity, allow graphs with too many edges to be quickly determined to be nonplanar. These theorems are based on a formula, derived by Leonhard Euler, relating the number of nodes and the number of links in a planar graph.

Theorem 8.3 (Euler's formula)   *For a planar embedding of a connected planar graph with N nodes and K edges, the equality*

$$N - K + F = 2 \tag{8.18}$$

holds, where F is the number of faces (regions bounded by edges, including the outer, infinitely large region).

It is easy to prove this theorem by induction. See, for instance, the proof in Ref. [47]. Euler's formula has many important consequences.

Theorem 8.4 (Corollary 1)   *In a plane graph with N nodes and K links, if every face is a k-cycle (i.e. the boundary of each face, including the exterior face, is a k-cycle), then:*

$$K = k\frac{N - 2}{k - 2}. \tag{8.19}$$

This is an immediate consequence of Euler's formula. Since every face of the graph is a k-cycle, each of the F faces in the graph has k edges, and each edge is on two faces. Hence, we can write a relation between the number of faces and the number of links: $kF = 2K$. When this relation is substituted in Eq. (8.18) it gives the result in Eq. (8.19).

A second corollary has to do with maximal planar graphs: planar graphs in which no link can be added without losing planarity.

Theorem 8.5 (Corollary 2)   *If G is a maximal plane graph with N nodes and K links, every face is a triangle and $K = 3N - 6$. If G is a plane graph in which every face is a 4-cycle, then $K = 2N - 4$.*

The corollary is proved by substituting, respectively, $k = 3$ and $k = 4$ in Eq. (8.19), and by observing that the maximum number of links in a planar graph occurs when each of the faces is a triangle (see Problem 8.3b).

Finally, making use of Corollary 2 we obtain a necessary condition for planarity of a graph in terms of its number of links.

Theorem 8.6 (Corollary 3: necessary condition for planarity)   *For a planar graph with N nodes and K edges, if $N \geq 3$, then $K \leq 3N - 6$.*
Furthermore if $N > 3$, and there are no cycles of length 3, then $K \leq 2N - 4$.

By using this result we can immediately rule out some graphs with too many links. For instance, we can easily prove that the complete graph with five nodes, and the complete 3×3 bipartite graph of the "three houses, three wells" puzzle, are not planar.

Example 8.14   *($\mathbb{K}_5$ and $\mathbb{K}_{3,3}$ are nonplanar)*   A complete graph with five nodes has $N = 5$ and $K = 10$. Since $K > 3N - 6 = 15 - 6 = 9$, by Corollary 3, $\mathbb{K}_5$ cannot be planar. The same result holds for $\mathbb{K}_{3,3}$. It has $N = 6$ nodes and $K = 9$ links. For this graph, the first necessary condition for planarity $K \leq 3N - 6$ is verified. However, since the graph is bipartite, by definition there cannot be cycles of length 3: in order to close a triangle there should be in the graph a direct link between two houses, or between two wells (see the figure in Example 8.13). Therefore, we have to check whether the second condition of Corollary 3 is satisfied. The answer is negative, since for $\mathbb{K}_{3,3}$ we have $K = 9$ and $2N - 4 = 8$, hence the graph is nonplanar.

After this detour in the world of planar graphs and planarity, we are now ready to go back to our main purpose, that is, counting the number of cycles in plane graphs of different world cities. We will compare each city with a plane graph with the same distribution of nodes in the square and with the maximum possible number of links. From Theorem 8.5 we know that the maximum number of edges, K_{max}, that can be accommodated in a graph with N nodes, without breaking the planarity, is equal to $K_{\text{max}} = 3N - 6$. A natural reference graph for a city with N nodes could then be the planar graph with $3N - 6$ links which minimises the total sum of link lengths. This is known in the literature as the *Minimum Weight Triangulation (MWT)*. Since no polynomial time algorithm is known to compute the MWT, we thus consider the so-called *Greedy Triangulation (GT)*, that is based on connecting pairs of nodes in ascending order of their distance provided that no edge crossing is introduced [60, 93]. The GT is easy to compute and leads to a maximal connected planar graph, while minimising as far as possible the total length of edges considered.

Example 8.15   *(Construction of the Greedy Triangulation)*   The GT induced by a spatial distribution of nodes $\{x_i, y_i\}_{i=1,\ldots,N}$ in the unit square can be constructed by means of the following brute force algorithm. First, all pairs of nodes i and j, representing all the possible edges in the complete graph, are sorted out by ascending order of their Euclidean distance $d_{ij}^{\text{Eucl}} = \sqrt{(x_i - x_j)^2 + (y_i - y_j)^2}$. The length of a possible edge between node i and j is here taken to be equal to such distance. Then, we browse the ordered list of pairs in ascending order of distance, and we add a link if it does not intersect other edges already in the graph. The graph obtained by this procedure has less than $3N - 6$ edges. This is due to the fact that all the edges are considered as being straight lines connecting the two end-nodes, and the nodes at the border of the square are often aligned. For such a reason, some of edges connecting the nodes on the border of the unit square cannot be placed without causing edge crossings. However, the number of such edges is usually small compared to the total number of edges, and can then be neglected in large GT graphs.

We have constructed the GT corresponding to each of the twenty cities in the data set of Box 8.1. An example of the results is shown in Figure 8.6 for the city of Savannah. Notice that comparing a city with its greedy triangulation is also meaningful in terms of the possible evolution of a city. The most primitive forms of street patterns are in fact close to trees, while more evolved forms involve the presence of cycles. For each city and the corresponding GT, we have counted the cycles of length 3, 4 and 5. In Table 8.2 we report $n(\mathbb{C}_k)/n^{GT}(\mathbb{C}_k)$, the ratio between $n(\mathbb{C}_k)$ and the number of cycles in the corresponding GT, for $k = 3, 4$ and 5. As expected, in most of the samples we have found a rather small value of $n(\mathbb{C}_3)/n^{GT}(\mathbb{C}_3)$, as compared, for instance, to $n(\mathbb{C}_4)/n^{GT}(\mathbb{C}_4)$, denoting that triangles are not common in urban street networks. This is a proof that the clustering coefficient C alone, measuring the number of triangles, would not be a good metric to characterise such networks. Walnut Creek, Los Angeles, Savannah and San Francisco are the cities with the smallest value of $\mathbb{C}_3/\mathbb{C}_3^{GT}$, while San Francisco, New York, Washington, Savannah and Barcelona are the cities with the largest value of $\mathbb{C}_4/\mathbb{C}_4^{GT}$ (larger than 0.1). Brasilia, Irvine 1 and Irvine 2 are the only cities with a prevalence of triangles with respect to squares. Concerning $\mathbb{C}_5/\mathbb{C}_5^{GT}$ we can distinguish three classes of cities. First of all, we have cities such as Ahmedabad, Cairo, Seul and Venice with $\mathbb{C}_3/\mathbb{C}_3^{GT} \simeq \mathbb{C}_5/\mathbb{C}_5^{GT}$. Then, we have other two groups of cities. Cities in the first group, such as Brasilia, Irvine and Paris have $\mathbb{C}_3/\mathbb{C}_3^{GT} > \mathbb{C}_5/\mathbb{C}_5^{GT}$, while cities in the second group, as Los Angeles, Savannah and Vienna, have $\mathbb{C}_3/\mathbb{C}_3^{GT} < \mathbb{C}_5/\mathbb{C}_5^{GT}$.

Another measure of the importance of cycles in a planar graph, which takes into account cycles of any length is the so called *meshedness coefficient* M [60]. The meshedness coefficient is defined as $M = F/F_{max}$, where F is the number of faces in a given plane graph with N nodes and K edges, and F_{max} is the maximum possible number of faces in a plane graph with the same number of nodes and $K_{max} = 3N - 6$ edges. By using the Euler's formula we have $F_{max} = 2N - 4$. The meshedness coefficient can vary from 0, in a tree

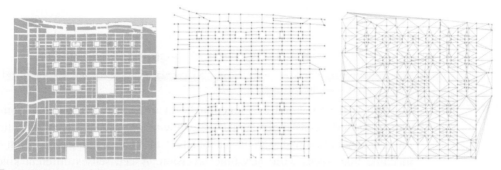

Fig. 8.6   The 1-square-mile map of Savannah, together with the associated spatial graph and the corresponding GT. Figure taken from Ref. [69].

Table 8.2 Cycles in urban street patterns. For each 1-square mile graph, we report the number $n(\mathbb{C}_k)$ of cycles of length $k = 3, 4, 5$ in the graph, normalised to the number of cycles, $n^{GT}(\mathbb{C}_k)$, in the corresponding GT. We also report the meshedness coefficient M.

CITY	M	$n(\mathbb{C}_3)/n^{GT}(\mathbb{C}_3)$	$n(\mathbb{C}_4)/n^{GT}(\mathbb{C}_4)$	$n(\mathbb{C}_5)/n^{GT}(\mathbb{C}_5)$
Ahmedabad	0.26	0.023	0.042	0.020
Barcelona	0.28	0.019	0.101	0.019
Bologna	0.21	0.015	0.048	0.013
Brasilia	0.15	0.029	0.027	0.012
Cairo	0.25	0.020	0.043	0.019
Irvine 1	0.08	0.035	0.022	0.005
Irvine 2	0.01	0.007	0.004	0.001
Los Angeles	0.21	0.002	0.075	0.011
London	0.25	0.011	0.060	0.020
New Delhi	0.15	0.011	0.020	0.011
New York	0.35	0.024	0.136	0.028
Paris	0.24	0.028	0.063	0.016
Richmond	0.28	0.034	0.068	0.022
Savannah	0.32	0.002	0.111	0.026
Seoul	0.25	0.021	0.051	0.021
San Francisco	0.31	0.003	0.148	0.003
Venice	0.15	0.016	0.030	0.010
Vienna	0.24	0.007	0.063	0.018
Washington	0.29	0.026	0.132	0.022
Walnut Creek	0.08	0.000	0.011	0.003

structure, to 1, in a maximally connected planar graph, as in a triangulation. The meshedness coefficient is also reported in Table 8.2. Three are the cities with a value of meshedness larger than 0.3, namely New York, Savannah and San Francisco. These represents the most complex forms of cities in terms of cycles. On the other hand, Irvine and Walnut Creek with a value of M lower than 0.1, have a tree-like structure. Notice that both the first and the second group of cities are examples of planned urban fabrics. Self-organised patterns such as Ahmedabad, Cairo and Seoul also exhibit high values of meshedness, which means the presence of a fairly large number of cycles.

8.4 Transcription Regulation Networks

The cell is the structural and functional unit of all living organisms. Some organisms, such as bacteria, consist of a single cell, while other organisms are multicellular: humans, for instance, are estimated to have about 10^{14} cells! Every single cell is an extraordinary machine that works with incredible precision. A cell senses different kinds of signals from the environment and from other cells, absorbs nutrients, transforms the nutrients into energy, performs specialised functions and reproduces itself. To accomplish all such tasks, the cell produces *proteins*. Each protein has its specific function, and the cell has to produce the right amount of a protein, at the right time. For instance, in bacteria, when sugar is sensed in the environment, the cell starts to produce proteins to take the sugar into the cell and to use it. Analogously, when a cell is damaged, it produces the best proteins to repair the particular damage that has occurred. The main information to produce proteins is stored in DNA segments called *genes*, and the process that turns the information carried by a gene into proteins is called *gene expression* [159]. Gene expression consists of two main steps: *transcription* and *translation*. In the first process of *transcription*, the information stored in a gene's DNA is transferred to another molecule, the so-called messenger RNA (mRNA) which carries the message for making a protein out of the cell nucleus. The second step, *translation*, takes place in the cytoplasm where the mRNA interacts with a specialised molecular machine called ribosome, which reads the sequence of mRNA and builds the protein, one aminoacid at a time. In order for the cell to obtain the proper amount of products when they are needed, the expression of a gene can be regulated, either promoted or inhibited, by a special class of proteins, the so-called *transcription factors*. Such process of *gene regulation* is fundamental for the adaptability of living organisms, and is a very complex biological mechanism schematically illustrated in the following example.

Example 8.16 *(Gene transcription regulation)*   Each gene is a portion of DNA whose sequence encodes the information needed for the production of a protein. The gene is transcribed into mRNA, which is then translated into the final product, the protein. Each gene is preceded by a regulatory DNA region, called the *promoter*, providing a binding site for RNA polymerase and also for the *transcription factors*. The RNA polymerase is the enzyme that in practice performs the transcription of genetic information from DNA to mRNA. *Transcription factors* are a special type of proteins that regulate the rate at which proteins encoded by genes are produced. When activated by specific signals in response to internal or external demands, the transcription factors bind to their specific sites on the promoter, promoting or blocking in this way the recruitment of RNA polymerase. The regulation process of transcription of a gene *Y* by a transcription factor protein *X* is sketched in the figure below.

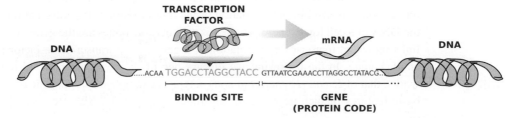

Basically, the coding of a protein is activated by one or more transcription factors, which bind to the binding site of the gene, allowing RNA polymerase to attach to the portion of DNA coding for that protein, and to start producing mRNA for that gene, which can then be used to synthesise the protein.

Of course, only a fraction of all cell proteins play the role of transcription factors. Transcription factors regulate the rate at which proteins encoded by genes are produced. Each transcription factor only acts on a specific subset of genes. A given transcription factor is

Box 8.2	DATA SET 8 BIS: The Transcription Network of *E. coli*

E. coli is a few microns long and is a rod-shaped bacterium that inhabits the intestinal tract of humans and other warm blooded animals. In 2002 Uri Alon, a physicist and biologist from the Weizmann Institute of Science in Israel, and his research group compiled a database of the interactions between transcription factors and the *operons* that they regulate in the *E. coli* [284]. Operons are groups of contiguous genes that are transcribed into the same mRNA molecule. The database contains 577 interactions and 424 operons, involving 116 different transcription factors, and the network provided in this data set is a directed graph with $N = 424$ nodes and $K = 519$ arcs.

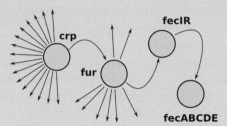

As sketched in the figure, the nodes of the network represent genes or operons, while a directed link from node i to node j means that the product of operon i is a transcription factor that binds the promoter of operon j to control the rate of transcription of j. The distribution of node out-degree is heterogeneous, since most of the genes are transcription factors for just one or two proteins, while a few genes (including *crp*, and *fur* shown in the figure) produce transcription factors for a lot of proteins, and can be thus responsible for relatively long chains of reactions.

also encoded by a gene, which is in turn regulated by other transcription factor proteins, which are regulated by other transcription factors, and so on. This intricate ensemble of interactions involved in the regulation of the transcription of the information contained in the DNA of a cell form a network. In this network, the nodes are the genes and the directed links represent transcription regulation of a gene by the transcription factor produced by another gene. In the following we discuss one of the best-characterised regulation networks, that of *E. coli* (*Escherichia coli*), a typical example of prokaryotic organism[1]. The basic features of such network are illustrated in the DATA SET Box 8.2. We denote the

[1] Based on cell type, living organisms can be divided in *eukaryotes* and *prokaryotes*. The main difference is that the cells of eukaryotes contain a distinct membrane-bound nucleus within which the genetic material is carried, while prokaryotes do not have a nucleus. Fungi, animals and plants are examples of species that are eukaryotes, while since 1990 the prokaryotes are divided into two separate domains, *bacteria* and *archaea*. The bacteria are single-celled organisms that inhabit virtually all environments, including water, soil and our

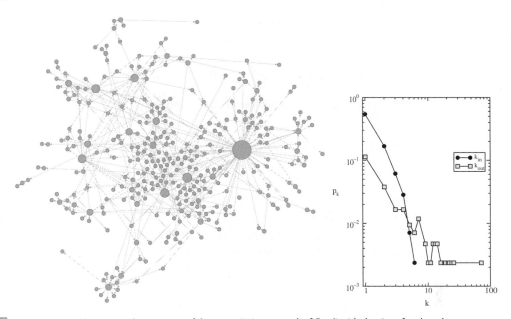

Fig. 8.7 The largest weakly connected component of the transcription network of *E. coli*, with the size of each node representing its out-degree (a). The in- and out- degree distributions of the entire network (b).

transcription network of *E. coli* as DATA SET 8 BIS, because in this chapter, at variance from the previous ones, we have introduced not one but two different databases. The network consists of 34 weakly connected components. The largest one has 329 nodes and 456 arcs, and is shown in Figure 8.7(a). The size of the nodes is proportional to their out-degree. Therefore, the big circles represent genes regulating the transcription of a large number of other genes. For instance, the best example of global regulator is gene *crp*, corresponding to the node with the largest out-degree in the network, namely $k^{out} = 72$. This is indeed an important gene, which regulates most of the genes involved in the energy metabolism of the *E. coli*. The in- and out-degree distributions of the network are reported in Figure 8.7(b). We notice that the out-degree distribution is more heterogeneous than the in-degree distribution. For instance, in this network we have 81 nodes with $k^{in} = 0$ and 230 nodes with $k^{in} = 1$, while the gene with the largest in-degree is gene *sodA*, with a k^{in} only equal to 6. Hence, the regulatory system is organised in such a way that, while each gene is regulated by only few other genes, a few key genes are responsible for the regulation of plenty of other genes.

Let us now go back to the main topic of this chapter. Already from a visual inspection of the network in Figure 8.7(a), it seems that the transcription network of *E. coli* does not have directed cycles. We can use the methods of Section 8.1 to count the precise number of cycles in the entire network. And indeed we find that the network, remarkably, has no

bodies. Archea are organisms living in extreme environments. Like all bacteria, they do not have a nucleus. However, from biochemical characteristics and DNA sequence analysis it has been found that there are numerous differences with respect to other bacteria, so that archea are now classified as a separate domain in the three-domain system [316].

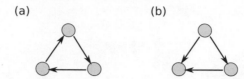

Fig. 8.8 A directed triangle and another small directed graph of three nodes known as a feed-forward loop.

directed cycles of any length. This might well be due to the fact that the network is very sparse. To verify this, we can evaluate the expected number of cycles in an ensemble of random graphs with the same number of nodes and links as the original network. Let us focus on the simple case of directed cycles with three nodes, as the one shown in panel (a) of Figure 8.8. We can still use Eq. (3.21) of Chapter 3, bearing in mind that now we have to deal with directed graphs. We recall that the expected number of directed triangles depends only on the average degree $\langle k \rangle$ and is equal to:

$$\overline{n}(\mathbb{C}_3) \sim \langle k \rangle^3/3 = (519/424)^3/3 \sim 0.6$$

(for more details see Example 8.7 and Problem 8.4a). As stated in Theorem 3.1, the number of triangles in an ensemble of ER random graphs follows a Poisson distribution. Consequently, the standard deviation on the number of triangles is equal to the square root of the average number $\overline{n}(\mathbb{C}_3)$. In conclusion, the fact that we have found no triangles in the transcription network of E. coli is perfectly consistent with the number 0.6 ± 0.8 of occurrences expected in an ensemble of random graphs. We might object that the ensemble of random graphs we have considered does not have the same single-node characteristics of the E. coli network, and the results might be affected by that. For a more fair comparison we can consider an ensemble of random graphs such that each node in a randomised network has the same number of incoming and outgoing edges as in the original network. This corresponds, in practice, to a generalisation of the configuration model of Section 5.3 in which both in-degree and out-degree sequences are assigned. The number of expected occurrences of directed triangles in such ensemble of random graphs is reported in Table 8.3, and is even smaller than that found for ER random graphs. This is again a confirmation that directed triangles are functionally not important in the regulation network of E. coli, since they occur with about the same frequency as in a random graph.

The question we ask now is whether there are other subgraphs of three nodes that occur in E. coli more often than in random networks. This is exactly the same question that initially motivated in 2002 Uri Alon and coworkers at the Weizmann Institute, and that finally led to the definition of *network motifs* that will be the subject of the next section. A good subgraph candidate to look at in the E. coli is the so-called *feed-forward loop*, shown in panel (b) of Figure 8.8. In fact, such a small graph describes a relatively common situation in a transcription network: a transcription factor X regulates a second transcription factor Y, and X and Y both bind the regulatory region of a third target gene Z, and jointly modulate its transcription rate. This mechanism can serve to produce useful redundancy in a biological system, or to speed up the response time of the target gene Z expression [284, 214]. Since this structural pattern is functionally relevant we should find its fingerprint in the structure of the E. coli. A direct search of the subgraph in the transcription network of

Table 8.3 The actual number of directed triangles and feed-forward loops in the transcription network of *E. coli* is compared to the expected number in ER random graphs and random graphs with the same degree sequence.

Network	directed triangles	feed-forward loops
Transcription network of *E. coli*	0	40
ER random graphs	0.6 ± 0.8	1.7 ± 1.3
Random graphs with the same degree sequence	0.2 ± 0.6	7 ± 5

E. coli tells us that the network contains 40 feed-forward loops (see Appendix A.16 for an algorithm to find all the 3-node subgraphs of a network). Again, to draw a definitive conclusion, we need to compare this number with the number of occurrences in the randomised version of the original graph. The result, first found by Alon and collaborators, and published in [284, 219], is unquestionable: as shown in Table 8.3 there are 40 feed-forward loops against an average of 2 or 7 respectively in the two ensembles of randomised graphs considered. The message is clear. The fact that the feed-forward loop appears at frequencies much higher than expected at random, indeed confirms that this particular pattern plays important specific functions in the information processing performed by the transcription regulation network.

Example 8.17 *(Functional meaning of the feed-forward loop)* Additional insights on the functional meaning of a subgraph can be gained by means of a mathematical description of their dynamics. In Ref. [284], Alon and collaborators have considered a set of differential equations to describe the concentration of transcription factors Y and Z at time t, in response to the concentration of X. In particular, they have studied a feed-forward loop circuit where X and Y act, as shown in the figure, with a so-called *AND-gate* to control operon Z. Let us see this more in detail. By indicating as $x(t)$, $y(t)$, and $z(t)$, the concentrations of the three transcription factors at time t, the dynamics can be written as:

$$\frac{dy}{dt} = \beta_y \Theta(x - k_{xy}) - \alpha_y y$$

$$\frac{dz}{dt} = \beta_z \Theta(x - k_{xz})\Theta(y - k_{yz}) - \alpha_z z.$$

The first equation tells us that the production of Y occurs at a rate β_y, when $x(t)$ exceeds the activation threshold k_{xy}, while its degradation takes place at a rate α_y. The Heaviside step function, $\Theta(x) = 0$ when $x < 0$ and $\Theta(x) = 1$ when $x > 0$, accounts for this step-like activation phenomenon. The second equation mimics the *AND-gate*: the production of Z with a rate β_z is only possible when both $y(t)$ exceeds the activation threshold k_{yz} *AND* $x(t)$ exceeds the activation threshold k_{xz}. In order to exemplify how the AND-gate works, we couple those two equations with a pulse-shaped signal $x(t)$, in which X is set equal to 1 for a certain time, and then suddenly removed, as shown panel (a) of the figure. The numerical integration of the equations shows that the circuit can reject rapid variations in the activity

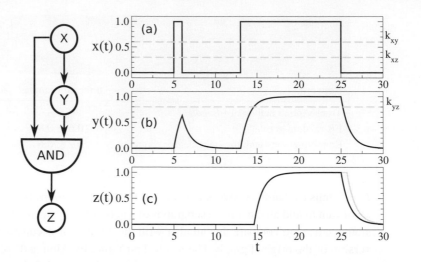

of the input X, and responds only to prolonged activation profiles. This is because, when X is on, the signal is transmitted to the output Z by two different pathways: a direct one from X and a delayed one through Y. Let us consider first the case where the activation of X is short-lived, e.g. when $x(t)$ is a short pulse like the one starting at $t = 5$ in panel (a). In this case, Y is activated immediately, since $x(t) > k_{xy}$ (we have set here $k_{xy} = 0.6$, represented by the top dashed line in panel (a)), and $y(t)$ starts increasing exponentially. However, the input signal is too short to activate also Z. In fact, although $x(t)$ is larger than $k_{xz} = 0.3$, the AND circuit remains inactive, since $y(t)$ does not have enough time to become larger than $k_{yz} = 0.8$ (dashed grey line in panel (b)). Conversely, when $x(t) = 1$ for a long enough time (like in the pulse starting at $t = 13$ shown in panel (a)), so that Y levels can build up and $y(t)$ becomes larger than k_{yz}, then Z will be activated as well. On the other hand, once X is deactivated, Z shuts down promptly, as shown by the solid black line in panel (c). In particular, $z(t)$ starts decreasing exponentially as soon as $x(t)$ is switched off. This kind of behaviour can be useful for making robust and quick decisions based on fluctuating external signals. A similar rejection of rapid fluctuations can be achieved also by the much more simple three-node cascade, $X \to Y \to Z$. However, in that case Z has a slower shutdown than the feed-forward loop (see the grey line in panel (c)), since $z(t)$ will start decreasing only as soon as $y(t)$ falls below k_{yz}. This introduces a delay in the deactivation of Z. The dynamical analysis suggests that the feed-forward loop can act as a circuit that rejects transient activation signals from a generic transcription factor and responds only to persistent signals, while allowing a rapid shutdown.

Even more interesting is the fact that the same subgraph, the feed-forward loop, has not only been found in other prokaryotes, but also in eukaryotic organisms such as the unicellular budding yeast *Saccharomyces cerevisiae*[2] or the nematode *C. elegans*, the

[2] The yeast that we use in baking and brewing, the eukaryotic model organism in molecular and cell biology, is much like *Escherichia coli* as the model prokaryote.

worm whose neural network we have introduced in DATA SET 4 and studied in Chapter 4. The presence of the feed-forward loop in various organisms from bacteria to humans suggests that this subgraph is a basic building block of all transcription networks. We will come back to this point in the next section.

Example 8.18 *(Signed interactions)*   One important feature of transcription networks is the sign of the interactions. In fact, the presence of a transcription factor can either promote or inhibit the production of another transcription factor. By convention, if a transcription factor A directly promotes the production of another transcription factor B, then the link from A to B is assigned a positive sign, represented by an arrow from A to B, while if A inhibits the production of B, the corresponding link is assigned a negative sign, represented by a perpendicular bar. If we take into account the sign of the interactions we have $2^3 = 8$ different instances of the feed-forward loop. As shown in the figure, the eight different configurations can be divided into two classes, namely *coherent* and *incoherent* loops.

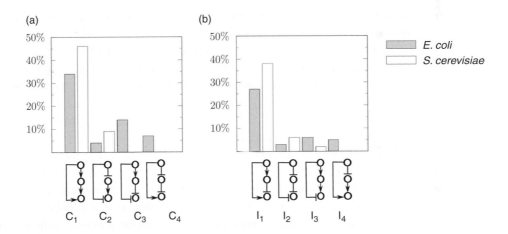

A loop is said to be coherent if the two directed paths in the loop have the same sign (the sign of a path consisting of two edges is just the usual algebraic product of the signs of the two edges), and is said to be incoherent otherwise. The coherent feed-forward loop types are denoted C_1 through C_4, while the incoherent types I_1 through I_4. There exists a fundamental functional difference between the two classes. In fact, as we have shown in Example 8.17, coherent loops normally work as low-pass filters, i.e. they avoid rapid variations in the input signal X to be transferred directly to Z, and introduce a delay in the activation of Z. Conversely, incoherent loops usually accelerate the expression of the target gene Z, and can either suppress or enhance the response to spikes in the control signal [214]. In the figure we report the relative abundance of the eight feed-forward loop types, i.e. the fraction of each type relative to the total number of feed-forward loops in the network of *E. coli* (grey bars) and *S. cerevisiae* (white bars). It is clear that two of the eight types are dominant with respect to the others, respectively the positive feed-forward loop C_1 in which there is no inhibition link, and the incoherent feed-forward loop I_1, where the direct path ($X \rightarrow Z$) is positive and the indirect one ($X \rightarrow Y \rightarrow Z$) is negative.

The transcription regulation we have studied in this section is one of the most important mechanisms taking place in a living cell. However, the transcription factors are only about 3 per cent of all protein-coding genes in *S. cerevisiae* (5 per cent in the *C. elegans*, and 10 per cent in humans). There are several other important biological processes in the cell that can be represented as networks. In addition to transcription networks, we therefore have the possibility to study other networks of interactions, such as *signal-transduction networks*, *protein–protein* interaction networks and *metabolic* networks.

8.5 Motif Analysis

In this section we introduce a systematic way to search for *network motifs*, i.e. the recurrent structures representing the building blocks of a given network. In the most general terms, the problem can be stated as follows. Given a directed or undirected graph $G = (\mathcal{N}, \mathcal{L})$ and any possible small connected graph $F_{n,l}$ with n nodes and l links, we want to find out if F is a "statistically significant" subgraph of G. In order to do this, we first need to count the number n_F of times graph $F_{n,l}$ appears in G, which means finding the number of subgraphs of G isomorphic to graph $F_{n,l}$. Then, we have to assess the statistical significance of the number of occurrences of F as a subgraph of G.

When selecting the small connected graphs $F_{n,l}$ to search for in G, the simplest non-trivial situation is when $n = 3$. In fact, there are only two different subgraphs of three nodes we can look for if graph G is undirected, namely the two connected graphs shown in Figure 8.9(a). The situation is much richer if we deal instead with directed graphs, such as the transcription regulation networks introduced in the previous section. It is in fact possible to show that there are 13 different ways to connect three nodes with directed links. All the possible 3-node weakly connected directed graphs are shown in Figure 8.9(b), and labelled with symbols from F1 to F13. In particular, the feed-forward loop and the directed triangle studied in the previous section are here indicated respectively as graph F5 and graph F9. The number of different possibilities rapidly increases if we consider larger subgraphs. For example, there are 199 different 4-node weakly connected directed graphs, and over 9000 weakly connected directed graphs with five nodes. Due to the finite size of real-world networks, in order to assess the statistical significance of a given directed subgraph in a real network G, we need to limit the study to graphs $F_{n,l}$ with only a few nodes, usually three or four. In this way, we have a non-negligible probability of finding $F_{n,l}$ in the network under study, so that we can count n_F, the actual number of times a given graph F appears in G, and we can compare this number to the expected number $\bar{n}_F^{\text{rand}}$ in an appropriate randomised version of the network. The concept of network motif, detailed in the following definition, was originally introduced by Uri Alon and coworkers in an article appeared in the journal *Science* in 2002 [219].

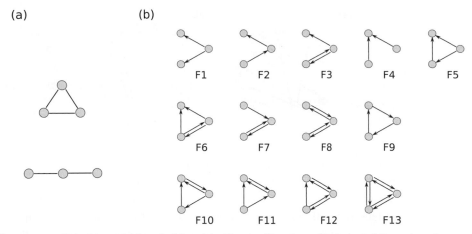

(a)   (b)

Fig. 8.9   The two types of 3-node connected graphs (a), and the 13 types of 3-node weakly connected directed graphs, sorted as in Ref. [219] and labelled as graphs F1 to F13 (b).

> **Definition 8.6 (Network motifs)**   *An undirected (directed) subgraph $F_{n,l}$ with n nodes and l links is said to be a* network *motif if it occurs in an undirected (directed) graph G at a number significantly higher than in randomised versions of the graph, i.e. in graphs with the same number of nodes and links, and degree distribution as the original one, but where the links are distributed at random.*

In practice, the search for network motifs is based on matching algorithms counting the total number of occurrences of each subgraph $F_{n,l}$ in the original graph G and in the randomised ones. The idea is sketched in the following example, while the details of an algorithm to find all the 3-node subgraphs of a network are discussed in Appendix A.16.

Example 8.19   *(Motif analysis)*   Suppose we are given the directed network G with $N = 16$ nodes and $K = 19$ arcs shown in panel (a) of the figure below, and we want to find out which of the 13 3-node directed graphs in Figure 8.9 is a network motif. The first step is to count the number of times each graph F appears as subgraph of the network. Let us focus for instance on the feed-forward loop F5. As shown in the figure, such a graph is found $n_{F5} = 5$ times in the network (the bold solid lines indicate links that participate in at least one such loop). The next step is to assess the statistical significance of the result. In practice, we need to compare the number n_{F5} to the expected number of occurrences, $\overline{n}_{F5}^{\text{rand}}$, of the feed-forward loop in randomised versions of the network with the same number of nodes and links. Such a comparison accounts for the fact that a given subgraph can appear by chance, i.e. due to the presence of a certain number of links in the network, or because some of the nodes have larger degrees than others. This is the reason why the usual procedure is to consider an ensemble of random graphs with the same number of incoming and outgoing links for each node as in the original network. In panels (b) and (c) we report two

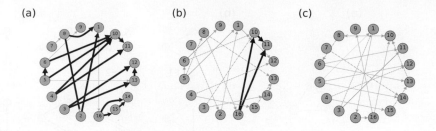

typical samples of the random graph ensemble corresponding to the network in panel (a). We notice that the graph in panel (b) contains just one feed-forward loop, while the graph in panel (c) contains none. These two samples reflect quite accurately the actual abundance of feed-forward loops in the ensemble. In fact, by sampling a large number of graphs from that ensemble, we obtain $\overline{n}_{F5}^{rand} = 0.7$, which is much smaller than the number of actual counting $n_{F5} = 5$ in the original graph. This result confirms the importance of the feed-forward loop as a building block of the network, i.e. as a network motif.

From the previous example it is clear that there is still an important point to specify in Definition 8.6, namely how to establish that the number of occurrences of $F_{n,l}$ in G is significantly higher than the expected number of occurrences in randomised versions of the graph. A complete treatment of this problem would require the use of appropriate statistical tests. The simplest approach to quantify the relevance of $F_{n,l}$ as a subgraph of G is based on the evaluation of the so-called *Z-score*, defined as follows [219].

Definition 8.7 (Subgraph Z-score) *The Z-score of $F_{n,l}$ in graph G can be written as:*

$$Z_F = \frac{n_F - \overline{n}_F^{rand}}{\sigma_{n_F}^{rand}} \tag{8.20}$$

where n_F is the number of times the subgraph $F_{n,l}$ appears in G, and $\overline{n}_F^{rand}$ and $\sigma_{n_F}^{rand}$ are, respectively, the mean and standard deviation of the number of occurrences in an ensemble of graphs obtained by randomising G.

The absolute value of Z_F measures the distance between the observed number of occurrences and the expected number of occurrences in an appropriate ensemble of randomised networks, in units of standard deviations. We are mainly interested here to positive values of Z_F, which indicate that n_F is above the mean. For instance, a value $Z_F = 2$, i.e. an observed number of occurrences that is two standard deviations larger than $\overline{n}_F^{rand}$, implies that the observed number of occurrences can be trivially due to fluctuations only in 2.14% of the cases (if we assume that the number of occurrences in the ensemble of randomised networks is Gaussian distributed). Analogously, a value of $Z_F = 3$ can be attributed to random fluctuations in only 0.13% of the cases. Therefore, finding values of Z_F larger than 2 or 3 is a good indication that F is a graph *motif*.

Notice now that in the graph in Example 8.19, the feed-forward loop F5 has a Z-score equal to 6.11, so F5 is a motif for that graph. Analogously, going back to real-world networks, we find that the Z-score of the feed-forward loop F5 in the transcription network of the *E. coli* is equal to 10.89 (40 occurrences in the real network, against 7.5 ± 3 in the randomised ensemble), while all the other possible 3-node directed subgraphs allowed in a directed acyclic graph (DAG), namely F1, F2, F4 have a negative and large Z-score, so they cannot be considered motifs. The same result holds for the transcription network of the *S. cerevisiae*, where the feed-forward loop has $Z = 14$ (70 occurrences in the real network, against 11 ± 4 in the randomised ensemble). By using the same approach it is also possible to explore 4-node motifs. For instance, the principal 4-node motif in transcription networks is the so-called bi-fan, consisting in two nodes X and Y both pointing to the same pair of nodes W and Q. The bi-fan has $Z = 13$ in *E. coli* (203 occurrences against 47 ± 12 in the randomised graph), and $Z = 41$ in the *S. cerevisiae* (1812 occurrences against 300 ± 40 in the randomised graph) [219].

Having developed a systematic method to extract the building blocks of a given network, it seems natural to compare the motif structure of different biological, social and man-made complex networks. In principle, we expect that systems performing different functions will also exhibit different sets of motifs in their networks. And this is indeed what we find. As a first example let us go back to the largest man-made network we have introduced in this book, the World Wide Web. If we study motifs in the network samples of the World Wide Web of DATA SET 5, we find that the 3-node motifs with the highest Z-score in the graph of the World Wide Web is not the feed-forward loop, but the complete graph indicated as F13 in Figure 8.9. This is a clear indication that in networks such as the World Wide Web, where the links have no cost, more tightly packed triads of Web pages are preferred [219]. A nice way to compare the various motifs emerging in networks from different fields can be obtained by constructing a *network motif profile* as described below. First of all let us label as $F_\ell, \ell = 1, 2, \ldots$ all the graphs $F_{n,l}$ with a fixed value of n. For instance, in the case of the thirteen directed graphs of order $n = 3$ in Figure 8.9(b), the label ℓ goes from 1 to 13. The motif profile of a given network is constructed by evaluating the Z-score of each graph F_ℓ, and by plotting the *normalised Z-score* S_{F_ℓ} defined as [220]:

$$S_{F_\ell} = \frac{Z_{F_\ell}}{(\sum_\ell Z_{F_\ell}^2)^{1/2}} \quad \ell = 1, 2, \ldots \tag{8.21}$$

as a function of ℓ. The normalisation in Eq. (8.21) emphasises the relative significance of all the possible subgraphs of a given order, rather than their absolute significance. The normalised Z-score is more suited than the non-normalised Z-score in Eq. (8.20) when it comes to compare real networks of different order and sizes, because motifs in large networks tend to display higher value of the Z-scores than motifs in small networks.

In Figure 8.10 we report the score profile of the thirteen 3-node subgraphs obtained for various real networks. Notice that we are here following Ref. [220], and the thirteen graphs on the x-axis of Figure 8.10 are sorted in a way different from that used in Figure 8.9(b) and in Ref. [219]. The real networks reported are transcription, signal-transduction, neuronal, World Wide Web, social and word co-occurrence. Something magic happens: it emerges

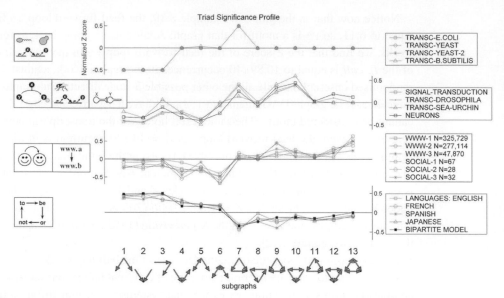

Fig. 8.10 Normalised score profile of the thirteen 3-node weakly connected directed graphs for various real-world networks. Figure taken from Ref. [220].

clearly that the networks can be divided into a few groups, each group being characterised by a given motif profile. The first profile in the figure includes all the transcription networks that control gene expression in bacteria and yeast in response to external stimuli. All such systems show one strong 3-node motif, the feed-forward loop, as we already know. And this is the only motif appearing with positive normalised score. Notice also that the profile admits three subgraphs with negative scores, namely 1, 2 and 3, corresponding respectively to F1, F4 and F2 in Figure 8.9(b). These subgraphs are therefore *anti-motifs*, i.e. significantly under-represented subgraphs, corresponding to patterns that are rarely used. The second group comprises three types of biological networks: a signal-transduction network, two developmental transcription networks that guide the development in the fruit fly and in the sea-urchin, and the neuronal network of the *C. elegans*. All these networks shows triads 7, 9 and 10 (respectively F5, F11 and F6) with a positive normalised score, and triads 1, 2, 4 and 5 (respectively F1, F4, F7 and F3) with negative scores. In contrast to the sensory transcription networks of microorganisms, these networks display two-node feedbacks that regulate or are regulated by a third node, such as triads 9 and 10 (F11 and F6), and are less biased against cascades corresponding to triad 3 (F2). The third group includes three directed social networks, in which directed links represent positive sentiments from one individual to another of a social group, and three graphs of the World Wide Web. All such networks display the same profile with a large tendency towards transitive interactions and transitive triads, such as motif 13 and 9 (namely F11 and F13). Finally, there is a fourth group including word-adjacency networks from texts written in different languages. In such networks each node represents a word and a directed connection occurs when one word directly follows the other in the text. Interestingly, the same profile appears in languages as different as English and Japanese.

The stunning results reported in Figure 8.10 appeared for the first time in a second work on motifs by Uri Alon and coworkers published in the journal *Science* in 2004. The same authors also introduced the term *network superfamily* to indicate a set of networks with the same motif profile [220]. A superfamily can contain networks of systems of very different nature and size that share a similar motif profile. The result indicate that different systems may have developed similar key circuit elements because they have evolved to perform similar tasks. This opens up new avenues of research for a better understanding of the relations between structure and function in networked systems, and leads to very intriguing connections between different research fields.

8.6 What We Have Learned and Further Readings

When it comes to characterising the structure of a complex network at a microscopic level it can prove extremely useful to look at the prevalence of cycles of different lengths and of other small subgraphs. In this chapter we have therefore learnt how to enumerate the cycles of a graph, and we have introduced a novel concept, that of *network motif*, indicating any pattern, usually of three or four nodes, which is statistically over-represented in a real network. We have then proposed a method, the so-called *motif analysis*, to assess the relevance of each small graph in a given network by a comparison to an appropriate randomisation of the network. In order to show the usefulness of counting cycles in a network, we have introduced and studied a data set with the *urban street networks* of different cities from all around the world. Namely, we have used short cycles, such as triangles, squares and pentagons, but also the so-called *meshedness coefficient*, which takes into account cycles of any length, to characterise the structure of a city. One of the most important things we have learnt by dealing with street networks is that these are a very special type of networks. They are in fact embedded in space and, like any other spatial network, carry signs of the spatial embedding in their structure. For a modern and complete overview on how to study and model spatial networks we recommend the review paper by Marc Barthélemy [26], while for a detailed account on urban street networks we suggest the book by Philippe Blanchard and Dimitri Volchenkov [39]. We also encourage the reader to have a look at the paper by Emanuele Strano et al. for an empirical study on how urban street networks grow over time [294].

This is also the first chapter where we have introduced two data sets instead of only one. The second data set made available is the *transcription regulation network of* E. coli, which is also the second data set of a biological system of the book. This is a nice example that helps us to understand why some graphs, such as the *feed-forward loop*, emerge as motifs in biological networks. They have an important functional role, so they are predominant in the structure of the system and thus represent its basic building blocks. More details about transcription networks and the various other networks at work in the cell can be found in the wonderful textbook on Systems Biology by Uri

Alon [8] and in a more recent handbook containing a collection of papers on biological networks [42].

Problems

8.1 Counting Cycles

(a) Find a relation between the Katz centrality defined in Eq. (8.3) and the α-centrality introduced in Section 2.3.

(b) Prove that $\mathbb{C}_3$ is k-cyclic for every $k \geq 3$ except $k = 4$.

(c) Prove that the property $c_k(\mathbb{C}_k) = 2k$ is valid for every $k \geq 3$. See in the text the proof that $c_3(\mathbb{C}_3) = 6$.

(d) Consider a graph G with N nodes K links and adjacency matrix $A = \{a_{ij}\}$. Prove that the number of subgraphs of G isomorphic to graphs H_1, H_2, H_3, H_4 and H_5, shown in Figures 8.1, 8.2 and 8.3, are respectively:

$$n_G(H_1) = K$$

$$n_G(H_2) = \sum_i k_i(k_i - 1)/2$$

$$n_G(H_3) = 1/2 \sum_{i,j} a_{ij}(k_i - 1)(k_j - 1) - 3n_G(\mathbb{C}_3)$$

$$n_G(H_4) = \sum_i \binom{k_i}{3}$$

$$n_G(H_5) = 1/2 \sum_i a_{ii}^{(3)}(k_i - 2)$$

where $a_{ij}^{(3)} = (A^3)_{ij}$ denotes the i, j component of matrix A^3.

(e) Prove that Eq. (8.11) and Eq. (8.12) can be rewritten in terms of the different powers of the adjacency matrix as [36]:

$$n_G(\mathbb{C}_4) = \frac{1}{8}\left[\sum_i a_{ii}^{(4)} - 2\sum_i a_{ii}^{(2)}a_{ii}^{(2)} + \sum_i a_{ii}^{(2)}\right]$$

$$n_G(\mathbb{C}_5) = \frac{1}{10}\left[\sum_i a_{ii}^{(5)} - 5\sum_i a_{ii}^{(2)}a_{ii}^{(3)} + 5\sum_i a_{ii}^{(3)}\right].$$

8.2 Cycles in Scale-Free Networks

(a) Prove that, in an undirected graph G, the sum of products of all pairs of eigenvalues ($\lambda_i\lambda_j$, with $i < j$) is equal to $-K$, while the sum of product of triples of eigenvalues ($\lambda_i\lambda_j\lambda_k$, with $i < j < k$) is twice the number of triangles (3-cycles) $n_\triangle$ in G. Such two properties are related to Theorem 8.1, and are obtained from the relation $Tr(A^l) = \sum_i^N \lambda_i^l$, together with $(\sum_i^N \lambda_i)^l = 0$.

(b) Consider an undirected graph with N nodes. Show that the subgraph centrality defined in Eq. (8.5) can be calculated as:

$$c_i^S = \sum_{j=1}^{N} (u_{j,i})^2 \, e^{\lambda_j}$$

where $\mathbf{u}_1, \mathbf{u}_2, \ldots, \mathbf{u}_N$ is an orthonormal basis of $\mathbb{R}^N$ composed by the eigenvectors of A associated with the eigenvalues $\lambda_1, \lambda_2, \ldots, \lambda_N$, and $u_{j,i} \equiv (\mathbf{u}_j)_i$ indicates the ith component of the jth eigenvector $\mathbf{u}_j$

8.3 Spatial Networks of Urban Streets

(a) Following the chord argument given in Example 8.13, prove that a complete graph with five nodes, $\mathbb{K}_5$, is nonplanar. Namely, there are five chords in a 5-cycle of $\mathbb{K}_5$, while there could be at most only four chords, two inside and two outside, without breaking the planarity.

(b) Prove that in a plane embedding of a maximal planar graph each face is a triangle.

(c) Verify Euler's formula for the five platonic solids (see Problem 1.7c).

(d) The *Minimum Spanning Tree (MST)* is the shortest tree which connects every nodes into a single connected component. See Definition 10.15 in Section 10.5. By construction the MST is an acyclic graph with $K_{min} = N - 1$ edges. This is the minimum possible number of edges to have all the nodes belonging to a single connected component. Compute and draw the MST for the city of Savannah given in DATA SET 8.

8.4 Transcription Regulation Networks

(a) Evaluate the average number of cycles in an ensemble of directed random graphs. Can Eq. (3.21) of Chapter 3 still be used for directed graphs? What is the value of a_F in this case?

(b) Compute number and sizes of weakly connected components in the transcription network of *E. coli* given in DATA SET 8 BIS. What is the size of the largest strongly connected component in the network?

8.5 Motif Analysis

(a) Show that there are only thirteen different 3-node weakly connected directed graphs. What would be the number of different 3-node directed graphs if we release the assumption of connectedness?

(b) Perform a 3-node motif analysis of the *Scientometrics* citation network in DATA SET 6. As discussed in Section 6.1, a citation network is a directed acyclic graph (DAG), so that some of 13 graphs shown in Figure 8.9 are in principle not allowed. What are the only possible motifs in a DAG?

9

Community Structure

Real-world networks present interesting *mesoscopic structures*, meaning that they carry important information also at an intermediate scale: a scale that is larger than that of the single nodes, but smaller than that of the whole network. In fact, their nodes are often organised into *communities*, i.e. clusters of nodes such that nodes within the same cluster are more tightly connected than nodes belonging to two different clusters. In such cases we say that the networks have a *community structure*. The most important point is that nodes in the same network cluster usually share common features. For instance, we will see that communities in the *Zachary's karate club network* coincide with real social groupings, communities in brain networks identify areas of the brain with different functions, while tightly connected groups of nodes in the World Wide Web correspond to pages on common topics. This is the reason why, by finding the communities of a network, we can learn a lot about the way the network works. In this chapter we will consider various methods to find communities, starting with two traditional approaches, namely spectral partitioning and hierarchical clustering, and then focusing on more recent methods specifically introduced by network scientists to find community structure in networks. We will present the *Girvan–Newman* approach that is based on the removal of the high-centrality edges, and then we will define a quality function, the so-called *modularity*, that quantifies the quality of a given partition of the nodes of the network. We will show that communities can be extracted directly by optimising the modularity over the set of possible graph partitions. Finally, we will discuss the *label-propagation algorithm*, a local and fast method to detect communities which can be used for very large graphs. The study of network community structure is now considered a research field by itself, and is an area of network science that is still rapidly expanding in different directions, with important contributions also from computer scientists and software engineers. Needless to say, it is very difficult to keep pace with the most recent approaches and algorithms. The choice of the topics of this chapter is therefore mainly didactic, and we have included in Section 9.8 a few pointers to some of the most advanced methods for community detection.

9.1 Zachary's Karate Club

In 1977 Wayne Zachary, an anthropologist at Temple University Philadelphia, published a study on conflict and divisions in small social groups [320]. For his PhD Zachary observed for a period of three years, from 1970 to 1972, the social interactions in a US university karate club. During this period, the club activities included regularly scheduled karate lessons as well as social events such as parties and banquets. In his paper Zachary constructed the network of social relationships in the club, including all the members of the club, the part-time karate instructor, Mr Hi, and also Mr John A., the president and chief administrator of the club. The basic properties of *Zachary's karate club network* and the corresponding graph are illustrated in DATA SET Box 9.1. As usual, the network can be downloaded from the book webpage. The most interesting aspect of Zachary's work is that

Box 9.1	DATA SET 9: Zachary's Karate Club Network

The karate club friendship network provided in this data set was constructed by Wayne Zachary and consists of $N = 34$ nodes and $K = 78$ edges. An edge is drawn if two individuals consistently were observed to interact outside the normal activities of the club during a period of three years. Node 1 and node 34 designate respectively the instructor of the club, Mr Hi, and the president/administrator Mr John A., while all the remaining nodes represent club members. A look at the network immediately indicates the presence of two groups, with about half of the nodes belonging to a group that contains node 1, and the remaining ones belonging to a group that contains node 34.

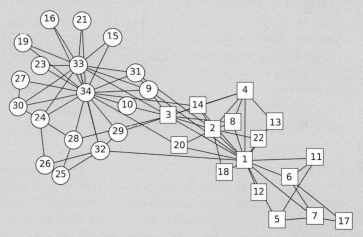

This is certainly due to the fact that, from the beginning of the observations, there was an incipient conflict between the president and Mr Hi over the price of karate lessons, which finally led to the formation of a second karate club by Mr Hi. Hence, this network comes with additional important information on the nodes. We know in fact to which of the two clubs each individual will go to after the split. For such a reason, the $N_1 = 18$ nodes that will stay with the president (node 34) are shown in the figure as circles, while the $N_2 = 16$ following the instructor (node 1) are shown as squares.

the social system under study experienced strong conflicts between the instructor and the administrator which led, after the observed period, to a division of the club. In fact, on the one hand, Mr Hi wished to raise prices, claiming the authority to set his own lesson fees, since he was the instructor. On the other hand, Mr John A., being the club's chief administrator, wanted to stabilise prices. As time passed, the entire club became divided over this issue, tension intensified and finally Mr John A. fired the instructor for attempting to raise lesson fees. The supporters of Mr Hi reacted by resigning and forming a new organisation headed by Mr Hi, thus completing an effective fission of the club into two clubs. The network shown in DATA SET Box 9.1 represents the pattern of relations shortly before the fission, while the shapes of the nodes (circles or squares) indicate their affiliation to one or the other club. In his paper Zachary raised the question whether it was possible to predict the actual split of the club from the structural properties of the graph before the splitting. This is one of the main questions we will try to address in this chapter. However, before starting to study some of the methods to find meaningful modules in a graph, we need to introduce the most basic definitions, those of *node partition* and *graph partition*.

Definition 9.1 (Node partition) *Given a graph $G \equiv (\mathcal{N}, \mathcal{L})$ and an integer number $M \geq 1$, a node partition $\mathcal{P}_\mathcal{N} = \{C_1, C_2, \ldots, C_M\}$ is a division of the elements of set $\mathcal{N}$ into M sets $C_1, C_2, \ldots, C_M$ such that $C_1 \cup C_2 \cup \ldots \cup C_M = \mathcal{N}$, and $C_i \cap C_j = \emptyset \; \forall i \neq j$.*

Definition 9.2 (Graph partition) *Given a graph $G \equiv (\mathcal{N}, \mathcal{L})$ and a node partition $\mathcal{P}_\mathcal{N} = \{C_1, C_2, \ldots, C_M\}$ we consider the set of subgraphs $\{g_1, g_2, \ldots, g_M\}$ induced by the node partition $\{C_1, C_2, \ldots, C_M\}$ (namely, for each $m = 1, \ldots, M$, graph g_m is the graph induced by the nodes in C_m.) Such a set of subgraphs induces a partition of the graph links $\mathcal{P}_\mathcal{L} = \{E_1, E_2, \ldots, E_M, E_{inter}\}$ where E_m, $m = 1, \ldots, M$, is the set of links of graph g_m, while E_{inter} is the set of links of $\mathcal{L}$ which do not belong to any subgraph g_m, $m = 1, \ldots, M$. A graph partition $\mathcal{P}_G$ is therefore given by the node partition $\mathcal{P}_\mathcal{N}$ and the associated link partition $\mathcal{P}_\mathcal{L}$.*

In practice, a node partition is a division of the graph nodes into M disjoint sets. For instance, going back to the Zachary's karate club network, node partition $\mathcal{P}_\mathcal{N} = \{C_1, C_2\}$, with C_1 containing all the nodes with odd label, and C_2 containing all the nodes with even label, is one of the possible partitions of the nodes of the karate club into $M = 2$ sets. This is shown in Figure 9.1, where the two different tones of grey indicate the two sets in the partition. Such a node partition induces a graph partition, since it produces a partition of the links into $\{E_1, E_2, E_{inter}\}$, with E_1 and E_2 containing respectively 20 and 19 links, and E_{inter} containing 39 links. Note that such a partition is an arbitrary one, and in fact the number of links connecting nodes in one set to nodes in the other set is comparable to the number of links within each set. For such a reason, this is certainly not a very useful partition of the graph. In fact, as shown in Figure 9.1, it puts together pairs of distant nodes, and indeed turns out to be very different from the real splitting of the system. A good partition of Zachary's network should instead be able to assign correctly the nodes to the two resulting clubs, based on the analysis of the network structure. Notice that a node partition $\mathcal{P}_\mathcal{N} = \{C_1, C_2\}$, with C_1 containing all the square nodes, and C_2 containing all the circle nodes, produces a partition of the links with only 10 links in the set E_{inter}. Hence, a meaningful division of the nodes of the karate club graph produces a partition of the links with most of the links internal to the two

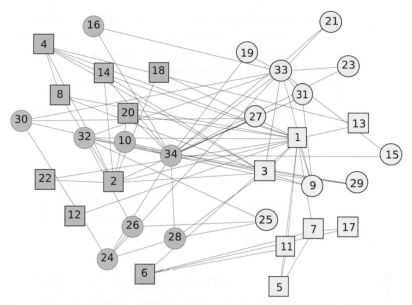

Fig. 9.1 The two different tones of grey indicate an arbitrary partition of the Zachary's karate club network into odd (light grey) and even (dark grey) nodes, while circles and squares show the two factions into which the club actually split during the course of the study.

sets, and only a few links connecting nodes of one set to nodes of the set. More in general, we can say that a graph has a good community structure if it admits a partition in which most of the links are internal to the communities. A first possibility to find the communities is therefore to search for the *cliques* in the network, as shown in the following example.

Example 9.1   *(Communities as cliques)*   In graph theory a *clique* is a complete subgraph, i.e. a subset of mutually adjacent nodes. A possible way to find communities in a graph is to look for cliques. For example, the graph shown in the figure below can be partitioned into one clique of five nodes, two cliques of three nodes and one clique with a single node. In this way we have 16 links internal to the communities, and only four links between communities. However, defining a community as a complete subgraph is a bit too

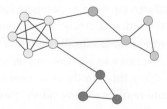

restrictive. Since the probability of finding large cliques in real networks is in general extremely small, a partition into cliques would produce only small communities. For instance, the largest clique in the karate club has only five nodes. Conversely, a large group of nodes can constitute a community without being a complete subgraph, as happens in the karate club. In practice, operational methods to extract communities in real networks must be based on more relaxed definitions of community.

The definition of a community as a clique adopted in Example 9.1 can be extended by weakening the requirement of adjacency between pairs of nodes into a requirement of reachability, or by reducing the number of other nodes to which each node must be connected. However, the concept of clique and the other derived from this, are all *local* definitions, meaning that in order to find a community to which a node belongs to, they only analyse a small portion of the graph in the neighbourhood of the node. In the following sections of this chapter we will instead study various methods to find meaningful partitions of a graph into communities which make use of *global* information.

9.2 The Spectral Bisection Method

A good community detection method should be able to identify automatically the two modules of the karate club in Box 9.1. With such a purpose, we start by considering a class of methods known, in the computer science literature, as methods for *graph partitioning* [116, 261]. Graph partitioning consists in dividing a network into two or, in general, into some fixed number M of subgraphs of roughly equal size, such that the number of edges lying between subgraphs $R = |E_{inter}|$, usually named the *cut size*, is minimal. The specification of the number of modules of the partition is a necessary input in graph partitioning. The problem of graph partitioning arises in many practical situations. A typical example is the allocation of computations across the processors of a cluster of computers.

Example 9.2 *(Graph partitioning in computer science)* Imagine a program performing a lengthy computation is composed of several small processes which can be run in parallel. Each process is responsible for a certain kind of computation and exchanges data with other processes. In this case, the whole computation can be represented as a graph whose N nodes correspond to processes while the edges indicate the data exchange between processes. If you have just one processor (CPU) available, then the different parts of your program will effectively have to compete for processor time, since a CPU can normally run only one instruction at a time. Now, imagine that you have $1 < M \le N$ processors available. In order to speed up the computation, it would be desirable to assign a set of roughly N/M processes to each of the different CPUs so that at least M processes can run in parallel at any time. In the best possible scenario, in which the number of CPUs is equal to the number of processes, all the processes of your program can run at the same time. However, allocating processes to more than one CPU introduces delays due to the transmission of data between different CPUs, which is usually much slower than memory access. An example is shown in the figure below for a program with $N = 9$ processes to be

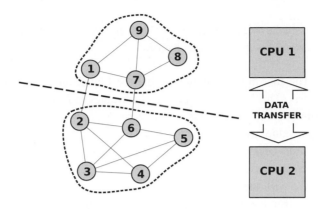

allocated to $M = 2$ processors. If we would like to distribute the processes (nodes) almost equally across the two available CPUs, and in such a way that delays due to inter-CPU data transfer are minimised, we should partition the nodes of the graph in two groups as shown in the figure, and assign nodes 1, 7, 8 and 9 to one processor, and nodes 2, 3, 4, 5 and 6 to the other processor. This is a partition with a minimal *cut size*, namely $R = 2$ edges connecting the two groups of nodes. Other classical applications of minimal cut size partitions can be found in the design of electronic circuits, where one usually wants to optimise the spatial embedding of the circuit components by dividing them into groups in such a way as to minimise the total wiring length and power consumption.

The most elementary partitioning of a graph is the bisection, which is a partitioning in $M = 2$ subgraphs. In addition to being the simplest partitioning, the bisection is also the first step in a hierarchical partitioning. Ideally, the best bisection is obtained by considering all possible partitions into two subgraphs of the same order, and selecting the one with the smallest cut size. However, the exhaustive search of the optimal bisection into two subgraphs of the same order has to consider $\begin{pmatrix} N \\ N/2 \end{pmatrix}$ partitions, which is a prohibitively large number. For instance, we have 252 different partitions for $N = 10$ nodes, and more than 10^{14} different partitions for $N = 50$. For this reason, we need to look for alternative methods that provide good partitions at a reasonable cost. One of such methods is the *spectral bisection method*, a technique to divide the graph into two subgraphs based on the spectral properties of the *Laplacian matrix*. In the case of an undirected graph, this is a matrix obtained from the adjacency matrix A, by placing on the diagonal the degrees of the vertices and by changing the signs of the other elements.

Definition 9.3 (Laplacian matrix) *The* Laplacian matrix, *or simply* Laplacian, *L, of a N-nodes undirected graph, is a $N \times N$ matrix defined as:*

$$L_{ij} = \begin{cases} -1 & \text{if } i \neq j \text{ and } (i,j) \in \mathcal{L} \\ k_i & \text{if } i = j \\ 0 & \text{otherwise.} \end{cases}$$

> *It can be written as $L = D - A$, where A is the adjacency matrix and D is the $N \times N$ diagonal matrix of node degrees.*

The first thing to notice is that the Laplacian of an undirected graph is a symmetric matrix. This implies that all the eigenvalues of L are real numbers. Moreover, all rows and columns of the Laplacian sum to zero: $\sum_i L_{ij} = \sum_j L_{ij} = 0$. Hence L has at least one zero eigenvalue, with eigenvector $\mathbf{1} = (1, 1, 1, \ldots, 1, 1)$. In addition to this, all the eigenvalues are non-negative, since L is a positive semi-definite matrix.[1] We can therefore sort the eigenvalues of L in increasing order as $0 = \lambda_1 \leq \lambda_2 \leq \ldots \leq \lambda_N$.

We will now show that there exists a direct relation between the presence of community structure and some peculiar features of the eigenvalues and the eigenvectors of L. Let us consider first the simplest possible case in which a graph consists of a single connected component. In this case, the zero eigenvalue is a simple root of the characteristic polynomial of L. This is so because, in a symmetric matrix, the number of independent eigenvectors associated with an eigenvalue is equal to its multiplicity, and it is possible to prove that $\mathbf{1}$ is the only eigenvector of L associated with the eigenvalue zero.

Example 9.3   Suppose we have a connected graph, and let L denote its Laplacian. We shall prove that the components of an eigenvector of L corresponding to the zero eigenvalue are all the same. In fact, let $\mathbf{u}$ be such an eigenvector. Then one has:

$$(L\mathbf{u})_i = ((D - A)\mathbf{u})_i = k_i u_i - \sum_j a_{ij} u_j = 0 \quad \forall i = 1, 2, \ldots, N$$

from which it follows $u_i = (\sum_j a_{ij} u_j)/k_i$, i.e. u_i is equal to the average component of the first neighbours of i, and this is true for each node i. Let ℓ denotes the index of the maximum component of $\mathbf{u}$. The relation $u_\ell = (\sum_j a_{\ell j} u_j)/k_\ell$ can be rewritten as:

$$\frac{\sum_j a_{\ell j}(u_\ell - u_j)}{k_\ell} = 0.$$

Now, since we have $u_\ell \geq u_i \ \forall i$, the above equality can be satisfied only if $u_j = u_\ell \ \forall j \in \mathcal{N}_\ell$. Repeating the same argument for the neighbours of ℓ we can also prove that the components of $\mathbf{u}$ corresponding to the second neighbours of ℓ are the same. By keeping repeating the argument, since the graph is connected, we explore the whole graph and show that all the components of $\mathbf{u}$ are equal.

Suppose now that the graph contains M communities. In the ideal situation in which the division into communities is perfect, each community is a distinct component of the graph.

[1] A matrix A is positive semi-definite, iff $\mathbf{x}^\top A \mathbf{x} \geq 0 \ \forall \mathbf{x} \neq \mathbf{0}$. This is true for the Laplacian matrix, since, making use of the property $\sum_i L_{ij} = \sum_j L_{ij} = 0$, we can write $\mathbf{x}^\top L \mathbf{x} = \sum_{ij} L_{ij} x_i x_j = -1/2 \sum_{ij} L_{ij}(x_i - x_j)^2$, which is larger than or equal to zero. Since a symmetric matrix is diagonalisable, we can write $L = U^\top \Lambda U$ with Λ diagonal and $U^\top U = I$ (see Appendix A.5.1). We therefore have $\mathbf{x}^\top L \mathbf{x} = \mathbf{x}^\top U^\top \Lambda U \mathbf{x} = \mathbf{u}^\top \Lambda \mathbf{u} = \sum_i |u_i|^2 \lambda_i$, where $\mathbf{u} = U\mathbf{x}$, and $\lambda_i, i = 1, \ldots, N$ are the eigenvalues of the Laplacian. It therefore follows that $\lambda_i \geq 0, \forall i = 1, \ldots, N$.

We can then divide the N nodes of the graph into M sets $C_1, C_2, \ldots, C_M$, respectively with $N_1, N_2, \ldots, N_M$ nodes ($N_1 + N_2 + \ldots + N_M = N$), so that each set is a component. In this way, there will be no edges between nodes of different sets. Hence, the vertices can be ordered in such a way that the Laplacian L displays M square blocks (of size $N_1, N_2 \ldots N_M$) along the diagonal, with entries different from zero, whereas all other elements vanish. Each block of size N_i ($i = 1, 2, \ldots, M$) is the Laplacian of the corresponding subgraph, the component of order N_i. Such a $N_i \times N_i$ Laplacian L_i has only one zero eigenvalue, and the corresponding eigenvector has all N_i components equal to 1. Therefore, L has M degenerate eigenvectors $\mathbf{u}_i$ ($i = 1, 2, \ldots, M$) with eigenvalue 0. Each of the eigenvectors has equal non-vanishing entries in correspondence to the nodes of a block, whereas all other entries are zero. Summing up, if a graph has M components, the multiplicity of the 0 eigenvalue is equal to the number of connected components M, and we can immediately identify the components by using the associated eigenvectors mentioned above.

Finally, let us consider a graph which separates well, but not perfectly, into communities. For instance, a connected graph consisting of M subgraphs with a few edges between each others. In this case, the Laplacian can no longer be written in a perfect block-diagonal form, because a few entries of the Laplacian matrix will lie outside the blocks. As illustrated in the following example, the spectrum of L will have only one zero eigenvalue and other $M - 1$ eigenvalues close to zero. The corresponding eigenvectors will approximately be linear combinations of the eigenvectors $\mathbf{u}_i$ defined above. Hence, by looking for eigenvalues of the graph Laplacian only slightly greater than zero and by taking linear combinations of the corresponding eigenvectors, one should in theory be able to find the communities themselves, at least approximately.

Example 9.4   *(A graph with three communities)*   Consider a graph G with $N = 20$ nodes, $K = 68$ links, and three well-defined communities. Assume the three communities are complete subgraphs with respectively $N_1 = 4$, $N_2 = 6$ and $N_3 = 10$ nodes and, in addition to the internal links, there are two inter-community links, namely one link between communities 1 and 2, and one link between communities 2 and 3. The graph Laplacian has the following eigenvalues with multiplicity reported in parentheses: 0.0 (1), 0.123 (1), 0.406 (1), 4.0 (2), 4.515 (1), 6.0 (3), 6.635 (1), 7.213 (1), 10.0 (8), 11.11 (1). The entire spectrum is shown in the upper part of the figure below. Hence, we have the eigenvalue 0 with multiplicity 1, and two other eigenvalues, namely 0.123 and 0.406, which are very close to zero if compared to the remaining ones.

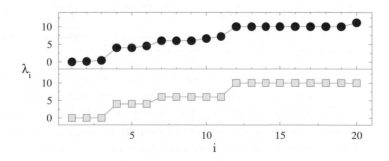

Notice that, in the case of three perfectly separate components, i.e. in a graph G' obtained from G by removing the two inter-community links, the eigenvalues with their multiplicity would read: 0.0 (3), 4.0 (3), 6.0 (5), 10.0 (9). We would therefore have the eigenvalue 0 with multiplicity three. The spectrum of G', shown in the lower part of the figure, does not look very different from that of G.

A special case is when there are only two blocks, as in the graph of the karate club. Since all eigenvectors associated with distinct eigenvalues of a real symmetric matrix are orthogonal, it is clear that all eigenvectors other than the eigenvector $\mathbf{1} = (1, 1, 1, \ldots, 1, 1)$ corresponding to the eigenvalue 0 must have both positive and negative elements. And for the case of two weakly coupled communities there will be one eigenvector with eigenvalue slightly greater than zero and elements all positive for one community and all negative for the other, since all elements are (nearly) equal within a community. Thus, we can divide the network into its two communities by looking at the eigenvector corresponding to the second lowest eigenvalue, and separating the vertices according to whether the corresponding element in this eigenvector is greater than or less than zero. This is the so-called *spectral bisection method*. The second lowest eigenvalue λ_2 is known as the *algebraic connectivity* of the graph, while the associated eigenvector is named *Fiedler vector* in honour of Miroslav Fiedler who first used the eigenvalues of the Laplacian matrix to extract information on the structure of a graph.

Example 9.5 *(The rationale of the spectral bisection method)*   Given a graph with an even number of nodes N, let us consider a partition of the set of nodes into two sets C_1 and C_2, respectively with N_1 and N_2 nodes. Any partition into two sets can be represented by a vector $\mathbf{x}$ whose component x_i is $+1$ if vertex i is in one set and is -1 if i is in the other set. Notice that $\sum_i x_i^2 = N$. A good way to evaluate how well the chosen bisection $\mathbf{x}$ represents a natural division of the graph is to evaluate the cut size of the partition R, i.e. to count the number of edges connecting nodes in C_1 to nodes in C_2. The quantity R can be written in terms of the Laplacian matrix L as:

$$R = \frac{1}{4}\mathbf{x}^\top L\mathbf{x}, \tag{9.1}$$

where $\mathbf{x}^\top$ is the transpose of vector $\mathbf{x}$. We have, in fact:

$$\mathbf{x}^\top L\mathbf{x} = \mathbf{x}^\top D\mathbf{x} - \mathbf{x}^\top A\mathbf{x} = \sum_{i=1}^{N} k_i x_i^2 - 2\sum_{\{i,j\}\in\mathcal{L}} x_i x_j = \sum_{\{i,j\}\in\mathcal{L}} (x_i - x_j)^2.$$

In the last term we observe that if i and j belong to the same set of nodes then $(x_i - x_j)^2 = 0$, while if i and j belong to different sets then $(x_i - x_j)^2 = 4$. Therefore, the last summation gives four times the number of links between the two sets, namely $4R$. Notice that, when looking for the minimum cut size, we have to add some constraints on the relative size of the two sets C_1 and C_2, in order to avoid trivial subdivisions such as those with a set with 0 or 1 nodes. For instance, we can impose that C_1 and C_2 have exactly the same number of nodes. In this way our problem is equivalent to the minimisation of the quadratic form $\mathbf{x}^\top L\mathbf{x}$ over vectors $\mathbf{x}$ with components $x_i = \pm 1$, with the further condition $\sum_{i=1}^{N} x_i = 0$. This is equivalent to the enumeration of all the graph partitions into two subsets of equal size, which is known to be a NP-complete problem (see Appendix A.2). We can find an approximate solution in the following way. Vector $\mathbf{x}$ can be written as $\mathbf{x} = \sum_i a_i \mathbf{u}_i$, where $\mathbf{u}_i$, $i = 1, \ldots, N$ are the eigenvectors of the Laplacian, chosen to form an orthonormal basis. Notice that $\sum_i x_i = 0$ means that vector $\mathbf{x}$ is orthogonal to $\mathbf{u}_1 = 1/\sqrt{N}$. Therefore, $a_1 = 0$. Hence, R can be written as:

$$R = \frac{1}{4} \sum_{i=2}^{N} a_i^2 \lambda_i, \tag{9.2}$$

where λ_i is the eigenvalue of the Laplacian associated with eigenvector $\mathbf{u}_i$. Observe that $\sum_{i=2} a_i^2 = \sum_i x_i^2 = N$. Finding the minimum R with the only constraint $\sum_{i=2} a_i^2 = N$ is equivalent to minimising expression (9.1) with $\mathbf{x} \in \mathbb{R}^N$, and the constraints $\sum_i x_i^2 = N$ and $\mathbf{x}$ orthogonal to $\mathbf{u}_1$. In this case the minimum is obtained by choosing $\mathbf{x}$ parallel to the Fiedler vector $\mathbf{u}_2$. This is because $R = \frac{1}{4} \sum_{i=2}^{N} a_i^2 \lambda_i \geq \frac{1}{4} \sum_{i=2}^{N} a_i^2 \lambda_2 = \frac{1}{4} N \lambda_2$, the latter being obtained by choosing $a_2^2 = N$ and $a_i = 0$ for $i > 2$. However, if we want $x_i = \pm 1$, then the vector $\mathbf{x}$ cannot be perfectly parallel to $\mathbf{u}_2$. So, the best we can do is to match the signs of the components and set $x_i = +1$ (-1) if the ith component of $\mathbf{u}_2$ is positive (negative), which corresponds to the spectral bisection method. Notice that N_1 and N_2 is fixed by the number of positive and negative components of $\mathbf{u}_2$.

The spectral bisection, based on the signs of the Fiedler vector, provides a subdivision of the nodes in two sets of automatically determined sizes N_1 and N_2. If one wants to control the relative size of the two sets, for example if one wants to fix $N_1 = N_2$, the best strategy is to order the components of the Fiedler vector from the lowest to the largest values and to put in one set the vertices corresponding to the first N_1 components from the top (or from the bottom), and the remaining vertices in the second set. Since, in general, the top or bottom procedures produce two different partitions, we will choose the one with the smallest cut size.

In Figure 9.2 we report the results of the spectral bisection method on the karate club network. In this case we find an algebraic connectivity $\lambda_2 = 0.469$. The corresponding eigenvector $\mathbf{u}_2$ has 19 positive components and 15 negative ones. As indicated in the figure, the algorithm works extremely well. It produces two communities, respectively with $N_1=19$ and $N_2 = 15$ nodes, which reproduce almost perfectly the known split of the network into two factions, with the exception of just one node, node 3, which is placed in the wrong group of nodes. Notice, however, that node 3 has the same number of links to nodes of the two communities, so that it can reasonably be considered an ambiguous case.

The spectral bisection method works very well in cases where the graph really does split nicely into two communities, as in the case of the karate club. Partitions into more than two modules are usually attained by iterative bisectioning. For instance, if we deal with the graph in Example 9.4, the existence of the eigenvalue 0 and other two eigenvalues close to 0 suggests that the natural partition of the graph is into three subgraphs. By looking at the

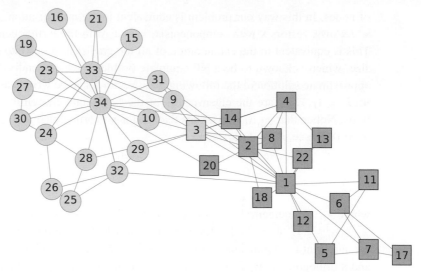

Fig. 9.2 The two different tones of grey of the nodes indicate the two communities of the Zachary's karate club network found by spectral bisection, while circles and squares indicate the two factions into which the club actually split during the course of the study. Only node 3 is classified incorrectly.

positive and negative components of the Fiedler eigenvector one performs a first bisection. One of the two subgraphs we obtain is the complete graph with 10 nodes, while the other subgraph will be composed of the two complete graphs with 4 and 6 nodes joined by one link. Then, we can perform a second bisection on the latter subgraph, finding that this splits into two subgraphs with respectively 4 and 6 nodes. Finally we get the three communities with $N_1 = 4$, $N_2 = 6$ and $N_3 = 10$ nodes. In conclusion, spectral bisection can also provide some information about the number and size of the various communities. However, if we want to perform a hierarchical clustering of the nodes of a graph, other methods are more suitable and will be discussed in the next section.

9.3 Hierarchical Clustering

We switch now, for a while, from the problem of finding the communities in a graph to the more general problem of organising a collection of data or objects into similar groups or *clusters*. Such a problem is known in the literature as *clustering* and can be stated as follows. Suppose we are given a set of N elements, being for instance distinguishable objects. Each object is characterised by a set of parameters which quantitatively describe its properties. We want to group together similar objects so that each cluster is therefore a collection of objects with similar properties. Notice that in the general problem of clustering, the objects to cluster do not need to be the nodes of a graph, neither needs the similarity between them to be based on graph properties. Some practical applications are illustrated in the following example.

Example 9.6 *(Clustering)*   The typical case of objects to cluster are the data resulting from a set of measurements. A classic example, which has also been thoroughly used to discriminate alternative clustering methods, is the iris data set studied for the first time by Ronald Fisher in 1936 [117]. This data set consists of $N = 150$ samples, 50 from each of the three species of iris flowers, namely *Iris setosa*, *Iris virginica* and *Iris versicolor*. Four features were measured from each sample, namely the width and the length of petal and sepal. Therefore with each sample i, with $i = 1, \ldots, N$, we can associate a four-component vector $\mathbf{x}_i$. The main goal of Fisher was to determine the species of the flower from the knowledge of the four features. We will come back to this point in the Example 9.8.

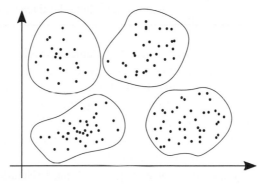

Locations of earthquakes, or their temporal occurrences, are other examples of measurements that might be useful to cluster [204]. In the figure above we sketch the geographical distribution of N earthquakes. The location of earthquake i is described by a two-dimensional vector $\mathbf{x}_i$ giving the latitude and longitude of its epicentre, and is shown as a point in the plane. Here the similarity criterion to adopt is geographical distance: two or more earthquakes belong to the same cluster if they are close in space. For instance, in the case in the figure we can identify four clusters, four dangerous seismic areas, into which the epicentres can be divided.

When it comes to cluster N elements, the first thing we need to evaluate is the number of different ways in which we can group the N elements. It is simple to understand that $N = 2$ objects can be partitioned in two different ways, namely either the two objects go together, or they are in two different sets. In the following example we show that $N = 3$ objects can be partitioned in five different ways. Unfortunately, as we will see, the number of partitions to be considered grows very fast with N.

Example 9.7 *(Partitioning three objects)*   Let us consider $N = 3$ objects and label them with the three letters a, b, and c. The three objects can be grouped into sets in the following five different ways: 1) a, b and c are all together in a single set; 2) a and b are together, and c is separate; 3) a and c are together, and b is separate; 4) b and c are together, and a is separate; or 5) a, b and c are in three separate sets. Hence, we have five different partitions to consider in this case. Notice that counting the number of different partitions of N objects is formally equivalent to counting the number of distinct ways of placing N distinguishable balls into one or more indistinguishable boxes. This is exactly what we have done for the cases $N = 3$. We have labelled as a, b, and c the three balls because they are distinguishable. And since the boxes cannot be distinguished from each others, there are indeed five different ways of putting the balls into the boxes. Namely: 1) all three balls go into one box (since the boxes are indistinguishable, this is only considered one combination); 2) ball c goes into one box, a and b go into another box; 3) ball b goes into one box, a and c go into another box; 4) ball a goes into one box, balls b and c go into another box; 5) the three balls go into three distinct boxes. In practice, we have one way to distribute the three balls into three boxes, one way to distribute the three balls into one box, and three different ways to distribute the three balls into two boxes.

The number of ways a set of N elements can be partitioned into nonempty subsets is called the *Bell number* and is usually denoted as B_N [30]. As illustrated in Box 9.2, the Bell number B_N can be obtained as a sum of the Stirling numbers of the second kind $S_{N,k}$ for $k = 1, \ldots, N$. In Table 9.1 we report such numbers for the first few values of N. The Bell numbers for $N = 1, 2, \ldots, 10$ are respectively 1, 2, 5, 15, 52, 203, 877, 4140, 21147, 115975. It is possible to prove that the Bell number grows at least exponentially in the number of nodes N. Indeed, the problem of optimising a certain quantity over all the possible partitions of a graph is NP-hard (Non-deterministic Polynomial-time hard: see Appendix A.2 for details) [56]. This means that, when N is large, the exhaustive exploration of all the possible partitions is out of reach. We therefore need a criterion to identify the most relevant partitions.

In 1967 Stephen C. Johnson from Bell Labs proposed a method, known as the *hierarchical clustering method*, to produce a series of partitions running from N clusters, each containing a single object, to a single cluster containing all objects [170]. The starting point of the method is the definition of an $N \times N$ distance matrix $\mathcal{D}$, or equivalently of an $N \times N$ similarity matrix $\mathcal{S}$. This is usually done from the values of the parameters quantifying the properties of the objects. The basic process of (agglomerative) hierarchical clustering consists then in the following four steps.

Table 9.1 Stirling numbers $S_{N,k}$

N \ k	0	1	2	3	4	5	6	7
1	0	1						
2	0	1	1					
3	0	1	3	1				
4	0	1	7	6	1			
5	0	1	15	25	10	1		
6	0	1	31	90	65	15	1	
7	0	1	63	301	350	140	21	1

| Box 9.2 | The Number of Distinct Partitions |

The number of distinct partitions of a set of N elements, the so-called *Bell number,* can be understood as the number of distinct ways we can place N *distinguishable* balls into one or more *indistinguishable* boxes. The number of ways to partition N objects into a fixed number k of non-empty subsets is known as the *Stirling number of the second kind* and is indicated as $S_{N,k}$. Therefore, the Bell number B_N can be obtained as a sum of the Stirling numbers, namely we can write:

$$B_N = \sum_{k=0}^{N} S_{N,k} \quad where \quad S_{N,k} = kS_{N-1,k} + S_{N-1,k-1} \ 1 \leq k < N.$$

The recurrence relation that allows the Stirling numbers to be calculated can be easily derived from the following argument. When it comes to partition N objects into k boxes, we have two possibilities. We can partition the first $N-1$ objects into k boxes and then add the last object into one of the k boxes (this can be done in k ways), or we can partition the first $N-1$ objects into $k-1$ boxes and then add the last object into a new box. By iterating the recurrence relation with the conditions $S_{N,1} = S_{N,N} = 1$ and $S_{N,0} = 0$ we get the numbers reported in Table 9.1. The Bell number B_N can then be obtained by summing the elements of line N of the table. See also Problem 9.3 for an alternative way to define and calculate Bell numbers.

1 Start by assigning each object to a cluster so that you have N clusters, each containing just one object. Let the distance (similarity) between two clusters be the same as the distance (similarity) between the two objects they contain.

2 Find the closest (most similar) pair of clusters and merge them into a single cluster, so that now you have one cluster fewer.

3 Compute distances (similarities) between the new cluster and each of the old clusters.

4 Repeat steps 2 and 3 until all objects are clustered into a single cluster of size N.

Notice that step 3 can be implemented in different ways, which is what distinguishes different methods, respectively known as single-linkage, complete-linkage and average-linkage. In *single-linkage* clustering, the distance between two clusters is equal to the smallest distance from any member of one cluster to any member of the other cluster. Of course, if we are working with similarities instead of distances, the similarity between two clusters is equal to the largest similarity from any member of one cluster to any member of the other cluster. In *complete-linkage*, the distance between two clusters is defined as the largest distance from any member of one cluster to any member of the other cluster. In *average-linkage* clustering, the distance is equal to the average distance from any member of one cluster to any member of the other cluster. The result of hierarchical clustering is a graphical structure known as *dendrogram*, or *hierarchical diagram*, which shows the clusters produced at each step of the agglomeration.

Example 9.8 *(Hierarchical clustering of iris flowers)* We report below the results of a complete-linkage hierarchical clustering for the database of the three species of iris flowers mentioned in Example 9.6. For the sake of simplicity we have only considered 30 of the 150 samples in the database. The shape of each node indicates that the corresponding sample

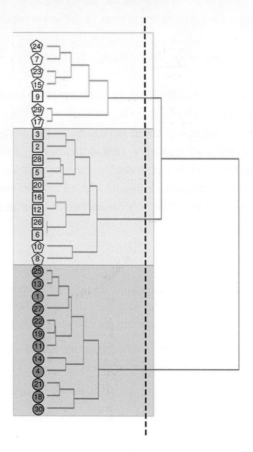

belongs to one of the three species: *Iris setosa* (circles), *Iris virginica* (squares) and *Iris versicolor* (pentagons). Since a sample i is characterised by a vector of four numbers $\mathbf{x}_i$, as the distance between two samples i and j we have used the Euclidean distance based on the four features:

$$\mathcal{D}_{ij} = \sqrt{\sum_{\alpha=1}^{4}\left[(\mathbf{x}_i)_\alpha - (\mathbf{x}_j)_\alpha\right]^2}.$$

The dendrogram obtained represents, from left to right, the hierarchy of the joins between different samples. The cross-section of the dendrogram shown by the vertical dashed line corresponds to a division into three clusters. Notice that only three samples, namely sample 8, 9 and 10, are not classified correctly by the method. Results indicate that it is quite reasonable to classify iris flowers solely based on the dimensions of petals and sepals.

We are now ready to come back to our original problem: that of finding the communities in a graph. We want to make use of the hierarchical clustering method. Of course, the key point of the method is how to define the distance or similarity/dissimilarity between the objects to cluster. When the objects are the N nodes of the graph, the $N \times N$ distance matrix $\mathscr{D}$ (or equivalently the similarity matrix $\mathcal{S}$) can be naturally constructed directly from the graph structure. Notice that for the distance matrix $\mathscr{D}$ we have used here a symbol different from $\mathcal{D}$ defined in Section 3.6 from the length of the shortest paths from i to j. As shown in the following example, $\mathscr{D}$ can in fact be constructed in several alternative ways.

Example 9.9 *(Similarity between the nodes of a graph)* The similarity or dissimilarity (distance) between two nodes of a graph can be defined by making use of the concept of structural equivalence, or alternatively by taking into account the number of distinct walks between the two nodes. *Structural equivalence* is a key concept in social networks analysis. Basically, two nodes of the graph are structurally equivalent if they have the same pattern of relationships with other nodes. For instance two individuals in a friendship network are structurally equivalent if they have the same friends [308, 278]. There are different ways to measure structural similarity. Structurally equivalent nodes have identical entries in their corresponding entries of the adjacency matrix A. Hence, we can define a distance matrix $\mathscr{D}$ by setting the entry $\mathscr{D}_{ij} = (\mathscr{D})_{ij}$ as:

$$\mathscr{D}_{ij} = \sqrt{\sum_{\substack{k=1 \\ k \neq i,j}}^{N} (a_{ik} - a_{jk})^2}. \tag{9.3}$$

In this way we say that two nodes i and j are structurally equivalent, and we indicate this here with the symbol $i \sim j$, if the entries in their respective rows of the adjacency matrix A are identical, and thus $\mathscr{D}_{ij} = 0$. Such distance has the properties of a metric distance: i) $\mathscr{D}_{ij} \geq 0 \ \forall i,j$; ii) $\mathscr{D}_{ij} = 0 \ \forall i \sim j$; iii) $\mathscr{D}_{ij} = \mathscr{D}_{ji} \ \forall i,j$, and can be used in a hierarchical clustering algorithm to cluster graph nodes.

An alternative possibility to construct a similarity matrix $\mathcal{S}$ is to define the similarity between two nodes from the Pearson correlation coefficient between the two corresponding rows (columns) of A:

$$s_{ij} = \frac{\sum_{k=1}^{N} (a_{ik} - \bar{a}_{i.})(a_{jk} - \bar{a}_{j.})}{\sqrt{\sum_{k=1}^{N} (a_{ik} - \bar{a}_{i.})^2} \sqrt{\sum_{k=1}^{N} (a_{jk} - \bar{a}_{j.})^2}}, \tag{9.4}$$

where by $\bar{a}_{i.}$ we have denoted the mean of the values in row i. The entries s_{ij} range from -1 to 1. In this case, if two nodes are perfectly structurally equivalent, then the correlation coefficient between their respective rows in A will be equal to 1.

Another different way to define the similarity between two nodes is to consider distinct walks that run between them. However, because the number of walks between any two vertices increases with the length of the walk, we weight walks of length l by a factor α^l. This is exactly the same idea used by Leo Katz to define his centrality index [174] (see Example 8.1). The similarity matrix S can be expressed in terms of the adjacency matrix A of the graph as:

$$S = \sum_{l=1}^{\infty} \alpha^l A^l = (I - \alpha A)^{-1} - I \tag{9.5}$$

where we know that, for the sum to converge, we must choose $|\alpha| < 1/\rho(A)$, where $\rho(A)$ is the *spectral radius* of the adjacency matrix A, i.e. the largest eigenvalue in absolute value.

Once we have defined a distance matrix $\mathcal{D}$ or a similarity matrix S based on the graph topology, the hierarchical method can be straightforwardly applied to find the communities in a graph. We show how the method works in practice for the karate club network. The obtained hierarchical diagram is reported in the left-hand side of Figure 9.3. We have considered a single-linkage agglomerative hierarchical clustering based on the distance matrix in Eq. (9.3). The tree represents, from left to right, the hierarchy of the joins between nodes. The horizontal length of the branching points in the tree are indicative only of the order in

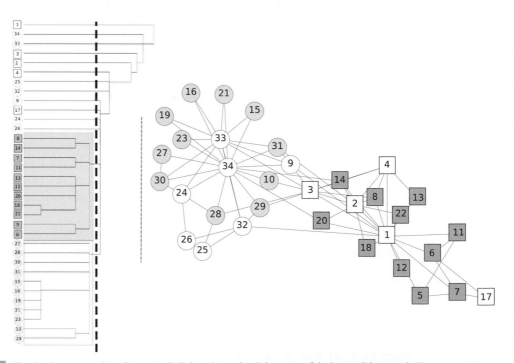

Fig. 9.3 The dendrogram resulting from a single-linkage hierarchical clustering of the karate club network. The cross-section indicated by the vertical dashed line corresponds to the community division reported in the network. Two large communities respectively formed by nodes in light and dark grey are found. Nodes in white are those not assigned to either of the two principal communities.

which the joins between nodes take place. When the lengths of some joins coincide, this reveals that the nodes joined at that level have identical similarities. Although hierarchical clustering methods do not require any a priori knowledge of the size and number of groups, they do not tell which is the best network partition, i.e. at which level the dendrogram should be considered. For instance, the vertical dashed line shown in the figure indicates the partition which most closely resemble the known division of the club. The corresponding communities are those shown in the right-hand side of Figure 9.3 with different tones of grey. We see that the method produces two large communities, and classifies correctly a substantial number of nodes. However, there is also a large number of nodes, shown as empty circles and squares not assigned to either of the two communities. For instance the method does not assign such relevant nodes as 1, 33, and 34 to any major group. Such a problem, which is typical of hierarchical clustering, makes this method unsatisfactory for the analysis of many large real networks.

9.4 The Girvan–Newman Method

To overcome the shortcomings of spectral bisection and hierarchical clustering methods, Michelle Girvan and Mark Newman came up in 2002 with an alternative approach to the problem of finding the communities in a graph [131]. Rather than constructing communities by progressively joining together similar nodes, as in the hierarchical clustering method of Section 9.3, Girvan and Newman proposed to look at the hierarchy of divisions produced by progressively removing edges from the original graph. The main ingredient of the method is to remove edges in decreasing order of their edge betweenness, the natural extension to edges of the measure of node betweenness we studied in Chapter 2. The betweenness of an edge is defined as the number of shortest paths between pairs of vertices that run along it. If there is more than one shortest path between the same pair of vertices, each path is given equal weight such that the total weight of all the paths sums to one.

Definition 9.4 (Edge betweenness centrality)   *In a connected graph, the betweenness centrality of edge e is:*

$$c_e^B = \sum_{j=1}^{N} \sum_{\substack{k=1 \\ k \neq j}}^{N} \frac{n_{jk}(e)}{n_{jk}} \tag{9.6}$$

where n_{jk} is the number of geodesics from node j to node k, whereas $n_{jk}(e)$ is the number of geodesics from node j to k, containing edge e. The definition can be extended to unconnected graphs, assuming that the ratio $n_{jk}(e)/n_{jk}$ is equal to zero if j and k are not connected. A normalised betweenness centrality of edge e, defined as:

$$C_e^B = \frac{c_e^B}{N(N-1)} \tag{9.7}$$

takes values in the range [0,1].

The basic idea of the method proposed by Girvan and Newman is very simple and intuitive. If a graph has a community structure, there are usually only a few edges between two communities. Then, all shortest paths from nodes in one community to nodes in the other must contain one of those few edges. Consequently, the edges connecting communities will have high edge betweenness. By removing these edges, we separate groups from one another and so we unveil the underlying community structure of the graph. In practice, Girvan and Newman proposed an algorithm that works as follows. One first calculates the betweenness for all the edges in the graph. Then, the edge with the highest betweenness is removed, and a component analysis is performed on the obtained graph, with each component representing a community. If there is more than one edge with the highest value of betweenness, ties are broken by selecting one of these edges at random with uniform probability. Notice that every time we remove an edge the number of components can remain the same or can increase by one. The edge betweenness is recalculated in the new graph and the same procedure is repeated until no edges remain and the graph is divided into N components, each containing a single node. Due to the possible existence of more than one edge with the highest betweenness, different runs of the algorithm can produce slightly different divisions of the network. More details on the GN algorithm can be found in Appendix A.17. The result of this procedure is a dendrogram which is similar overall to those produced by the hierarchical clustering method, with the only difference being that in this case the dendrogram is obtained by successive divisions rather than by agglomeration. Figure 9.4 (a) shows a dendrogram obtained by running the GN algorithm on the Zachary's karate club network. We observe that the first splitting into two communities produced by the algorithm and indicated by the black vertical dashed line, matches closely the real division of the club. As denoted by the shapes and colours (the two tones of grey) of the vertices, only one node, namely node 3, is incorrectly classified. This is certainly a much better result than that obtained by the agglomerative hierarchical clustering in the previous section. However, also in this case, the choice of where to cut the dendrogram is completely arbitrary. We will come back to this point in the next section. An important thing to notice is that, every time a link is removed, it is essential to recalculate the betweenness of all the graph links. The right panel of Figure 9.4 shows the results obtained by a variation of the method without recalculation of betweennesses after each edge removal. It is clear from the hierarchical diagram produced that this version of the algorithm fails to find the split into the two correct factions.

In cases such as Zachary's karate club, when the real splitting is a priori known, it is possible to evaluate the quality of an algorithm by comparing the partition it produces with the real partition. One possibility is to make use of the so-called *normalised mutual information*, a measure based on information theory that allows the similarity between two partitions $\mathcal{P}_{\mathcal{N}_1}$ and $\mathcal{P}_{\mathcal{N}_2}$ of the nodes of a graph to be quantified [90].

Definition 9.5 (Normalised mutual information) *Given two partitions $\mathcal{P}_{\mathcal{N}_1}$ and $\mathcal{P}_{\mathcal{N}_2}$ of the nodes of a graph respectively with M_1 and M_2 sets, the normalised mutual information between $\mathcal{P}_{\mathcal{N}_1}$ and $\mathcal{P}_{\mathcal{N}_2}$ is:*

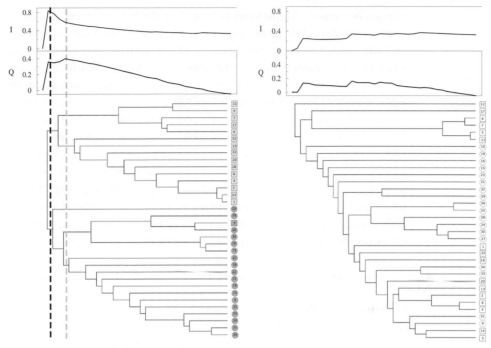

The dendrograms representing a division of the karate club network obtained by the Girvan–Newman method (left), and by a modification of the method in which the edge betweennesses are not recalculated after each edge removal (right). For each partition of the network we also report the value of normalised mutual information *I* and that of modularity *Q*, a quality function introduced in Section 9.6.

$$
I(\mathcal{P}_{\mathcal{N}_1}, \mathcal{P}_{\mathcal{N}_2}) = \frac{-2 \sum_{m=1}^{M_1} \sum_{m'=1}^{M_2} N_{mm'} \log\left(\frac{N_{mm'} N}{N_m^1 N_{m'}^2}\right)}{\sum_{m=1}^{M_1} N_m^1 \log\left(\frac{N_m^1}{N}\right) + \sum_{m'=1}^{M_2} N_{m'}^2 \log\left(\frac{N_{m'}^2}{N}\right)} \tag{9.8}
$$

where $N_{mm'}$ is the number of nodes in common between set C_m^1 of partition $\mathcal{P}_{\mathcal{N}_1}$ and set $C_{m'}^2$ of partition $\mathcal{P}_{\mathcal{N}_2}$, while N_m^1 and $N_{m'}^2$ are respectively the number nodes in set C_m^1 and in set $C_{m'}^2$.

The quantity $I(\mathcal{P}_{\mathcal{N}_1}, \mathcal{P}_{\mathcal{N}_2})$ takes its maximum value equal to 1 when the two partitions $\mathcal{P}_{\mathcal{N}_1}$ and $\mathcal{P}_{\mathcal{N}_2}$ are identical, while it tends to zero when the two partitions are totally independent. In our case, we can take as $\mathcal{P}_{\mathcal{N}_1}$ the real division of the karate club consisting in the group of the club members following the instructor after the split, and in the group of those remaining. We can then take as $\mathcal{P}_{\mathcal{N}_2}$ any of the partition found by the Girvan–Newman algorithm. Figure 9.4 shows the normalised mutual information between the real division of the network and each partition found by the Girvan–Newman algorithm. In this way it is possible to follow and compare the partitions obtained at different levels of the hierarchical diagram. As expected, the partition into two communities found by the

Girvan–Newman algorithm and indicated by the black vertical dashed line in the figure, corresponds indeed to the highest value of mutual information, namely $I = 0.836$.

Node roles. Once the best partition of the network has been identified, we can ask whether it is somehow possible to quantify the role played by each node in the overall network organisation into communities. For instance, we can measure the contribution of a given node to the community it belongs to by counting the number of links from the node to other nodes within the same community. Suppose that the best partition of the graph corresponds to a division of the nodes into the M disjoint sets $C_1, C_2, \ldots, C_M$, which respectively induce subgraphs $g_1, g_2, \ldots, g_M$. Consider now a generic node i. We indicate as k_i its degree, and as k_{i,C_m} the number of links connecting node i to nodes in the set C_m, with $m = 1, \ldots, M$. In particular, we denote respectively as g^* and C^* the subgraph and the set of nodes to which i belongs. Therefore, k_{i,C^*} is the so-called *within-module degree* of node i, that is the number of links connecting node i to other nodes in the same community C^*. Since the number k_{i,C^*} depends on the number of nodes in C^* and on the number of links in g^*, we can get a measure of the contribution of i to the intra-modular connectivity by comparing k_{i,C^*} to the average within-community degree of the nodes in C^* [146, 147].

Definition 9.6 (Normalised within-module degree)   *The* normalised within-module degree z_i *of a node i belonging to community C^* is defined as:*

$$z_i = \frac{k_{i,C^*} - \langle k_{i,C^*} \rangle}{\sigma_{k_{i,C^*}}} \tag{9.9}$$

where k_{i,C^*} is the number of links connecting node i to other nodes in community C^*, while $\langle k_{i,C^*} \rangle$ and $\sigma_{k_{i,C^*}}$ are respectively average and standard deviation of k_{i,C^*} over all nodes in community C^*.

In practice, the quantity z_i is a *Z-score*, since it measures how many standard deviations the within-module degree of node i deviates from the mean. With such a definition, z_i will be large and positive (large and negative) for nodes that are particularly tightly (loosely) connected to the other nodes in the same community.

Another feature that is important to capture is how the $k_i - k_{i,C^*}$ external links of node i are distributed to the other modules. Together with the normalised within-module degree, we can therefore characterise the inter-modular connectivity of a node i by evaluating the following quantity.

Definition 9.7 (Participation coefficient)   *The* participation coefficient p_i *of node i is defined as:*

$$p_i = 1 - \sum_{m=1}^{M} \left(\frac{k_{i,C_m}}{k_i} \right)^2 \tag{9.10}$$

where k_i is the total degree of node i, and k_{i,C_m} is the number of links connecting node i to nodes in the C_m module g_m.

The participation coefficient p_i is equal to zero if node i is linked exclusively to other nodes in its own module. On the other hand, the value of p_i will tend to one if the links of i are homogeneously distributed to all communities. The two-dimensional space $\{p, z\}$ defined by the normalised within-module degree and the participation coefficient, can be used to assign roles to the nodes of the network. We will adopt the classification proposed by Roger Guimerá and Luís A. Nunes Amaral in Refs. [146, 147]. The hubness of a node is defined by its within-module degree: if a given node i has a large value of z_i, usually one assumes $z_i > 2.5$, it is classified as a *hub*, otherwise as a *non-hub*. Then, both hubs and non-hubs are more finely classified by using the values of the participation coefficient. In particular, for non-hubs we can define four different classes: if a non-hub i has value $0 < p_i < 0.05$, the node i is classified as an *ultra-peripheral node* (a node with all its links within its module); $0.05 < p_i < 0.62$ corresponds to a *peripheral node* (a node with most of its links within its module); $0.62 < p_i < 0.80$ corresponds to a *connector node* (a node with many links to other modules); finally $0.80 < p_i < 1.0$ is a *kinless node* (a node whose links are homogeneously distributed among all modules). For hubs, $0 < p_i < 0.30$ corresponds to a *provincial hub* (a hub with most of its links within its module); $0.30 < p_i < 0.75$ corresponds to a *connector hub* (a hub with many links to most of the other modules); and $0.75 < p_i < 1.0$ is a *kinless hub* (a hub with links homogeneously distributed among all modules). These different categories allow the classification of the nodes according to their topological function in the community structure of the network. For example, a provincial hub is a hub with intra-modular connectivity larger than inter-modular connectivity, thus having a pivotal role for its community, whereas a connector hub plays a central role in transferring information from its module to the rest of the network.

Example 9.10 *(The cartography of the karate club)*   We have computed the participation coefficient p_i and the within module degree z_i for each node $i = 1, \ldots, 34$ of the karate club network. As a reference partition we consider the division into the two communities found by the Girvan–Newman algorithm and corresponding to the first vertical dashed line in Figure 9.4). We show in the figure the scatter plot obtained by reporting for each node a point in a $\{p, z\}$ plane. Node 34, i.e. the administrator of the club, is the node with the largest value of z_i, immediately followed by node 1, the karate instructor. These two nodes are the hubs of the graph, but they are provincial hubs

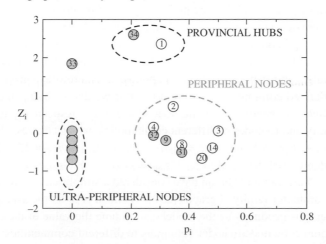

because they have most of the their links within their own community ($0 < p_i < 0.30$), with node 1 being the one of the two with the largest value of p_i, denoting a slightly higher tendency to have links also to the other community. There is a third node, namely node 33, which has also a quite large value of z_i, although not in the hubs' range. Such a node has almost the same number of links as node 34, and all its links are within its module ($p_i = 0$). All the remaining nodes are definitely non-hub nodes. Nine of them, namely nodes 2, 3, 4, 8, 9, 14, 20, 31, 32, are peripheral nodes (with $0.05 < p_i < 0.62$), while the others have $p_i = 0$ and are ultra-peripheral nodes.

9.5 Computer Generated Benchmarks

Finding a real network such as that of Zachary's karate club, where the true division in communities is a priori known from empirical observations, is an uncommon and very lucky situation. In order to test how well a community detection method performs, it is thus of fundamental importance to be able to generate synthetic networks with a given community structure. These are usually known as *community structure benchmarks*. The simplest and most famous computer generated benchmark is the so-called *planted partition model* [85].

Definition 9.8 (The planted partition model)   *Let N and M be two positive integers, with N multiple of M, and $0 \leq p_{in} \leq 1$, $0 \leq p_{out} \leq 1$ be two tunable parameters, such that $p_{out} < p_{in}$. The model constructs graphs with N nodes and M communities by arranging the N nodes in M sets of equal size $n = N/M$, and connecting by an edge two nodes in the same set with probability p_{in}, and two nodes in different sets with probability p_{out}.*

Since the links are placed independently at random between vertex pairs, with probability p_{in} for an edge to fall between vertices in the same set, and probability p_{out} to fall between vertices in different set, the subgraphs induced by each set of nodes are ER random graphs with linking probability p_{in}. Furthermore, given the constraint $p_{in} > p_{out}$, the intra-set edge density exceeds the inter-set density and the graphs produced have a well-defined community structure. By tuning the values of the two control parameters of the model, p_{in} and p_{out}, it is possible to control separately the number of links within communities and those between communities, and therefore to produce graphs with a more or less defined community structure. In fact, if we denote by z_{in} and z_{out} the expected number of links connecting a generic node of the graph respectively to nodes of the same community and to nodes of different communities, we can express these quantities in terms of the model probabilities: $z_{in} = p_{in}(n - 1)$, and $z_{out} = p_{out}n(M - 1)$. Finally, the expected node degree will be equal to $z_{in} + z_{out} = p_{in}(n - 1) + p_{out}n(M - 1)$.

Newman and Girvan have considered a simplified benchmark model that produces one-parameter family of graphs [237]. The idea is to fix the average node degree $\langle k \rangle$ for all the graphs produced by the model, and to tune the value of the expected number of links that connect a node to nodes belonging to different communities.

Definition 9.9 (The Newman–Girvan benchmark)   *Let N and M be two positive integers, with N multiple of M, and z a non-negative real number. Let z_{out} be a tunable real parameter such that $0 \leq z_{out} \leq z$. The model considers graphs with N nodes and M communities of equal size $n = N/M$, where all nodes have expected degree equal to z, and an expected number z_{out} of links connecting to nodes of different communities.*

In practice, this benchmark is a simplified version of the planted partition model, and can be obtained by opportunely choosing the values of p_{in} and p_{out}. In fact, if we fix the value of the average node degree equal to z, the relation $z = p_{in}(n-1) + p_{out}n(M-1)$ makes one of the two parameters p_{in} and p_{out} of the planted partition model depending on the other. This means that we can fix the number of nodes N, the number of communities M, and the value of z, and we can produce a family graphs with a more or less defined community structure by varying only one parameter of the planted partition model, namely p_{out}, and choosing the other accordingly: $p_{in} = 1/(n-1)[z - p_{out}n(M-1)]$. All the graphs have, on average, the same total number of links, and a variable number of links, $z_{out}N/2$, between different communities, so that with the Newman–Girvan benchmark we can test in a controlled way up to which vale of z_{out} a given algorithm is able to recognise and extract the communities.

To be more explicit, let us test the GN algorithms on Newman–Girvan benchmark graphs with $N = 128$ vertices, average degree $\langle k \rangle = z = 16$, and $M = 4$ communities of $n = 32$ vertices each. Since the average node degree is set to 16, we can tune p_{out} from 0 to 1 producing graphs with a tunable value of z_{out}, and with $z_{in} = 16 - z_{out}$. As shown in Figure 9.5, when z_{out} is very small there is a clear community structure as most edges will join vertices of the same community, whereas when $z_{out} = 5$ and $z_{out} = 8$ the division in community starts to look less defined. In this way, we can check how the GN algorithms works in recognising correctly the communities under different degrees of mixing between the communities, up to a certain value of z_{out} at which the division in communities disappears. This value is simple to calculate. In fact, the condition $p_{in} = p_{out}$ at which the communities disappear leads to:

$$z_{out} = \frac{n(M-1)}{n(M-1) + n - 1}\langle k \rangle.$$

In the case in which the number of nodes in each community is large we can approximate $n - 1 \approx n$ and the expression above reduces to $z_{out} \simeq \langle k \rangle(M-1)/M$. For our graph

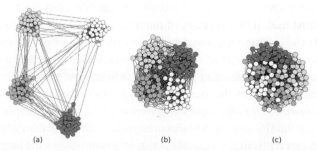

(a)   (b)   (c)

Fig. 9.5   A network with $N = 128$ nodes, $z = 16$ and a tunable division into $M = 4$ communities. The three pictures correspond respectively to $z_{out} = 1$ (a), $z_{out} = 5$ (b) and $z_{out} = 8$ (c).

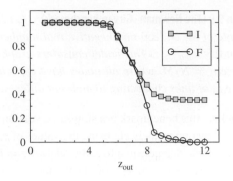

Fig. 9.6 Normalised mutual information *I* and fraction *F* of correctly classified nodes for the GN algorithm in Newman and Girvan benchmark graphs with $N = 128$ nodes, $\langle k \rangle = 16$, four communities of $n = 32$ nodes each, and a variable value of z_{out}. The value $z_{out} = 8$ is the point at which vertices have as many connections outside their own community as inside it, while $z_{out} = 12$ is where the communities disappear. Each point of the curve has been obtained as an average over 100 graph realisations.

with $M = 4$ communities and $\langle k \rangle = 16$, we get $z_{out} \simeq 12$, so that, in principle, the four communities are visible up to a value of z_{out} approximately equal to 12. However, even if the communities are there, it can be very hard if not impossible to detect them already before this limit since the benchmark graphs may look indistinguishable from random graphs because of the fluctuations. In Figure 9.6 we report the results of a test of the GN algorithm on this benchmark. We have adopted the following procedure. For each value of z_{out} we have generated an ensemble of 100 graphs. For each of these graphs we have measured the similarity between the results produced by the GN algorithm and the correct partition in the four equal-sized sets used to construct the benchmark. Since the GN algorithm produces a dendrogram, i.e. an entire collection of divisions of the graph, we have decided to store the highest value of similarity between each partition of the dendrogram and the natural partition of the graph. We have then repeated the procedure for each of the 100 graphs of the ensemble, and what we plot in the figure is the average value of (the highest) similarity as a function of z_{out}.

We have considered two different measures of similarity. The first measure is the mutual information *I* we have defined in Eq. (9.8). Since we know exactly the real communities in the benchmark model, we can define another measure of similarity as follows. We say that a vertex is correctly classified by the GN method if it is placed in the same set with at least half, in this case 16, of the nodes in the same built-in group in the benchmark [119]. If the partition has one set given by the union of two or more natural groups, all vertices in this set are considered incorrectly classified. In order to have a final number between 0 and 1, as for the case of the mutual information, the number of correctly classified nodes is then divided by the total number of nodes in the graph. We call this number the *fraction of correctly classified nodes* and we indicate it by *F*. The results shown in the figure indicate that the GN algorithm works pretty well, with more than 90 per cent of all vertices classified correctly from $z_{out} = 0$ all the way to around $z_{out} = 6$. For $z_{out} > 6$ the classification begins to deteriorate rapidly, with the algorithm correctly classifying the community structure in the network almost only up to the point $z_{out} = 8$ at which each vertex has on average the same number of connections to vertices outside its community as it does to those inside.

In the previous two benchmark models, all the communities have exactly the same number of nodes by construction. Moreover all the nodes have approximately the same degree. This is in contrast with what usually observed in real systems. For this reason, Andrea Lancichinetti, Santo Fortunato and Filippo Radicchi have proposed a benchmark model that takes into account the heterogeneity of node degrees and community sizes [197]. Namely, the model considers graphs with a power law degree distribution with exponent γ, and where the size of communities is distributed as a power law with an exponent τ.

Definition 9.10 (The Lancichinetti–Fortunato–Radicchi (LFR) benchmark)   *Let γ and τ be two positive real numbers (typical values are $2 \leq \gamma \leq 3$ and $1 \leq \tau \leq 2$). Let $0 \leq \mu \leq 1$ be a tunable parameter. The model considers graphs with M communities where the number of nodes n in each community is distributed as a power law $p_n \sim n^{-\tau}$ with exponent τ, while the degree of each node is distributed as a power law $p_k \sim k^{-\gamma}$ with exponent γ. Each node has a fraction μ of its k edges connected to nodes of other communities, and a fraction $1 - \mu$ connected to other nodes in the same community.*

In practice, the networks can be generated by an appropriate modification of the configuration model of Chapter 5. First of all, the order of the M communities is generated by sampling M random numbers $n_1, n_2, \ldots, n_M$ from a power law distribution with exponent τ. This also fixes the number of nodes in the network $N = \sum_{m=1}^{M} n_g$. Then, the degree sequence of the graph is generated by sampling N natural numbers $k_i, i = 1, \ldots, N$ from a degree distribution as in Eq. (5.10). Each nodes is then assigned a number of stubs equal to its degree. Of the k_i stubs of node i, a fraction $(1 - \mu)k_i$ are considered internal stubs, while the remaining μk_i are external. Then, for each set of nodes, all the internal stubs are matched in pairs at random with uniform probability. Finally, all the external stubs of each node i are matched to nodes of different communities, until all the edges of the graph are formed. This benchmark again allows a one-parameter family of networks with tunable community structure to be constructed. The control parameter is in this case the so-called mixing parameter μ. An example of the graphs produced by the LFR benchmark model is reported in Figure 9.7 which shows the case of a network with 500 nodes, power law degree distribution with exponents $\gamma = 2$, power law distribution of the community size with exponent $\tau = 1$, and a mixing parameter $\mu = 0.1$.

9.6 The Modularity

In the case of Zachary's karate club or in the computer-generated benchmarks of the previous section we have a priori knowledge of the true division in communities. Is it possible to derive a measure of the quality of a partition even without such a priori information? This is an interesting question with important practical outcomes. In fact, as shown in the dendrogram of Figure 9.4, the GN algorithm produces a large number of different

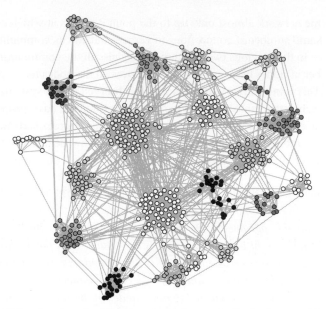

Fig. 9.7 A realisation of the LFR benchmark graphs with 500 vertices and $\mu = 0.1$. The distributions of the node degree and that of the community size are both power laws, respectively with exponents $\gamma = 2$ and $\tau = 1$.

possible partitions to choose from and, in principle, all of them would be equivalent if we did not know that the real Zacharys karate club finally split into two communities after the observation period. Finding a *quality function*, i.e. a quantitative criterion to evaluate how good a partition is, would also provide an operative way to better choose among the various partitions in the dendrogram. We will concentrate here on how to derive a proper quality function.

Given a graph G and a partition $\mathcal{P}_\mathcal{N} = \{C_1, C_2, \ldots, C_M\}$ of its nodes into M sets, in order to describe how well the nodes from one set are connected to the nodes of another set, let us construct a $M \times M$ symmetric matrix $E \equiv \{e_{mm'}\}$ as follows. For $m \neq m'$, the entry $e_{mm'}$ of E is set to be equal to one-half of the number of graph edges that connect vertices in set C_m to those in set $C_{m'}$, so that the total number of such edges is $e_{mm'} + e_{m'm} = 2e_{mm'}$ because we are considering undirected graphs. If $m = m'$, e_{mm} is set to be equal to the number of edges joining two nodes in set C_m, i.e. the number of links internal to set C_m. With such a definition of E, the quantities $e_{mm'}$ obey the normalisation condition:

$$\sum_{m=1}^{M} \sum_{m'=1}^{M} e_{mm'} = K \tag{9.11}$$

where K is the total number of links in the graph G. We can therefore define, for each values of m and m', the quantities:

$$p_{mm'} = \frac{e_{mm'}}{K} \tag{9.12}$$

which can be interpreted as probabilities. In fact $p_{mm'} + p_{m'm} = 2p_{mm'}$ is equal to the probability that a randomly chosen link connects nodes in set C_m with nodes in set $C_{m'}$, while p_{mm} is the probability that a randomly chosen link connects nodes in the same set C_m. The normalisation $\sum_m \sum_{m'} p_{mm'} = 1$ holds, while the summation $\sum_{m'} p_{mm'}$, with $m = 1, 2, \ldots, M$, is equal to one-half of the sum of the degrees of all the nodes in set m. The quantities $e_{mm'}$ and $p_{mm'}$ we have just introduced are similar to the quantities $e_{kk'}$ and $p_{kk'}$, respectively representing the number of edges between nodes of degree k and nodes of degree k', and the probability that a randomly chosen link goes from a node of degree k to a node of degree k', defined in Section 7.2 when we treated graphs with degree–degree correlations. Here, however, are the sets of the partition $\mathcal{P}_N$ to define node classes, and not the degrees of the nodes.

Example 9.11   Consider the graph with $N = 6$ nodes and $K = 6$ links shown in the figure, and the partition $\mathcal{P}_N = \{C_1, C_2\}$ in the two sets indicated. This graph has somehow an organisation into two communities since there is only one link connecting a node in C_1 to a node in C_2.

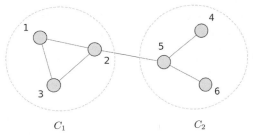

$$C_1 \qquad\qquad\qquad C_2$$

Matrix E can be easily constructed. We get:

$$E = \begin{pmatrix} 3 & 1/2 \\ 1/2 & 2 \end{pmatrix}.$$

Notice that the sum $\sum_{m'=1}^{2} e_{1m'}$ of the first row of E is equal to one-half of the sum of the degrees of the nodes in the first set, namely 7/2, while the sum $\sum_{m'=1}^{2} e_{2m'}$ of the second row of E is equal to 5/2, which is one-half of the sum of the degrees of the nodes in the second set. Also we have $\sum_{m=1}^{2} \sum_{m'=1}^{2} e_{mm'} = 6$, i.e. the sum of all the elements of E is equal to the number of links K in the graph, in agreement with the normalisation in Eq. (9.11). We can then normalise the entries of matrix E to construct the probabilities $\{p_{mm'}\}$ as in Eq. (9.12), getting the following matrix:

$$\begin{pmatrix} 3/6 & 1/12 \\ 1/12 & 2/6 \end{pmatrix} = \begin{pmatrix} 6/12 & 1/12 \\ 1/12 & 4/12 \end{pmatrix}.$$

We now have: $\sum_{m=1}^{2} \sum_{m'=1}^{2} p_{mm'} = 1$, $a_1 = \sum_{m'=1}^{2} p_{1m'} = 7/12$, $a_2 = \sum_{m'=1}^{2} p_{2m'} = 5/12$. The quantities a_1 and a_2 are useful, as we shall see in the following, to define the modularity Q_P in Eq. (9.16), and can be seen as the fraction of all edge ends that are attached to nodes in set C_1 and C_2 respectively. In our graph we have a total of $2K = 12$ edge ends, and 7 of them are attached to nodes in C_1, while the remaining five are attached to nodes in C_2.

Clearly, matrix E depends both on the graph structure and on the partition chosen. The total fraction $F_{\mathcal{P}}$ of edges that connects nodes in the same sets is:

$$F_{\mathcal{P}} = \sum_{m=1}^{M} p_{mm}. \tag{9.13}$$

By definition, the maximum possible value this sum can take is 1. This would happen in the ideal case in which each set of the partition is a graph component, so that there are no links between nodes in two different sets. We could therefore think to use this sum as a measure of community structure, and say that the partition $\mathcal{P}_{\mathcal{N}}$ is a good division of the network into communities if the corresponding $F_{\mathcal{P}}$ is as close as possible to 1. However, this is not a good strategy in general. It may work if we have to find the best partition into a fixed number of modules, for instance into two modules, as seen in Section 9.2, but certainly $\mathcal{P}_{\mathcal{N}}$ is not a good quality function when we want to consider and compare partitions with a different number of sets. For instance, we would find that the best partition is that where all the nodes of the graphs are placed into a single set. In fact, for such a partition the sum above would take its maximal value of 1.

A better quality function can thus be obtained by considering, for each set in the partition, the difference between the actual fraction of internal links and the expected fraction of internal links in an appropriate *null model*:

$$Q_{\mathcal{P}} = F_{\mathcal{P}} - F_{\mathcal{P}}^{\text{exp}} = \sum_{m=1}^{M} \left(p_{mm} - p_{mm}^{\text{null}} \right). \tag{9.14}$$

The idea behind this quality function is that a partition of a graph into any number of sets is a good partition when the actual number of edges between nodes in the same set is smaller than the number expected purely by chance. To evaluate the expected fraction of internal links, $F_{\mathcal{P}}^{\text{exp}}$, we need to consider the case in which graph links are placed between nodes independently from the sets the nodes belong to. We can find an explicit expression for the term p_{mm}^{null}. We have to take into account that some of the sets will have a larger number of links simply because they contain a larger number of nodes. Therefore, let us first calculate the fraction a_m of all edge ends that are attached to vertices in set C_m in the original graph G. This fraction can be expressed in terms of the quantities $p_{mm'}$ as:

$$a_m = \sum_{m'=1}^{M} p_{mm'}. \tag{9.15}$$

In other words, the probability a_m is the marginal probability associated with the probability $p_{mm'}$, as in Section 7.2 q_k was the marginal probability associated with the joint probability $p_{kk'}$. Now, in a randomised network with no community structure, the product $a_m a_{m'}$ with $m \neq m'$ will be equal to half of the expected fraction of links between nodes of set C_m and nodes of set $C_{m'}$. If instead $m = m'$, the product a_m^2 will be the fraction of links between pairs of nodes both in set C_m. Therefore, in the null model the fraction of edges that connect vertices within sets will be equal to $F_{\mathcal{P}}^{\text{exp}} = \sum_m a_m^2$. We finally have the following quality function:

$$Q_{\mathcal{P}} = \sum_{m=1}^{M} \left(p_{mm} - a_m^2 \right). \tag{9.16}$$

This quantity was proposed for the first time in Ref. [237] by the same two authors who developed the GN algorithm and the benchmark model given in Definition (9.9), and is known today as the *modularity* of a graph partition. Expression (9.16) can be rewritten in a more convenient way. Notice in fact that we can write $p_{mm} = K_{mm}/K$, where K_{mm} is equal to the quantity e_{mm} we have previously defined, and denotes the number of links joining nodes in set C_m. Also, we can easily find a more compact expression for the fraction a_m of edges attached to vertices in set C_m. In fact we can write $a_m = K_m/2K$, where by K_m we denoted the sum of the degrees of all the nodes in set C_m. We can finally give the following definition.

Definition 9.11 (Modularity) *Given a graph G and a partition $\mathcal{P}_{\mathcal{N}} = \{C_1, C_2, \ldots, C_M\}$ of its nodes into M sets, the* modularity $Q_{\mathcal{P}}$ *of partition $\mathcal{P}_{\mathcal{N}}$ can be written as:*

$$Q_{\mathcal{P}} = \sum_{m=1}^{M} \left[\frac{K_{mm}}{K} - \left(\frac{K_m}{2K} \right)^2 \right] \tag{9.17}$$

where K_m is the degree of set C_m, defined as the sum of the degrees of all the nodes in set C_m, K_{mm} is the total number of links joining nodes in set C_m, and $2K$ is the total number of links in G.

This expression of the modularity of a graph partition is in all equivalent to Eq. (9.16). In fact, the first term of each summand in Eq. (9.17) is the fraction of edges of the graph inside set C_m, whereas the second term is the expected fraction of edges inside set C_m in a random graph with the same degree for each set of the partition as in the original graph. In such a case, a node can equally link any other node of the graph, and the probability of a link between two sets C_m and $C_{m'}$ is proportional to the product of the degree K_m and $K_{m'}$ of the two sets. Thus, in a random graph, the probability of a link between nodes in the same set C_m is proportional to K_m^2. Nodes in set C_m of the partition form a real community of the graph if the number of edges inside the set is larger than the expected number in the null model. If this is the case, then the nodes of C_m are more tightly connected than expected. Basically, if each summand in Eq. (9.17) is non-negative, the corresponding set is a real community. Besides, the larger the difference between real and expected edges, the more modular the set C_m. Finally, large positive values of $Q_{\mathcal{P}}$ are expected to indicate a good partition where all the sets of nodes are good communities.

Example 9.12 *(Range of variability of the modularity)* The modularity is by definition always smaller than one, and can assume negative values as well. Consider the simplest partition we can imagine, consisting in a single set formed by all the nodes in the graph. As expected, the modularity of such a partition is zero, because the two terms of the only summand in Eq. (9.17) are in this case equal to 1 and opposite. More in general, a value $Q_{\mathcal{P}}$ close to 0 indicates that the partition $\mathcal{P}$ is not meaningful, that is the edges of the graph are placed independently from the fact the nodes belong to one or another of the sets in partition $\mathcal{P}$. Conversely, values of $Q_{\mathcal{P}}$ different from zero represent deviations from randomness. In particular, if the values of $Q_{\mathcal{P}}$ are large and positive, then the sets in partition $\mathcal{P}$ represent meaningful communities of the network. Let us consider for instance the graph with $N = 6$ nodes and $K = 6$ links in Example 9.11. Does such graph have a community structure? In particular, is the partition $\mathcal{P}_{\mathcal{N}} = \{C_1, C_2\}$ shown in the figure of Example 9.11 a good division in communities? We can answer such questions by evaluating the modularity of the partition. Using Eq. (9.17) we get:

$$Q_{\mathcal{P}} = \left[\frac{6}{12} - \left(\frac{7}{12} \right)^2 \right] + \left[\frac{4}{12} - \left(\frac{5}{12} \right)^2 \right] = \frac{46}{144} \approx 0.32$$

which confirms indeed that the division of the graph into the two communities is meaningful.

Notice that the modularity can also be negative. For instance, in any graph, the partition in which each vertex is a set has a negative modularity, since each term in the summation (9.17) is either zero or negative. The fact that the modularity can assume negative values is interesting. In fact, the existence of partitions with large negative modularity values indicates the existence of sets of nodes with very few internal edges and many edges lying between them. For instance, in a bipartite graph (see Definition 1.6), the partition into two sets corresponding to the two types of nodes is the one with the largest negative modularity.

The modularity can be written alternatively as a sum over node pairs, instead of a sum over sets. This requires us only to count the links that are in the same set as in the following expression:

$$Q_{\mathcal{P}} = \frac{1}{2K} \sum_{i=1}^{N} \sum_{j=1}^{N} \left(a_{ij} - e_{ij}^{\text{null}} \right) \delta(C^{(i)}, C^{(j)}). \tag{9.18}$$

Here the sum runs over all pairs of vertices, including the cases in which index i is equal to index j. The quantity a_{ij} is the (i, j) entry of the adjacency matrix of the graph, e_{ij}^{null} is the expected number of links between i and j in an appropriate null model, and we denote by $C^{(i)}$ the set containing node i. The Kronecker δ yields the value 1 if vertices i and j are in the same set, and zero otherwise. Hence, the only contributions to the sum come from node pairs belonging to the same set. As *null model* we could take ER random graphs with the same number of nodes and links as the original graph, which implies setting $e_{ij}^{\text{null}} = p \ \forall i, j$ with a linking probability p equal to $2K/N(N - 1)$. However, a much better choice is to consider randomised graphs with the same degree sequence as the original graph. This is equivalent to set e_{ij}^{null} equal to the expression of $\overline{e}_{ij}^{\text{conf}}$ in the configuration model. Making use of Eq. (5.31) derived in Chapter 5, and assuming that the graph is large enough so that $(2K - 1) \simeq 2K$, we obtain the following useful expression of the modularity as a sum over node pairs:

$$Q_{\mathcal{P}} = \frac{1}{2K} \sum_{i=1}^{N} \sum_{j=1}^{N} \left(a_{ij} - \frac{k_i k_j}{2K} \right) \delta(C^{(i)}, C^{(j)}). \tag{9.19}$$

The nice thing of writing the modularity as in Eq. (9.18) is that, being such an expression defined at the level of the nodes rather than at the level of group of nodes as Eq. (9.14), it allows more freedom in the selection of the most appropriate null model to adopt. We can therefore derive expressions of modularity different from that in Eq. (9.19) just by changing the null model in Eq. (9.18). For instance, we can decide to use more information from the system we are studying, and consider, as a null model, graphs with the same degree–degree correlations of the original network. In such a case, the modularity would be obtained by plugging $e_{ij}^{\text{null}} = k_i p_{k_j|k_i}$ into Eq. (9.18), where $p_{k_j|k_i}$ is the conditional degree distribution constructed from the original graph. However, in this book we will focus on the definitions of modularity $Q_{\mathcal{P}}$ given in Eq. (9.17) and in Eq. (9.19).

The program `modularity`, which can be downloaded from the book website `www.complex-networks.net`, takes as input a graph G and a partition $\mathcal{P}_{\mathcal{N}}$, and computes the corresponding value of the modularity $Q_{\mathcal{P}}$ from Eq. (9.17). Having a function, such as $Q_{\mathcal{P}}$ to quantify the quality of a partition can be very useful. For instance, we can now come back to the GN algorithm and use the modularity as a quality function to select the best partition found by the algorithm. As shown in Figure 9.4 left, the modularity Q exhibits two local maxima in the dendrogram obtained for the Zachary karate club. Notice that the first local maximum, $Q = 0.360$, corresponds exactly to the split into two communities indicated by the black vertical dashed line that we have already discussed in Section 9.4. However, it is the second local maximum, $Q = 0.401$, that is higher than the first one. This value of modularity corresponds to the partition into five communities denoted by the grey vertical dashed line in Figure 9.4 left. Such partition is reported with five different tones of grey (from white to black) in Figure 9.8, and consists of four large groups, respectively with 6, 12, 10 and 5 nodes, and one group with a single vertex, namely vertex number 10. In practice, the algorithm breaks into two or more pieces the two true factions (indicated as usual by circles and squares) into which the club actually split during the course of the study.

Given that higher values of modularity correspond to better community divisions, we can attempt to avoid the expensive recalculation of betweenness required by the GN after the removal of each edge, and try instead to find the best partition by directly maximising Q. Since an exhaustive exploration of all the partitions is impossible in large graphs, as we have seen in Section 9.3, we have to rely on approximate optimisation methods. A particularly simple and efficient method has been proposed by Mark Newman and is based on a standard "greedy" optimisation algorithm [226]. The algorithm starts with a partition with N sets, each containing one of the N nodes of the graph. Then, for each pair of sets $C_{m'}$ and $C_{m''}$ connected by at least one edge, the change in modularity ΔQ obtained by merging the two sets is evaluated. This is simply done by using the formula:

$$\Delta Q = p_{m'm''} + p_{m''m'} - 2a_{m'}a_{m''} \tag{9.20}$$

illustrated in the following example.

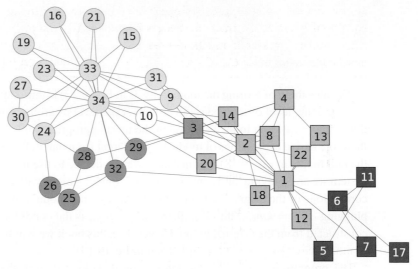

Fig. 9.8 The five different tones of grey show the partition of the Zachary's karate club corresponding to the highest value of modularity, $Q = 0.401$, found in Figure 9.4 by the GN algorithm.

Example 9.13   Let us evaluate the change in modularity obtained by merging the two sets $C_{m'}$ and $C_{m''}$ into a single set $C_{m'''} = C_{m'} \cup C_{m''}$. Such change can be written as:

$$\Delta Q = Q^{\text{after}} - Q^{\text{before}}$$

where Q^{before} and Q^{after} are respectively the modularity before and after the merging. By isolating the contributions to the modularity in Eq. (9.16) due to the two sets involved in the merging we can write:

$$Q^{\text{before}} = Q_{m'} + Q_{m''} + \sum_{\substack{m \neq m' \\ m \neq m''}} Q_m$$

$$Q^{\text{after}} = Q_{m'''} + \sum_{\substack{m \neq m' \\ m \neq m''}} Q_m$$

where we have defined Q_m, the contribution to the summation due to set C_m, as $Q_m = p_{mm} - a_m^2$. Therefore we have:

$$\Delta Q = Q_{m'''} - Q_{m'} - Q_{m''}$$

where:

$$Q_{m'} = p_{m'm'} - a_{m'}^2$$
$$Q_{m''} = p_{m''m''} - a_{m''}^2$$
$$Q_{m'''} = p_{m'm'} + p_{m''m''} + p_{m'm''} + p_{m''m'} - (a_{m'} + a_{m''})^2.$$

Notice that the fraction $p_{m'''m'''}$ of edges internal to set $C_{m'''}$ in the expression of $Q_{m'''}$ is equal to the sum of the fraction of edges internal to $C_{m'}$ and to $C_{m''}$, plus the fraction of edges connecting a node in $C_{m'}$ to a node in $C_{m''}$. Similarly, the fraction of edges $a_{m'''}$ attached to nodes in set $C_{m'''}$ is equal to $a_{m'} + a_{m''}$. We finally get Eq. (9.20).

The two sets $C_{m'}$ and $C_{m''}$ producing the highest value of ΔQ are merged together, and the matrix $\{p_{mm'}\}$ is updated accordingly. This means that the row (and column) corresponding to $C_{m'}$ is replaced by the sum of the rows (columns) corresponding to $C_{m'}$ and $C_{m''}$, and then we delete the row (and column) associated with $C_{m''}$. The procedure is then repeated with the new matrix $\{p_{mm'}\}$, until all the nodes have been merged in a single set. In this way we obtain a dendrogram, and of the produced partitions we choose the one corresponding to the maximum value of modularity. Since there can be at most K pairs of sets connected by at least one edge, where K is the number of links in the graph, at each iteration we need to calculate the quantity ΔQ for at most K different mergings. Notice that mergings of two sets not connected by edges cannot produce an increase of the modularity, so that such mergings do not need to be checked. Additionally, the merging of two sets requires at most $2N$ operations. Thus the total number of steps required by the algorithm scales in the worst case as $(K+N \times N)$. Even if this algorithm is pretty simple to explain, an efficient implementation requires a dedicated data structure and clever procedures to update the matrix E and to evaluate the maximum of ΔQ. Such efficient implementation was proposed in Ref. [79] and runs essentially linearly in the number of nodes in a sparse graph. All the details on the implementation of this algorithm can be found in Appendix A.18.

When used on the Zachary's karate club network, the algorithm gives a partition of the network into three communities of respectively 17, 9 and 8 nodes each, corresponding to a value of modularity $Q = 0.381$. Notice that even this is not the highest possible value of modularity that can be found. Today it is in fact well accepted that the best partition in terms of modularity of Zachary's network is achieved by a division into four groups. Such a partition corresponds to a value $Q = 0.419$ and is shown in Figure 9.9.

9.7 A Local Method

We discuss here a *local* method to detect communities based on the usual assumption that nodes belonging to a community have more links to other nodes in the same community than to nodes in the rest of the graph. The method, known as *label-propagation* and proposed by Usha Raghavan, Réka Albert and Soundar Kumara is simple, time-efficient and does not optimise any specific measure [266]. In addition to this, the method is easily parallelisable, and can be therefore used for large graphs. The algorithm works as follows. We assume that each node carries a label denoting the set of nodes to which it belongs to. Initially, we start with N different labels, one for each of the nodes in the graph. At each iteration (epoch), each node i adopts the label of the set to which the maximum number of its neighbours belong. If two or more labels are present in the neighbourhood of i with

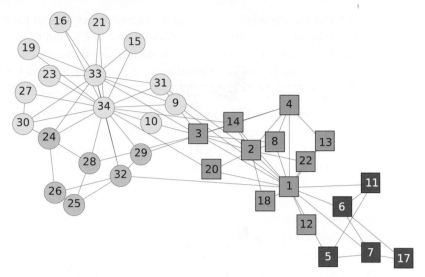

Fig. 9.9 The four different tones of grey indicate the best known partition of Zachary's karate club in terms of modularity, corresponding to a value $Q = 0.419$.

the same number of occurrences, i will adopt one of them with equal probability. As the method is iterated, the labels propagate throughout the network, with densely connected groups of nodes quickly reaching consensus on a unique label. Such groups continue to expand outwards until it is possible to do so. Communities will be then defined as sets of nodes sharing the same label. There are two possible label update strategies. The first one consists in computing the new labels for all the N nodes of the graph, and then updating all the labels at once. The second strategy considers a single node, computes its new label and updates it before proceeding to the next node. The first approach, the so-called synchronous update, can give rise to cases in which the algorithm indefinitely cycles through the same sequence of label configurations. We will therefore adopt an asynchronous update rule, where the order in which the N nodes are considered and updated is chosen at random at each epoch. The process ends when, for every node i having label C_m, the following condition holds:

$$k_i^{C_m} \geq k_i^{C_{m'}} \quad \forall m' \tag{9.21}$$

where $k_i^{C_m}$ is the number of neighbours of node i with label C_m. Notice that, even by using an asynchronous update, it is not possible to completely eliminate situations in which the algorithm indefinitely cycles through the same sequence of label configurations. We will discuss how to deal with this problem in Appendix A.19. Due to the random order in which nodes are considered in the asynchronous update, and the stop criterion adopted, the label propagation algorithm can produce different partitions at each run. For instance, in Figure 9.10 we report four different outcomes of the algorithm on the Zachary's karate club network. In two out of four cases (the two top panels in the figure) the label-propagation method finds four communities. It is not a surprise that, if we compute the modularity for such partitions, we find two large values, namely $Q = 0.416$ and $Q = 0.415$ respectively.

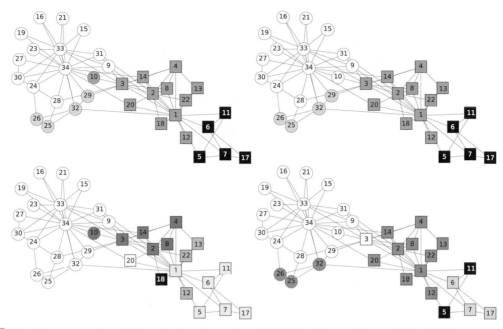

Fig. 9.10 The different tones of grey of the nodes show the partitions of the Zachary's karate club network found by the label-propagation method in four different runs of the algorithm.

The other two partitions given by the algorithm have respectively seven and four communities, and smaller values of modularity, namely $Q = 0.343$ and $Q = 0.330$. We notice that in all the four cases the algorithm finds two large communities, one around node 34 and one around node 1. Also, there are some critical nodes, such as nodes 3, 10 and 20, which are sometimes associated with the community of node 1 and sometimes with that of node 34. In one case, node 10 forms a community by itself.

The label-propagation method is extremely efficient. In fact, at each epoch the number of operations is a linear function of the number of links K, and in practice, the algorithm converges within a few epochs. See details in Appendix A.19. Therefore, the algorithm can be used to find communities in large graphs. In the following we consider some applications to find communities in other real networks of increasing order. The first network we consider is the neural system of the *C. elegans* we introduced in Section 4.2. The network has $N = 282$ nodes which are divided by the label-propagation algorithm into two different communities as shown in Figure 9.11 (a). Notice that the positions of the nodes in the figure correspond to the real positions of the neurons in the neural system of the *C. elegans*, even if the relative proportions of head, body and tail have been rescaled to better visualise the large number of neurons placed in the head. The two communities have roughly equal sizes, and correspond quite closely to the neurons in the head (empty circles) and to the neurons in the tail and in the ventral cord (full circles). This partition has a modularity $Q = 0.347$. In Figure 9.11(b) we report for comparison the results found by using the GN algorithm, which divides the network in three large communities, respectively with 139 (grey circles), 100 (empty circles) and 35 nodes (full circles), and some isolated nodes. In

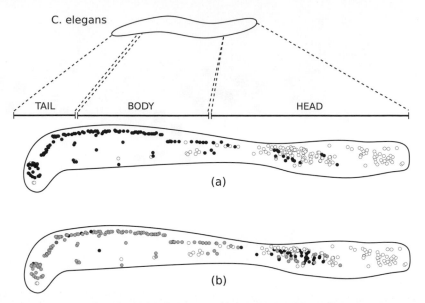

Fig. 9.11 The partition in communities found by the label-propagation method for the neural network of the *C. elegans* (a) is compared to that obtained by the GN algorithm (b). The placement of the nodes corresponds to their positions along the body of the nematode, but the proportions of the head, body and tail have been readjusted to facilitate visualisation.

this case, the nodes in the head are placed into two different communities, and the partition has a modularity $Q = 0.367$. However, while the label propagation algorithm runs in a few milliseconds on this graph, the GN algorithm requires more than half a minute on the same machine.

The second network we consider is one of the scientific coauthorship networks introduced in Section 3.5, namely that obtained from papers posted in the cond-mat directory of ArXiv. Such a network has $N = 16726$ nodes and a largest connected component of 13861 nodes. The best partition found by running the label-propagation algorithm 1000 times consists of 1177 communities and corresponds to a value of modularity $Q = 0.722$. The most striking result is the presence in the partition of communities of very different order. For instance, while the largest community contains about 1 per cent of the nodes, the smallest one has 3 nodes only. In Figure 9.12 we plot the order distribution $p(s)$, i.e. the probability of finding a community of order s, which is well fitted by a power-law $p(s) \sim s^{-\tau}$ with an exponent $\tau = 2.5$. Notice that, if we run on the same network the greedy modularity optimisation algorithm introduced in Section 9.6, we obtain a partition with only 168 communities and with a higher value of modularity, namely $Q = 0.771$. This means that the communities found by greedy modularity optimisation contain in general a larger number of nodes than those discovered through label-propagation. In particular, the three largest communities found have order 2053, 1846 and 1751, and contain about 40 per cent of the nodes of the graph, which is very different from what was found by the label-propagation method. The presence of discrepancies between the results obtained by running different algorithms of community detection on the same network is the norm rather than an exception. As highlighted in Box 9.3 the key question is which partition

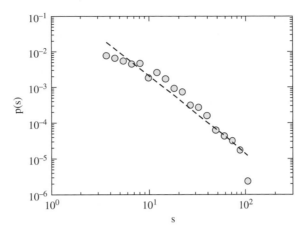

Fig. 9.12   Order distribution of the communities found by the label-propagation algorithm for the ArXiv cond-mat coauthorship network.

Box 9.3   **Which Partition Is the Best One?**

The problem of community detection has interested researchers with diverse backgrounds (from mathematics to computer science, from social science to biology) who have tackled the problem from various angles and have contributed refining existing methods or developing new ones. As a consequence, we can nowadays choose among hundreds of different algorithms. The main problem is that each algorithm produces a partition that is usually different from those found by the other algorithms, and there is no *a priori* criterion to decide which of the partitions is *the correct one*. Only in a very few exceptions, as in the case of the Zachary's karate club network, we know a so-called *ground truth* partition, so that the performance of different community detection algorithms can be assessed on the basis of the matching of the proposed partitions with the reference one. In most of the situations, instead, we perform community detection on a network because we do not know already whether the graph admits a meaningful partition in communities, and how such a partition looks like. In all such cases, community detection algorithms can only be compared on the basis of the values of a *quality function*, implicitly assuming that higher values of the quality function correspond to better partitions in communities. There is no universal agreement about which quality function should be used to compare algorithms for community detection, although modularity still remains among the most reliable ones.

---

to choose among those found by different methods, and this is probably one of the main reasons why the problem of detecting community structure in a network has attracted so much attention in the last decade.

9.8 What We Have Learned and Further Readings

In this chapter we have learned that complex networks, apart from having interesting structures at the microscopic scale, such as the presence of small cycles and motifs, can also carry important information at the mesoscopic scale, namely they can have a *community structure*. Finding the communities can be useful not only to group together the nodes with similar functions in a network, but also to coarse-grain a large network so that is possible to draw it in a meaningful way. In the last ten years the availability of large networks with billions of nodes has stimulated the development of efficient methods for community detection, and hundreds of different algorithms have appeared in the literature. For comprehensive and updated reviews see the paper written by Santo Fortunato [119], and that by Mark Newman [241]. In particular, great efforts have been devoted to the optimisation of the modularity: various approximate methods to find the maximum of the modularity, such as simulated annealing, genetic algorithms and local greedy schemes, have been proposed over the years. Notwithstanding the wide use and the large number of applications of the modularity, there are some caveats which concern this measure. First, it is not possible to directly compare the optimal values of modularity for two different graphs. Second, the modularity suffers of a resolution limit. In fact, it has been proven by Santo Fortunato and Marc Barthélemy that the modularity increases when subgraphs smaller than a characteristic size are merged together, and this implies that communities smaller than a certain size cannot be resolved by modularity optimisation [120]. A method that allows for a multiple resolution analysis of the modules in a network has been proposed by Alex Arenas, Alberto Fernández, and Sergio Gómez [11]. Finally, a systematic study of modularity performed in Ref. [140] revealed that this quality function is normally affected by extremely high degeneracy levels. Typically, in any network with community structure there is an exponentially high number of different partitions associated with very high values of modularity, while there is not a single partition associated with a value of modularity which is remarkably larger than that of similar partitions. This means that looking for the partition with the largest possible modularity is not of the utmost importance, since for each network there is in general a large set of partitions of comparable quality which can provide meaningful information about the community structure of the network. Other generalisations of the modularity and alternative quality measures can be found for instance in Refs. [271, 195, 276].

Today the state of the art in the analysis of community structure, in terms of efficiency and accuracy, is mainly represented by two algorithms. The first one is known as the *Louvain algorithm* because it was proposed in 2008 by Vincent Blondel, Jean-Loup Guillaume, Renaud Lambiotte and Etienne Lefebvre, four scientists from the Université Catholique de Louvain in Belgium [40]. This was historically the first algorithm able to handle networks with hundred millions of nodes and billions of edges. The algorithm uses a simple local scheme to optimise modularity, and allows the existence of modules at different scales to be revealed. The second algorithm, by Martin Rosvall and Carl Bergstrom, is called *Infomap*. It is not based on the modularity but on an information-theoretic framework. According to this algorithm, the best partition in communities of a network corresponds to the one requiring the minimum amount of information to be represented [274]. Both the Louvain algorithm and Infomap require a computational time which is a linear function of the number of edges in a graph. Even if these algorithms are very efficient, they are both non-deterministic and can produce slightly different partitions at each run. This problem can be solved by using a recent method which combines the results of several runs of the same algorithm in a single meaningful partition [196].

Finally, a few more words about network models. The benchmark models considered in Section 9.5 do not take into account the possible growth of the network over time. There are in the literature examples showing how communities can arise as a result of the growth dynamics of a network. These are abstract models, as in Ref. [221], or models specific to a given system, such as for instance, the model in Ref. [168] which mimics the growth mechanism of a social network. In the latter model, when the new node connects to node which belongs to a community C, this event will enhance the probability of creating other links from the new node to other nodes in C, so that ultimately the new node becomes itself a member of the community C.

Problems

9.1   Zachary's Karate Club

(a) We have defined a *partition* as a division of the graph nodes into sets, with each node assigned to only one set. Such a condition can be relaxed in various way. For instance, a division of the graph nodes into sets, such that a node can be assigned to one or to more than one set, is called a *cover*. Consider the Zachary's karate club network and construct the set of nodes at distance $d \leq 2$ from node 1, and the set of nodes at distance $d \leq 2$ from node 34. Show that these two sets form a cover of the network.

(b) Are you able to check that there are no cliques of six nodes in the Zachary's network? HINT: consider all nodes with at least five links. Then, find a partition and a cover of the nodes of the network in cliques.

9.2   The Spectral Bisection Method

(a) Suppose you have a graph with $N = 8$ nodes. How many different partitions can you produce? How many of these partitions are bisections, and how many of them are perfect bisections, i.e. divisions into two clusters with four nodes each?

(b) Consider the graph in Example 9.1. Apply the spectral bisection method to partition the graph. Comment on the results.

9.3   Hierarchical Clustering

(a) Prove that the Bell number B_N representing the number of distinct partitions of a set of N elements can be computed by the following recurrence formula:

$$B_{n+1} = \begin{cases} 1 & \text{if } n = 0 \\ \sum_{k=0}^{n} \binom{n}{k} B_k & \text{if } n \geq 1. \end{cases}$$

HINT: The proof can be done by induction on n.

(b) Implement a complete-linkage hierarchical clustering of the karate club network based on the distance matrix of Eq. (9.3). Construct the dendrogram and compare it with that in Figure 9.3.

9.4   The Girvan–Newman Method

(a) Prove that the shortest-path betweennesses of the nodes and the edges of a graph are related by:

$$c_i^B = \frac{1}{2} \sum_{l=1}^{N} a_{il} c_{e_{il}}^B - 2(N-1) \qquad \forall i = 1, \ldots, N$$

where by e_{il} we have indicated the edge connecting nodes i and l. This means that the betweenness of a node i is equal to half the sum of the betweenness of all the edges incident in i minus $2(N-1)$. Notice that the term $2(N-1)$ comes from the fact that node i is $N-1$ times the first node of a shortest path that contribute to the edge betweenness, and $N-1$ times the last node.

(b) Due to the way ties are broken in the selection of the edge to be removed at each step, the Girvan–Newman algorithm can in general return a different partition at each run. Run several times the program gn discussed in Appendix A.17 on the Zachary's karate club, and plot the distribution of the values of maximum modularity Q obtained. Compute also the value of mutual information I between the division of the network in two factions shown in Box 9.1 and the partitions of maximum modularity obtained at each run. Plot the distribution of I.

9.5   Computer Generated Benchmarks

(a) Consider the Girvan–Newman benchmark with $N = 1024$ nodes, M groups, $z = 20$ and a tunable value z_{out} of the expected number of links connecting a node of the graph to nodes in different groups. Evaluate the critical value of z_{out} at which the division in communities disappears for a number of groups M respectively equal to 4, 8, 16 and 32.

(b) Test the GN algorithm on the LFR benchmark. HINT: plot the normalised mutual information I and the fraction F of correctly classified nodes as a function of the mixing parameter μ.

9.6   The Modularity

(a) The modularity in Eq. (9.17) is defined in terms of number of links within modules. Can you write a quality function (similar to the modularity) in terms of links between different modules, instead of links within modules? In such a case the resulting quality function has to be minimised.

(b) Consider the graph with $N = 6$ nodes and $K = 6$ links, and the partition $\mathcal{P}_N = \{C_1, C_2\}$ in two sets shown in the figure of Example 9.11. Evaluate the modularity $Q_{\mathcal{P}}$ of the partition using the definition in Eq. (9.19). Compare the result obtained to the value found in Example 9.12 using Eq. (9.17).

(c) Can you show the relation between the two expressions of modularity in Eq. (9.17) and in Eq. (9.19)? HINT: Starting from Eq. (9.19), express the sum

over pairs of nodes as a sum over the sets of the partition and get rid of the
Kronecker δ.

9.7 A Local Method

Run the label propagation algorithm on the four network samples of the World Wide
Web introduced in Box 5.1 for $E = 100$ epochs, and compute for each case the distribution $p(s)$ of the size of the communities found. Assuming that those distributions
are power-laws $p(s) \sim s^{-\tau}$, provide an estimation of the exponent τ in each of the
four networks.

10

Weighted Networks

In the networks studied in the previous chapters, for each pair of nodes we can either have a link or not. In actual fact, real networks can display a large variety in the strength of their connections. Examples are the existence of strong and weak ties between individuals in any type of social system, or unequal capacities in infrastructure networks such as the Internet or a transportation system. When we have access to information about the intensity of interactions in a complex system, the structure of such a system can certainly be better described in terms of a *weighted network*, i.e. a network in which each link is associated with a numerical value, in general a positive real number, representing the strength of the corresponding connection. In this chapter, we extend and generalise to weighted networks the concepts and methods we have introduced in the previous chapters of the book. We will start introducing some basic measures to characterise and classify a weighted network. Next, we will discuss how to perform a motif analysis and how to detect community structures in weighted networks. The results of our empirical studies will demonstrate that purely topological models are often inadequate to explain the rich and complex properties observed in real systems, and that there is also a need for models to go beyond pure topology. We will then introduce some models of weighted networks which can reproduce the broad scale distributions and the correlations between topology and weights found empirically. Finally, as an application, we will show what we can learn about financial systems by describing correlations among *stocks in a financial market* in terms of a weighted network.

10.1 Tuning the Interactions

There are plenty of cases where (unweighted) graphs are a poor representation of real-world networks. As a concrete example, let us come back to the scientific collaboration networks studied in Chapter 3. When we constructed such graphs in Section 3.5 we linked pairs of scientists who have coauthored at least one paper. However, it is clear that scientists who have written many papers together are expected to know each other better than those who have coauthored only one paper. A reasonable way to account for this is to weight links by the frequency of collaboration, so that a link between two scientists having coauthored many papers together is given a weight larger than that of a link between two authors with only one paper in common.

Example 10.1 *(Weighting scientific collaborations)* In Ref. [234, 233] Mark Newman introduced a weighted version of coauthorship networks. The idea is to take into account the number of papers coauthored by a given pair of scientists, together with the number of coauthors involved in each paper. Namely, the weight w_{ij} of the interaction between two collaborators i and j, with $j \neq i$, is defined as:

$$w_{ij} = \sum_p \frac{\delta_i^p \delta_j^p}{n_p - 1}$$

where the index p runs over all papers, δ_i^p is 1 if author i has contributed to paper p and 0 otherwise, and n_p is the number of authors of the paper p. In this way, we sum over all the papers written together by the authors i and j. However, for each paper, it is reasonable to consider that two scientists whose names appeared together with many other coauthors know each other less well on average than two who were the only two authors of the paper. To account for this, we weight the contribution of each paper by the factor $1/(n_p - 1)$, assuming that each author divides his time equally with the other authors.

There are also many examples of man-made networks, such as various communications systems or transport infrastructures, whose links have very different capacities. All such networks are better described in terms of *weighted graphs*, graphs in which each link carries a numerical value measuring the intensity of the connection. The precise definition of a weighted graph is given below.

Definition 10.1 (Weighted graph) *A weighted or valued graph, $G^{\mathrm{w}} \equiv (\mathcal{N}, \mathcal{L}, \mathcal{W})$ consists of a set $\mathcal{N} = \{n_1, n_2, \ldots, n_N\} \neq \emptyset$ of nodes, a set $\mathcal{L} = \{l_1, l_2, \ldots, l_K\}$ of links and a set of weights or values $\mathcal{W} = \{w_1, w_2, \ldots, w_K\}$, positive real numbers associated with the links.*

In matricial representation, a weighted graph G^{w} can be described by the so-called *weight matrix W*, also known as the *weighted adjacency matrix* of the graph, defined in the following way.

Definition 10.2 (Weight matrix) *The weight matrix, or weighted adjacency matrix, W of a weighted graph is a $N \times N$ square matrix whose entry w_{ij} is equal to the weight of the link connecting node i to node j, while $w_{ij} = 0$ if the nodes i and j are not connected. As usual, $w_{ii} = 0 \ \forall i$.*

From now on, unless otherwise specified, we will only consider the case of positive and symmetric weights $w_{ij} = w_{ji} \geq 0$. As for the case of unweighted graphs, we can also define the *adjacency matrix* of a weighted network, which simply indicates which pairs of nodes are connected. This is a $N \times N$ matrix A such that $a_{ij} = 1$ if $w_{ij} \neq 0$, and $a_{ij} = 0$ if $w_{ij} = 0$.

The first weighted graph we analyse in this book describes the network of aeroplane connections in the USA. The network is described in DATA SET Box 10.1. The most

The weighted network provided in this data set was constructed from publicly available data by Vittoria Colizza, Romualdo Pastor-Satorras and Alessandro Vespignani, and represents air travel connections among the 500 airports with the largest amount of traffic in the USA [84]. The network has $N = 500$ nodes and $K = 2980$ edges. Two airports are connected by an edge if there is a non-stop flight in both directions. The graph is undirected, and the weight $w_{ij} = w_{ji}$ associated with the link (i, j) is the number of available seats on the connection between node i and node j on a yearly basis. Weights range from a minimum value of nine to a maximum equal to 2,253,992. A sketch is shown in the figure below, with the width of the edges representing their weights.

With about 95 million passengers per year, the Hartsfield Jackson International Airport in Atlanta is the busiest airport in the world. It has flight connections to almost all the other airports in the USA, and the corresponding node in the airline transportation data set is the one with the highest number of links. However, not all the routes from Atlanta are equally popular, and the system is thus better represented as a weighted graph.

basic thing we need to look at in a weighted network is how the weights are distributed among the edges. In the network under study here, weights are positive integers, because they represent the numbers of available aeroplane seats, and range from a minimum value of nine to a maximum equal to 2,253,992. We can then count, for each value of w, the number of edges $E(w)$ of weight w, and define the probability $q(w) = E(w)/K$ that an edge has weight equal to w. However, due to the high variability of the weights, and to the relatively small number of links, $K = 2980$, we expect the plot of $q(w)$ versus w to show large fluctuations, especially in the tails. In order to avoid this problem, we can bin the x-axis and count how many of the links have weights in a given interval around a value equal to w. However, selecting the binning can be a tricky issue, and can be convenient to use a logarithmic binning as we have done in Section 5.2. Another possibility, that we will adopt here, is to construct and work with the *cumulative weight distribution* function $Q(w) = \sum_{w'=w}^{+\infty} q(w')$, representing the probability of finding an edge of weight larger than or equal to w, instead that with $q(w)$. Notice that, even if in this case the weights are integer numbers, we have decided to use the symbols $q(w)$ and $Q(w)$, instead of the more appropriate q_w and Q_w, to remind the reader that in general the variable w is continuous. Figure 10.1a reports, as a function of w, the cumulative weight distribution $Q(w)$ we have obtained for the network in DATA SET 10. We have adopted a double-logarithmic scale to point out that the distribution does not appear as a power law. However, we find that 10 per cent of the links have a weight larger than 400,000, and only 1 per cent of the links have a weight larger than 1,160,000.

We can now focus on node properties. As for any unweighted graph, the first step is to compute the degree of each of the nodes and plot the degree distribution. Figure 10.1b shows the cumulative degree distribution P_k. This is a power law over two decades, with a few airports directly connected to more than hundreds, which is a remarkable result considering that the network has only $N = 500$ nodes. Of course, what is crucial to a given airport is not only the number of different direct connections to other airports, i.e. the degree of the corresponding node, but also the number of seats available for each connection. In a weighted graph, the natural generalisation of the degree k_i of a node i is the so-called *node strength*, that combines the information on the number of links incident in i with that on their weights [22, 247]. In the case of the US transport network, the node strength gives

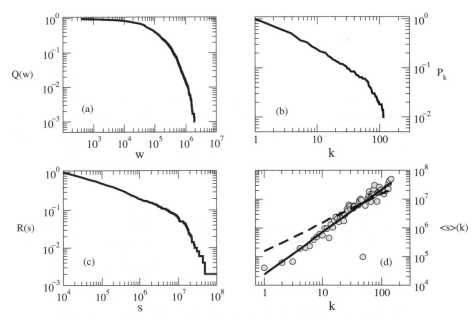

Fig. 10.1 Basic properties of the US air transport network. Cumulative distributions of link weights (a), node degrees (b) and node strengths (c). Average strength $\langle s \rangle (k)$, of vertices of degree k as a function of k (d): circles represent the result obtained for the real network, the solid line is a fit with a curve Ak^β giving $\beta = 1.5$, while the dashed line is an average over an ensemble of 500 networks obtained by reassigning the weights to the links.

the total number of seats available from a given airport to all other possible destinations, so it better captures the importance of the airport than just the node degree.

Definition 10.3 (Node strength)   *The strength s_i of a node i is the sum of the weights of the edges incident in i:*

$$s_i = \sum_{j=1}^{N} w_{ij}. \tag{10.1}$$

If the graph is directed, the strength of the node has two components: the sum of weights of outgoing links s_i^{out}, referred to as the out-strength *of the node, and the sum of weights of ingoing links s_i^{in}, referred to as the* in-strength *of node i:*

$$s_i^{out} = \sum_{j=1}^{N} w_{ij} \qquad s_i^{in} = \sum_{j=1}^{N} w_{ji}. \tag{10.2}$$

The total strength of the node is then defined as $s_i = s_i^{out} + s_i^{in}$.

Having given the definition of node strength, we can now look at how the strength is distributed among the nodes of the network. Also, since the strength is in this case a positive integer number, we can construct the probability $r(s)$ that a node has a strength equal to s. As for the case of the weight distribution, it is better to work with the *cumulative strength distribution* function $R(s) = \sum_{s'=s}^{+\infty} r(s')$ representing the probability that a vertex has strength larger or equal to s. Figure 10.1c shows $R(s)$ as a function of s for the US air transport network. We observe again a power-law behaviour, extending in this case over three decades. Although the node strength is related to the node degree, we will now show that the heavy-tailed $R(s)$ distribution in the network under study does not follow trivially from the power law P_k. In fact, if the link weights were independent from the topology, the quantities w_{ij} in Eq. (10.1), on average, would not depend on i and j, and could be well approximated as $w_{ij} \simeq \langle w \rangle$, where $\langle w \rangle = \sum_{i,j} w_{ij} / \sum_{i,j} a_{ij}$ is the typical link weight in the network. In other words, if the weights were not correlated to the topology, the strength of a node i of degree k_i would factorise as $s_i \simeq \langle w \rangle \, k_i$, and in such a case the strength distribution $R(s)$ would carry no more information than the degree distribution P_k. However, as we will see, in the US airport transport network the strength is non-trivially related to the degree, since the weights depend on the topology. To shed light on the relationship between node strength and node degree, we plot in Figure 10.1(d) the average strength $\langle s \rangle(k)$ of vertices of degree k, as a function of k. This quantity can be written as:

$$\langle s \rangle(k) = \frac{1}{N_k} \sum_{i=1}^{N} s_i \delta_{k_i k} \tag{10.3}$$

where N_k is the number of nodes of degree k, and $\delta_{k_i k} = 1$ when $k_i = k$, while it is zero otherwise. Circles in the figure are the results obtained for the air transport network, while the dashed line is the result of an average over 500 randomised versions of the network generated by redistributing the weights across the edges. The figure shows that real and randomised networks exhibit different behaviours. In fact, while the points for the randomised networks are, as expected, in perfect agreement with a linear behaviour, $\langle s \rangle(k) \simeq \langle w \rangle \cdot k$, in the case of the real US air transport network the circles are instead well fitted by a curve $\langle s \rangle(k) \simeq Ak^{\beta}$, with an extracted exponent $\beta = 1.5$. The result of such a fit is reported as a solid line. A value of the exponent larger than 1 implies that the strength of nodes grows faster than their degree. This means that the weights of edges belonging to highly connected nodes have higher values than those expected if the weights were assigned at random. In practice such a strong correlation between weighted and topological properties in the air transport system means that the higher the number of direct connections of an airport, the more traffic the airport can handle for each direct connection.

Now, for a given node i of degree k_i and strength s_i, different situations may occur. All weights w_{ij} from i to its neighbours j can be approximately equal to s_i/k_i or, in contrast, only one or a few weights can dominate over all the others. Various different quantities can be used to discriminate between these two opposite situations. The measure we adopt here was proposed in Ref. [28] to quantify the disparity in the weights of a node, and conceptually is not different from the node participation coefficient defined in Eq. (9.10) to evaluate the contribution of a node to the different communities of a network.

> **Definition 10.4 (Node disparity)**   *The disparity Y_i of the weights of node i is defined as*
>
> $$Y_i = \sum_j \left[\frac{w_{ij}}{s_i} \right]^2 , \qquad (10.4)$$
>
> *where w_{ij} are the link weights, and s_i is the strength of node i.*

Notice that, for each node i, the quantities $w_{ij}/s_i, j = 1, \ldots, k_i$ are non-negative numbers summing up to 1. Therefore, by summing the squares of such numbers we obtain a quantity, the disparity Y_i, that is always a number smaller than or equal to 1. The maximum value of disparity $Y_i = 1$ is obtained when the weight of the node i is concentrated in a single link. Conversely, it can be easily shown that the minimum value of Y_i is obtained when all the edges have the same weight. In this case we can write $w_{ij} = (\sum_l w_{il})/k_i = s_i/k_i \; \forall j$, and Y_i reduces to $1/k_i$. Given the implicit dependence of Y_i on the node degree k_i, when we want to characterise an entire network, it can be useful to compare the quantity $\langle Y \rangle(k)$, i.e. the disparity averaged over all nodes with degree k, to the curve $1/k$. If $\langle Y \rangle(k)$ scales

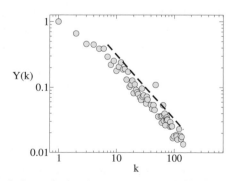

Fig. 10.2   Disparity as a function of the node degree for the US air transport network. The dashed line corresponds to a $1/k$ behaviour.

as $1/k$, this means that in practice all the links have the same weight. If instead the weight of a single link dominates for each node, then $\langle Y \rangle(k) \approx 1$, or in other words $\langle Y \rangle(k)$ is independent of k. As an example, we have computed the disparity for the nodes of the US aeroplane network. Figure 10.2 reports the resulting $\langle Y \rangle(k)$ as a function of the degree k, showing that the edges of a node in the aeroplane network have comparable weights, since $\langle Y \rangle(k)$ scales indeed as $1/k$. In practice, this means that airports usually distribute uniformly their traffic over all their possible direct connections. The use of the disparity can have more exotic applications, as shown in the following example.

Example 10.2   *(The best football team)*   In 2006 Italy won the World Cup in Berlin, beating France on penalty kicks. Italy's triumph will be remembered as a team effort with ten different Azzurri players scoring at least one goal. The final match between Italy and France can be analysed by looking, for each team, at the number of passes from every player to each of his teammates. The first figure shows the network of passes. For the sake of clarity we report for each player only the three teammates he passed to most frequently. This means that each node in the figure has only three out-going links. The width of the arrows is proportional to the number of passes. In the figure below we report the disparity for each of the 28 players that played the match. The players of the two teams are ranked according to their value of disparity, from the highest to the smallest value. Although the average value of node disparity for Italy, 0.17 ± 0.06, is only slightly smaller than the average value for France, 0.19 ± 0.06, the curves of the rankings show that the disparity of each Italian player is systematically smaller than that of the French. This result indicates that the Italian network has a more

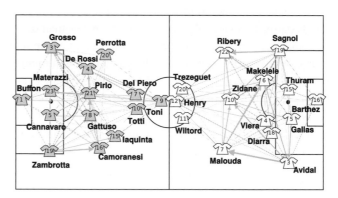

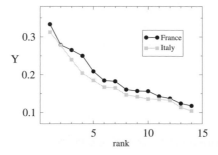

balanced distribution of the weights incident in each node. Is this an indication that the Italian team played better?

10.2 Basic Measures

In this section we show how to generalise to weighted networks some of the basic concepts previously introduced for unweighted networks. The most interesting aspect of the metrics that we will discuss is that they combine structural features of a weighted graph together with information on the edge weights.

The first measure we want to extend to weighted graphs is the clustering coefficient introduced in Chapter 4. The definition proposed in Eq. (4.3) does not consider that, in a weighted network, some neighbours can be more tightly connected than others. One possible definition of *weighted clustering coefficient* which takes into account the edge weights, was proposed by Alain Barrat, Marc Barthélemy, Romualdo Pastor-Satorras and Alessandro Vespignani [22, 28].

Definition 10.5 (Weighted clustering coefficient) *The weighted clustering coefficient, c_i^w, of node i is defined as:*

$$c_i^w = \begin{cases} \frac{1}{s_i(k_i-1)} \sum_{l,m} \frac{(w_{il}+w_{im})}{2} a_{il}a_{lm}a_{mi} & \text{for } k_i \geq 2 \\ 0 & \text{for } k_i = 0,1. \end{cases} \qquad (10.5)$$

The weighted clustering coefficient C^w of a weighted graph is the average of c_i^w over all graph nodes:

$$C^w = \langle c^w \rangle = \frac{1}{N} \sum_{i=1}^{N} c_i^w. \qquad (10.6)$$

By definition, we have $0 \leq c_i^w \leq 1$ $\forall i$ and $0 \leq C^w \leq 1$.

It is simple to understand the idea behind Eq. (10.5). The unweighted node clustering coefficient c_i in Eq. (4.2) and in Eq. (4.4) counts the number of triangles containing node i. Instead, c_i^w takes into account, for each triangle containing node i, the average weights of the two links adjacent to i. In this way, the weighted clustering coefficient captures the relative weight of triangles in the neighbourhood of node i, with respect to the total strength of node i. The normalisation factor $s_i(k_i - 1)$ in Eq. (10.5) ensures that $0 \leq c_i^w \leq 1$ $\forall i$. In fact, in the best possible case in which all pairs of neighbours of node i are linked, the term $\sum_{l,m} \frac{(w_{il}+w_{im})}{2} a_{il}a_{lm}a_{mi}$ reduces to:

$$\sum_l \sum_{m\neq l} \frac{(w_{il}+w_{im})}{2} a_{il}a_{im} = \sum_l \sum_{m\neq l} w_{il}a_{il}a_{im}$$

$$= \sum_l w_{il}a_{il} \sum_{m\neq l} a_{im} = (k_i - 1) \sum_l w_{il}a_{il} = (k_i - 1)s_i.$$

Notice also that the definition of c_i^w reduces to the topological node clustering coefficient c_i of Eq. (4.2) when all weights are equal to 1. It is therefore interesting to compare, in a weighted graph, the quantity C^w with the topological clustering coefficient C of Eq. (4.3). Finding $C^w \neq C$ implies that there are correlations between topology and weights. In particular we can have two different cases. If $C^w > C$, we are in the presence of a network where triangles are more likely formed by links with larger weights. On the contrary, $C^w < C$ indicates that the clustering is generated by edges with low weights. This can be easily understood by considering the difference:

$$c_i^w - c_i = \frac{1}{s_i(k_i - 1)} \sum_{l,m} (w_{il} - \overline{w}_i) a_{il} a_{lm} a_{mi}$$

where $\overline{w}_i = s_i/k_i$ is the average weight of links incident in i. Since the sum above considers only the links of node i which belong to triangles, if $c_i^w - c_i > 0$ then the weights of such links are greater than average, thus indicating a positive correlation between weights and triangles. If instead $c_i^w - c_i < 0$, then the weights of the links belonging to triangles are smaller than average. Following is an example of a node of a weighted graph with $c_i^w < c_i$. This means that, in the neighbourhood of node i, strong connections occur on edges not belonging to triangles.

Example 10.3 *(Weighted vs unweighted measures)* When we have a weighted graph, it can be very instructive to compute and compare topological and weighted quantities. Figure below reports two typical cases of local configurations whose topological and weighted measures are different [22]. In the weighted graph shown in panel (a), node i (black circle) has four neighbours. If we do not take into account edge weights we find a clustering coefficient $c_i = 0.5$, since only three of the possible six links are present in the graph induced by the neighbours of i. Notice, however that node i has a strong link of weight $w = 5$ to one of its neighbours, namely the node which contributes less to the clustering coefficient. The weighted clustering coefficient captures

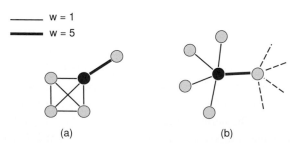

(a)   (b)

more precisely the effective level of cohesiveness of the neighbourhood of i due to the actual interaction strength. We obtain in fact $c_i^w = 0.25$, a value smaller than c_i, which indicates an accumulation of the weights on edges not contributing to triangles.

In panel (b) we focus on degree–degree correlations. In the graph shown, node i (black circle) has five neighbours, four of degree $k = 1$ and one of degree $k = 5$. If we compute the average nearest neighbours' degree of node i as defined in Eq. (3.36) we obtain $k_{nn,i} =$

$9/5 = 1.8$, since all five neighbours of i are taken into account in the same way. However, node i has a strong link of weight $w = 5$ to the node of degree $k = 5$ so that, by evaluating the weighted average nearest neighbours' degree in Eq. (10.6), we obtain $k^{\mathrm{w}}_{\mathrm{nn},i} = 3.2 > k_{\mathrm{nn},i}$, which better captures the actual pattern of interaction weights.

Let us now come back to the US air transport network introduced in the previous section. If we evaluate the graph clustering coefficient of such a network both in the unweighted and in the weighted version we find $C^{\mathrm{w}}/C \approx 1.1$. This means that larger weights contribute preferentially to connections between airports more involved into triangles. We can explore this effect more in detail by focusing on the node quantity c_i, and by looking at the clustering coefficient of nodes with different degree k. In practice, we group nodes according to their degrees and, for each value of k, we define the average unweighted node clustering coefficient $c(k)$ as in Eq. (4.5). Analogously, we can construct the weighted node clustering coefficient $c^{\mathrm{w}}(k)$ by averaging the quantity c^{w}_i in Eq. (10.5) over all nodes of degree k. In Figure 10.3 we report the two quantities $c(k)$ and $c^{\mathrm{w}}(k)$ as a function of k. The first thing we notice is that $c(k)$ decays with k. This is because airports with large k usually provide nonstop connections to very distant destinations and these destinations are often not interconnected among them, which implies a low clustering coefficient for the hubs. We found that the weighted clustering coefficient $c^{\mathrm{w}}(k)$ decreases with k as well. It can also be useful to compare the values of $c^{\mathrm{w}}(k)$ to those of $c(k)$ by taking in mind that we expect $c^{\mathrm{w}}(k) \approx c(k)$ in a randomised network where the existing weights are randomly redistributed over the network edges. Figure 10.3 shows that $c^{\mathrm{w}}(k) > c(k)$ for any value of k, and that the range of variability of $c^{\mathrm{w}}(k)$ with k is more limited. This indicates that high-degree airports have a progressive tendency to use larger weights for the links used in the fewer and fewer triangles present, thus partially balancing the reduced unweighted clustering coefficient.

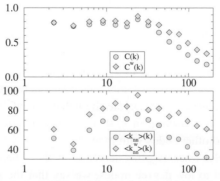

Fig. 10.3 Clustering coefficient and average nearest neighbours' degree of nodes of degree k for the US air transport network. Both weighted and unweighted quantities are reported. (a) The weighted clustering coefficient is larger than the unweighted one for each value of k. (b) The network shows a disassortative behaviour for $k > 15$, with values of $\langle k^{\mathrm{w}}_{\mathrm{nn}} \rangle(k)$ larger than $\langle k_{\mathrm{nn}} \rangle(k)$ for each value of k.

Degree Correlations

In a weighted network it is also possible to investigate degree–degree correlations by taking into account the fact that links may have different weights. For instance, we can define the average degree of the nearest neighbours of a given node, similarly to what we did for unweighted networks in Definition 3.8, by weighting this time each neighbour by the weight of the connection [22]. And then, we can construct a weighted average nearest neighbours' degree function $\langle k_{nn}^{w} \rangle (k)$.

Definition 10.6 (Weighted average nearest neighbours' degree)   *The weighted average degree $k_{nn,i}^{w}$ of the nearest neighbours of node i is defined as:*

$$k_{nn,i}^{w} = \frac{1}{s_i} \sum_{j=1}^{N} a_{ij} w_{ij} k_j \tag{10.7}$$

where w_{ij} is the weight of the link (i,j) and s_i is the strength of node i.
The weighted average nearest neighbours' degree function $\langle k_{nn}^{w} \rangle (k)$ *is then obtained by averaging $k_{nn,i}^{w}$ over all nodes i with k links:*

$$\langle k_{nn}^{w} \rangle (k) = \frac{1}{N_k} \sum_{i=1}^{N} k_{nn,i}^{w} \, \delta_{k_i k} \tag{10.8}$$

where N_k is the number of nodes of degree k.

In practice, in Eq. (10.7) we take into account the presence of the link weights by weighting each neighbour of i by the normalised weight of the connecting edge: w_{ij}/s_i. In this way, the degrees of nodes more tightly connected to node i contribute more to $k_{nn,i}^{w}$. As in the case of the clustering coefficient, comparing weighted and unweighted measures can provide useful information. In fact, finding $k_{nn,i}^{w} > k_{nn,i}$ indicates that the edges with the larger weights are pointing to the neighbours with larger degree, while $k_{nn,i}^{w} < k_{nn,i}$ denotes the opposite case. Panel (b) of the figure in Example 10.3 reports a case of a graph where $k_{nn,i}^{w} > k_{nn,i}$. In fact, node i (black circle) is connected by the largest weight to the node with the largest degree, so that in this case we get that the weighted average nearest neighbours' degree is two times the unweighted average nearest neighbours' degree.

We can now make use of the weighted average nearest neighbours' degree of nodes with k links to characterise whether a weighted network is assortative or disassortative. For this reason we need to investigate the behaviour of $\langle k_{nn}^{w} \rangle (k)$ as a function of k. If $\langle k_{nn}^{w} \rangle (k)$ is an increasing function of k, that is if nodes with high degree tend to have the strongest weights to other nodes with many links, we say that the weighted network is *assortative*. Conversely, when $\langle k_{nn}^{w} \rangle (k)$ decreases with k, that is if nodes with many links tend to be connected to low degree nodes, we say that the network is *disassortative*. Again, it is useful to compare $\langle k_{nn}^{w} \rangle (k)$ to its unweighted counterpart $\langle k_{nn} \rangle (k)$. Figure 10.3 shows the results obtained for the US air transport network. The first thing we notice is that $\langle k_{nn}^{w} \rangle (k) > \langle k_{nn} \rangle (k) \; \forall k$, i.e. the values of the weighted quantity are always larger than those of the unweighted one. Hence, for each value of k, the edges with the larger weights from a node are pointing to the neighbours with larger degree. Furthermore, both the topological and the weighted quantities do exhibit a dependence on the degree for $k > 15$, revealing a disassortative behaviour. This means that, in this network, high-degree airports are mainly connected to low-degree airports, and also that the major part of the traffic (in terms of number of seats) of high-degree airports is directed to low-degree airports.

The Rich Club

The rich-club behaviour can be detected and quantified in unweighted networks by plotting, as a function of k, the normalised rich-club coefficient $\rho(k)$ defined by Eq. (7.22) and Eq. (7.23). As a first step, we can use the same approach to extend the concept of *rich club* to weighted graphs, such as the US air transport network. However, this means to neglect the information contained in the weights or, equivalently, to assume that all edge weights are equal. The results obtained by treating the US air transport system as an unweighted network are shown in Figure 10.4(a). We observe that $\rho(k)$ is approximately constant and equal to 1, with the only exception of a few nodes with k larger than 100. Thus, if we analyse the US air transportation system as an unweighted network, the message we get is that the system does not exhibit a rich-club behaviour. Such a conclusion can just be a trivial consequence of the fact that we did not take into account the link weights in our analysis. A more effective approach is to extend the concept of *rich club* to weighted networks in order to measure the tendency of nodes with large degree to be connected by links of large weights. We will follow here the work of Tore Opsahl, Vittoria Colizza, Pietro Panzarasa and José Ramasco [249]. To be as general as possible, let us indicate as r the *richness* of a node. The richness can be a structural property of a node, such as its degree k, its strength s, or any other centrality measure, but it can also be an intrinsic node property measuring its importance. The basic idea to detect a weighted rich-club behaviour is to measure, as a function of r, how much of the weights in the network belong to links connecting nodes of richness larger than r. For such a reason, let us define respectively as $E_{>r>r}$ and $W_{>r>r}$ the number of links and the sum of weights of links among nodes of richness larger than r. We will also need to rank the K links of the network in decreasing

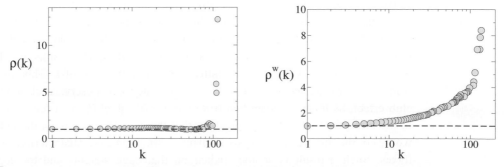

Fig. 10.4 The unweighted (a) and weighted (b) normalised rich-club coefficients are plotted as function of k for the US air transport network. To normalise the rich club coefficient in the weighted case we have considered an ensemble of 100 networks obtained by randomisations of the weights.

order of their weights, so that $w_l^{\text{rank}} \geq w_{l+1}^{\text{rank}}$ with $l = 1, 2, \ldots, K$. We are now ready for the following definition.

Definition 10.7 (The weighted rich-club coefficient) *The weighted rich-club coefficient $\phi^{\text{w}}(r)$ is defined as:*

$$\phi^{\text{w}}(r) = \frac{W_{>r>r}}{\sum_{l=1}^{E_{>r>r}} w_l^{\text{rank}}} \tag{10.9}$$

where $W_{>r>r}$ is the sum of the weights of the $E_{>r>r}$ links among nodes with richness larger than r, and the denominator is the sum of the weights of the $E_{>r>r}$ links in the network with the highest weights.

In practice, for each value of r, the weighted rich-club coefficient in Eq. (10.9) is equal to the ratio between the weights shared by the rich nodes, i.e. those with richness larger than r, and the maximum possible weights they could share if they were connected by the strongest links of the network. Hence, we have $\phi^{\text{w}}(r) \leq 1$ $\forall r$ by definition. As for the unweighted case, in order to assess the presence of a rich-club behaviour, we need to compare the weighted rich-club coefficient of a real weighted network to that of a *network null model*, obtained for instance by randomising the weights associated with the links. We can finally define a properly normalised weighted rich-club coefficient as:

$$\rho^{\text{w}}(r) = \frac{\phi^{\text{w}}(r)}{\phi_{\text{null}}^{\text{w}}(r)} \tag{10.10}$$

and use this quantity to extend the concept of rich-club behaviour to weighted networks as follows.

Finding values of the normalised weighted rich-club coefficient $\rho^{\text{w}}(r)$ larger than 1 for large values of r, and increasing as a function of r indicates the presence of a *weighted rich club* in a network.

We are now ready to perform again a rich-club analysis of the US air transport network, this time taking into account the influence of the link weights through the use of Eq. (10.10). We have adopted the node degree as the simplest possible measure of richness of a node, i.e. we set $r_i \equiv k_i$ for each node i. The results are shown in Figure 10.4(b), where we plot the quantity $\rho^{\text{w}}(k)$ as a function of k. This time, differently from when we treated the network as unweighted, we observe a clear weighted rich-club effect, as it can be identified from the growth of $\rho^{\text{w}}(k)$ as a function of the degree of the airports. This means, in practice, that the connections among the hubs of the network, i.e. the airports with flights to many destinations, are characterised by large travel fluxes. Such a result is another indication that edge weights and topology of the air transport network are nontrivially related, as we found already in Figure 10.1 when we plotted the strength as a function of the degree of the nodes. In particular, the US air transport system is an example of a weighted network with disassortative degree–degree correlations which exhibits a weighted rich club behaviour. These results further support our discussion in Section 7.3, showing that the rich club is not necessarily related to assortativity.

10.3 Motifs and Communities

In this section we will concentrate on how to detect significant structural features in weighted networks, both at the microscopic and at the mesoscopic scale. Namely, we will discuss how to extend the motif analysis of Chapter 8 and some of the community detection methods of Chapter 9 to take into account the edge weights.

Motif Analysis

Concerning the definition of *motifs*, it is easy to understand that in a weighted network we need to do more than just counting the number of times a given small graph appears in the network. Let us see this in a practical case. Suppose we are interested in undirected subgraphs of three nodes and three links, and in particular in triangles. If we have to evaluate whether triangles are relevant in a weighted network, what matters is not only the number of triangles in the network, but also the values of the weights of the three links in all the triangles we find. There are many alternative ways to combine together numbers and weights of triangles, and to implement this into a significance score. We will follow here the work of Jukka-Pekka Onnela, Jari Saramäki János Kertész and Kimmo Kaski, a group of researchers from the Helsinki University of Technology who proposed to evaluate two different quantities, namely the *intensity* and the *coherence*, for each subgraph found [248]. The intensity of a subgraph measures the typical value of its edge weights, while the coherence quantifies the diversity of the weights. The definitions of subgraph intensity and coherence, and the two associated significance scores are quite complex, so it is good to start from scratch.

Suppose we have a weighted undirected (or directed) graph $G^W = (\mathcal{N}, \mathcal{L}, \mathcal{W})$ and we indicate as $G = (\mathcal{N}, \mathcal{L})$ the unweighted graph associated with G^W. Our final goal is to assess the statistical significance of each possible small connected unweighted graph $F_{n,l}$, with n nodes and l links, as a subgraph of G^W. For instance, for $n = 3$ this means that we should consider, one by one, the two types of three-nodes undirected graphs, or the thirteen types of directed graphs shown in Figure 8.9. Assume now that G^W is undirected, and that the graph $F_{n,l}$ we are interested in is the undirected triangle. Once $F_{n,l}$ is fixed, we need to consider all the topologically equivalent weighted undirected triangles $F_{n,l}^W(\mathcal{N}_F, \mathcal{L}_F, \mathcal{W}_F)$ in the weighted network G^W. Instead of just counting the number of such triangles, we can sum the intensity of each of them, defined as the average weight of the three links in the triangle. Also, we can sum the coherence of each of them, which measures the diversity of the three links. The precise definitions of subgraph intensity and coherence in the most general case are given below.

Definition 10.8 (Subgraph intensity and coherence)   *The* intensity $I(F^w)$ *and the* coherence $Q(F^w)$ *of a weighted subgraph* $F^w \equiv F^w_{n,l}$ *with n nodes and l links in a weighted graph* G^w *are respectively defined as:*

$$I(F^w) = \left(\prod_{(ij) \in \mathcal{L}_F} w_{ij} \right)^{1/l}, \qquad Q(F^w) = \frac{I(F^w)}{\frac{1}{l} \sum_{(ij) \in \mathcal{L}_F} w_{ij}} \qquad (10.11)$$

where w_{ij} is the weight of the link (i, j).
The intensity I_F *and the* coherence Q_F *of graph* $F \equiv F_{n,l}$ *as a subgraph of* G^w *are then defined as:*

$$I_F = \sum_{F^w} I(F^w), \qquad Q_F = \sum_{F^w} Q(F^w) \qquad (10.12)$$

namely as the sum of the intensity and of the coherence of all the weighted subgraphs $F^w_{n,l}$ of G^w topologically equivalent to $F_{n,l}$.

Notice that the *intensity* $I(F^w)$ in Eq. (10.11) is defined as the geometric and not as the arithmetic mean of the weights in F^w. The geometric mean is best used when the product of the values, more than their sum, is significant. With such a definition, the subgraph intensity $I(F^w)$ may be equally low because all the weights are low, but also if only one of the weights is very low. The introduction of the second quantity in Eq. (10.11), the *coherence* $Q(F^w)$, equal to the ratio between the geometric and the arithmetic mean, allows us indeed to distinguish between these two cases. In fact, since the geometric mean of non-negative quantities is always less than or equal to their arithmetic mean, we have $Q(F^w) \leq 1 \ \forall F^w$. In particular, when all quantities are equal, geometric and arithmetic mean coincide so that $Q(F^w) = 1$. Thus, $Q(F^w) \simeq 1$ if the subgraph weights do not differ too much, that is the subgraph weights are internally coherent, while $Q(F^w) \simeq 0$ if one of the weights is very low. Once we know how to calculate the intensity and coherence of a given weighted subgraph $F^w_{n,l}$ in the weighted graph G^w, we can obtain the *intensity* and *coherence* of graph $F_{n,l}$ in G^w by summing, as in Eqs. (10.12), the intensity and coherence of all the weighted subgraphs $F^w_{n,l}$ topologically equivalent to $F_{n,l}$.

Example 10.4   *(Computing motifs intensity and coherence)*   Let us consider the weighted graph G^w shown below:

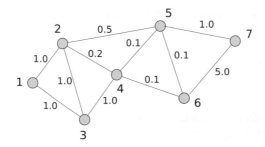

Subgraph	$I(F^w)$	$Q(F^w)$
(1,2,3)	1.00	1.00
(2,3,4)	0.58	0.79
(2,4,5)	0.22	0.81
(4,5,6)	0.10	1.00
(5,6,7)	0.79	0.39
	I_F	Q_F
	2.69	3.99

which has $N = 7$ nodes and $K = 11$ edges, and let us compute intensity and coherence of triangles in this graph. This means that, as small graph $F_{n,l}$, with $n = 3$ and $l = 3$, we take the triangle. Graph G^W has five weighted triangles, namely $(1,2,3)$, $(2,3,4)$, $(2,4,5)$, $(4,5,6)$ and $(5,6,7)$. The values of intensity $I(F^W)$ and coherence $Q(F^W)$ of each triangle are reported in the table. Triangle $(1,2,3)$ has the largest intensity, $I(F^W) = 1.0$, and the largest coherence, $Q(F^W) = 1.0$, since its weights are all equal to 1.0. Triangle $(4,5,6)$ has the minimal value of intensity, $I(F^W) = 0.1$, since its edges have all weight equal to 0.1, but also the largest coherence $Q(F^W) = 1.0$. The minimal value of coherence is instead associated with triangle $(5,6,7)$, whose edges are characterised by the largest heterogeneity. It is interesting to compare these results with those obtained on the weighted graph below,

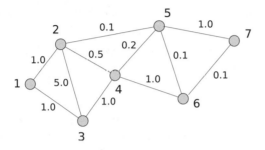

Subgraph	$I(F^w)$	$Q(F^w)$
(1,2,3)	1.70	0.73
(2,3,4)	1.36	0.63
(2,4,5)	0.22	0.81
(4,5,6)	0.27	0.62
(5,6,7)	0.22	0.54
	I_F	Q_F
	3.77	3.33

having the same topology and set of weights of the previous one, but with a different association between weights and edges. Most of the triangles have a relatively larger value of intensity in this case, while the coherence is in general smaller than in the other graph, mainly because no triangle has edges with identical weights. For instance triangle $(1,2,3)$ had $I(F^W) = 1.0$ and $Q(F^W) = 1.0$ in the first graph while, due to the rearrangement of edge weights, its intensity in the second graph is relatively larger, $I(F^W) = 1.7$, at the expenses of a smaller coherence, $Q(F^W) = 0.73$.

Summing up, when we have a weighted network, instead of simply counting the number of times n_F that a given pattern $F_{n,l}$ appears in the graph, we get two values, namely I_F and Q_F, respectively quantifying the mean and the diversity of the weights used in such patterns. As we did in Section 8.5, in order to assess the significance of finding a given value of I_F and a given value of Q_F, we need to compare graph G^W to an appropriate randomised version of the graph. The standard approach is to generalise Eq. (8.20) to the case of weighted networks, and to define two *Z-score* values by replacing the number of subgraphs respectively by their intensity and by their coherence.

Definition 10.9 (Subgraph intensity and coherence Z-scores)   *The* intensity Z-score ZI_F and *the* coherence Z-score ZQ_F of $F_{n,l}$ in weighted graph G^W can be written as:

$$ZI_F = \frac{I_F - \overline{I}_F^{\text{rand}}}{\sigma_{I_F}^{\text{rand}}} \qquad ZQ_F = \frac{Q_F - \overline{Q}_F^{\text{rand}}}{\sigma_{Q_F}^{\text{rand}}} , \tag{10.13}$$

where I_F (Q_F) is the intensity (coherence) of subgraph $F_{n,l}$ in G^w, while $\bar{I}_F^{\text{rand}}$ and $\sigma_{I_F}^{\text{rand}}$ ($\bar{Q}_F^{\text{rand}}$ and $\sigma_{Q_F}^{\text{rand}}$) are, respectively, mean and standard deviation of the intensity (coherence) in an ensemble of graphs obtained by randomising G^w.

The ensemble of random graphs usually proposed as a *null model* when looking for weighted motifs is the usual configuration model where the K edge weights of the original graph G^w are assigned at random [248]. This randomisation is different from the one used in Figure 10.1(d) or from that adopted as null model for the normalised rich-club coefficient, because here we randomise both the weight assignment and the topology of the network (keeping the degree sequence fixed). There is a reason for this. Notice in fact that, due to this choice of the randomisation adopted, when the link weights are all equal, both ZI and ZQ in Eq. (10.13) reduce to the unweighted Z-score Z in Eq. (8.20).

Similarly to the case of unweighted networks, we are here interested in subgraphs with an intensity significantly higher than in randomised versions of a weighted network. We can therefore measure the relevance of a subgraph F as a *motif* of a weighted network by the value of the intensity score ZI_F. For each motif F, we can then estimate, by means of the coherence score ZQ_F, if its weights are internally more or less coherent than in randomised versions of the graph. For instance, finding a Z-score equal to 2 or 3, means that the observed value of intensity (or coherence) is two or three standard deviations larger than that expected if the same links and weights were organised at random, i.e. that the observed I_F (Q_F) can trivially be due to fluctuations only in 2.14% or 0.13% of the cases respectively. Therefore a value of ZI_F larger than 2 or 3 is a good sign that F is a graph *motif* in terms of intensity, while a value of QI_F larger than 2 or 3 indicates that the weights of the motif are coherent.

Example 10.5   *(Computing intensity and coherence Z-scores)*   We can now evaluate the Z-scores of the two graphs shown in Example 10.4. If we consider 200 randomisations of the graphs, the average value of intensity of triangles in the null model is equal to $\bar{I}_F^{\text{rand}} = 2.34$, while the standard deviation is $\sigma_{I_F}^{\text{rand}} = 0.76$. Hence, the intensity score of triangles in the first graph is $ZI_F = 0.47$, while for the second graph we get $ZI_F = 1.89$. As a result, the intensity of triangles in those graphs is nor particularly different from the typical values observed in the null model. Conversely, the average coherence of triangles in the null model is $\bar{Q}_F^{\text{rand}} = 2.42$, with a standard deviation $\sigma_{Q_F}^{\text{rand}} = 0.61$. This yields a value of the coherence score equal to $ZQ_F = 2.57$ for the first graph and to $ZQ_F = 1.49$ for the second graph. We can conclude that the coherence of triangles in the first graph is noticeably different from that expected in the corresponding null model.

Finding community structures

Concerning the presence of *community structures* and their detection in weighted networks, the ideas and definitions introduced in Chapter 9 can be easily generalised also to take the edge weights into account [229, 12]. In particular, we can extend to weighted networks the definition of modularity of a partition given in Eq. (9.17).

Definition 10.10 (Weighted modularity) *Given a weighted graph G^w and a partition $\mathcal{P}_\mathcal{N} = \{C_1, C_2, \ldots, C_M\}$ of its nodes into M sets, the weighted modularity $Q_\mathcal{P}^w$ of partition $\mathcal{P}_\mathcal{N}$ can be written as:*

$$Q_\mathcal{P}^w = \sum_{m=1}^{M} \left[\frac{W_{mm}}{W} - \left(\frac{W_m}{2W} \right)^2 \right] \tag{10.14}$$

where W_m is the strength of set C_m, defined as the sum of the strengths of all the nodes in set C_m, W_{mm} is the sum of the weights of the links joining nodes in set C_m, and $2W$ is the total strength in G^w.

In such a definition of *weighted modularity* the summation is over the M sets of the partition. Notice that, since $2W = \sum_i s_i$ is the total strength in the network, the first term of each summand in Eq. (10.14) is the fraction of the weights of graph G^w inside a set, whereas the second term is the expected fraction of weights that would be inside the set in a randomised graph having the same strength W_m for each set m, $m = 1, \ldots, M$, as the original graph G^w. Hence, according to Eq. (10.14), a set of nodes C_m is a real community of a weighted network if the sum of weights inside the set is larger than the expected sum of weights in an appropriate null model conserving the internal strengths of the sets in G^w. Finally, a positive and large value of $Q_\mathcal{P}^w$ indicates that the partition $\mathcal{P}$ is a good division of the network into communities. Notice that the definition of weighted modularity is perfectly consistent with that of modularity introduced in Eq. (9.17). In fact, in an unweighted graph, W_{mm}/W is simply equal to the fraction of edges of the graph joining nodes in set C_m, $(W_m/2W)^2$ is the expected fraction of edges inside set C_m (in a random graph with the same degree for each set of the partition as in the original graph), and Eq. (10.14) reduces to Eq. (9.17).

As in the case of unweighted networks, it can be useful to rewrite the weighted modularity as a sum over node pairs instead than a sum over sets of the partition. In this way, the modularity can be expressed directly in terms of the weighted adjacency matrix w_{ij}, where i and j are node labels and not set labels, and of the node strengths $s_i = \sum_j w_{ij}$, with $i, j = 1, 2, \ldots, N$. We thus obtain:

$$Q_\mathcal{P}^w = \frac{1}{2W} \sum_{ij} \left(w_{ij} - \frac{s_i s_j}{2W} \right) \delta(C^{(i)}, C^{(j)}). \tag{10.15}$$

To count only links between nodes belonging to the same set we denote as $C^{(i)}$ the cluster containing node i, and we make use of the Kronecker delta-function $\delta(C^{(i)}, C^{(j)})$ that yields the value 1 if node i and j are into the same cluster, and zero otherwise. The modularity of a given partition is then the fraction of weight of internal links in the network under study, minus the expected fraction of weight of internal links in a network with the same number of nodes, where the edges are placed at random preserving the strength of each node. For unweighted networks Eq. (10.15) reduces to Eq. (9.19). In fact, if all weights are equal to 1, w_{ij} coincides with a_{ij}, s_i is equal to the degree of node i, and W is the total number of links in the network.

Summing up, since the larger the modularity the best the partition, we can use Q_P^w in the form of Eq. (10.14) or in the form of Eq. (10.15) as the quality function to optimise when looking for the best partition of a weighted network. The two following examples show that in some cases the communities found in a weighted graph can be very different from those obtained by neglecting the weights.

Example 10.6 *(Weighted communities in Zachary's karate club)*   In his empirical observations, Wayne Zachary was able to associate an integer number between 0 and 8 to each pair of club members. Such value represented the number of different social contexts (more details can be found in the original paper by Zachary [320]) in which the corresponding interaction took place. It is therefore possible to consider a weighted version of the karate club network. To find communities in this network, we used the *Louvain algorithm*, since this algorithm can be easily extended to optimise the weighted modularity in Eq. (10.14), thus allowing a direct comparison between partitions obtained in the weighted to those of the unweighted graph. In both cases the algorithm finds $M = 4$ communities, respectively corresponding to a value of modularity $Q = 0.419$ and of weighted modularity $Q^w = 0.445$. The partition found for the unweighted network is indeed the best partition in terms of modularity of the Zachary's network that we reported in Figure 9.9. Interestingly, the partition obtained by taking weights into account is almost identical, except for the position of a single node, node 10, which is assigned to a different community (this is the same node that was assigned to a single-vertex community by the GN algorithm in Figure 9.8). As a result, the value of the normalised mutual information (see Eq. (9.5)) between the two partitions is very high, namely $\mathcal{I} = 0.923$.

Example 10.7 *(Weighted communities in the US air transport network)*   Notice that the community structure of a weighted network and that of its underlying unweighted counterpart can in general be very different. An interesting example is that of the US

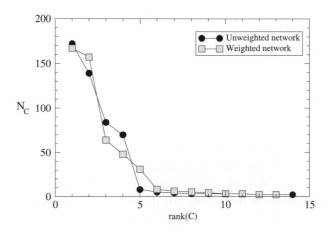

air transport network of DATA SET 10. Again we have made use of a modification of the *Louvain algorithm* for weighted networks. In the figure above we plot the order of each of the communities found in the weighted (grey squares) and in the unweighted (black circles) network of US air transport, ranked from the largest to the smallest. In this case, although the best partition of the weighted graph consists of $M = 14$ communities, only one less than the $M = 15$ communities found in the unweighted network, the sizes of the communities of the two networks differ substantially. More importantly, there are considerable differences in the respective composition of the communities in the two cases, as confirmed by a value of the normalised mutual information equal to $\mathcal{I} = 0.549$.

10.4 Growing Weighted Networks

The simplest possible way to construct a weighted graph is to first sample a random graph with a given degree distribution p_k, for instance by using the configuration model studied in Chapter 5, and then to assign weights to the edges as random independent variables sampled from a given weight distribution $q(w)$. With this procedure we can obtain graphs which are random both in their topology and in the weight assignment. Although it is possible to control the two distributions p_k and $q(w)$, clearly in this way we will produce graphs in which the weights are independent from the topology. As seen with the air transport network studied in Section 10.1, the most interesting situation is instead when the link weights are coupled to the topology, for instance to quantities such as the node degrees. Furthermore, the method outlined above is static, while most of the weighted networks in the real world are the result of a growing process in which the number of nodes and links, but also the link weights can change over time.

   Of course, there is a wide range of different ways to model growing networks whose weights are active variables. Here, we will discuss only a few of the models that have been proposed over the last ten years, and we will sort them in increasing order of complexity. We start with what can certainly be considered the minimal model to grow weighted scale-free networks, the so-called *Antal–Krapivsky (AK)* model. In this model, the structural growth of the network is coupled to the edge weights through a generalisation of the BA preferential attachment rule [10].

Definition 10.11 (The AK model) *Given three positive integer numbers N, n_0 and m (with $m \leq n_0 \ll N$), at time $t = 0$, one starts with a complete graph with n_0 nodes and $l_0 = \binom{n_0}{2}$ edges with equal weights. The graph grows by iteratively repeating at time $t = 1, 2, 3, \ldots, N - n_0$ the following three steps:*

(1) A new node, labelled by the index n, being $n = n_0 + t$, is added to the graph. The node arrives together with m edges.

(2) The m edges link the new node to m different nodes already present in the system. The probability $\Pi_{n \to i}$ that a new link connects the new node n to node i (with $i = 1, 2, \cdots, n - 1$) at time t is:

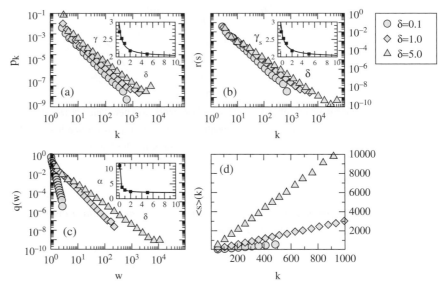

Fig. 10.6 Basic properties of the BBV model for $N = 10000$, $m = 2$, $w_0 = 1$ and three values of δ. Distributions of node degrees (a), node strengths (b) and link weights (c). Average strength $\langle s \rangle (k)$ of vertices of degree k as a function of k (d). Insets in the first three panels show agreement of the exponents with analytical predictions.

of generality, we can set $w_0 = 1$, so that the model depends on a single parameter δ, that is the fraction of weight which is induced by the new edge onto the others. Details about the algorithm we have used to grow graphs with the BBV model can be found in Appendix A.21. Figure 10.6 shows the results obtained by simulating the model for $N = 10000$, $m = 2$, and three different values of δ, namely $\delta = 0.1, 1.0$ and 5.0. For each value of the parameter, curves reported are averages over 500 realisations of the model. In panels (a) and (b) we plot respectively the node degree distributions p_k and the strength distributions $r(s)$, while in panel (c) we show the edge weight distribution $q(w)$. We immediately notice that all the curves look like straight lines in a double-logarithmic scale. In the limit $N \to \infty$ of very large networks, it is in fact possible to prove analytically that the BBV model produces power-law distributions for both vertex degree and vertex strength, $p_k \sim k^{-\gamma}$ and $r(s) \sim s^{-\gamma_s}$. Moreover, the two power laws have the same exponent $\gamma = \gamma_s = (4\delta + 3) / (2\delta + 1)$ [24]. This result is confirmed by our numerical simulations. The two insets in panel (a) and (b) report, as a function of δ, the values of γ and γ_s obtained by data fitting (full squares), and show the excellent agreement with the analytic expressions (solid lines). In addition to this, we also have a power-law distribution for the weights, $q(w) \sim w^{-\gamma_w}$ with an analytically predicted exponent $\gamma_w = 2 + 1/\delta$. Also this expression is in perfect agreement with the results of the numerical simulations, as shown in the inset in panel (c).

Summing up, the BBV model produces scale-free networks with a tunable exponent $\gamma \in [2, 3]$. In particular, when $\delta \to 0$, the model is equivalent to the BA model, since the weight distribution approaches a Dirac delta centred at 1, meaning that all link weights are

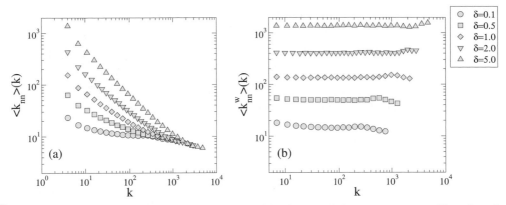

Fig. 10.7 Average nearest neighbours' degree function for the BBV model with $N = 10000, m = 2, w_0 = 1$ and five values of δ. Both unweighted (a) and weighted (b) quantities are reported.

equal to 1, and we have $p_k \sim k^{-\gamma}$ with $\gamma = 3$. For larger values of δ, the power law p_k gets broader, and the exponent γ tends to the value 2 in the limit $\delta \to \infty$. Interestingly, the model also produces correlations between topology and weights. In fact, as shown in Figure 10.6(d), the node strength scales as a function of the degree as $\langle s \rangle(k) \simeq Ak^{\beta}$, with $\beta = 1$ and $A \neq \langle w \rangle$ [24, 23]. Finally, the model exhibits disassortative degree–degree correlations. Figure 10.7(a) shows that the average nearest neighbour degree, $\langle k_{nn} \rangle(k)$, is a decreasing function of k, with a slope depending on the value of the parameter δ. More precisely, for small δ, the average nearest neighbours' degree function $\langle k_{nn} \rangle(k)$ is almost flat as in the BA model, while a disassortative behaviour clearly emerges as δ increases, giving rise to a power-law behaviour of $\langle k_{nn} \rangle(k)$ as a function of k. Conversely, the weighted average nearest neighbour's degree function $\langle k^{w}_{nn} \rangle(k)$ shows no dependence on k for any value of δ. However, for almost any value of k, we observe $\langle k^{w}_{nn} \rangle(k) > \langle k_{nn} \rangle(k) \; \forall k$, i.e. the values of the weighted quantity are always larger than those of the unweighted one, meaning that, for each value of k, the edges of larger weights from a node are pointing to the neighbours with larger degree [23].

An alternative model leading to results similar to those of the BBV is the *Dorogovtsev–Mendes (DM)* model [96]. While in the first model, high strength nodes attract new edges and then, after the attachment, the weights of their links are modified, in the DM model are the links with large weights which attract new connections, as illustrated in Figure 10.8, and increase their weights. In other words, while in the first case the attachment is to strong vertices, in the second case, the attachment is to high weight links. Of course, both growth mechanisms describe situations that can occur in real systems. For instance, in coauthorship networks it can happen that a successful collaboration between two authors, i.e. a high weight link, more than a high strength node, attracts new collaborators, and this is perfectly reproduced by the main ingredient of the DM model.

Definition 10.13 (The DM model)   *Let N be a large positive integer number. At time $t = 0$ one starts with a single link of weight 1. The graph grows by iteratively repeating at time $t = 1, 2, 3, \ldots, N - 2$ the following steps:*

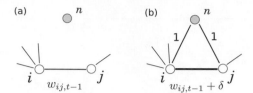

Fig. 10.8 Basic mechanisms of network growth in the DM model. (a) A new node n arrives at time t and a link (i,j) is chosen with a probability proportional to its weight $w_{ij,t-1}$. (b) Then, new node n is attached to both i and j with links of weight equal to 1, while the weight $w_{ij,t-1}$ is increased by a quantity equal to δ.

(1) A link (i,j) in the graph is chosen with a probability proportional to its weight, and such weight is increased by a constant $\delta > 0$:

$$w_{ij,t} = w_{ij,t-1} + \delta \tag{10.20}$$

(2) A new node labelled by the index n, being $n = 2 + t$, is added to the graph and is attached to both i and j by links of weight equal to 1.

Also this model can be solved analytically [96]. It is possible to prove that, in the limit of large N, the distribution of edge weights, node degrees and node strengths of the resulting network are all power laws, with exponents respectively equal to $\gamma_w = 2 + 2/\delta$ and $\gamma = \gamma_s = 2 + 1/(1 + \delta)$.

We finally discuss a model proposed by Jouko Kumpula, Jukka-Pekka Onnela, Jari Saramäki, Kimmo Kaski and Janos Kertèsz, and directly inspired to the growth of social systems. We have seen in Section 4.3 that *transitivity*, or *triadic closure*, i.e. the tendency to connect to friends of friends and produce triangles, is a basic mechanism of link formation in social networks. There is, however, another important mechanism, known as *focal closure* [114], in which links form because of the presence of common activities. Two people can create a link, independently of their distance on the social network, just because they live in the same neighbourhood, frequent a particular place, or work for the same company. These activities and places are in fact all "social focal points" that enhance the probability of interactions and link creation [184, 91]. The *Kumpula–Onnela–Saramäki–Kaski–Kertèsz (KOSKK)* model considers a network with a fixed number of nodes and implements both mechanisms of triadic and focal closure mentioned above, producing weighted networks with tunable community structure [191].

Definition 10.14 (The KOSKK model) *Let N be a large positive integer number, $0 < w_0$ and $0 \leq \delta \leq w_0$. Let $0 \leq p_\triangle \leq 1$, $0 \leq p_r \leq 1$ and $0 \leq p_d \leq 1$. Starting with N isolated nodes, the graph changes by iteratively repeating the following three steps:*

(1) A node i is selected at random. If i has at least one neighbour, one of its neighbours, say node j, is chosen with probability w_{ij}/s_i. If the selected node j has other neighbours apart from i, one of them, say k, is chosen with probability $w_{jk}/(s_j - w_{ij})$. If the link (i,k) already exists, its weight is increased by an amount δ, as shown in Figure 10.9(a). If instead there is no link between i and k, the link (i,k) is created with a local attachment

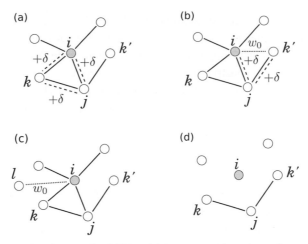

Fig. 10.9 Basic mechanisms of network evolution in the KOSKK model. In step 1 a node i and one of its neighbours j are selected. Then, one neighbour of j is selected. (a) If the neighbour chosen is k, the weight of the already existing link (i, k) is increased by δ. (b) if the node chosen is k', a new link (i, k') of weight w_0 is created with probability $p_\triangle$. In both cases also the weights of the two links (i, j), and either (j, k) or (j, k') are increased by the quantity δ. (c) In step 2, with probability p_r, node i creates a link of weight w_0 to a random node l. (d) In step 3, with probability p_d a randomly chosen node i has all its edges removed.

probability $p_\triangle$ and $w_{ik} = w_0$, as shown in Figure 10.9(b). In both cases, w_{ij} and w_{jk} are also increased by δ.

(2) *If node i selected in step 1 has no links, or otherwise with* global attachment probability p_r, *it creates a link of weight* w_0 *to a randomly chosen node l, as shown in Figure 10.9(c).*

(3) *With a* deletion probability p_d *a node is randomly chosen and all its links are removed, as shown in Figure 10.9(d).*

This model is among the most complicated we have considered in the book. It implements quite a number of different processes, and consequently has various control parameters. Notice, first of all, that the number of nodes N in the network is a quantity fixed as a model input, and remains constant over time. The number of links can increase in the first two steps of the process, while it can decrease in the third step. The first step of the model corresponds to a local search and to the formation of a new triangle or the reinforcement of an existing one, and stands for the *triadic closure* mechanism. The second step mimics *focal closure* with the creation of a new link to a randomly chosen node in the network. This is accounted for without the explicit introduction in the model of focal points, but simply allowing the formation of links between two nodes, independently of their distance on the network. Finally, the third step is in all equivalent to deleting a randomly selected node and all its links, and replacing it with a new node initially with no links, so as to keep N constant and avoid, at the same time, an unlimited growth of the number of links in the graph.

The most interesting property of the KOSKK model is that, when triadic and focal clo-sure are implemented together, they are able to produce networks with non-trivial structures both at a microscopic and at a mesoscopic scale. We will now show that is possible, for instance, to construct networks with different types of community structures. We discuss some of the numerical simulations reported in Ref. [191] and obtained by setting $w_0 = 1$ and exploring different values of δ, namely $\delta = 0, 0.1, 0.5, 1$. Also, the input probabili-ties chosen are $p_r = 5 \times 10^{-4}$ and $p_d = 10^{-3}$, while the value of p_Δ has been adjusted to keep the average degree $\langle k \rangle \approx 10$ equal for each value of δ considered. Under this selection of the input parameters, after an initial transient, the system reaches a stationary state where macroscopic properties of the network such as number of links and clustering coefficient stabilise over time. The largest connected components of the networks obtained for the four different values of δ are shown in colours in Figure 2 of the original paper [191]. Link colours represent link weights, with green standing for weak links, yellow for medium weight links and red for strong links. Notice that the weights enter the KOSKK model through the value of the parameter δ. In particular for $\delta = 0$, all the links have weight $w_0 = 1$ and we obtain unweighted networks with no community structure. When $\delta > 0$ we have instead the emergence of communities. This happens because of the local formation and reinforcement of triangles around randomly selected nodes, which act as nuclei for the formation of communities. This is clearly visible in the last two panels of Figure 2 of Ref. [191], where the division into communities becomes more pronounced as we increase the value of δ, and triangles with a large number of strong links, shown in red, appear inside each community. Notice also that the links connecting different communities are green links, i.e. small weight links. This observation immediately leads us back to the discussion of Example 4.10 on the importance of weak ties in bridging different parts of social networks.

Example 10.8 *(The strength of weak links)* We have introduced in Example 4.10 the work of Mark Granovetter on the importance of weak ties. In his paper, Granovet-ter showed that *weak ties* play a crucial role in bridging parts of social networks otherwise inaccessible through strong ties [142]. Here we can quantify this effect in

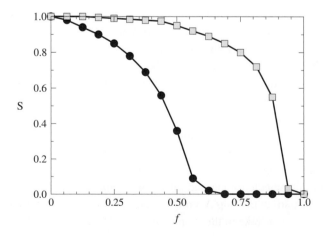

networks produced by the KOSKK model. We have in fact a natural way to define weak ties as those links with small weights. Now, if links between communities are weak, while links within communities are strong, then the network will fragment faster when links are removed in increasing rather than in decreasing order of weight. In order to show this, we follow again Ref. [191] and we remove links in both orders. We consider the same set of input parameters w_0, p_r, p_d, p_Δ as in the previous simulations and the case $\delta = 1$. The figure above reports the relative order of the largest connected component $S = s_1/N$ as a function of the fraction f of deleted links, defined as $f = K_{\text{del}}/K$ where K_{del} is the number of removed links. When links are removed in decreasing order of their weight (grey squares), the size of the largest connected component of the network remains almost unchanged ($S \simeq 1$) even when as much as 50 per cent of the links have been removed, and we still have $S \simeq 0.5$ when more than 85 per cent of the links have been removed. Conversely, when links are removed in increasing order of their weight (black circles), the network gets more fragmented and S decreases more quickly. In particular, the removal of slightly more than 50 per cent of the links with the smallest weights is sufficient to completely fragment the system ($S \sim 0$). This implies that weak links indeed play an important role in maintaining the connectivity of the entire system in the KOSKK model.

The KOSKK model not only gives rise to the emergence of communities, but also reproduces other important features of real-world networks. In fact, it allows the construction of assortative graphs with high clustering coefficients and small characteristic path lengths.

10.5 Networks of Stocks in a Financial Market

In this final section we show that networks can also be derived from time series. In particular, we will construct and study networks describing correlations among a set of stocks traded in a financial market. However, our arguments are pretty general and can be applied to other time series, for instance those derived from brain imaging as discussed in Box 4.3. We will take as an example the work of Giovanni Bonanno, Fabrizio Lillo and Rosario Mantegna, three researchers from the Observatory of Complex Systems of the University of Palermo, who studied the relations among 100 highly capitalised stocks traded in the New York Exchange Market from January 1995 to December 1998 [215, 54]. We will follow here the same approach and reproduce the analysis for a more recent period, from January 2012 to December 2014, using only a subset of $N = 62$ stocks of the original 100 ones, namely those still exchanged in the New York Exchange Market in the new period. We start from the most basic information which is the set of N time series of the stock prices. The temporal information can be very detailed: high-frequency data today allows price variations to be studied with a time resolution of seconds. However, we will focus here on price changes over the timescale of one day. Let us denote by $S_i(t)$, $i = 1, \ldots, N$, the value of the stock price i at time t, namely the value at the end of each trading day t. When dealing with financial data, it is usual to consider the logarithm of the stock values,

follow Ref. [54]. In our network the smallest distance, $\ell = 0.656$, corresponding to the couple of series with the highest correlation coefficient, $\rho = 0.785$, is obtained for the pair of stocks US Bancorp (USB) and Wells Fargo (WFC). The value of correlation for the pair of stocks Bank of America Corporation (BAC) and JP Morgan Chase (JPM) reported in Figure 10.10 is instead equal to $\rho = 0.712$ and gives a distance $\ell = 0.758$. The network contains all $N(N - 1)/2 = 1891$ edges, of course with different weights. To reduce the number of edges, we can choose to maintain only the links with weights w_{ij} larger than a given threshold w_{thresh} (with distance ℓ_{ij} smaller than ℓ_{thresh}). In this way the resulting graph will depend on the value of the adopted threshold. A better method to simplify the information contained in the matrix of weights (or in the distance matrix), is to reduce the number of edges to the smallest possible number, $K = 61$, to still have a connected network. Following Ref. [215, 54] we construct here the minimum spanning tree connecting the 62 stocks. This will give an immediate representation of the most relevant correlations, and also the ability to visualise the outcome, N being not too big in this case, since a tree is by definition a planar graph, so it can be drawn on a plane with no link intersections.

> **Definition 10.15 (Minimum spanning tree (MST))**   *Given a connected undirected weighted graph, a* spanning tree *of the graph is a subgraph that is a tree and connects all the nodes. A* minimum (distance) spanning tree *(MST) is a spanning tree such that the sum of its link lengths is smaller than or equal to the sum of the link lengths of any other spanning tree.*

In practice, a minimum spanning tree is a tree that connects all the nodes in the graph and has the shortest possible length. Similarly to the minimum distance spanning tree, if we work with the link weights instead of the link distances, we can define a *maximum weight spanning tree* as a spanning tree such that the sum of its link weights is greater than or equal to the sum of the link weights of any other spanning tree. Although the definition of minimal spanning tree is pretty simple, devising an efficient algorithm to find minimal spanning trees is a complicate task. One of the simplest algorithms consists of enumerating all the possible trees of G and selecting the one having minimal total weight. However, as shown in the following example, this is not a viable solution.

Example 10.9   *(Number of spanning trees)*   The *Matrix-Tree Theorem* is a formula for the number of spanning trees of a graph in terms of the determinant of a certain matrix. There are different versions of the theorem [55]. In its eigenvalue version, the theorem states that the total number of distinct spanning trees $t(G)$ of a weighted connected graph with N nodes can be written as [51]:

$$t(G) = \frac{1}{N} \prod_{i=2}^{N} \lambda_i$$

where $\lambda_2, \lambda_3, \ldots, \lambda_N$ are the $N - 1$ non-null eigenvalues of the Laplacian matrix L (see Definition 9.3) of the underlying unweighted graph, that is the graph in which all the edges have unitary weight. Let us consider for instance the weighted graph with $N = 6$ nodes and $K = 10$ edges shown below. The Laplacian matrix L of the underlying

$$L = \begin{pmatrix} 2 & -1 & 0 & -1 & 0 & 0 \\ -1 & 4 & -1 & -1 & 0 & -1 \\ 0 & -1 & 4 & -1 & -1 & -1 \\ -1 & -1 & -1 & 5 & -1 & -1 \\ 0 & 0 & -1 & -1 & 2 & 0 \\ 0 & -1 & -1 & -1 & 0 & 3 \end{pmatrix}$$

unweighted graph has five non-null eigenvalues, namely $\lambda_2 = (7 - \sqrt{13})/2$, $\lambda_3 = (7 - \sqrt{5})/2$, $\lambda_4 = (7+\sqrt{5})/2$, $\lambda_5 = (7+\sqrt{13})/2$, $\lambda_6 = 6$. Consequently, the number of distinct spanning trees in this graph is $t(G) = 99$. A large number for such a small graph. Indeed, the number of spanning trees grows pretty fast with order and size of the network. For instance, a complete graph of six nodes will have $t(G) = 1296$, while the complete graph of 62 stocks has about 3.5×10^{107} distinct spanning trees. In fact, it is easy to show that, in a complete graph of order N, we have $t(G) = N^{N-2}$ since the Laplacian matrix has $N - 1$ identical non-null eigenvalues $\lambda_2 = \lambda_3 = \ldots = \lambda_N = N$. Just to make another example, the number $t(G)$ of spanning trees in the US air transport network ($N = 500$, $K = 2980$) is about 5.6×10^{318}. An exhaustive search of all spanning trees of such graphs to find the minimum ones is therefore beyond our computational capacity.

To construct the minimum spanning tree for the set of 62 stocks under investigation we have made use of an algorithm proposed by Joseph Kruskal in 1956 [190]. The algorithm is explained in detail in Appendix A.20. In Figure 10.11 we report the minimum spanning tree we have obtained. Each of the stocks belongs to one of 12 different economic sectors

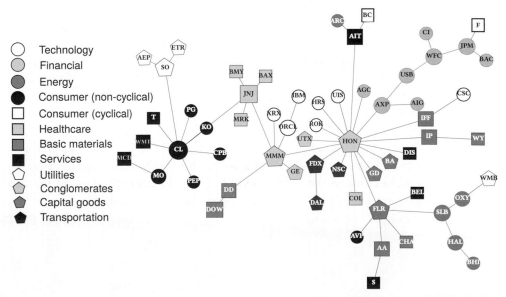

Fig. 10.11 Minimum spanning tree obtained from the stock correlation network. Each node is one of the $N = 62$ stocks and is labelled by the corresponding symbol. Node size represents the degree, while shape and colour distinguish the different economic sectors.

namely: technology, financial, energy, consumer/non-cyclical, consumer/cyclical, health-care, basic materials, services, utilities, conglomerates, capital goods and transportation. The economic sectors are indicated with shades and shapes with the code as reported in the left-hand side of the figure. Notice that the division in groups of different economic sectors and the relations between them hidden in the correlation matrix are now clearly visible from the minimum spanning tree. In particular, it is easy to identify well-defined groups, such as the group of financial stocks, the group of consumer non-cyclical and those of energy, conglomerates and healthcare stocks. The different technology stocks are instead connected through conglomerates, and in particular through Honeywell (HON) and 3M (MMM). The same thing happens for transportation stocks. Instead, basic materials and services are more fragmented in different parts of the tree. We can also focus our attention on some local properties of the nodes, for instance, their degree. The minimum spanning tree exhibits a large number of stocks with only one link, and some stocks with several links. Some of the high-degree stocks are prevalently linked to other stocks of the same sector, and act as local centres of such stocks. Examples are Colgate–Palmolive (CL), Johnson & Johnson (JNJ), JP Morgan Chase (JPM) and Southern Company (SO), respectively for consumer stocks, healthcare, financial and utilities. Some others connect instead stocks of different sectors. The most notable example is Honeywell (HON), the node with the largest degree in the tree, whose behaviour is correlated with 17 stocks belonging to six different economic sectors, including American Express (AXP), FedEx (FDX), International Paper (IP) and Boeing Corporation (BA). In order to show explicitly the distance at which the different parts of the tree are merged together in Kruskal's algorithm, we report in Figure 10.12 the associated dendrogram. We have already seen examples of dendrograms in Chapter 9. In the case reported here, each vertical line corresponds to a stock, while each horizontal line stands for a connection between different clusters and its height is equal to the distance at which the stocks are connected. Notice that the position of the stocks from left to right is arbitrary, so that we have adopted in the figure an ordering which emphasises the clustering structure of the system. We highlighted in the figure the sequences of two

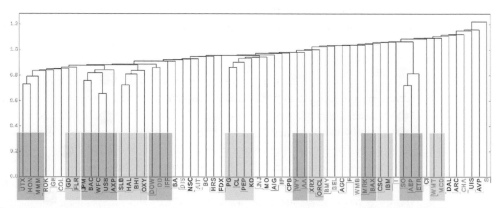

Fig. 10.12 Dendrogram of the set of 62 stocks obtained from the stock correlation network. The shaded regions identify sequences of two or more stocks belonging to the same economic sector that are successively merged by Kruskal's algorithm.

or more stocks belonging to the same economic sector which are successively merged by the Kruskal's algorithm. Again we observe from the dendrogram that several groups are clearly identified and correspond to the economic sectors.

In conclusion, in this section we have shown how to extract important information on a financial market by constructing the correlation network of the stocks from the associated time series of one-day logarithm price differences. Since the obtained network is complete we have constructed its minimum spanning tree. This is certainly a drastic filtering of the complete information contained in the correlation matrix. Still, it allows us to characterise clusters of stocks of different economic sectors, and also to investigate the relation between them.

10.6 What We Have Learned and Further Readings

A large number of databases of weighted networks have become available in the last decade, together with novel ways to characterise their properties and to model them. In this chapter we have introduced and studied two different data sets, namely the weighted network of the *US air transport system* and a network of *financial stocks*. However, what we have learnt about data, measures and models for weighted networks, and on their possible applications is only a small fraction of what is now available in the literature. The interested reader can find more on various aspects of weighted networks, including characterising shortest paths [246, 135] and spanning trees [213, 283], and on applications of weighted networks to the most diverse systems. See, for instance, the work by Diego Garlaschelli et al. on market investments [129], by Giorgio Fagiolo et al. on the World Trade Web [110], by Marco Baiesi and Maya Paczuski on earthquake networks [14], and by Paolo Tieri et al. on the human immune cell network [297]. Finally, concerning network models, some intriguing variations and generalisation of the models considered here can be found in Refs. [318, 321, 161, 6, 33, 322, 13, 149, 2].

Among other things, in this chapter we have extended the concept of rich club to weighted networks. As further reading we suggest the work by Athen Ma and Raúl Mondragón on a method to define the rich core of a network and extract the nodes belonging to it [211]. The proposed algorithm works both for unweighted and weighted networks [212].

Finally, in the last section we discussed how to construct a network from a set of time series. We considered, in particular, a complete weighted network obtained from the correlations among N time series of stocks traded in a financial market, and we focused on the problem of how to extract a subset of representative links from such a complete network of N nodes. We studied the most extreme approach, that of retaining only $N - 1$ links and producing a spanning tree, i.e. a tree connecting all the N nodes. Among all the possible spanning trees, we worked with the ones with the shortest length, the so-called *minimum spanning trees*. There are other ways to produce relevant spanning trees. See, for instance the work of Salvatore Scellato, Alessio Cardillo, Vito Latora and Sergio Porta to learn how to construct *maximum centrality spanning trees* [277]. There are also various ways to tune the link filtering and keep more than $N - 1$ links. We recommend a method introduced by

Michele Tumminello, Tomaso Aste, Tiziana Di Matteo and Rosario Mantegna to produce graphs that not only preserve the hierarchical organisation of the minimum spanning trees, but also contain cycles [299]. The approach based on correlations we have considered in this book is only one of many possible ways to build a network from one or more time series. As an example of an interesting method to transform *a single* time series into a network, the so-called *visibility graph*, see the work by Lucas Lacasa, Bartolo Luque and colleagues [193, 192].

Problems

10.1 Tuning the Interactions

(a) Prove that the definition of weights adopted for scientific collaboration networks in Example 10.1 implies that the strength s_i of node i is equal to the number of papers author i has coauthored with others.

(b) Show that the disparity Y_i of the weights of a node i in a weighted graph can be written as:

$$Y_i = \frac{1}{k_i} \left(1 + \frac{\sigma_i^2}{\langle w \rangle_i^2} \right)$$

where $\langle w \rangle_i$ and σ_i^2 are respectively the average and the variance of the weights of node i, while k_i is the degree of node i.

10.2 Basic Measures

(a) Show that the weighted clustering coefficient introduced in Definition 10.5 reduces to the unweighted definition of Section 4.3 in the case in which all weights are equal.

(b) Compute the clustering coefficients of the nodes of the Zachary's karate club network both in the unweighted and in the weighted case, and rank the nodes accordingly. In order to compare the two rankings, evaluate the Spearman's or the Kendall's rank correlation coefficient defined in Section 2.4. Finally, compute the graph clustering coefficient both in the unweighted and in the weighted case, and evaluate the ratio C^w/C. Explain the implications of the results found.

10.3 Motifs and Communities

(a) Prove that the geometric mean of $N = 2$ non-negative quantities is less than or equal to their arithmetic mean. Are you able to generalise this result to $N > 2$?

(b) Consider the network of the Zachary's karate club. Evaluate intensity, coherence and the relative Z-scores for the two types of three-nodes undirected graphs.

10.4 Growing Weighted Networks

Sort the links of the Zachary's karate club network according to their weights. Perform, as in Example 10.8, an analysis of the largest connected component as a function of the fraction f of links removed. Consider removing links in decreasing or in increasing order of their weights. In which case, does the network disintegrate faster? Comment on the results found.

10.5 Networks of Stocks in a Financial Market

(a) The percentage difference $\bar{R}(a, b)$ between two values a and b is the quantity usually calculated when you want to know the difference in percentage between the two values. It is defined as:

$$\bar{R}(a, b) = 2\frac{b - a}{a + b}.$$

Consider now $\bar{R}_i(t, \Delta t) = \bar{R}(S_i(t), S_i(t - \Delta t))$, and show that the logarithm price difference introduced in Section 10.5 is also directly related to $\bar{R}_i(t, \Delta t)$. In other words, prove that:

$$Y_i = \bar{R}_i + \frac{1}{12}(\bar{R}_i^3) + O(\bar{R}_i^5)$$

where $Y_i = Y_i(t, \Delta t)$ is the logarithm price difference defined in Eq. (10.21).

(b) Consider the urban street networks introduced in DATA SET 8. Compute, draw and compare the minimal spanning trees corresponding to the two cities of Bologna and San Francisco. See Ref. [277].

Appendices

A.1 Problems, Algorithms and Time Complexity

We introduce here the elementary concepts of computer science and complexity theory which will be useful in understanding the material of the other appendices. This appendix is primarily intended for those readers who do not have a background in computer science. Its main aim is that of presenting basic notions about computational problems, the axiomatic definition of algorithms, the standard methods to represent algorithms, the concept of time complexity and a few useful tools to estimate the time complexity of simple algorithms. Whenever possible, the discussion has been intentionally left informal and all the unnecessary technicalities have been discarded, in order to allow the reader to focus on the essential concepts without being distracted by definitions and theorems. For a more formal treatment of the material presented in this appendix we encourage the interested reader to refer to any classical book on algorithms, such as the trilogy by Donald Ervin Knuth [181, 182, 183] or in Complexity Theory [250].

A.1.1 What Is a "Problem"?

Formally, a *problem* $\mathcal{P}$ is a pair (D, Q) of an abstract description D and a question Q requiring an answer. A simple example of a problem in graph theory is the Graph Connectivity Problem: "Given a graph $G(\mathcal{N}, \mathcal{L})$ where $\mathcal{N}$ denotes a set of vertices and $\mathcal{L}$ is the set of edges among vertices in $\mathcal{N}$, is the graph G connected?" In this example, the description provides the context of the problem, which is usually represented as a class of objects (a generic graph $G(\mathcal{N}, \mathcal{L})$), while the question to be answered ("is G connected?") is a precise enquiry about a specific property of the class of objects under consideration.

The definition above is quite general and is given for an entire class of objects, without any specific connection to a particular object of that class. Conversely, an *instance* of a problem includes a full specification of one particular object of a class, which we indicate as x, and the solution $\mathcal{P}(x)$ is the answer to the question Q *for the specific object x* under consideration. For example, the sentence "given the graph G with nodes 1, 2, 3 and 4 and the edges $(2, 3)$ and $(2, 4)$, is graph G connected?" is a proper instance of the Graph Connectivity Problem, and the answer to the question "is the graph G connected" will be solely related to the particular graph with four nodes and two edges under consideration. We call *input* the specific object to which an instance of a problem is related, while the answer $\mathcal{P}(x)$ to an instance of a problem $\mathcal{P}$ defined on the specific input x is called a *solution* of the problem for the input x.

Problems can be divided into two classes, namely *decision problems* and *function problems*. The difference is that the answer to a decision problem is simply YES or NO, while the answer to a function problem is more complex than just YES/NO. The Graph Connectivity Problem stated above is a typical example of a decision problem, since it admits an answer that is either YES or NO (a graph can be either connected or disconnected). Usually, any decision problem has a corresponding function problem. The function problem corresponding to the Graph Connectivity Problem is the so-called Graph Connected Components problem, which can be formulated as follows: "Given a graph $G(\mathcal{N}, \mathcal{L})$, which are the connected components of G?" A proper answer to this problem consists of the list of all the connected components of the graph G, and of a list of vertices that belong to each connected component.

A.1.2 Algorithms and Decidability

Given a problem $\mathcal{P}$, it is interesting to ask whether there exists a procedure such that, for each possible input x, one can obtain the corresponding solution $\mathcal{P}(x)$ in a finite number of steps. We define an *algorithm* for a given problem $\mathcal{P}$ as a finite ordered sequence $\mathcal{A}$ of elementary operations (also called *instructions*), such that by applying this sequence of operations on an input x one obtains the corresponding solution of the problem $\mathcal{P}(x)$ in a finite number of steps. Given an algorithm $\mathcal{A}$ associated with the problem $\mathcal{P}$, we informally say that $\mathcal{A}$ *solves* $\mathcal{P}$, and we call *output* of $\mathcal{A}$ the solution of the instance $\mathcal{P}(x)$ provided by $\mathcal{A}$. Notice that there are two fundamental pre-requisites for an algorithm $\mathcal{A}$ to be correctly defined, i.e. that $\mathcal{A}$ should be able to provide a solution to the corresponding problem $\mathcal{P}$: *(i)* for any possible input x, and *(ii)* in a finite number of steps. Given a problem $\mathcal{P}$ we say that the problem is *decidable* if there exists at least one algorithm $\mathcal{A}$ that solves $\mathcal{P}$. If such an algorithm does not exist, then $\mathcal{P}$ is said *undecidable*. All the problems discussed in this book belong to the class of decidable problems, so that each of them has associated at least one algorithm that can provide a solution to any instance of the problem.

A.1.3 Algorithm Representation and Pseudocodes

The definition of algorithm is so general that it practically includes any kind of process that can be expressed as an ordered list of elementary operations, executed on an input and producing a certain output. The simple steps needed to reply to an email, the set of operations performed to shut down the turbine of an electrical power plant, the method devised by Euclid to find the greatest common divisor of two numbers, and the set of computer instructions used to encode an image in JPEG format, are just some examples of algorithms. The generality of the concept of algorithm implies that there are many different ways of expressing the same list of instructions, and that a unique standard to represent algorithms does not exist. A pretty simple way to express an algorithm consists in using statements in a natural language, i.e. English or French. Natural languages have the advantage to be easily understood by humans, so that a human who is given an input and a list of instructions in English can read and *execute* them in the given order to obtain the output. As an example, we consider a simple function problem that everybody has faced at primary school, namely the Greatest Common Divisor (GCD) problem. This problem can be stated as follows: "Given two integer numbers a and b, what is the greatest number c that exactly divides both a and b ?". The first algorithm that comes to mind consists in computing the set of all the common divisors of a and b and then finding the maximum of such set. A possible description of this algorithm in natural language reads:

1 **input:** two integer numbers, a and b
2 compute the set A of divisors of a
3 compute the set B of divisors of b
4 find the set C given by the intersection of A and B
5 find the maximum element m of set C.
6 **output:** m

Any human who reads this description and can understand English should be able to compute $GCD(a, b)$, for any pair of integers (a, b) given as input, by *executing* each instruction in the given order. Indeed, the proposed algorithm makes the assumption that finding all the divisors of a number, computing the intersection of two sets and finding the maximum element in a set are *elementary* instructions, and that the interpretation of these instructions is *unique* and *unambiguous*. However, whenever we deal with natural languages, the risk of mistakes and misunderstandings is always around the corner. In fact, virtually every human language includes words with multiple (sometimes even contradictory) meanings, and this would make the specification and interpretation of an algorithm pretty fuzzy and therefore disputable. In practice, it is fundamental to always have a precise and unambiguous description of an algorithm. The main reason is that correctly defined algorithms, i.e. appropriate sequences of instructions having a unique, unambiguous interpretation, can be also executed by an automatic calculator, like a digital computer, which is much faster than humans at executing repetitive tasks. Computers are not able to understand and execute algorithms described in natural language. They can only manage data encoded as strings of binary digits (bits) and execute sequences of instructions written in a particular language, so-called machine code, which is practically unreadable by humans. Therefore, we need a way to fill the existing gap between algorithms expressed in natural language (easy for humans but unusable for computers) and the corresponding machine code (unreadable to humans but executable on a computer).

This gap is filled in by *computer languages* (see Figure A.1), like FORTRAN, C, Java, etc., which are artificial languages specifically designed to communicate a list of instructions from a human to a computer. Computer languages (also called *programming languages*) help humans describing algorithms in an unambiguous way, and provide tools to transform an algorithm into a *program*, i.e. a corresponding list of instructions in machine code to be executed on a computer. Consequently, programming languages represent a trade-off between natural languages and machine code. An algorithm expressed in a computer language is relatively easy to understand for a human who knows that programming language, and is ready to be translated into machine code as needed.

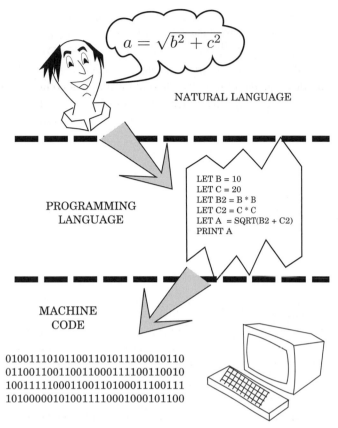

$$a = \sqrt{b^2 + c^2}$$

NATURAL LANGUAGE

PROGRAMMING
LANGUAGE

LET B = 10
LET C = 20
LET B2 = B * B
LET C2 = C * C
LET A = SQRT(B2 + C2)
PRINT A

MACHINE
CODE

0100111010110011010111000010110
0110011001100110001111100110010
1001111100011001101000011100111
1010000010100111100010000101100

Fig. A.1 Digital calculators cannot directly process algorithms expressed in natural language. Thus, in order to be executed on a computer, an algorithm must first be written in an intermediate programming language, and then translated from the programming language into the corresponding machine code.

A description of an algorithm given in a specific programming language is called an *implementation* of the algorithm. Among the hundreds of different programming languages created in the last sixty years, we have decided to use the C language to implement all the algorithms discussed in this book. This choice was motivated by the wide availability of C programming environments for different hardware platforms and operating systems, and by the well-known efficiency of C programs.

In order to compare C programs to algorithms expressed in natural language, in Figure A.2 we present, side by side, Euclid's algorithm to compute the greatest common divisor of two numbers expressed in natural language (current English), and one of the possible implementations of this algorithm in C.

Notice that in the implementation of the algorithm in C we find some extra instructions (like "#include <stdio.h>", "printf()", "int a, b, c", etc.) that are apparently not strictly related to the algorithm expressed in natural language. However, these instructions allow the program to communicate with the user (i.e. receiving inputs and returning outputs) and are needed to ensure that the implementation is unambiguous (i.e. we are working with integer numbers and not with real numbers).

1: consider two integer numbers, a and b
2: if b is different from zero, go to step 3, otherwise go to step 5
3: assign the value of b to the variable c
4: compute the remainder of the division a/c, and assign it to b
5: assign the value of c to a and repeat from step 2
6: the result is a

```c
#include <stdio.h>

int main(){
    int a, b, c;
    printf("Insert a: ");
    scanf("%d", &a);
    printf("Insert b: ");
    scanf("%d", &b);
    while(b!=0){
        c = b;
        b = a % c;
        a = c;
    }
    printf("%d\n", a);
}
```

(a) Natural language   **(b)** C programming language

Fig. A.2   The Euclid's algorithm to compute the greatest common divisor between two integer numbers a and b, expressed in natural language (left panel) and in the C programming language.

Due to the relatively large amount of additional instructions needed to provide a working C implementation of an algorithm, it is usually more convenient to express algorithms in a hybrid language, so-called *pseudocode*, which discards all the ancillary (or too elementary) instructions needed to make the program runnable on a computer. In Algorithm 1 we show the pseudocode of Euclid's algorithm.

Algorithm 1 Euclid's algorithm for greatest common divisor

Input: $a, b \in \mathbb{N}$
Output: gcd(a,b)
  1: **while** b **is not** 0 **do**
  2:     $c \leftarrow b$
  3:     $b \leftarrow a \bmod b$
  4:     $a \leftarrow c$
  5: **end while**
  6: **return** a

Pseudocode comes in handy when one wants to produce an accurate description of an algorithm, focusing only on the *logic* and discarding unnecessary implementation details. All the algorithms discussed in the following sections are given in a C-like pseudocode, while the corresponding programs written in C are available for download from the website www.complex-networks.net.

A.1.4 Time Complexity

Sometimes a decidable problem can be solved by more than just one single algorithm. An example is the $GCD(a, b)$ problem, for which we have already mentioned at least two algorithms: the first one based on the factorisation of a and b, and Euclid's algorithm which uses successive divisions. Both algorithms provide a solution to the $GCD(a, b)$ problem for any pair of integers (a, b), but for practical reasons one could ask which of the two requires the smallest number of operations. In general, if we have the option to choose among different computer programs to solve the same problem, then we would prefer the one that requires fewest resources in terms of processing time (number of elementary operations) and/or required amount of memory. It is usually not easy to compute the exact number of elementary operations performed by an algorithm *on any possible input*. Then, one possibility is to compare two algorithms by considering the number of operations needed on average (where the average is computed over the set of all possible inputs), or the number of operations performed in the worst possible case. However, when the set of possible inputs is not bounded, as in the case of integer numbers for $GCD(a, b)$, and for the input domains of the vast majority of algorithms, the number of operations performed on average or in the worst case is not defined, so that we need some other method to compare the relative efficiency of two algorithms.

The concept of *time complexity* solves this problem by allowing the relative efficiency of two different algorithms to be assessed, even when the exact number of operations performed by each algorithm is not easy to compute *a priori*. Before giving a precise definition of time complexity, we need to introduce here some useful asymptotic notation:

Definition A.1 (O (big-O) notation)   *Given two functions $f, g : \mathbb{N} \longrightarrow \mathbb{N}$, we say that $f(n)$ is $O(g(n))$, to be read "$f(n)$ belongs to big-O of $g(n)$", if there exist $n_0 \in \mathbb{N}$ and a constant c such that $f(n) \leq cg(n) \; \forall n > n_0$.*

Definition A.2 (Ω notation)   *Given two functions $f, g : \mathbb{N} \longrightarrow \mathbb{N}$, we say that $f(n)$ is $\Omega(g(n))$ if there exist $n_0 \in \mathbb{N}$ and a constant c such that $f(n) \geq cg(n) \; \forall n > n_0$.*

Definition A.3 (Θ notation)   *Given two functions $f, g : \mathbb{N} \longrightarrow \mathbb{N}$, we say that $f(n)$ is $\Theta(g(n))$ if $f(n)$ is $\Omega(g(n))$ and $f(n)$ is $O(g(n))$.*

The O-notation sets an asymptotic upper bound for a given function defined on integers. If we say that a function f is $O(n^4)$, then there exists a constant c such that f is asymptotically smaller than cn^4. This means that $f(n)$ does not grow faster than cn^4, when n goes to infinity. Conversely, the Ω-notation sets an asymptotic lower bound for f, such that when we say that f is $\Omega(n^2)$ we mean that f grows at least as fast as cn^2 when n goes to infinity. Finally, by saying that f is $\Theta(g(n))$ we mean that f grows *exactly* as $g(n)$ when n goes to infinity. We are now ready to define the time complexity of an algorithm $\mathcal{A}$:

Definition A.4 (Time complexity of an algorithm)   *The time complexity of an algorithm $\mathcal{A}$ solving a problem $\mathcal{P}$ is the smallest upper bound on the amount of time taken by $\mathcal{A}$ to run (in terms of number of elementary instructions), as a function of the size n of the input to the problem $\mathcal{P}$.*

In practice, the time complexity of an algorithm (also called *order* of the algorithm) tells us how quickly the running time of the algorithm grows as the length of the input increases. Operationally, given an algorithm $\mathcal{A}$, we compute its time complexity by considering the maximum number of steps $f(n)$ performed by $\mathcal{A}$ on any possible input of a given length n, as a function of n. The time complexity of $\mathcal{A}$ is the smallest possible upper bound $g(n)$ of $f(n)$. Once we have determined such $g(n)$, we say that the time complexity of $\mathcal{A}$ is $O(g(n))$.

Given a problem $\mathcal{P}$ and two algorithms $\mathcal{A}'$ and $\mathcal{A}''$ whose time complexity are respectively $O(g(n))$ and $O(h(n))$, we say that $\mathcal{A}'$ is *more efficient* than $\mathcal{A}''$ in solving $\mathcal{P}$ if:

$$\lim_{n \to \infty} \frac{g(n)}{h(n)} = 0. \tag{A.1}$$

For example, an algorithm that runs in $O(n^3)$ is more efficient than an algorithm running in $O(2^n)$, since all polynomial functions in n are asymptotically smaller than any exponential function of n. Notice that, when $g(n)$ is $\Theta(h(n))$, then the two algorithms have the same time complexity, but still $\mathcal{A}'$ might require less time than $\mathcal{A}''$ to provide a solution, for any value of the input size n. In Table A.1 we list some of the most common time complexity classes, with corresponding examples of graph algorithms. There exist algorithms having *constant* time complexity (which we denote as $O(1)$), such as the algorithm to select one element at random from an array of length n. Other algorithms are *logarithmic* (time-complexity $O(\log n)$), like the binary search algorithm on a sorted array. Other algorithms run in *linear time* (complexity $O(n)$), like the simple sequential scan used to find an element in an unsorted array. The most efficient sorting algorithms are *linearithmic*, i.e. have time complexity $O(n \log n)$, and a vast class of problems admit *polynomial-time* algorithms, i.e. $O(n^k)$ for a given value k. There exist also problems for which only *exponential* algorithms are known, i.e. algorithms running in $O(b^n)$, for $b > 1$, while some combinatorial problems have algorithms with *factorial* time complexity, i.e. $O(n!)$. The following chain of inequalities shows the relations existing among time complexity classes:

$$O(1) < O(\log n) < O(n) < O(n \log^\ell n) < O(n^k) < O(b^n) < O(n!) \tag{A.2}$$

where $\ell, b, k \in \mathbb{R}^+$, $\ell \geq 1$, $b > 1$, $k > 1$. For most practical uses, algorithms having a time complexity which is a combination of polynomials and logarithmic functions are considered *efficient algorithms*, while algorithms running in exponential (or factorial) time are considered *inefficient algorithms*.

A.1.5 Determining the Time Complexity of an Algorithm

We discuss here some properties of the O-notation that are useful in assessing the time complexity of an algorithm. The first property is that the O-notation is invariant under positive constant multiplication. This means that, for any constant $k \in \mathbb{R}^+$, if f is $O(kg(n))$ then we can simply write f is $O(g(n))$. Since algorithms are designed to produce a specified output when fed with a given input, it is usually possible to construct more sophisticated algorithms through combinations of simpler algorithms, i.e. by appropriately using the output of an algorithm as the input of another algorithm. Sometimes an algorithm $\mathcal{A}$ can be expressed as a sequence of simpler algorithms $\{\mathcal{A}_1, \mathcal{A}_2, \ldots, \mathcal{A}_L\}$ (usually called *functions* or *subroutines*) executed in a certain order, so that the output of algorithm $\mathcal{A}_{i-1}$ is fed as input to algorithm $\mathcal{A}_i$. In this case, which is depicted in Figure A.3(a), the total number of

Name	Running time	Example algorithm
Table A.1 Common time complexity classes.		
Constant	$O(1)$	select a random element from an array
Logarithmic	$O(\log n)$	binary search on sorted arrays
Linear	$O(n)$	single-source shortest paths on a graph
Linearithmic	$O(n \log(n))$	sampling of a Watts–Strogatz small-world graph
Quadratic	$O(n^2)$	betweenness centrality of all the nodes
Cubic	$O(n^3)$	shortest paths in a weighted graph (Floyd's)
Exponential	$2^{O(n)}$	enumerating all the cycles of a graph
Factorial	$O(n!)$	enumerating all the paths of a complete graph

a)   b)

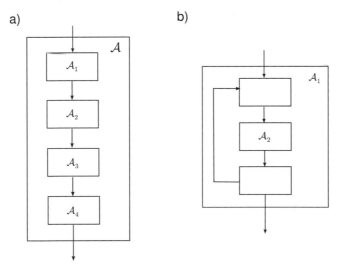

Fig. A.3 Sophisticated algorithms can be obtained by proper combinations of simpler algorithms. Some algorithms can be decomposed as a linear concatenation of subroutines (panel a), while in general more complicated algorithms can involve loops and recursive calls to other subroutines (panel b).

operations performed by $\mathcal{A}$ is equal to the sum of the operations executed by each subroutine. Let us denote as $f_i(n)$ the complexity of algorithm $\mathcal{A}_i$. The following two properties of the O-notation allow us to determine the time complexity of $\mathcal{A}$:

Definition A.5 (Sum rule)   *If $f_1(n)$ is $O(g_1(n))$ and $f_2(n)$ is $O(g_2(n))$, then $f_1(n) + f_2(n)$ is $O(g_1(n) + g_2(n))$*

Definition A.6 (Dominant term rule)   *If a function $f(n)$ can be written as the finite sum of other functions, then only the fastest growing term determines the time complexity of f.*

The sum rule can be derived directly from the definition of O-notation, while the dominant term rule reminds us that when $n \rightarrow \infty$ all the other terms of a sum of functions become negligible with respect to the fastest growing one. In general, we say that the time complexity of an algorithm is *dominated* by the sub-algorithm whose time complexity grows fastest. In the example depicted in Figure A.3(a), the algorithm $\mathcal{A}$ is a sequence of four subroutines $\{\mathcal{A}_1, \mathcal{A}_2, \mathcal{A}_3, \mathcal{A}_4\}$. Let us suppose that the time complexity of these subroutines is such that f_1 is $O(n^5)$, f_2 is $O(n^3)$, f_3 is $O(\log^2 n)$ and f_4 is $O(n)$. In this case, the time complexity $f = (f_1 + f_2 + f_3 + f_4)$ of the composition of the four subroutines is $O(n^5) + O(n^3) + O(\log^2 n) + O(n)$. By using the sum rule, we can say that f is $O(n^5 + n^3 + \log^2 n + n)$, and by using the dominant term rule, we can simplify it further, concluding that f is $O(n^5)$. Thus, the time complexity of $\mathcal{A}$ is $O(n^5)$. In Figure A.4 we plot the four functions f_1, f_2, f_3, f_4 and their sum f. Notice that, in practice, all the terms of the sum very soon become negligible with respect to n^5, so that the sum of the four functions becomes indistinguishable from n^5 as n increases. Indeed, not all algorithms can be expressed as simple sequences of subroutines. In fact, some algorithms have *loops*, i.e. sequences of instructions that are executed multiple times during the life of a program. In principle, an instruction in a loop can itself consist of a call to another subroutine. In Figure A.3(b) we show an example of an algorithm $\mathcal{A}_1$ which consists of a loop involving the usage of another algorithm $\mathcal{A}_2$. If an algorithm $\mathcal{A}_2$ performs $f_2 \in O(g_2(n))$ operations, and it is repeatedly called inside a loop of another algorithm $\mathcal{A}_1$ which is executed $f_1 \in O(g_1(n))$ times, we can intuitively conclude that the time complexity of $\mathcal{A}_1$ should be equal to the product of f_1 and f_2. The following rule allows us to simplify the expression of time complexity for products of functions:

Definition A.7 (product rule) *If $f_1(n)$ is $O(g_1(n))$ and $f_2(n)$ is $O(g_2(n)) \Rightarrow f_1(n) \cdot f_2(n)$ is $O(g_1(n) \cdot g_2(n))$*

For instance, if an algorithm $\mathcal{A}$ performs $f_1 \in O(n)$ times a loop which involves the usage of a subroutine whose time complexity is $f_2 \in O(n \log n)$, then the time complexity of $\mathcal{A}$ is equal to $f_1 f_2 \in O(n \times n \log n) = O(n^2 \log n)$.

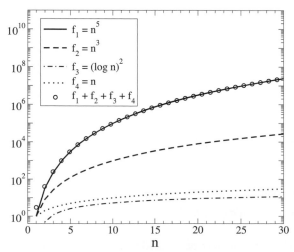

Fig. A.4 The subroutine $\mathcal{A}_i$ of an algorithm $\mathcal{A}$ having the fastest growing time complexity dominates the time complexity of $\mathcal{A}$.

A.1.6 Asymptotic Notation for Multiple Variables

When we gave the definitions of O, Ω and Θ notations, we made the simplifying assumption that the input of an algorithm is *mono-dimensional*, so that its length can be expressed by a single variable n. In general, the input to an algorithm can be multidimensional, and in the particular case of graph algorithms the time complexity could depend either on the number of nodes N, or on the number of links K, or on a combination of the two. It is therefore necessary to extend the definitions of asymptotic notation for the more general case of multidimensional functions.

Definition A.8 (Multidimensional O-notation)   *Given two functions $f, g : \mathbb{N}^k \longrightarrow \mathbb{N}$ we say that $f(\boldsymbol{n})$ is $O(g(\boldsymbol{n}))$ if there exists $M \in \mathbb{N}$ and a constant C so that $|f(\boldsymbol{n})| \leq C|g(\boldsymbol{n})|$ for all $\boldsymbol{n}$ such that $n_i > M \;\; \forall i = 1, 2, \cdots, k$.*

Definition A.9 (Multidimensional Ω-notation)   *Given two functions $f, g : \mathbb{N}^k \longrightarrow \mathbb{N}$ we say that $f(\boldsymbol{n})$ is $\Omega(g(\boldsymbol{n}))$ if there exists $M \in \mathbb{N}$ and a constant C so that $|f(\boldsymbol{n})| \geq C|g(\boldsymbol{n})|$ for all $\boldsymbol{n}$ such that $n_i > M \;\; \forall i = 1, 2, \cdots, k$.*

Definition A.10 (Multidimensional Θ-notation)   *Given two functions $f, g : \mathbb{N}^k \longrightarrow \mathbb{N}$ we say that $f(\boldsymbol{n})$ is $\Theta(g(\boldsymbol{n}))$ if $f(\boldsymbol{n})$ is $O(g(\boldsymbol{n}))$, and $f(\boldsymbol{n})$ is $\Omega(g(\boldsymbol{n}))$.*

Also the sum, product and dominant term rules can be extended in a similar way to the case of multidimensional inputs. These definitions, together with the rules discussed in the previous section, will be sufficient to determine the time complexity of all the algorithms described in this appendix.

A.2 A Simple Introduction to Computational Complexity

In Appendix A.1 we have given a formal definition of problems and algorithms, and we have defined the concept of time complexity, discussing how to compute and compare the time complexity of different algorithms. As we mentioned in Section A.1.2, all the problems discussed in this book are decidable, meaning that there exists at least one algorithm to solve each one of them. However, there are some classes of problems which seem more difficult to solve than others. This is not because it is more difficult to find an algorithm to solve them, but instead because all the possible algorithms to solve them are *inefficient*, i.e. have an exponential time complexity.

As an example, we consider two classical problems in graph theory, namely the *Eulerian trail* (ET) problem, that we saw in Section 1.5 and the *Hamiltonian path* (HP) problem. These two problems are defined as follows:

Definition A.11 (Eulerian trail problem)   *Given a graph $G(\mathcal{N}, \mathcal{L})$, find an Eulerian trail, i.e. a trail which visits each and every edge of the graph exactly once.*

Definition A.12 (Hamiltonian path problem)   *Given a graph $G(\mathcal{N}, \mathcal{L})$, find a Hamiltonian path, i.e. a path which visits each and every node of the graph exactly once.*

At a first glance, these two problems look quite similar, the only difference being that a ET is a visit to all the edges of the graph while a HP is a visit to all the nodes. Our intuition could even suggest that, once we have found an algorithm $\mathcal{A}$ to solve one of them, let us say the ET problem, it is sufficient to slightly modify $\mathcal{A}$ in order to solve the HP problem with not much additional effort. Unfortunately, this is a typical example of how much our intuition can be misleading when we try to assess the difficulty of a computational task. In fact, while for the ET problem there exist different algorithms having a polynomial time complexity, all the known algorithms for the HP problem have exponential time complexity. This fact suggests that these two problems are intrinsically different, as if they belonged to two different *classes*. In particular, it seems that the HP problem should be essentially *harder to solve* than the ET problem, since given a graph G, even the most efficient algorithm to solve the former will require exponentially more time than the standard algorithms to solve the latter.

Now the question is: how can we compute the relative difficulty of two problems? There is an entire branch of computer science, called Computational Complexity Theory, whose aim is to define the difficulty of a problem and to classify computational problems according to their relative difficulty. It is important to notice that computational complexity and time complexity are two distinct concepts, even if they are somehow related. In particular, time complexity gives information about the *amount of time* needed for a specific *algorithm* to produce an output, and is useful to compare the relative efficiency of different algorithms which solve the same problem. Instead, computational complexity tries to assess the *difficulty* of a given *problem*, independently of the particular algorithm that solves it, and proposes a classification of problems based on their relative difficulty. The connection

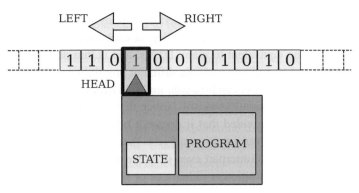

Fig. A.5 A Deterministic Turing Machine reads one symbol at a time from a tape of infinite length, and executes a program, which consists in a sequence of instructions.

between computational complexity and time complexity consists in the observation that all the problems falling in the same *complexity class*, i.e. having comparable difficulty, can usually be solved by algorithms having comparable time complexity.

A.2.1 Turing Machines

An accurate and comprehensive discussion of the foundations of computational complexity theory requires the introduction of a considerable amount of new concepts and definitions, and is far beyond the scope of this book. However, we believe that the reader should be somehow exposed to the basic notions of *computational complexity classes* without entering in the mathematical details behind those concepts.

The basic tools of computational complexity theory are abstract computational devices, called *machines*, rigorously defined through a set of specifications which describe the type of operations that each device can execute. The most famous example of such abstract devices is the so-called Deterministic Turing Machine (DTM), an ideal device introduced by Alan Turing in 1937 which is able to read inputs from a tape, execute a certain number of operations and produce the corresponding output on the same tape [300]. In Figure A.5 we show a schematic representation of a deterministic Turing machine. This machine consists of four elements, namely a *tape*, a *state register*, a *head* and a *program*. At each step, the machine reads the symbol at the position in the tape pointed by the head and executes the instruction of the program which corresponds to the value of the symbol read and to the current value of the state register. Each instruction consists of three operations, namely: *(1)* writing a new symbol in the tape at the current position of the head (or leaving it unmodified); *(2)* moving the head by one position, either to the right or to the left (or leaving the head at the same position); *(3)* updating the value of the state register. A schematic representation of a DTM is shown in Figure A.5. Without loss of generality, we can assimilate a DTM to a modern digital computer, where the tape is represented by the central memory, the state register is one of the internal registers of the Central Processing Unit (CPU), the head is a pointer to the location of memory currently read or written, and the program is a set of instructions in machine code.

Another remarkable example of ideal machines is the "Non-deterministic Turing Machine" (NTM). A non-deterministic Turing machine is similar to a conventional DTM,

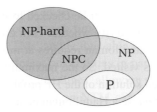

Fig. A.6 A picture of what we currently know about the relations among the complexity classes **P**, **NP**, **NP**-complete (**NPC**), and **NP**-hard. The class of **NP** problems contains the class of polynomial problems **P**, but it is still unknown whether **P** = **NP** or **P** ≠ **NP**. The set of all the problems that are **NP**-hard and belong to **NP** form the class of **NP**-complete problems. However, there are **NP**-hard problems that are not in **NP**. If **P**=**NP** then all the **NP**-complete problems will be solvable in polynomial time.

In Figure A.6 we show a tentative diagram which summarises the known relationships between the different computational complexity classes introduced in this appendix. Notice that in general a problem can be **NP**-hard even if it is not in **NP**. For instance, all the problems that can be solved in exponential time on an NTM are **NP**-hard but, by definition, they are not in **NP**.

A.3 Elementary Data Structures

As we have seen in Appendix A.1, a computational problem can be usually solved by more than just one algorithm. There are at least two factors contributing to the selection of one specific algorithm, namely the time complexity and the difficulty of the implementation. The first aspect is usually more important. In practice, even if we know that a given problem on a graph is in **P**, meaning that it can be solved by algorithms with polynomial time complexity, using an algorithm with time complexity $O(N^2)$ instead of another algorithm with time complexity $O(N^3)$ will make a substantial difference in terms of the time needed to obtain a solution.

Quite often the efficiency of an algorithm is mainly due to the way in which input data is stored and processed, and in particular to the specific data structure chosen to represent the information. Let us consider the problem of computing the degrees of all the N nodes of a graph. If the graph is stored as an $N \times N$ adjacency matrix, whose (i, j) entry is equal to 1 if node j is a neighbour of node i, then the simplest algorithm to compute the node degrees will have time complexity $O(N^2)$, since for each of the N nodes we have to check whether each of the N elements of the corresponding row of the matrix is equal to 1, which amounts to N^2 operations. Conversely, as we will see in Appendix A.4, sparse matrix representations allow the degree sequence of a graph to be computed in $O(K)$, in the case of the so-called *ij-form*, or even in $O(N)$ if we use a compressed row storage form.

In this appendix we review a few elementary data structures which will be useful for the efficient implementation of the algorithms described in the following sections. In general, the efficiency of a data structure is measured in terms of the time complexity required to perform operations on it, e.g. accessing (i.e. reading or writing) a specific element, searching for a certain element (i.e. checking whether it is contained in the structure or not), adding a new element, removing an existing one, and so forth. We will discover that some of those data structures are specifically designed to optimise certain operations. For instance, *Heaps* allow the maximum or the minimum element of a set to be accessed in $O(1)$, while *Binary Search Trees* guarantee that insertion, deletion, and search have all time complexity $O(\log N)$, on average. In the following, we will mainly refer to the time complexity of the *average case* of each of the operations supported by a given data structure already containing N elements. For some applications, instead, it might be more informative to look at the time complexity of one or more of those operations in the *worst case*. The interested reader can find more detailed information about data structures and their time complexity in computer science textbooks such as in Ref. [86].

A.3.1 Arrays

The most basic data structure is the `array` (or vector), which is a contiguous region of memory able to store a number of variables of the same type, often called *components* or *elements* of the array. An array is normally identified by a symbolic name, and each of the components are addressed using an integer index, indicated in square brackets. In

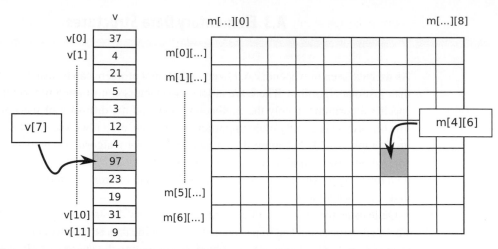

Fig. A.7 An array is a contiguous area of memory able to contain N elements of the same type, e.g. integer numbers. In the C language, array indexing starts at 0. As a consequence, the value 97, contained in the eighth position of the array v[], can be accessed using the notation v[7]. Similarly, the seventh element of the fifth row of the two-dimensional array m[][] will be denoted by m[4][6].

Figure A.7 we show the schematic representation of an array of N integers. In the C programming language, the first component of an array of length N has index 0 and the last one has index $N-1$. So, if we have an array v[] with $N = 12$ elements, v[0] indicates the first element of the array and v[11] indicates the last one. It is in general possible to define and use arrays with more than one dimension, which are useful to represent matrices and tensors. The indexing for higher-order arrays works as shown in Figure A.7, so that if m is a two-dimensional array, i.e. a matrix, with 7 rows and 9 columns, then m[4][6] indicates the element placed in the seventh element of the fifth row of the matrix. Similarly, if t is an order-3 tensor, t[2][6][9] is the tenth element of the seventh row of the third slice of the tensor.

Arrays are extremely simple structures to deal with. For instance, the time complexity of reading the value of a specific component of an array, once its position in the array is known, is $O(1)$, i.e. constant and independent on the size of the array. However, both adding an element at the end of an existing array of size N (thus incrementing its size to $N+1$), or removing one of the elements of the array (thus decreasing its size to $N-1$) in general have average time complexity $O(N)$. In fact, if we want to add a new element to the array v[], we could ask the operating system to extend the memory reserved to the array so that we can accommodate one more element after the current last element of the array. Depending on the programming language, on the operating system in use, and on the current state of the memory of your computer, this operation might be performed either in $O(1)$ or in $O(N)$. For instance, if the memory after the last entry of your array is free, i.e. it is not used to store other data of your program or by other programs, then most operating systems can "extend" the size of an existing array in $O(1)$. Conversely, if this is not possible, the operating system will need to find another contiguous area of memory large enough to contain an array of $N+1$ elements, and then copy the values of the existing array in the new location, which has time complexity $O(N)$. On the other hand, the deletion of an element of the array, say v[k], requires, in the worst case, to compact the array by copying v[k+1] into v[k], then v[k+2] into v[k+1], and so forth, again with average time complexity $O(N)$. Similarly, a naïve algorithm to search a generic array of size N for a specific element has average time complexity $O(N)$, i.e. linearly increasing with the length of the array. In fact, if we want to check whether a generic array v[] contains the value x, we could employ the so-called *linear search algorithm*, which consists into looping over all the elements of the array, starting with the first one, until we find an element v[k] = x. If x is not contained in v[], then we have to check all the N elements of v, while if x is an element of v[] then we will have to perform, on average, $N/2$ checks. In fact, in the latter case x can be placed with equal probability in each of the N positions, so on average we will find x after we have looped over $\frac{1}{N}\sum_{i=1}^{N} i = \frac{N+1}{2}$ elements of v. As a result, the time complexity of linear search is $O(N)$.

Arrays are mostly used when the number of elements is fixed *a priori* or does not change much during the computation, and when the typical operation is reading or writing elements of the array using their index. However, some of the shortcomings of arrays, specifically those related with linear time complexity for checking for the presence of a specific element, can somehow be alleviated by *sorting* the array, i.e. by rearranging its elements so that either v[i] $\leq$ v[i+1] or v[i] $\geq$ v[i+1] $\forall i = 0, 1, \ldots, N-1$. In fact, it is possible to efficiently search for an element of a sorted array through the so-called *binary search algorithm*, whose time complexity is $O(\log N)$ [86]. Algorithm 2 if the pseudocode for the binary search algorithm.

Algorithm 2 binary_search()

Input: x, N, v[] (sorted in ascending order)
Output: position of x, or -1 if x is not in v[]

1: high $\leftarrow$ N-1
2: low $\leftarrow$ 0
3: cur $\leftarrow$ $\lfloor$(low + high)/2$\rfloor$
4: **while** low $<$ high **do**
5:   **if** x $>$ v[cur] **then**
6:     low $\leftarrow$ cur + 1
7:   **else**
8:     **if** x $\leq$ v[cur] **then**
9:       high $\leftarrow$ cur
10:    **end if**
11: **end if**
12: **end while**
13: **if** x $=$ v[cur] **then**
14:   **return** cur
15: **else**
16:   **return** -1
17: **end if**

The idea behind binary search is simple: if we want to check whether a sorted array $v[]$ with N elements contains a certain value x, we first compare x with the median element of the array, i.e. the one placed at position $cur = \lfloor (N-1)/2 \rfloor$. If the elements of $v[]$ are sorted in ascending order of their values and x is smaller than $v[cur]$, then we can restrict our search to the lower half of the array $v[]$, since all the elements $v[k]$ with $k > cur$ will be larger than x, by definition. By iterating this procedure, at each step we restrict our search to a chunk of the array that is half the size of the chunk considered at the previous step, until we finally end up with an array chunk consisting of just one element. If that element is not equal to x, then we are certain that $v[]$ does not contain the value x. Since binary search is based on successively dividing the original array in halves, and a number N can be successively divided in two equal halves at most $\log_2 N$ times, the time complexity of binary search is $O(\log N)$.

But all improvements come at a cost, and the possibility of using binary search is bound to the possibility of obtaining a sorted array in the first instance. There are several algorithms to efficiently sort arrays according to a given relation, but one of the most widely used is the *Quick-sort* algorithm, whose pseudocode is provided in Algorithm 3. Let us assume, without loss of generality, that our array contains numeric values and we want the elements of the array to be sorted in ascending order, from the smallest (to be placed at the first position) to the largest (to be placed in the last position of the array). The Quick-sort algorithm is a recursive procedure which consists of three steps, namely:

Algorithm 3 `quick_sort()`

Input: A, `low`, `high`
Output: the array A, sorted in-place
1: **if** `low` $<$ `high` **then**
2:     `p` $\leftarrow$ `partition(A, low, high)`
3:     `quick_sort(A, low, p-1)`
4:     `quick_sort(A, p+1, high)`
5: **end if**

- at each step, select an element of the (sub-)array A, called *pivot*;
- partition the (sub-)array into two sub-arrays, each containing, respectively, the elements of A which are smaller or larger than the pivot;
- recursively apply the Quick-sort algorithm to the two sub-arrays obtained.

In practice, the array is usually partitioned by appropriately swapping its elements such as all the elements smaller than the pivot will be placed in the lower positions of the array, followed by the pivot, placed at position p, and then by all the elements larger than the pivot. Notice that after the function `partition()` returns, the pivot is placed in the position it will occupy when the array is sorted, and the subsequent application of the Quick-sort algorithm will work on sub-arrays whose length will in general be smaller than that of the original array. Ideally, an appropriate selection of the pivot element can guarantee that the two sub-arrays are roughly of the same size, so that the function `partition()` (which requires $O(N)$ operations) will be called at most $O(\log N)$ times.

It is possible to prove that, by using an appropriate procedure to select the pivot element at each iteration (for instance, by selecting the pivot as close as possible to the median of the subarray) and with an accurate implementation of the function `partition()`, the Quick-sort algorithm has average time complexity $O(N \log N)$. More importantly, it is possible to prove that the average time complexity of *any* general-purpose sorting algorithm cannot be smaller than $O(N \log N)$. In other words, it is not possible to sort a generic array with a time complexity smaller than $O(N \log N)$, on average. However, these proofs are beyond the scope of the present book, and the interested reader can find more details in Ref. [86]. Although Quick-sort is used in some of the algorithms described in this appendix, we decided to not include a custom implementation of the algorithm, and we instead preferred to use the implementation provided by the function `qsort()` of the standard ANSI C library.

Since arrays can be efficiently sorted in $O(N \log N)$, it is convenient to implement element searching through binary search whenever the elements of the array do not change too often, and in particular when one performs a large number of searches between two consecutive insertion/deletion operations. A typical example is the use of arrays to store the sorted lists of neighbours for each node of a graph, which do not change at all once the graph is given, so that it is possible to check if a certain node is a neighbour of another node in $O(\log N)$. Conversely, if your algorithm modifies the elements of the array frequently, the benefits of having a logarithmic-time search might easily be overcome by the necessity to sort the array again after each insertion/deletion.

A.3.2 Stacks

Like arrays, stacks are data structures that are able to contain a homogeneous collection of elements, but they have a peculiar way to access their elements. In fact, as the name suggests, the elements of a stack are arranged like a pile of books or a stack of plates at a restaurant: at each time you can get only the plate placed at the top of the stack (or the book at the top of the pile), and if a new plate is added to the stack, it becomes the new top. The access policy implemented by stacks is usually called LIFO, which stands for Last In First Out, indicating that the element that can be accessed at each time is always the last one that has been inserted on the stack. This access policy is particularly useful for maintaining a record of the progress made by an algorithm, and is actually used by modern operating systems to keep track of the calls to subroutines and functions.

As a consequence of the LIFO access policy, just two functions are sufficient to operate with a stack, namely `push()`, which inserts a new element at the top of the stack, and `pop()`, which removes the element at the top of the stack and returns it to the user. The most direct way to implement a stack is by means of an array `s` of size N plus a variable `top`, which contains the index of the top of the stack. By convention, when the stack is empty the variable `top` is set to -1. In Figure A.8 we show how a stack implemented as an array works in practice. In panel (a) we show an empty stack `s` for which `top = -1`. If we use the `push()` function to insert one element onto the stack, e.g. the number `23`, the element will be placed in the first available position of the stack, which is `s[0]`, and the variable `top` is set to `0` (panel b). After we successively call `push()` to insert the

the recursive procedure `bst_search()` to search an element in a BST, whose pseudocode is presented in Algorithm 4.

Algorithm 4 `bst_search()`

Input: node, t
Output: {TRUE | FALSE}
  1: **if** key[node] = t **then**
  2:     **return** TRUE
  3: **else**
  4:     **if** key[node] > t **then**
  5:       **if** node has `left` child **then**
  6:         `bst_search(left[node], t)`
  7:       **else**
  8:         **return** FALSE
  9:       **end if**
10:     **else**
11:       **if** node has `right` child **then**
12:         `bst_search(right[node], t)`
13:       **else**
14:         **return** FALSE
15:       **end if**
16:     **end if**
17: **end if**

Notice that at each step the function `bst_search()` discards an entire subtree and focuses the search to a smaller section of the tree. This mechanism is similar to that used by the binary search algorithm on sorted arrays (see Appendix A.3.1), which was shown to have time complexity $O(\log N)$. In the case of Binary Search Trees, the time complexity of `bfs_search()` depends substantially on the shape of the tree, and in particular on whether the subtree discarded at each step is of the same size of the subtree on which the search continues. Just for comparison, we present in Figure A.10 two Binary Search Trees generated by two different sequences with the same set of elements of the BST shown in Figure A.9. We immediately notice that in the BST shown in Figure A.10(a) only the root and the node with label 27 have both a left and a right child, while in the BST in Figure A.10(b) the root (label 18) and both its left and right children have in turn both a left and a right child. As a result, the first BST looks more stretched, while the second one looks rather compact. This difference has a substantial effect on the time needed to search an element in each of the two BSTs. In fact, the search for an element in a BST will require, on average, a number of iterations of `bfs_search()` equal to the average path length from the root of the tree to any other node in the tree. In the case of the BST in Figure A.10(a), a call to `bst_search()` will require on average 2.63 iterations (at most five iterations in the worst case), while for the BST shown in panel (b) of the same figure each call to `bst_search()` requires on average 1.38 iterations (at most 3 iterations in the worst case).

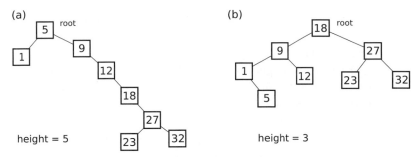

Several different Binary Search Trees can be associated with the same set of elements. The two BSTs shown have the same set of elements as the BST in Figure A.9. However, the BST in panel (a) looks more stretched, while that in panel (b) seems more compact. The reason is that the height of the BST in panel (a) is much larger than the height of the BST in panel (b).

The differences between two Binary Search Trees with respect to the time required to search their elements can be quantified in terms of the *height* of the trees. Given a node of a Binary Search Tree, its *height* is defined as the length of the (unique) path connecting that node to the root of the tree. For instance, the node with label 27 in Figure A.9 has height equal to 1, while the same node in the BST in Figure A.10(a) has height equal to 4. The height of a Binary Search Tree is defined as the maximum of the heights of its nodes. Consequently, the height of the BST shown in Figure A.9 is 3, and the heights of the BSTs shown in Figure A.10 are respectively equal to (a) 5 and (b) 3. It is easy to show that the time complexity of `bfs_search()` is $O(h)$, where h is the height of the Binary Search Tree.

A Binary Search Tree is said to be *balanced* if its height is $O(\log N)$, where N is the number of elements in the BST. It is possible to show that access, insertion, deletion and searching in balanced Binary Search Trees have time complexity $O(\log N)$, on average (see Ref. [86] for details). As a consequence, balanced Binary Search Trees are a very good choice to store data which can be accessed and updated frequently, and are used in many graph algorithms. Specific techniques exist to ensure that a BST remains balanced. However, these techniques, as well as the specific algorithms used to insert and delete elements from Binary Search Trees, are beyond the scope of the present book. The interested reader can find more details in Refs. [86, 279].

We have included an implementation of Binary Search Trees in the software library distributed with the programs available at www.complex-networks.net, which is used, not least, by the programs to sample random graphs from Erdős and Rényi models and from the configuration model, and by the program cnm for the greedy optimisation of modularity.

A.3.4 Binary Heaps

In many applications it is necessary to know which of the elements of a set is the maximum or the minimum one. A typical example is Dijkstra's algorithm to compute the shortest paths in a weighted graph, where, at each step, one needs to select, among the unvisited nodes, the one having the shortest distance from the source node. Notice that, if we keep the list of unvisited nodes in an array, the search for the element with minimum distance will require, at each step, the whole array to be scanned, which would cost $O(N)$. We might instead decide to keep the array sorted in increasing order of distance, so that the first element of the array is always the one with the shortest distance from the source node. However, since Dijkstra's algorithm proceeds by updating the distances of all the unvisited nodes, we need to sort the array at each step, resulting in $O(N \log N)$ cost, which is even worse than the linear search on the unsorted array. The solution in these cases is to use a Binary Heap, a data structure that guarantees $O(1)$ time complexity to access and extract the minimum (or the maximum) element in a set, and requires $O(\log N)$ operations to keep the data structure updated. If a binary heap allows the maximum (the minimum) element of a set to be accessed in constant time, then it is called a Max-Heap (Min-Heap). Without loss of generality, in the following we focus on Max-Heaps.

A Binary Max-Heap is a tree satisfying all these properties at any time:

1 Each node of the tree is associated with a numeric `key`, not necessarily unique;
2 The tree has exactly one node labelled as `root`;
3 Each node of the tree, with the only exception of the node labelled as `root`, has one *parent node*, namely the neighbour which is closer to `root`;
4 Each node of the tree can have at most two *children nodes* respectively called `left` and `right`;
5 The tree is *complete*, meaning that all the levels of the tree are fully filled, with the possible exception of the last level, which is always filled from left to right;
6 (Max-Heap property) the key of any node is larger than or equal to the keys of all its children.

Notice that property 6 effectively enforces a (partial) order over the set of elements of the heap. The definition of a Binary Min-Heap is identical except for property 6, which establishes that the key of any node must be smaller than or equal to the keys of its children. In Figure A.11 we show three trees with the same set of elements. It is easy to verify that only the tree shown in Figure A.11(c) is a Binary Max-Heap. In fact, the tree in Figure A.11(a) is not a Binary Max-Heap, since it does not respect property 3 (node 9 is labelled as root but has a parent, while the node with key 5 does not have a parent), property 4 (the node with key 1 has three children), and property 6 (both node 1 and node 9 have children whose keys are larger than theirs). The tree in Figure A.11(b) respects properties 1–4, but falls short on property 5 (the tree is not complete, since the second level contains only three nodes) and on property 6 (node 9 has a child whose key is larger than its own, namely the node with key 12).

Notice that, due to property 5, the height of a heap containing N elements is always equal to $\lfloor \log_2(N) \rfloor$. Moreover, thanks to property 6, the root node of a Binary Max-Heap always contains the maximum element of the set, so that finding the maximum corresponds to reading the element stored in the root of the Max-Heap, which can be done in $O(1)$. Similarly, in a Binary Min-Heap the root will always contain the minimum of the set.

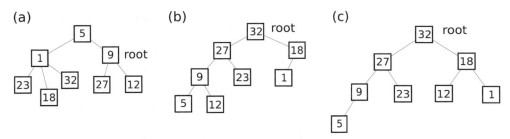

There are many different ways of organising a set of numbers in a tree. Of the three cases shown here, only the tree in panel (c) satisfies all the properties of a Binary Max-Heap.

Binary Heaps normally support at least two operations, namely:

- `heap_extract()`: remove the root of the Heap and rearrange the elements to ensure that the remaining structure is still a Heap (sometimes also called `delete()`);
- `heap_insert()`: inserts a new element in the Heap.

According to the specific way in which the Heap is implemented, some other functions might be needed. The easiest way to implement a Heap, which is the one we used in the programs distributed at `www.complex-networks.net`, is by means of an appropriately organised array. We start by noticing that any rooted binary tree can be represented through an array, if the elements of the array satisfy the following constraints:

- the left child of a node at position i in the array is placed at position $2 \times i + 1$;
- the right child of a node at position i in the array is stored at position $2 \times i + 2$;
- the parent of a node at position i in the array is stored at position $\lfloor (i - 1)/2 \rfloor$.

In Figure A.12 we show the array representation of the Max-Heap given in Figure A.11(c). In the figure we denote the links between a node and its left and right children respectively using solid and dashed lines. The root of the Heap (key 32) is placed at position 0 in the array. Then, its left child (key 27) is placed at position $2 \times 0 + 1 = 1$, while its right child is at position $2 \times 0 + 2 = 2$. Similarly, the node with key 1 is placed at position 6, and its parent (the node with key 18) can be found at position $\lfloor (6 - 1)/2 \rfloor = \lfloor 2.5 \rfloor = 2$. Notice that the array does not have any holes, in accordance with the fact that Heaps are by definition complete binary trees.

In Algorithm 5 we present the pseudocode of the function `heap_extract()`, which returns the current root of the Heap (corresponding to the maximum element in a Max-Heap, or to the minimum element in a Min-Heap), removes it from the Heap, and rearranges the Heap such that the new maximum is placed in the root. We assume that the Heap is stored in the vector `H[]`, and contains `N` elements. The function `heap_extract()` reads the maximum element of the Heap, which is stored, by construction, in the first position of the array `H` and puts it in the return variable `max_H`. Then, it puts the last element of `H` in the first position of the array, and calls the function `sift_down()`, which is responsible of restoring the Max-Heap property. The pseudocode of `sift_down()` is shown in Algorithm 6. In practice, `sift_down()` assumes that the left and right sub-trees of the node `current` satisfy the Max-Heap property, while the key of the node `current` itself

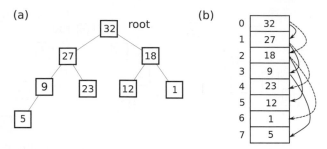

A Binary Max-Heap (a) can be represented by an appropriately organised array (b), where the root is placed at position 0 of the array, and the left and the right children of a node at position i are respectively placed at $2 \times i + 1$ and at $2 \times i + 2$. In panel (b) we denote respectively by solid and dashed arcs the relation between a node and its left and right child.

Algorithm 5 `heap_extract()`

Input: H, N
Output: H, N, max_H
1: `max_H ← H[0]`
2: `H[0] ← H[N-1]`
3: `N ← N - 1`
4: `sift_down(H, N, 0)`
5: **return** max_H

---

might actually be smaller than that of one of its children. If this is the case, the current node and the child having the largest key are swapped, and `sift_down()` is called recursively on the corresponding sub-tree, in order to let the `current` node literally "sift down" to its correct position in the Heap. This is needed because, after we have swapped the node `current` with its largest child, we need to ensure that the sub-tree rooted at `current` respects the Max-Heap property.

Let us consider, for instance, the Max-Heap shown in Figure A.13(a). When we call the function `heap_extract()`, the root is removed and replaced with the last node in the Heap, in this case the node with key 5 (panel b), which becomes the `current` node. Then, the key of `current` is compared with those of its children. In this case, the left child of `current` has a key equal to 27, which is larger than both the key of `current` (5) and the key of the right child of `current` (18). Thus, we swap `current` with its left child, and obtain the situation in Figure A.13(c). This time the key of the right child of the `current` node (which is equal to 23) is larger than both that of `current` and that of its left child (9). Hence we swap `current` with its right child, ending up in the situation shown in Figure A.13(d). In this case, the `current` node does not have any children, and the function `sift_down()` terminates. Notice that the Max-Heap in Figure A.13(d) satisfies the Max-Heap property. It is easy to verify that the function `sift_down()` has time complexity $O(\log(N))$, since it requires at most h operations, where h is the height of the Heap.

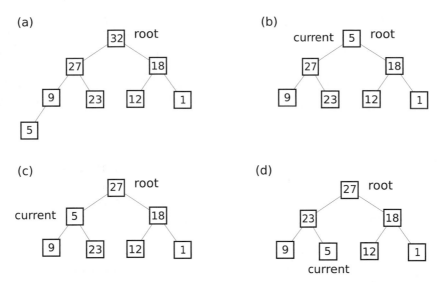

Fig. A.13 After we extract the maximum from a Max-Heap (panel a), it is necessary to rearrange to Heap in order to guarantee that the Max-Heap property is respected. This is done by the function `sift_down()`. It starts by placing the last element of the Heap as the root, and labelling it as `current`. Then, the `current` element is compared with its children and swapped with the largest of them (panels b and c). This procedure is recursively repeated, until the `current` node does not have any child (panel d).

Algorithm 6 `sift_down()`

Input: H, N, current
Output: H (with max-heap property respected)

 1: `largest ← current`
 2: `left ← 2 × current + 1`
 3: `right ← 2 × current + 2`
 4: **if** `left` < N **and** `H[left] > H[largest]` **then**
 5:    `largest ← left`
 6: **end if**
 7: **if** `right` < N **and** `H[right] > H[largest]` **then**
 8:    `largest ← right`
 9: **end if**
10: **if** `largest ≠ current` **then**
11:    `swap (H[current], H[largest])`
12:    `sift_down(H, N, largest)`
13:    **return**
14: **end if**

The insertion of a new element in a Max-Heap is performed by the function `heap_insert()`, whose pseudo-code is presented in Algorithm 7. In Figure A.14 we show schematically how `heap_insert()` works. Let us assume that we want to insert a node with key equal to 37 to the Max-Heap shown in Figure A.14. The new node is

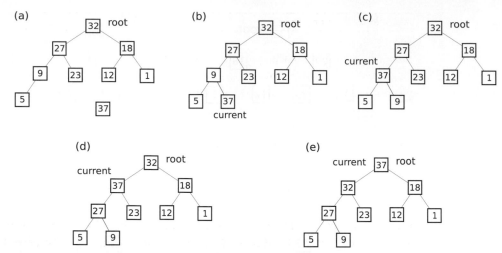

Fig. A.14 The insertion of an element in an existing Max-Heap (panel a) starts by putting the new element at the last position of the Heap (panel b). Then, the function `sift_up()` recursively compares the `current` element with its parent, and the two nodes are swapped if the key of `current` is larger than that of its parent (panels c, d, e). When `sift_up()` finishes (panel e), the new element has been placed in a position which guarantees that the Max-Heap properties is satisfied.

Algorithm 7 `heap_insert()`

Input: H, N, new_key

Output: N, H (with `new_key` added and max-heap property respected)

  1: N ← N + 1

  2: H[N−1] ← new_key

  3: sift_up(H, N, N)

  4: **return** H

---

labelled as `current` and placed at the first available position in the Heap, i.e. as the right child of the node with key 9 (see panel b). Then, the recursive function `sift_up()` (whose pseudocode is in Algorithm 8) compares the key of the `current` element with that of its parent, and if the parent has a smaller key, the two nodes are swapped. This is what happens in Figure A.14(b) since the parent of 37 has a smaller key (9). After the swap (see panel c), the sub-tree rooted at `current` satisfies the Max-Heap property, since the keys of the children of `current` are both smaller than 37. At this point, we need to recursively call `sift_up()` in order to let the `current` node find the correct position in the Max-Heap. Since the parent of the `current` node in Figure A.14 has key 27, we swap the two nodes to obtain the situation in Figure A.14(d). Again, `sift_up()` is called recursively, and the `current` node is then swapped with its parent (which has key 32), and we get to the final configuration shown in Figure A.14(e), which satisfies the Max-Heap property. Notice that the way in which `sift_up()` works is somehow dual to that of `sift_down()`, the main difference being that in `sift_down()` the element `current` flows from the root to the leaves of the heap, while in `sift_up()` the new node flows from the leaves up to the first

---

Algorithm 8 `sift_up()`

Input: `H, N, current`
Output: `H` (with max-heap property respected)
 1: `parent ← ⌊(current -1)/2⌋`
 2: **while** `H[parent] < H[current]` **do**
 3:     `swap(H[parent], H[current])`
 4:     `current ← parent`
 5:     `parent ← ⌊(current -1)/2⌋`
 6: **end while**
 7: **return** `H`

---

place in the Heap, which guarantees that the Max-Heap property is satisfied. Again, it is easy to show that the function `sift_up()` has time complexity $O(\log(N))$. This means that both insertion into and extraction from a Max-Heap can be performed in $O(\log(N))$ operations.

The library of utility functions distributed with the programs accompanying this book includes an implementation of Max-Heaps and Min-Heaps, together with an implementation of Priority Queues, a particular kind of Max-Heaps where the value of a key can be changed maintaining the Max-Heap property. This latter data structure is particularly useful for an efficient implementation of Dijkstra's algorithm to compute the shortest paths in a weighted network, as we will discuss in Appendix A.6.2, and for the implementation of the greedy modularity algorithm in Appendix A.18.

A.4 Basic Operations with Sparse Matrices

A matrix $A \in \mathbb{R}^{m \times n}$ is a table of $m \times n$ real numbers, placed in m rows and n columns. The elements of matrix A will be denoted by $a_{ij}, i = 1, \ldots, m, j = 1, \ldots, n$. One usually writes this in a compact form as: $A \equiv \{a_{ij}\}$. We shall indicate the entry i, j of matrix A as a_{ij}, or alternatively as $(A)_{ij}$. The pair (m, n) is the *order* of the matrix. If $m = n$, then A is a square matrix of order n.

The set of matrices of order (m, n) is a vector space. The operation of sum of elements of the vector space, and multiplication by a scalar are defined as

$$A, B \in \mathbb{R}^{m \times n}, \alpha, \beta \in \mathbb{R} \rightarrow C = \alpha A + \beta B \in \mathbb{R}^{m \times n}$$

with

$$c_{ij} = \alpha a_{ij} + \beta b_{ij}, \; i = 1, \ldots, m, \; j = 1, \ldots, n.$$

In addition, matrix multiplication is defined as

$$A \in \mathbb{R}^{m \times p}, \; B \in \mathbb{R}^{p \times n} \rightarrow C = AB \in \mathbb{R}^{m \times n}, \;\; c_{ij} = \sum_{l=1,p} a_{il} b_{lj}.$$

Note that, at variance with regular product, matrix product is not commutative, i.e. in general $AB \neq BA$.

If $A \in \mathbb{R}^{m \times n}$ is a $m \times n$ real matrix, its transpose, denoted with the symbol $A^\top$, is defined as:

$$(A^\top)_{ij} = (A)_{ji}$$

which means that the transpose of a matrix is obtained by exchanging the role of rows and columns. A real matrix equal to its transpose, i.e. such that $A = A^\top$, is said to be *symmetric*.

In this book we deal with matrices associated with graphs, like the adjacency matrix and the Laplacian matrix, which are $N \times N$ matrices where N is the number of nodes of the graph. Since matrices associated with real-world networks are usually sparse, meaning that the number of non-zero entries M of a matrix of order N is considerably smaller than N^2, we describe in the following two standard ways to represent a sparse matrix, namely the *ij-form* and the *compressed row storage format*, and we show how to switch from one representation to the other. We will also show that a sparse representation of a matrix allows a standard matrix operation to be performed, namely matrix-vector multiplication, more efficiently.

A.4.1 The *ij*-Form

The *ij-form* of an $N \times N$ matrix A uses three vectors, $\mathbf{i}, \mathbf{j}, \mathbf{s}$, whose length M is equal to the number of non-zero elements of the matrix. The three vectors store respectively the row and column indices of the non-zero elements of A, and their values. If A is the adjacency matrix of a weighted graph, then the vector $\mathbf{s}$ stores the weights of the links, while if A is the adjacency matrix of an unweighted graph, then all the entries of **s** would be equal, so that it is sufficient to assign **i** and **j** only. If the graph is directed, the length of vectors **i**, **j** (and, if necessary, **s**) is equal to the number of arcs K. If the graph is undirected, the matrix A is symmetric, and it would be sufficient to store only, say, the lower triangular part, i.e. the non-zero elements of the matrix for which $i > j$. However, it is more convenient to store the indices of all the non-zero elements of A. In fact, the disadvantage of using twice as much memory is counterbalanced by the fact that some operations, such as iterating over the neighbours of a node, are much easier. In this case, the length of vectors **i**, **j** is twice the number of edges, $M = 2K$.

Performing matrix-vector products with the *ij-form* is more efficient than using the full adjacency matrix. In fact, if $\mathbf{x} \in \mathbb{R}^N$ is a column vector, the product $\mathbf{y} = A\mathbf{x}$ can be computed in the following way:

Algorithm 9 Matrix vector multiplication (*ij-form*)

Input: i, j, s, x
Output: y $= A\mathbf{x}$
  1: **y** $\leftarrow \bar{0}$
  2: **for** $k = 0$ to $M - 1$ **do**
  3:     $y[i[k]] \leftarrow y[i[k]] + s[k] * x[j[k]]$
  4: **end for**

Notice that in the C-like pseudocode syntax used for the description of algorithms in the appendix, the indices of a vector **x** of size N are in the range $[0, N - 1]$, so that $x[0]$ is the first component of **x**, $x[1]$ is the second component and so forth. This notation is different from that adopted main text of the book (see for instance Section 1.6 and Section 2.3), where usually the indices of a vector **x** of size N range in $[1, N]$.

The algorithm performs M multiplications. Furthermore, it requires N additional operations to initially set all components of vector **y** to zero. Hence, the number of operations required is $M + N$, so that the time complexity of this algorithm is $O(M)$. Instead, the algorithm which implements matrix-vector multiplication using the full adjacency matrix A requires two nested cycles over N and has time complexity $O(N^2)$. Since in sparse matrices $M \ll N^2$, then matrix-vector multiplication in *ij-form* is more efficient.

A.4.2 The Compressed Row Storage (CRS) Format

Another common way of storing a sparse matrix is the *compressed row (or column) storage* (CRS). When using the CRS format, the matrix information is stored into three vectors, **r**, **j**, **s**. At variance with the *ij*-form of the matrix, now the three vectors have different lengths. In fact, **j** and **s** have M components, while **r** has $N + 1$ components. Vector **j** stores the column index of each non-zero component of the matrix, ordered by row index, while **s** stores the non-zero values of the matrix. Finally **r** is such that $r_i, i = 0, \ldots, N - 1$, is the index of the first non-zero element of the ith row, and r_N is set equal to M. In this way all the non-zero elements of the ith row of the matrix will be stored in positions $k = r_i, r_i + 1, r_i + 2, \ldots, r_{i+1} - 1$. Note that the CRS format is sometimes referred as the

---

Algorithm 10 Matrix vector multiplication (CRS-form)

Input: r, j, s, x
Output: y $= A$**x**
1: **y** $\leftarrow \vec{0}$
2: **for** $i = 0$ **to** $N - 1$ **do**
3:     **for** $k = r[i]$ **to** $r[i + 1] - 1$ **do**
4:         $y[i] \leftarrow y[i] + s[k] * x[j[k]]$
5:     **end for**
6: **end for**

Algorithm 11 From CRS to ij-format

Input: r
Output: i
1: **for** $\ell = 0$ **to** $N - 1$ **do**
2:     **for** $k = r[\ell]$ **to** $r[\ell + 1] - 1$ **do**
3:         $i[k] \leftarrow \ell$
4:     **end for**
5: **end for**

---

adjacency-list representation [86], since it is equivalent to storing the list of neighbours of each node. If a matrix A is stored in CRS format, then the matrix-vector product $\mathbf{y} = A\mathbf{x}$ is obtained by Algorithm 10.

Sometimes it is useful to switch from one sparse matrix representation to the other. Here we show how to switch from the CRS storage to the ij-representation of the matrix, and vice versa. It is trivial to switch from the compressed row storage to the ij-format; in fact, it is enough to define a vector **i** as shown in Algorithm 11.

The construction of the CRS representation from the ij-representation is slightly more complicated, mainly because graph links are stored in **i** and **j** vectors without any specific order. One possibility to switch from ij- to CRS format is to sort the edges (i, j) of the graph in increasing order of i, and then create a vector **r** by scanning the sorted vector **i**. However, there is a simpler and more efficient way to do it, which is shown in Algorithm 12. The algorithm works as follows. First we compute the number of occurrences of each node label i in **i** and we store it in $r_{i+1}, i = 0, \ldots, N - 1$. From this information, the vector **r** is iteratively computed as $r_0 \leftarrow 0, r_{i+1} \leftarrow r_{i+1} + r_i, i = 0, \ldots, N - 1$. Finally, vector **i** is scanned again. If the node label ℓ appears at position k of **i**, the corresponding value $j[k]$ is assigned to $j_{CRS}[r[\ell] + p[\ell]]$. The vector p is an auxiliary integer vector which keeps track of the number of neighbours of each node already found during this second scan of **i**.

The procedure can be better understood by looking at Figure A.15. Let us focus on the loop at lines 12–16 of the algorithm. Suppose that the first 11 entries of vector **i** (i.e. from $i[0]$ to $i[10]$) contained label 0 four times (so that $p_0 = 4$), label 1 three times ($p_1 = 3$), label 2 two times ($p_2 = 2$) and label 3 two times ($p_3 = 2$), and suppose that the vector **r**

Algorithm 12 From ij-to CRS format

Input: i, j

Output: $\vec{r}$, $\mathbf{j}_{CRS}$

1: $\vec{r} \leftarrow \vec{0}, \vec{p} \leftarrow \vec{0}$ {Set $\vec{r} \in \mathbb{N}^{N+1}$ and $\vec{p} \in \mathbb{N}^N$ to zero}

2: Count the number of elements for each row

3: **for** $k = 0$ **to** $M - 1$ **do**

4:    $r[i[k] + 1] \leftarrow r[i[k] + 1] + 1$

5: **end for**

6: Compute vector **r**

7: $r[0] \leftarrow 0$

8: **for** $l = 0$ **to** $N - 1$ **do**

9:    $r[l + 1] \leftarrow r[l] + r[l + 1]$

10: **end for**

11: Compute vector $\mathbf{j}_{CRS}$

12: **for** $k = 0$ **to** $M - 1$ **do**

13:    $\ell \leftarrow i[k]$

14:    $j_{CRS}[r[\ell] + p[\ell]] \leftarrow j[k]$

15:    $p[\ell] \leftarrow p[\ell] + 1$

16: **end for**

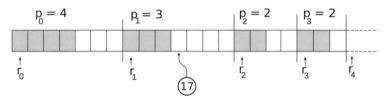

Fig. A.15   Creation of the $\mathbf{j}_{CRS}$ vector from vectors $\boldsymbol{i}$ and $\boldsymbol{j}$.

we have already generated at lines 7–10 of the algorithm reads $\mathbf{r} = (0, 7, 14, 18, 21, \ldots)$. This means that node 0 has $r_1 - r_0 = 7$ links, node 1 has $r_2 - r_1 = 7$ links, node 2 has $r_3 - r_2 = 4$ links and node 3 has $r_4 - r_3 = 3$ links. Figure A.15 shows how the vector $\mathbf{j}_{CRS}$ is modified when the 12th link is read. Suppose that $i[11] = 1$ and $j[11] = 17$. Then we insert the value 17 in $\mathbf{j}_{CRS}$ as indicated in the figure. Namely we set $j_{CRS}[r[1] + p[1]]$ equal to 17, and increase p_1 by one.

A.5 Eigenvalue and Eigenvector Computation

The problem of finding eigenvalues and eigenvectors of a matrix, the so-called *eigenvalue eigenvector problem*, is a classical problem in numerical analysis with numerous applications in science and engineering. Several methods have been designed to solve this problem. All such methods are based on iterations, since the eigenvalue problem is equivalent to the problem of finding the root of a polynomial, and there are no direct methods to solve algebraic equations of degree greater than four. We can distinguish between methods that are suited to compute one or few eigenvalues (and associated eigenvectors), and methods that provide all eigenvalues and eigenvectors. In the following we define the eigenvalue eigenvector problem, and we discuss the power method algorithm, which allows the leading eigenvalue and the associated eigenvector of a matrix to be computed.

A.5.1 Eigenvalues and Eigenvectors of a Matrix

Let $A \in \mathbb{R}^{N \times N}$ be a real square matrix. Suppose we want to find a vector whose direction remains unchanged under the action of the matrix, i.e. we look for a non-trivial vector $\mathbf{u} \in \mathbb{C}^N$ that is proportional to $A\mathbf{u}$. This means that we seek a vector such that:

$$A\mathbf{u} = \lambda\mathbf{u} \tag{A.3}$$

for some values of λ. If λ and $\mathbf{u}$ exist, they are called, respectively, *eigenvalue* and *right eigenvector* (or simply *eigenvector*) of matrix A. Note that the system in Eq. (A.3) may be written in the form:

$$(\lambda I - A)\mathbf{u} = 0$$

where I denotes the $N \times N$ identity matrix. Since this is a homogeneous system of linear equations, if $\det(\lambda I - A) \neq 0$ it admits the only solution $\mathbf{u} = 0$. If we look for a non-trivial solution, then we have to impose:

$$p(\lambda) \equiv \det(\lambda I - A) = 0 \tag{A.4}$$

where $p(\lambda)$ is called *characteristic polynomial* of matrix A. Therefore the eigenvalues of A are the roots of the polynomial $p(\lambda)$. Observe that $p(\lambda)$ is a polynomial of degree equal to N. The fundamental theorem of algebra states that $p(\lambda)$ has exactly N roots (in general, complex numbers), if each root is counted with the proper multiplicity. Once the eigenvalues $\lambda_1, \ldots, \lambda_N$ have been found, the eigenvectors can be computed by solving the linear homogeneous systems:

$$(\lambda_\alpha I - A)\,\mathbf{u}_\alpha = 0, \quad \alpha = 1, \ldots, N. \tag{A.5}$$

The set of eigenvalues $\lambda(A) = \{\lambda_1, \ldots, \lambda_N\}$ is called the *spectrum* of A. The largest absolute value of all the eigenvalues is called the *spectral radius* of the matrix and is denoted by $\rho(A) = \max_{1 \leq \alpha \leq N} |\lambda_\alpha|$.

Similarly, it is possible to define the *left eigenvectors* $\mathbf{v} \in \mathbb{C}^N$, such that:

$$\mathbf{v}^\dagger A = \lambda\mathbf{v}^\dagger \tag{A.6}$$

where the symbol $\dagger$ denotes the adjoint of a vector. Given a column vector $\mathbf{x} \in \mathbb{C}^N$, its adjoint $\mathbf{y} = \mathbf{x}^\dagger$ is a row vector such that $y_i = \bar{x}_i, i = 1, \ldots, N$, where $\forall x \in \mathbb{C}$, $\bar{x}$ denotes the complex conjugate of x. It is possible to show that left and right eigenvectors corresponding to different eigenvalues are orthogonal, i.e. $\mathbf{v}_\alpha^\dagger \mathbf{u}_\beta = 0$ if $\lambda_\alpha \neq \lambda_\beta$.

A.5.2 Diagonalisable Matrices

Real symmetric and, more in general, Hermitian matrices have special properties concerning eigenvalues and eigenvectors. For any matrix $A \in \mathbb{C}^{M \times N}$ we define its *adjoint* $A^\dagger \in \mathbb{C}^{N \times M}$ a matrix such that $(A^\dagger)_{ij} = \overline{(A)_{ji}}$. This means that $A^\dagger$ is the complex conjugate of the transpose of A. A square matrix $A \in \mathbb{C}^{N \times N}$ is said Hermitian if $A = A^\dagger$.

The first thing we need to observe is that the eigenvalues of real symmetric matrices, or more in general of complex Hermitian matrices, are real. In fact, taking the adjoint of the relation:

$$A\mathbf{u} = \lambda\mathbf{u}$$

one has:

$$\mathbf{u}^\dagger A^\dagger = \bar{\lambda}\mathbf{u}^\dagger. \tag{A.7}$$

Now, by left multiplying the first relation by $\mathbf{u}^\dagger$, right multiplying the second relation by $\mathbf{u}$ and subtracting each other, one obtains:

$$\mathbf{u}^\dagger (A - A^\dagger)\mathbf{u} = (\lambda - \bar{\lambda})\mathbf{u}^\dagger\mathbf{u}.$$

Let $\|\mathbf{u}\| \equiv (\sum_{i=1}^N |u_i|^2)^{1/2}$ denotes the *Euclidean norm* of a vector $\mathbf{u}$. Since $\mathbf{u}^\dagger\mathbf{u} = \|\mathbf{u}\|^2 > 0$, if $A = A^\dagger$ it follows $\lambda = \bar{\lambda}$, i.e. $\lambda \in \mathbb{R}$. Notice also that left and right eigenvectors of a real symmetric (or Hermitian) matrix coincide, as a consequence of Eq. (A.7), since $\lambda = \bar{\lambda}$. Moreover, since the eigenvalues of a real symmetric matrix are real, then also the eigenvectors are real, because they are solution of a real linear homogeneous system.

Because of the orthogonality property of the eigenvectors, it appears natural to normalise them in such a way that their Eucledian norm is equal to 1, i.e. we impose that:

$$\mathbf{u}_\alpha^\top \mathbf{u}_\alpha = 1, \quad \alpha = 1, \ldots, N.$$

In this way, if the eigenvalues are all distinct, i.e. if $\lambda_\alpha \neq \lambda_\beta \forall \alpha \neq \beta, \alpha, \beta = 1, \ldots, N$, we have:

$$\mathbf{u}_\alpha^\top \mathbf{u}_\beta = \delta_{\alpha\beta}, \quad \alpha, \beta = 1, \ldots, N.$$

In matrix form this can be written as $U^\top U = I$, where we denote by $U = (\mathbf{u}_1, \ldots, \mathbf{u}_N)$ a square matrix whose columns are the normalised eigenvectors of matrix A. This relation tells us that the eigenvectors are independent (since $\det(U) \neq 0$), and, furthermore, that the transpose of matrix U is equal to its inverse, i.e. that $U^{-1} = U^\top$. If a matrix is not symmetric, then it is not guaranteed that its eigenvalues are real. However, it may be shown that, if the eigenvalues are all distinct, i.e. if:

$$\lambda_\alpha \neq \lambda_\beta, \quad \alpha \neq \beta, \quad \alpha, \beta = 1, \ldots, N$$

then the N eigenvectors corresponding to the N eigenvalues are independent. This means that the set $\{\mathbf{u}_\alpha, \alpha = 1 \ldots, n\}$ forms a basis of $\mathbb{C}^N$ (see Chapter 8 of Ref. [198]). This is in fact a corollary of the following theorem stating that eigenvectors corresponding to distinct eigenvalues of a matrix are linearly independent.

Theorem A.1   *Let* $\lambda_1, \lambda_2, \ldots, \lambda_k$ *denote* $k \leq N$ *distinct eigenvalues of matrix* $A \in \mathbb{C}^{N \times N}$, *and let us indicate* $\mathbf{u}_1, \ldots, \mathbf{u}_k$ *the corresponding eigenvectors. The k eigenvectors* $\mathbf{u}_1, \ldots, \mathbf{u}_k$ *are linearly independent.*

For a proof of the theorem see, for example, Ref. [198]. An immediate consequence of the theorem is that if $k = N$ then the eigenvectors form a basis. Observe that the relation:

$$A\mathbf{u}_\alpha = \lambda_\alpha \mathbf{u}_\alpha, \quad \alpha = 1, \ldots, N$$

can be written, using matrix notation, as:

$$AU = U\Lambda$$

where $\Lambda = \mathrm{diag}(\lambda_1, \ldots, \lambda_N)$ is a diagonal matrix containing the eigenvalues. If the eigenvectors form a basis of $\mathbb{C}^N$, then $\det(U) \neq 0$, and one can write:

$$\Lambda = U^{-1}AU.$$

In this case we say that the matrix is diagonalisable.

If the eigenvalues are not all distinct, then it may or may not still be possible to find a basis of eigenvectors, depending on the structure of the matrix. Suppose that the distinct eigenvalues are $\lambda_1, \ldots, \lambda_m$, where the eigenvalue λ_i has multiplicity m_i, so that $\sum_{i=1}^m m_i = N$. Then, since the eigenvectors corresponding to different eigenvalues are linearly independent, and since there could be no more than m_i independent eigenvectors corresponding to a given eigenvalue λ_i, a necessary and sufficient condition for diagonalisability is that for each eigenvalue λ_i there are exactly m_i linearly independent eigenvectors. If this condition is satisfied, then the matrix is still diagonalisable[1]. If this is not the case, then the matrix cannot be reduced to a diagonal matrix by a similarity transformation. It can be shown, however, that in this case the matrix can be put in the so-called canonical Jordan form (see, for instance, Chapter 10 of [198]).

A.5.3 The Power Method

The power method and the Lanczos method are the two most common means of finding the largest eigenvalue of a matrix and its associated eigenvector. Although algorithms based on the Lanczos method are probably the most effective ones for the computation of eigenvalues and eigenvectors of sparse matrices, their description is beyond the scope of the book, and we address the interested reader to the book by Gene Golub and Charles Van Loan, which is an excellent reference for numerical linear algebra [137]. In the following we describe in detail the power method. Notice that an implementation

[1] If the matrix is real symmetric or Hermitian, it can be proved that the matrix is always diagonalisable [198].

of the method is provided by the program pm, available for download from the website www.complex-networks.net. From now on we shall deal with real matrices only. Let $A \in \mathbb{R}^{N \times N}$ be a real square matrix. Let us denote by $\lambda_1, \ldots, \lambda_N$ its eigenvalues sorted in decreasing order of absolute value, and assume $|\lambda_1| > |\lambda_2|$, i.e.

$$|\lambda_1| > |\lambda_2| \geq |\lambda_3| \geq \cdots \geq |\lambda_N|. \tag{A.8}$$

This implies, in particular, that λ_1 is simple and real. We can easily prove this by contradiction. In fact, let us assume that λ_1 has a non-zero imaginary part. Then, $\overline{\lambda_1} \neq \lambda_1$ would also be an eigenvalue because the characteristic polynomial of the matrix has real coefficients. In this case the two largest eigenvalues would coincide in absolute value, which contradicts assumption (A.8). If λ_1 is real, the corresponding eigenvector $\mathbf{u}_1$ is also real. Let us assume, furthermore, that A admits a basis of eigenvectors, denoted by $\mathbf{u}_i : A\mathbf{u}_i = \lambda_i\mathbf{u}_i$, $i = 1, \ldots, N$, and consider the sequence of vectors:

$$\boldsymbol{x}_{n+1} = A\boldsymbol{x}_n, \ n = 0, 1, \ldots, \tag{A.9}$$

where $\boldsymbol{x}_0$ is any arbitrary non-zero vector of $\mathbb{R}^N$, for instance $\boldsymbol{x}_0 = (1, 1, \ldots 1)^\top$. We will show below that:

$$\lim_{n \to \infty} \frac{\boldsymbol{x}_n}{\|\boldsymbol{x}_n\|} = \mathbf{u}_1 \tag{A.10}$$

i.e. that the normalised vector $\boldsymbol{x}_n/\|\boldsymbol{x}_n\|$ converges to the normalised eigenvector $\mathbf{u}_1$ associated with the dominant eigenvalue λ_1, and also that $\boldsymbol{x}_{n+1}$ tends to be proportional to $\boldsymbol{x}_n$ as $n \to \infty$, with proportionality constant equal to λ_1.

Since the eigenvectors $(\mathbf{u}_1, \mathbf{u}_2, \ldots, \mathbf{u}_N)$ of A form a basis of $\mathbb{R}^N$, then any vector of $\mathbb{R}^N$ can be expressed as a linear combination of these eigenvectors[2]. In particular, we have:

$$\boldsymbol{x}_0 = \sum_{i=1}^{N} \alpha_i\mathbf{u}_i$$

$$\boldsymbol{x}_1 = A\boldsymbol{x}_0 = A\sum_{i=1}^{N} \alpha_i\mathbf{u}_i = \sum_{i=1}^{N} \alpha_1 A\mathbf{u}_i = \sum_{i=1}^{N} \alpha_i\lambda_i\mathbf{u}_i$$

$$\boldsymbol{x}_2 = A\boldsymbol{x}_1 = A\sum_{i=1}^{N} \alpha_i\lambda_i\mathbf{u}_i = \sum_{i=1}^{N} \alpha_i\lambda_i A\mathbf{u}_i = \sum_{i=1}^{N} \alpha_i\lambda_i^2\mathbf{u}_i$$

and in general:

$$\boldsymbol{x}_n = \lambda_1^n \left(\alpha_1\mathbf{u}_1 + \sum_{i=2}^{N} \alpha_i \left(\frac{\lambda_i}{\lambda_1} \right)^n \mathbf{u}_i \right).$$

If we assume that $\alpha_1 \neq 0$, since by hypothesis $|\lambda_1| > |\lambda_i| \ \forall i \neq 1$, then we have:

$$\lim_{n \to \infty} (\lambda_i/\lambda_1)^n = 0 \quad \forall i \neq 1$$

[2] Strictly speaking, the eigenvectors form a basis if $\mathbb{C}^N$, and therefore of $\mathbb{R}^N$, but only if we allow complex coefficients in the expansion on such a basis.

which proves that the vector $\mathbf{x}_n$ will become proportional to the first eigenvector $\mathbf{u}_i$ as $n \to \infty$. Moreover, if we indicate by $x_{n,j}$ and $u_{i,j}$, respectively, the jth component of $\mathbf{x}_n$ and of $\mathbf{u}_i$, and we consider the ratio:

$$\frac{x_{n+1,j}}{x_{n,j}} = \lambda_1 \frac{\alpha_1 u_{1,j} + \sum_{i=2}^{N} \alpha_i (\lambda_i/\lambda_1)^{n+1} u_{i,j}}{\alpha_1 u_{1,j} + \sum_{i=2}^{N} \alpha_i (\lambda_i/\lambda_1)^n u_{i,j}} \tag{A.11}$$

we can conclude that, as long as $\alpha_1 u_{i,j} \neq 0$, the ratio $x_{n+1,j}/x_{n,j}$ will converge to λ_1, $\forall j$, because the fraction in the right-hand side of Eq. (A.11) will tend to 1 as $n \to \infty$. This means that by iterating Eq. (A.9) we obtain an approximation of the leading eigenvalue of A and of the corresponding eigenvector.

It is also possible to show that asymptotically the ratio $x_{n+1,j}/x_{n,j}$ converges to λ_1 as a geometric sequence of common ratio λ_2/λ_1, i.e. that:

$$\frac{x_{n+1,j}}{x_{n,j}} = \lambda_1 + O\left(\left(\frac{\lambda_2}{\lambda_1}\right)^n\right). \tag{A.12}$$

In fact, from Eq. (A.11) we obtain:

$$\begin{aligned}
\frac{x_{n+1,j}}{x_{n,j}} &= \lambda_1 \frac{\alpha_1 u_{1,j} + \alpha_2 u_{2,j}(\lambda_2/\lambda_1)^{n+1} + \sum_{i=3}^{N} \alpha_i (\lambda_i/\lambda_1)^{n+1} u_{i,j}}{\alpha_1 u_{1,j} + \alpha_2 u_{2,j}(\lambda_2/\lambda_1)^n + \sum_{i=3}^{N} \alpha_i (\lambda_i/\lambda_1)^n u_{i,j}} \\
&\simeq \lambda_1 \left(\frac{\alpha_1 u_{1,j} + \alpha_2 u_{2,j}(\lambda_2/\lambda_1)^{n+1}}{\alpha_1 u_{1,j} + \alpha_2 u_{2,j}(\lambda_2/\lambda_1)^n}\right) \\
&= \lambda_1 \left(\frac{1 + c(\lambda_2/\lambda_1)^{n+1}}{1 + c(\lambda_2/\lambda_1)^n}\right) \\
&= \lambda_1 \left(\frac{1 + cr^{n+1}}{1 + cr^n}\right)
\end{aligned}$$

where both in the numerator and in the denominator we considered only leading order terms in $(\lambda_2/\lambda_1)^n$ and neglected $(\lambda_i/\lambda_1)^n$ for $i > 2$, since these latter terms will converge to zero at least as fast as $(\lambda_2/\lambda_1)^n$. We have set $c = (\alpha_2 u_{2,j})/(\alpha_1 u_{1,j})$ and $r = \lambda_2/\lambda_1$. The last term in Eq. (A.13) can be rewritten as:

$$\lambda_1 \left(\frac{1 + cr^{n+1}}{1 + cr^n}\right) = \lambda_1 (1 + c\,r^{n+1})(1 - c\,r^n) + O(r^{2n}) = \lambda_1 \left(1 + c\,(r-1)\,r^n + O(r^{2n})\right)$$

where we used the approximation $(1 + x)^{-1} = 1 - x + O(x^2)$, with $x = c\,r^n$. Combining the last two expressions we obtain Eq. (A.12), meaning that the error on the estimation of λ_1 decreases by a constant factor $|\lambda_2/\lambda_1|$ at each iteration. Hence, the smaller the ratio $|\lambda_2/\lambda_1|$, the faster the convergence to λ_1 of the ratio $x_{n+1,j}/x_{n,j}$.

In the previous method it remains to decide which component j we want to consider to approximate λ_1. Rather than selecting one specific component, it is better to make use of all the information contained in $\mathbf{x}_n$ and $\mathbf{x}_{n+1}$, e.g. by weighting each component $x_{n+1,j}$ by a weight proportional to $x_{n,j}$. This can be obtained by introducing the quantity:

$$\sigma_n = \frac{\sum_j x_{n,j} x_{n+1,j}}{\sum_j x_{n,j} x_{n,j}} = \frac{\mathbf{x}_n^\top A \mathbf{x}_n}{\mathbf{x}_n^\top \mathbf{x}_n}$$

which is known as the *Rayleigh quotient*, and has the following remarkable property. Let A be a real symmetric matrix, let x be an approximation of an eigenvector associated with an eigenvalue, say λ, and let $\tilde{\lambda}$ be an approximation of λ. Consider the quantity $\eta = Ax - \tilde{\lambda}x$. This quantity is the *residual*, and measures how much the pair $(\tilde{\lambda}, x)$ fails to satisfy the eigenvalue equation. If $\tilde{\lambda}$ is an eigenvalue, and x is an eigenvector, then the residual is zero. It can be proven that for any approximation x of $\mathbf{u}_i$, the quantity $\tilde{\lambda}$ that minimises the residual is the Rayleigh quotient:

$$\sigma(x) = \frac{x^\top A x}{x^\top x}.$$

This suggests that if x approximates an eigenvector, then the Rayleigh quotient is an approximation of the corresponding eigenvalue.

Using an argument analogous to the one adopted above, it is possible to show that, in general:

$$\sigma_n = \lambda_1 + O\left((\lambda_2/\lambda_1)^n\right)$$

and in particular, if the matrix A is symmetric, the following asymptotic relation holds:

$$\sigma_n = \lambda_1 + O\left((\lambda_2/\lambda_1)^{2n}\right) \tag{A.13}$$

meaning that the rate of convergence of the Rayleigh quotient is *faster* than in the general case of non-symmetric matrices (for the same λ_2/λ_1 ratio). This can be easily proved as follows. If A is real and symmetric, then its eigenvectors $\{\mathbf{u}_i\}$ form an orthonormal basis of $\mathbb{R}^N$, such that:

$$x_n^\top A x_n = \sum_{i=1}^N \alpha_i \lambda_i^n \mathbf{u}_i^\top \sum_{j=1}^N \alpha_j \lambda_j^{n+1} \mathbf{u}_j = \sum_{i=1}^N |\alpha_i|^2 \lambda_i^{2n+1} ||\mathbf{u}_i||^2$$

where $\|\mathbf{u}_i\|^2 = \|\mathbf{u}_i\|_2^2 = \mathbf{u}_i^\top \mathbf{u}_i$ is the square of the Eucledian norm of vector $\mathbf{u}_i$. Hence we get:

$$\sigma_n = \lambda_1 \frac{|\alpha_1|^2 ||\mathbf{u}_1||^2 + \sum_{i=2}^N |\alpha_i|^2 (\lambda_i/\lambda_1)^{2n+1} ||\mathbf{u}_i||^2}{|\alpha_1|^2 ||\mathbf{u}_1||^2 + \sum_{i=2}^N |\alpha_i|^2 (\lambda_i/\lambda_1)^{2n} ||\mathbf{u}_i||^2}$$

which has exactly the same structure of Eq. (A.11). Consequently, when A is symmetric, using the same procedure adopted to get Eq. (A.12), we obtain the relation in Eq. (A.13).

Despite the description of the power method provided above being quite simple to understand, a naïve algorithm based on the iteration of Eq. (A.9) will not work in most cases. In fact, if $|\lambda_1| > 1$, then x_n will exponentially diverge, and will sooner or later produce a machine overflow. Conversely, if $|\lambda_1| < 1$, then x_n will exponentially converge to 0, yielding an underflow as soon as machine precision is reached. A simple strategy to avoid this problem is to re-normalise x_n at each step, guaranteeing that the norm of the vector remains bounded when n increases. For this reason, we define these new sequence of vectors:

$$y_n = \frac{x_n}{||x_n||}, \quad x_{n+1} = A y_n, \quad n = 0, 1, \ldots$$

If the norm used is the Eucledian norm, then x_n is always bounded by $\|A\|_2$, and will converge asymptotically to $|\lambda_1|$.

A.6 Computation of Shortest Paths

We discuss here algorithms to find shortest paths in a graph. We will consider separately the case of unweighted graphs and that of weighted ones. In particular, we will focus on the problem of constructing the shortest paths from a given node i to all other nodes in the graph. As we shall see, efficient algorithms consist in visiting all nodes of the graph, starting from node i, in increasing order of distance.

A.6.1 Unweighted Graphs

If we are only interested in the *distance* between nodes, i.e. in the *length* of the shortest paths connecting a pair of nodes, we can construct an algorithm based on the following argument. As shown in Theorem 8.1, the entry i,j of the lth power of the adjacency matrix $(A^l)_{ij}$ is different from zero if and only if there is a walk of length exactly equal to l from node i to node j. Therefore, the length d_{ij} of the shortest walk from i to j, which is necessarily a path from i to j, is the minimum value of l such that $(A^l)_{ij} \neq 0$. Notice that such a value of l is equal to 0 iff the two nodes coincide (we can imagine to set $A^0 = I$), it is 1 iff the two nodes are connected by an edge, it is 2 iff node j is a neighbour of a neighbour of i, without being a neighbour of i, and so on. If there is no value of $l \in [1, \ldots, N-1]$ such that $(A^l)_{ij} \neq 0$ this means that the graph is not connected and the nodes i and j belong to two different components. In principle, to find the distance from a given node i to all other nodes, we can iteratively compute $\mathbf{x}_l = A\mathbf{x}_{l-1}$ starting from $\mathbf{x}_0 = \mathbf{e}_i$, where $\mathbf{e}_i$ denotes the ith vector of the canonical basis of $\mathbb{R}^N$, i.e. the ith column of the $N \times N$ identity matrix. The distance d_{ij} is equal to the smallest l such that $(\mathbf{x}_l)_j \neq 0$. If we are interested only in the distance between i and j, then the algorithm stops as soon as $(\mathbf{x}_l)_j \neq 0$, or when $l = N$. The latter case happens if i is not connected to j. If a value $l < N$ has been found, j is marked as being connected to node i. Instead, if we are interested in the distance between a node i and all other nodes in the graph, then the algorithm stops when all nodes are marked as connected to node i, or when $l = N$. Although simple to understand, such an algorithm is quite inefficient, and it is therefore rarely used in practice for large graphs. In fact, in a connected graph, the time complexity of computing the distances from a node i to all the other nodes through successive matrix multiplication is $O(DK)$, where K is the number of links and D is the diameter of the graph. This is so, because the cost of the multiplication of a sparse matrix A times a vector equals the number on non-zero elements of A, which is K in a directed graph, and $2K$ in an undirected graph. This operation has to be repeated D times, so that all nodes can be reached. In the case of unconnected graphs, the time complexity of the algorithm is $O(NK)$.

More efficient algorithms are based on *visiting* the graph. Here we will describe an algorithm which is a variant of the classic *Breadth-First Search* (BFS) algorithm [234, 233]. Further details about the BFS and other algorithms for the visit of trees and graphs can be found in Ref. [279]. Given a node i, our algorithm allows us to compute the distances from i to all the other nodes, and also to keep track of all the shortest paths. Since there might be in general more than one shortest path from i to a given node j, then for each node j, we need to store the value d_{ij} of the distance from node i to node j, and also all the shortest paths of distance d_{ij} from i to j. A simple way to reconstruct all the shortest paths is to store, for each node j, a list of the *predecessors* of j, that is, the set of all the first neighbours of j which belong to at least one shortest path from i to j. In this way, by recursively visiting the list of predecessors of j, and of its predecessors, we can enumerate all the shortest paths from i to j.

Algorithm 14 BFS()

Input: G (unweighted graph), i
Output: distances from i to all the other nodes in G, and list of predecessors
  1: **for** j = 0 **to** N-1 **do**
  2:     dist[j] ← N
  3:     marked[j] ← N+1
  4: **end for**
  5: dist[i] ← 0
  6: marked[0]← i
  7: d ← 0
  8: n ← 0
  9: nd ← 1
 10: ndp ← 0
 11: **while** d<N **and** nd >0 **do**
 12:     **for** k=n **to** n + nd - 1 **do**
 13:         cur_node ← marked[k]
 14:         **for all** j **in** neigh[cur_node] **do**
 15:             **if** dist[j] = d+1 **then**
 16:                 add_predecessor(j,k,preds)
 17:             **end if**
 18:             **if** dist[j] = N **then**
 19:                 dist[j] ← d+1
 20:                 add_predecessor(j,k,preds)
 21:                 marked[n + nd + ndp] ← j
 22:                 ndp ← ndp + 1
 23:             **end if**
 24:         **end for**
 25:     **end for**
 26:     n ← n + nd
 27:     nd ← ndp
 28:     ndp ← 0
 29:     d ← d+1
 30: **end while**
 31: **return** dist, preds

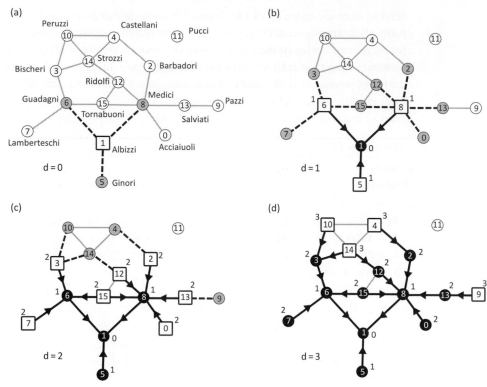

Fig. A.16 Computation of the shortest paths from node 1, representing the Albizzi family, to all other nodes. The states of the nodes, and their distances from node 1, are shown for the first four iterations of the algorithm. At each step we indicate unvisited nodes as white circles, all the nodes at the current distance d as grey circles, all the just-marked nodes as white squares (e.g. node 1 in panel (a), nodes 5, 6, 8 in panel (b)), and all the processed nodes as black circles.

The pseudocode of the procedure is presented in Algorithm 14, while in Figure A.16 we illustrate how the algorithm works on the network of Florentine families discussed in Section 2.1. Notice that we have labelled the 16 nodes of the graph starting from 0 and not from 1, since all the pseudocodes of this appendix available for download from the book website are written in the C language, where array indices start from 0. The distances d_{ij} are stored in a vector of size N, called dists, which is initialised so that dists[j] $= N$, $\forall j = 0, \ldots, N-1$. This is a good choice, since a finite distance in a graph with N nodes cannot be greater than $N-1$. When a node j is visited, the algorithm sets its distance to a value $d_{ij} < N$, so that all nodes whose final distance from i remains N at the end of the algorithm, do not belong to the connected component containing i. Another vector called marked[] is used to keep track of all the visited nodes. Since the number of predecessors is not fixed *a priori*, it is convenient to store all the predecessors of each node in a suitable vector structure, that we call preds[]. Notice that by simply allowing each node to have more than one predecessor, we are able to treat the case of multiple shortest paths from i to j.

In the example of the Florentine families, we have chosen $i = 1$, that is, we are computing all the shortest paths from the Albizzi family. At the beginning, the vector named `marked[]` contains only node 1, "Albizzi", and the current distance `d` is set equal to 0. The cycle beginning at line 12 scans all the nodes in `marked[]` which are at distance `d`. This is done by considering the nodes in `marked[]` at positions $\{n, n+1, \ldots, n+nd-1\}$, where `n` is the number of nodes visited so far, and `nd` is the number of nodes which were added to the array `marked` at the previous step. In the first iteration `cur_node` is set to the label of node "Albizzi", i.e. 1, and at line 14 we cycle over all the neighbours of node 1. In this case we find "Ginori", "Guadagni" and "Medici", respectively, nodes 5, 6 and 8. For each $j = 5, 6, 8$, the algorithm enters the if-statement at line 18, sets `dists[j]=1`, adds the node j to the `marked` vector, and adds node 1 to the list of the predecessors of j.

In the figure, we indicate unvisited nodes as white circles, all the nodes at distance `d` from i as grey circles, all just-marked nodes as white squares, and all processed nodes as black circles. In Figure A.16(a), i.e. when `d=0`, the only white square is node 1, and nodes 5, 6 and 8 are shown in grey. All the edges connecting a white square to a grey circle are indicated as dashed black lines. At line 22 we update the counter `ndp` of nodes at distance $d + 1$. When all nodes at distance `d` have been processed, `d` is incremented at line 29 and a new iteration starts. Notice that an already processed node, i.e. a node for which $dists[j] \leq d$, is never processed more than once. In Figure A.17 we plot the contents of `marked[]`, `dists[]` and `preds` at the end of each iteration, that is at line 29 of the algorithm. For instance, at the end of the first iteration, the `marked[]` vector contains node 1 ("Albizzi"), which is at distance 0, and three nodes at distance 1, namely "Ginori", "Guadagni" and "Medici", indicated in grey in Figure A.17(a). The values of the distances are given in vector `dists`, whose elements at positions 1, 5, 6, 8 are respectively set to 0, 1, 1, 1, while all other elements are equal to $N = 16$. The labels of the precedessors are added to the structure `preds`. This kind of structure is technically known as a *dictionary* and a simple implementation consists of a vector of vectors. In fact, for each node j we store a vector `preds[j]`, whose first element `preds[j][0]` indicates the number of predecessors of node j, while the remaining `preds[j][0]` elements are the labels of the predecessors of j in the shortest paths towards node i. Here, at the end of the first step, only nodes 5, 6 and 8 have one predecessor, which is node 1.

In Figure A.16(b) we illustrate the second iteration of the algorithm, corresponding to `d=1`. The three nodes at distance `d=1` from node 1, namely 5, 6 and 8, are indicated as white squares, and the edges belonging to the shortest paths to node 1 are indicated as thick black lines, whose arrows point to the predecessors. At the second step, the algorithm scans these three nodes, which are the last three nodes added in the `marked[]` vector. For each of them, the algorithm visits the neighbours j, as it did for node "Albizzi" in the first step. Node 5 has no new neighbours, node 6 has new neighbours, namely nodes 3, 7, 15, while node 8 has nodes 0, 2, 12, 13, 15 as new neighbours. For each node $j = 3, 7, 15, 0, 2, 12, 13$ the algorithm sets `dists[j]=2`, and adds the node j to the `marked[]` vector. These nodes are presented as grey circles in Figure A.16(b). For each of them, the algorithm adds the respective predecessor to the vector `preds[j]`, by calling the function `add_predecessor`. Note that the node "Albizzi" is not added again to the array

(a)

marked	dists		preds	
1	0	16	0	
5	1	0	0	
6	2	16	0	
8	3	16	0	
	4	16	0	
	5	1	1	1
	6	1	1	1
	7	16	0	
	8	1	1	1
	9	16	0	
	10	16	0	
	11	16	0	
	12	16	0	
	13	16	0	
	14	16	0	
	15	16	0	

d=0

(b)

marked	dists		preds		
1	0	2	1	8	
5	1	0	0		
6	2	2	1	8	
8	3	2	1	6	
3	4	16	0		
7	5	1	1	1	
15	6	1	1	1	
0	7	2	1	6	
2	8	1	1	1	
12	9	16	0		
13	10	16	0		
	11	16	0		
	12	2	1	8	
	13	2	1	8	
	14	16	0		
	15	2	2	6	8

d=1

(c)

marked	dists		preds		
1	0	2	1	8	
5	1	0	0		
6	2	2	1	8	
8	3	2	1	6	
3	4	3	1	2	
7	5	1	1	1	
15	6	1	1	1	
0	7	2	1	6	
2	8	1	1	1	
12	9	3	1	13	
13	10	3	1	3	
10	11	16	0		
14	12	2	1	8	
4	13	2	1	8	
9	14	3	2	3	12
	15	2	2	6	8

d=2

Fig. A.17 The three structures marked, dists and preds at the first three iterations of the BFS algorithm.

marked[], since its distance is smaller than the distance of "Guadagni", "Medici" and "Ginori". Note also that node "Tornabuoni" is included in marked[] only once. In fact, the first time this node is visited, that is as a neighbour of "Guadagni", its distance is set equal to 2. When "Tornabuoni" is visited again, this time as a neighbour of "Medici", the corresponding node is not added to marked[] again, although a new predecessor is assigned to "Tornabuoni". This is because we want to keep track of all the shortest paths from one node to another. Indeed, there are two shortest paths from "Tornabuoni" to "Albizzi", one through node "Guadagni" and the other through node "Medici". At the end of the second iteration, the new values of marked, dists, preds[j] are show in Figure A.17(b).

At the third iteration, corresponding to d=2, the algorithm goes through the new neighbours of the seven nodes at distance 2, namely nodes 4, 9, 10 and 14, and repeats the procedure. The seven nodes at distance 2 are indicated as white squares in Figure A.16(c), while the new neighbours are shown as grey circles. The values of marked[], dists[] and preds[j] at the end of the third iteration are shown in Figure A.17(c).

Finally, Figure A.16(d) shows the state of all the nodes at the beginning of the fourth step, when d=3. There are only three nodes at distance d=3, and no reachable node at distance $\geq$ d+1, so that no more nodes are added to the marked[] vector, and no predecessors are added to preds[]. Therefore the algorithm stops because nd=0, and the while cycle terminates. Notice that only node 11, the "Pucci" family, is unreachable from "Albizzi" since dists[11] has remained equal to $N = 16$, and is the only node still represented as a white circle in Figure A.16(d).

Once the algorithm has terminated, all the shortest paths from i to any other node j belonging to its component can be easily reconstructed by using the structure preds[] and going recursively from one node to its predecessor, up to node i. The procedure is illustrated in Figure A.18, where the dashed and dotted lines indicate, respectively, the two shortest paths connecting Albizzi to Strozzi. In particular, the path indicated by dotted lines

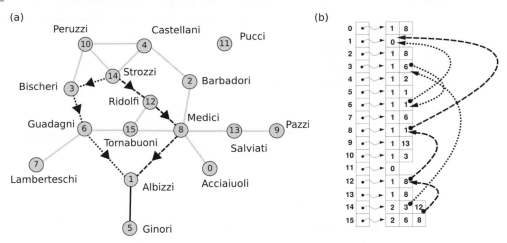

(a) The two shortest paths between Albizzi and Strozzi are shown in the graph respectively with dashed and dotted black lines, while their reconstruction from array `preds` is represented in panel (b).

runs through the nodes 1, 6, 3 and 14, while the path denoted through dashed lines includes nodes 1, 8, 12 and 14.

It can be proven that the BFS algorithm is optimal, i.e. that it is not possible to find a more efficient algorithm to visit all the nodes of a graph starting from a given initial node [279]. Its time complexity is $O(N + K)$ because, in order to run to completion, the algorithm needs to scan all edges and all nodes of the graph exactly once. This time complexity is much lower than $O(DK)$, the complexity of the *naïve* algorithm based on the powers of the adjacency matrix. By repeating the BFS procedure for each node of the network, the algorithm allows all pair distances to be computed in a time $O(N(N + K))$.

There are a few aspects to take into account for an efficient implementation of the BFS algorithm. First of all, notice that the usage of the `marked[]` vector improves the efficiency of each single step of BFS, since it automatically provides a list containing the visited nodes in increasing order of distance. Second, since for each node we need the list of neighbours, the adjacency matrix of the graph has been represented using the CRS format, as described in Appendix A.4.2. A final remark concerns the actual implementation of the dictionary `preds[]`. A straightforward initialisation of `preds[]` requires us to allocate a vector of size $k_j + 1$ for each node j, and assign it to `preds[j]`. Such initialisation requires an allocation $O(K + N)$ in terms of memory. In this case, the function `add_predecessor()` can be implemented as illustrated in Algorithm 15. Since the number of predecessors of each node is usually much lower than the number of neighbours, a more efficient usage of the memory may be obtained by reallocating the array `preds[j]` in the function `add_predecessor()` as needed.

The program `shortest`, available for download at www.complex-networks.net, is an implementation of Algorithm 14 which takes as input a graph and a node label i, and computes the distances (and the shortest paths) from i to all the other nodes of the graph. A similar program, called `shortest_avg_max_hist`, computes and prints in output the

---

Algorithm 15 add_predecessor()

1: $\ell \leftarrow$ preds[j][0]
2: preds[j][ℓ] $\leftarrow$ k
3: preds[j][0] $\leftarrow$ preds[j][0]+1

---

average and the maximum distance from node i to all the other nodes of the graph, together with the entire distribution of distances.

A.6.2 Weighted Graphs

Finding shortest paths in weighted graphs is slightly more complicated than in unweighted graphs. This is mainly because, if a graph is weighted, the distance between a node and one of its first neighbours is not always equal to the weight of the link connecting them. Suppose, for instance, we are given the weighted graph with $N = 6$ nodes and $K = 8$ links shown in Figure A.19, with the weights representing the length or cost of each link, and we want to calculate the shortest paths from node 1. It is easy to realise that node 2 is at distance $d_{1,2} = 3$ from node 1, and the corresponding shortest path contains only the edge $(1, 2)$, while node 0 is at distance $d_{1,0} = 8$, and the shortest path between node 1 and node 0 contains two edges, namely $(1, 2)$ and $(2, 0)$. Of course, if we discard all the weights and consider all graph links as having the same unitary cost, the BFS would assign to both pairs of nodes $(1, 2)$ and $(1, 0)$ the same distance $d_{1,2} = d_{1,0} = 1$.

The shortest paths from a node i to all the other nodes of a weighted graph can be efficiently computed by the Dijkstra algorithm [94, 86, 279]. The algorithm is illustrated in Figure A.20, and the corresponding pseudocode is presented in Algorithm 16. At each iteration, each node of the graph is in one of three possible states: *current*, *visited* or *unvisited*. Initially, all the nodes are marked as *unvisited* and are included in the set U of unvisited nodes: each node $j \neq i$ is assigned a tentative distance $d_{i,j} = \infty$ from node i, while the distance from node i to itself is set to 0 (lines 1–6). The algorithm makes use of the array dist[] to store the tentative distance of each node from node i, while the array prev[]

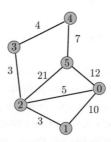

Fig. A.19   A weighted graph with $N = 6$ nodes and $K = 8$ links. Each link has an associated weight representing the *length* or the *cost* of the link. The shortest path between two nodes in a weighted graph is not the one with the minimal number of links; e.g. the shortest path between node 1 and node 0 has length $d_{1,0} = 8$ and contains two edges, namely $(1, 2)$ and $(2, 0)$.

Algorithm 16 `dijkstra()`

Input: G (weighted graph), i
Output: distances from i to all the other nodes in G, and list of predecessors

1: **for** j=0 **to** N-1 **do**
2:     dist[j] $\leftarrow \infty$
3:     prev[j] $\leftarrow$ -1
4: **end for**
5: dist[i] $\leftarrow$ 0
6: $U \leftarrow \mathcal{N}$
7: **while** U **is not** empty **do**
8:     cur_node $\leftarrow$ get_min(U)
9:     $U \leftarrow U \backslash \{cur_node\}$
10:    **if** dist[cur_node] = ∞ **then**
11:        **break**
12:    **end if**
13:    **for all** n **in** neigh[cur_node]) $\bigcap U$ **do**
14:        t_dist $\leftarrow$ dist[cur_node] + w[cur_node][n]
15:        **if** t_dist < dist[n] **then**
16:            dist[n] $\leftarrow$ t_dist
17:            prev[n] $\leftarrow$ cur_node
18:            update(U, n)
19:        **end if**
20:    **end for**
21: **end while**
22: **return** dist, prev

---

contains the label of the tentative predecessor of each node. At each step, we choose the node in U with the minimal distance from i, we remove this node from U and we mark it as *current*. The node marked as current at the first step is node i, since this is the only node having a finite tentative distance `dist[i]=0`. Then, for each neighbour j of the current node, we consider the shortest path from i to j which includes the current node. We set `t_dist` equal to the tentative distance of the current node from i plus the weight of the link connecting the current node to j. If `t_dist` is smaller than `dist[j]`, then `t_dist` becomes the new tentative distance from i to j, and the current node is set as predecessor of j (lines 13–17). When all its neighbours have been processed, the current node is marked as *visited*. Once a node is marked as *visited*, its distance from node i is final and minimal, and the node is not considered anymore. The algorithm stops when the set U is empty, or when the current node has infinite distance from i. In the latter case, all the other nodes contained in U do not belong to the same component of i, and are therefore unreachable.

Figure A.20 shows how the Dijkstra algorithm works in practice to compute distances from node 1 in the graph of Figure A.19. At each step, the current node is shown in grey, while unvisited and visited nodes are indicated, respectively, in white and in black. The numbers shown in grey, close to each node, represent their tentative distance to node 1. For

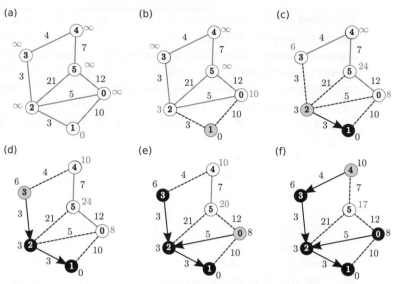

Fig. A.20 Computation of shortest paths from node 1 to all the other nodes of a weighted graph using the Dijkstra algorithm. At each step, we indicate the *current* node in grey, the *visited nodes* in black, and the *unvisited* nodes in white.

the sake of clarity, it is also convenient to consider different types of links: unvisited links are in grey, visited ones are shown as dashed black lines, while a solid black directed link is used to indicate the final predecessor of a node. Panel (a) presents the state of the system right after the initialisation: all nodes belong to the set U, they are marked as *unvisited* and coloured in white, and their tentative distance from node 1 is set to ∞, except for node 1 itself. At the first iteration, shown in panel b), node 1 is selected as *current*, since it is the only unvisited node with finite tentative distance, and is marked in grey. Then, all its unvisited neighbours, namely 0 and 2 are considered, and their tentative distances are respectively updated to dist[0]=10 and dist[2]=3. Also, node 1 is marked as tentative predecessor of these two nodes. In panel c) we display the state of the system at the second step. Node 1 is marked as *visited* and coloured black. Node 2 is selected as the new *current* node, since it is the unvisited node with the minimal tentative distance from node 1. A solid black directed link from node 2 to node 1 is used to indicate that 1 is the final predecessor of node 2 and that the distance 3 is final. Then, the tentative distances of the three unvisited neighbours of the current node 2, namely nodes 0, 3 and 5, are updated. Notice that the tentative distance of node 0 at the end of the first step was dist[0]=10, while at the end of the second step it becomes dist[0]=8, since the algorithm discovered a shorter path of length 8 passing through node 2. At the end of the second step, node 4 is the only unvisited node having infinite distance from 1. At the third step, shown in panel d, node 2 is marked as *visited* and node 3 is selected as the new current node, since it has the minimal tentative distance from 1 (dist[3]=6). The only unvisited neighbour of 3, namely node 4 is considered and its tentative distance is updated to dist[4]=10. The algorithm iterates over the remaining unvisited nodes of the network (panels e–f) and stops when the node selected as *current* has infinite tentative distance, or when all nodes have been marked as *visited*.

The Dijkstra algorithm looks similar to the Breadth-First Search. There are however a few important differences. The first is that, at the end of each iteration, Dijkstra adds only one node to the set of *visited* nodes, namely the *current* node, while the BFS adds all the unmarked neighbours of the current node to the array `marked`. The second difference is that the current node has to be selected from U in increasing order of distance from the initial node i, therefore it is important to store U using an efficient data structure. The time complexity of the single source Dijkstra algorithm can be expressed as $O(K \cdot up + N \cdot min)$, where *up* is the time complexity of the `update` function and *min* is the time complexity of `get_min`. Notice that both *up* and *min* crucially depend on the actual implementation of the set U. For instance, if U is implemented as an unsorted array, then each call to `get_min` requires $O(N)$ operations, while each call to `update` has $O(1)$ complexity, since no reordering of U is performed. In this case, the time complexity becomes $O(K + N^2)$, which is dominated by $O(N^2)$. However, it is possible to use more efficient structures for U to reduce the running time of `get_min` and `update`. In the code provided on the website, we implemented U as a Priority Queue (see Appendix A.3.4). In this case, both `get_min` and `update` have $O(\log N)$ running time, and the time complexity of the Dijkstra algorithm becomes $O((N + K) \log N)$, which is dominated by $O(K \log N)$. Notice that the best implementation of the Dijkstra algorithm is obtained by using a *Fibonacci heap* for U, and the corresponding time complexity is $O(K + N \log N)$. The interested reader can find detailed information on this subject, for example, in the book by Sedgewick [279] or in the book by Cormen, Leiserson, Rivest and Stein [86].

We provide an implementation of Algorithm 16 in the program `dijkstra`, available for download at `www.complex-networks.net`. The program takes as input a graph and a node label i, and computes the distances and shortest paths from i to all the other nodes of the graph. Notice that Algorithm 16 finds just one shortest path between each pair of connected nodes. In the case of degeneracy, i.e. when there is more than one shortest path between a given node pair, and one is interested in finding all of them, then a data structure for the storage of the predecessors similar to the one used for unweighted graphs is required. Such a data structure is indeed used in the code provided in the web site of the book.

A.7 Computation of Node Betweenness

In order to compute the betweenness centrality defined in Section 2.4 it can be useful to rewrite the expression for the node betweenness centrality of node i given in Definition 2.7 as:

$$c_i^B = \sum_{\substack{j=1 \\ j \neq i}}^{N} \sum_{\substack{k=1 \\ k \neq i,j}}^{N} \delta_{jk}(i) \qquad (A.14)$$

where $\delta_{jk}(i)$ is equal to the fraction $n_{jk}(i)/n_{jk}$, of shortest paths from j to k that passes through i, if j and k are connected, while $\delta_{jk}(i) = 0$ if j and k are not connected. In this way, Eq. (A.14) is valid both for connected and unconnected graphs. A direct use of Eq. (A.14) to compute the betweenness of a node i requires the calculation of all-pairs shortest paths, and $O(N^2)$ additional operations to perform the two nested summations. Hence, computing the betweenness of all the N nodes in a graph has time complexity equal to the time complexity of the all-pairs shortest path algorithm, plus $O(N^3)$. As shown in the previous section, the time complexity of the BFS and the Dijkstra algorithm are respectively $O(N(N + K))$ and $O(N(K + N \log N))$. Therefore, the complexity of computing the betweenness of all the nodes by directly using Eq. (A.14) is dominated by $O(N^3)$, for both unweighted and weighted graphs.

In this appendix we describe faster algorithms to compute node betweenness introduced by Ulrik Brandes [57] and based on the following two simple observations. The first thing to notice is that a vertex i lies on a shortest path between j and k if and only if:

$$d_{j,k} = d_{j,i} + d_{i,k}$$

This is known as the *Bellman criterion*. The second observation is that the number of shortest paths from j to i is equal to the sum of the number of shortest paths from j to each of the predecessors of i, where a node v is a *predecessor* of i, iff $(v, i) \in \mathcal{L}$ and (v, i) belongs to a shortest path from j to i. Therefore the number of shortest paths can be obtained recursively. This way of counting the number of shortest paths is called *combinatorial shortest path counting*. An example is shown in Figure A.21, where we plot the three predecessors v_1, v_2 and v_3 of node i, and all the possible shortest paths from j to i. We

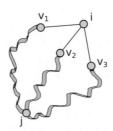

Fig. A.21   Predecessors of i along the shortest paths from j to i. Curly grey curves connecting two nodes indicate that there can be one or more shortest paths between the two nodes.

use curly grey curves to represent all the shortest paths connecting two nodes so that, in the case shown in the figure, there can be more than three shortest paths from j to i. The Bellman criterion allows the following characterisation of the set $P_j(i)$ of the predecessors of node i in a shortest path from j to i:

$$P_j(i) = \{v \in \mathcal{N} : (v, i) \in \mathcal{L}, d_{j,i} = d_{j,v} + \ell_{vi}\} \tag{A.15}$$

where ℓ_{vi} is the length or "cost" of edge (v, i), which is equal to 1 for all the links of an unweighted graph. Given a pair of nodes j and k, and making use of the definition of $P_j(i)$ given in Eq. (A.15), the combinatorial shortest path counting reads:

$$n_{jk} = \sum_{v \in P_j(k)} n_{jv}. \tag{A.16}$$

It is now useful to introduce another quantity, known as the *dependency* of a vertex j on another vertex i and defined as:

$$\delta_{j\bullet}(i) = \sum_{\substack{k=1 \\ k \neq i,j}}^{N} \delta_{jk}(i). \tag{A.17}$$

If the graph is a tree, there is exactly one shortest path connecting each pair of nodes. Therefore, the quantity $\delta_{jk}(i)$ is either one, if the shortest path from j to k passes by node i, or zero, if it does not. Consequently the dependency of vertex j on i is equal to the number of shortest paths from j that traverse i. In a graph with cycles the situation is different because there might be in general more than one shortest path connecting j to k, and $\delta_{jk}(i)$ is a real number between zero and one, representing the fraction of shortest paths from j to k which cross i. In this case, the dependency $\delta_{j\bullet}(i)$ is the sum of the fractions of shortest paths originating in j and traversing i.

By making use of Eq. (A.14), the betweenness of node i can be written as:

$$c_i^B = \sum_{\substack{j=1 \\ j \neq i}}^{N} \delta_{j\bullet}(i) \tag{A.18}$$

and can therefore be computed by summing the dependencies $\delta_{j\bullet}(i)$ over all the sources j. In this way, the evaluation of the betweenness of node i reduces to the task of computing the dependencies $\delta_{j\bullet}(i)$ of all nodes j. We will now show that simple recurrence relations can be derived for the computation of such dependencies, based on the introduction of the definition of *successor* of a node. A node w is said to be a successor of a node i in the shortest paths originating in j and traversing i, if and only if i is a predecessor of w, i.e. iff $i \in P_j(w)$. We denote by $S_j(i)$ the set of successors of node i for paths originating at node j.

When the graph is a tree, the recurrence relations for the dependencies are particularly simple. In fact, since there is only one shortest path from j to k, the quantity $\delta_{jk}(i)$ equals 1 if i lies on the shortest path from j to k, otherwise $\delta_{jk}(i) = 0$. Furthermore, as illustrated in Figure A.22, i lies on all the shortest paths from j which pass through the successors of i. Therefore the number $\delta_{j\bullet}(i)$ of shortest paths passing through i is equal to the number of

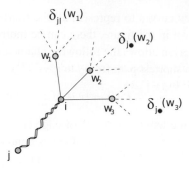

Fig. A.22 Successors of node i, and computation of the dependency of i on j.

shortest paths ending in a successor of i, plus the number of shortest paths passing through each of the successors of i. In formula one has:

$$\delta_{j\bullet}(i) = |S_j(i)| + \sum_{w \in S_j(i)} \delta_{j\bullet}(w) = \sum_{w \in S_j(i)} (1 + \delta_{j\bullet}(w)) \qquad (A.19)$$

where $|S_j(i)|$ indicates the number of nodes in set $S_j(i)$.

In the case of a generic graph with cycles, there can be more than one shortest path connecting two given nodes j and k. Therefore, not all the shortest paths passing through w also pass through its predecessor i. To compute $\delta_{j\bullet}(i)$ we have to sum the fraction of shortest paths originating in j, passing through i and ending in w, and the fraction of shortest paths originating in j, and passing through i and w. We thus obtain:

$$\delta_{j\bullet}(i) = \sum_{w \in S_j(i)} \frac{n_{ji}}{n_{jw}} + \sum_{w \in S_j(i)} \frac{n_{ji}}{n_{jw}} \delta_{j\bullet}(w) = \sum_{w \in S_j(i)} \frac{n_{ji}}{n_{jw}} (1 + \delta_{j\bullet}(w)). \qquad (A.20)$$

A formal derivation of such relation can be found in Ref. [57]. This formula, together with Eq. (A.18), is used to construct Algorithm 17 for the computation of the betweenness of all the nodes in an unweighted graph. The algorithm is a slight modification of the BFS algorithm for the computation of shortest paths that we discussed in Appendix A.6. For clarity, the lines that have been added to Algorithm 14 are shown in grey. The main difference is that, whenever the current node is set as a predecessor of node w, the number of shortest paths from j to w is updated (line 24 and 31). At the end of the while cycle (line 39), the wth component of the array nj contains the number n_{jw} of shortest paths starting from node j and arriving at node w. This allows the use of Eq. (A.20) to compute the dependencies of node j on each node w, and the corresponding contribution to the betweenness of w due to paths starting from j (lines 40–45). Notice that the cycle starting at line 40 has time complexity $O(K)$, since each pair of neighbours is considered at most once. This implies that, for each node j, we perform $O(N + K)$ operations to find the shortest paths starting from j (BFS) plus $O(K)$ operations to calculate dependencies and contributions to the betweenness. Since the computation of the betweenness of all the nodes requires to loop over all the N possible starting nodes j, the algorithm has time complexity $O(N(N + K))$ in an unweighted graph, which is more efficient than the $O(N^3)$ implementation of the definition in Eq. (A.14). A similar algorithm can be obtained for weighted graphs, and has the same

Algorithm 17 `brandes()`

Input: G

Output: betweenness centrality cB of all the nodes

1: **for** i = 0 **to** N-1 **do**
2:     cB[i] ← 0
3: **end for**
4: **for** j = 0 **to** N-1 **do**
5:     **for** i = 0 **to** N-1 **do**
6:         dist[i] ← N
7:         marked[i] ← N+1
8:         nj[i] ← 0
9:         delta[i] ← 0
10:     **end for**
11:     dist[j] ← 0
12:     nj[j] ← 1
13:     marked[0] ← j
14:     d ← 0
15:     n ← 0
16:     nd ← 1
17:     ndp ← 0
18:     **while** d<N **and** nd >0 **do**
19:         **for** i=n **to** n + nd - 1 **do**
20:             cur_node ← marked[i]
21:             **for all** w **in** neigh[cur_node] **do**
22:                 **if** dist[w] = d+1 **then**
23:                     add_predecessor(w,cur_node,preds)
24:                     nj[w] ← nj[w] + nj[cur_node]
25:                 **end if**
26:                 **if** dist[w] = N **then**
27:                     dist[w] ← d+1
28:                     add_predecessor(w,cur_node,preds)
29:                     marked[n + nd + ndp] ← w
30:                     ndp ← ndp + 1
31:                     nj[w] ← nj[w] + nj[cur_node]
32:                 **end if**
33:             **end for**
34:         **end for**
35:         n ← n + nd
36:         nd ← ndp
37:         ndp ← 0
38:         d ← d+1
39:     **end while**

```
40:     for k = n - 1 down to 1 do
41:         w ← marked[k]
42:         for all i in predecessors(w) do
43:             delta[i] ← delta[i] + nj[i]/nj[w]*(1 + delta[w])
44:         end for
45:         cB[w] ← cB[w] + delta[w]
46:     end for
47: end for
48: return cB
```

time complexity of finding all shortest paths using Dijkstra, i.e. $O(N(K + N \log N))$, which is again faster than $O(N^3)$. The computation of betweenness using Brandes algorithm is parallelisable in a straightforward way. In fact, if we have P processors available, we can assign to each of them the computation of the contributions to the betweenness due to all the shortest paths originating in a subset of $\sim N/P$ nodes.

We provide an implementation of Algorithm 17 in the program `betweenness`, available for download at `www.complex-networks.net`. The program can compute both the exact value of betweenness for all the nodes of the graph given as input, and also an approximation of the betweenness based only on the paths originating at a subset of nodes. We also provide the program `bet_dependency`, which computes the dependency of all the nodes of the graph due to the shortest paths originating at a subset of nodes. This program can be used to parallelise the computation of node betweenness in large graphs, by assigning a different set of starting nodes to each of the processors available, and then summing up the values of dependency computed by each processor to obtain the values of node betweenness. Finally, we will discuss in Appendix A.17 a slight modification of Algorithm 17 which allows the betweenness centrality of all the *edges* of a graph to be computed, according to Definition 9.4 of Chapter 9.

A.8 Component Analysis

As seen in Chapter 2, an undirected (directed) graph is connected (strongly connected) if and only if its adjacency matrix A is irreducible. According to Theorem 2.3, one way to check the irreducibility of A is to verify whether $(I + A)^{N-1} > 0$. We can therefore use the powers of matrix $(I + A)$ to find the connected components in a graph, as explained below. Let us consider first the case of an undirected graph. The positions of the non-zero entries of $\boldsymbol{u}_{N-1} = (I + A)^{N-1}\boldsymbol{e}_i$ correspond to the labels of all nodes that can be reached from node i. Hence, by computing $\boldsymbol{u}_{N-1}$ we can find all the nodes in the connected component to which node i belongs. Actually, we do not need to compute the $(N - 1)$-th power of matrix $I + A$, because in the worst case, it is sufficient to stop the computation at $\boldsymbol{u}_{D_c+1}$, where D_c is the diameter of the component c to which node i belongs. This is because all the nodes of such a component can be reached from i in at most D_c steps. During each of these steps, the number on non-zero components of vector $\boldsymbol{u}$ increases monotonically until $\boldsymbol{u}_{D_c}$, and then the number of non-zero components of $\boldsymbol{u}_{D_c+1}$ is the same as that of $\boldsymbol{u}_{D_c}$, indicating that D_c is indeed the diameter of the c component. Since the cost of a matrix-vector multiplication is equal to the number of non-zero entries of the matrix, the number of operations required to find all the components in an undirected graph is $\sum_{c=1}^{N_c}(D_c + 1)K \leq NK$. In the case of a directed graph, the number of operations required to find the strongly connected components by using powers of $(I + A)$ is even higher, because for each node i we have to compute the intersection of its out- and in-components.

The approach discussed above is rather inefficient, and is of no practical use for large graphs. More efficient algorithms to decide whether a graph is connected and to compute number and sizes of its connected components are based on the visit of the graph. For instance, the Breadth-First Search (BFS) algorithm used in Section A.6 to find all the shortest paths from a given node i gives, as a by-product, the connected component containing i, if the graph is undirected, or the out-component of node i, if the graph is directed. Such component is contained in the array `marked`. Therefore, we can find all the components in a graph with BFS in $O(N + K)$. However, if we are not interested in finding the shortest paths, we can use an alternative method to explore the graph, called *Depth-First Search (DFS)*, which is somehow simpler to implement. The DFS, in its basic implementation, requires a very simple data structure, namely a Boolean variable associated with each node to indicate whether the node has been visited or not, and a variable counting the number of visited nodes. DFS can be used to find the connected components of an undirected graph, and, with a suitable modification of the basic algorithm, the strongly connected components of a directed graph.

A.8.1 Undirected Graphs

Let us consider the undirected graph in Figure A.23(a). The $N = 9$ nodes in the graph are labelled by integer numbers, for convenience from 0 to 8. We identify each connected component g_ℓ by an integer $\ell \in [1, \ldots, N_c]$, where N_c is the number of connected components. In our case we have $N_c = 3$ components. Nodes $\{0, 3, 6, 7, 8\}$ belong to the first

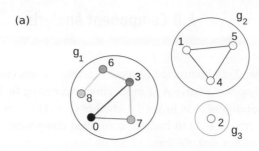

(a)

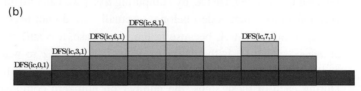

(b)

Fig. A.23 (a) An undirected graph with three connected components g_1, g_2 and g_3. DFS starting from node 0 visits component g_1 along a spanning tree. The nodes and edges of g_1 are coloured according to the visiting order of DFS. (b) Recursive calls to DFS to find the component of node 0 are represented in step-notation. The events are ordered chronologically from left to right.

component g_1, nodes $\{1, 4, 5\}$ belong to the second component g_2, while the third component g_3 consists of a single node, node 2. To find all the nodes of the graph reachable from a given node i we use the *Depth-First Search* (DFS) algorithm. The pseudocode of the DFS is shown in Algorithm 18. The algorithm makes use of vector ic[], of size N, whose null components identify the nodes which have not yet been visited. Given a node i, the DFS sets ic[i] = nc, where the value of nc denotes the currently visited component. Then DFS considers each neighbour j of i and, if j has not yet been visited (i.e. ic[j] = 0), it calls again DFS for node j, thus recursively proceeding towards nodes which are farther and farther away from i. Notice that in such a recursive procedure, DFS avoids loops, thus producing a tree whose root is the current node i. A call to DFS terminates only when all the branches of such a tree have been visited.

A simple way to represent recursive calls is shown in Figure A.23(b). Each step represents a call to the recursive function (in this case the function is DFS) with all its parameters. An upward step is added each time we make a recursive call, while a downward step is added whenever a function at a given level terminates. Steps are temporally ordered from left to right. In the example, DFS is initially called for node 0. As we shall see, DFS visits component g_1 along a spanning tree rooted at node 0. Let us suppose that, at the beginning, all the elements of ic are equal to zero, meaning that the nodes have not been assigned to any component. In order to find all the nodes reachable from node 0 in Figure A.23(a), we call first DFS(ic,i=0,nc=1). This call to DFS() is represented in Figure A.23(b) with the first upward step on the left-hand side. According to Algorithm 18, this call to DFS() sets ic[0]=1 and then scans all the neighbours of node 0, in this case nodes 3 and 7. Since node 3 has not yet been visited (ic[3] = 0), then DFS()

Algorithm 18 DFS()

Input: ic, i, nc, f
Output: s{size of the tree starting at i}
  1: ic[i] ← nc
  2: s ← 1
  3: **for all** j **in** *neigh*[i] **do**
  4:     **if** ic[j] = 0 **then**
  5:         s ← s + DFS(ic,j,nc,f)
  6:     **end if**
  7: **end for**
  8: f[time] ← i
  9: time ← time + 1
 10: **return** s

---

is recursively called starting from node 3. This call is represented in the figure by the first dark-grey step. During this call, DFS() sets ic[3] = 1, and scans all the neighbours of node 3. Since node 6 has not yet been visited, DFS is called again from node 6 (upward medium-grey step). Here we set ic[6] = 1, and, since 8 is the only unvisited neighbour of 6, DFS() is called from node 8. This is represented by the upward light-grey step. At this point we set ic[8] = 1 and, since node 8 has no unvisited neighbours, this call to DFS() terminates, and the control returns to the previous DFS() call, namely the one corresponding to node 6. This is represented by the first downward step in Figure A.23(b). Again, since the two neighbours of node 6, namely nodes 3 and 8, have already been visited, the DFS() call from node 6 terminates, and the control returns to the DFS() call from node 3 (downward medium-grey step). The call from node 3 continues by considering the next unvisited neighbour of node 3, namely node 7. Then, DFS() is called again from node 7 (second upward medium-grey step), then ic[7] = 1 and, since node 7 has no unvisited neighbours, the DFS() call from node 7 terminates, and the control returns again to node 3 (second downward medium-grey step). Now the whole subtree downstream of node 3 has been visited, and the control returns to the original node 0. Since node 0 has no unvisited neighbours, also the original call to DFS() from node 0 terminates.

The other vector used by DFS(), namely the vector f[], contains the labels of the visited nodes in ascending order of *finishing time*, i.e. the time step at which a recursive call to DFS terminates and the control returns to the previous DFS() call. Notice that, since the DFS() call from node 8 is the first one to terminate, then f[0] is set to 8. Analogously, f[1] = 6, f[2] = 7, f[3] = 3 and f[4] = 0. Actually, the vector f[] is not needed to find components in an undirected graph. Indeed, the two lines marked in grey at the end of DFS() have been included only to simplify the extension of the algorithm to the case of a directed graph, as we shall see in the next subsection. When DFS(ic,i=0,nc=1) terminates, all the nodes reachable from node 0 have been visited and marked as belonging to the first component of the graph. In particular, the connected component associated with node 0 contains all the nodes *i* for which ic[i]=1. The brightness of those nodes in Figure A.23 represents the depth in the spanning tree rooted at node 0, with darker shades of grey representing nodes placed closer to the root.

In order to find all the components of an undirected graph we have to call DFS() for all the nodes of the graph. This is done in Algorithm 19, which sets ic to zero and scans all the nodes. If a node i has not yet been visited, the number of connected components nc is increased by one, and DFS() is called from node i. Notice that a call to DFS() from a generic node j returns the size of the subtree rooted at j, so that when DFS(ic,i,nc,f) returns at line 10, s is the size of the component to which node i belongs.

Algorithm 19 components()

Input: G
Output: ic,nc,sizes, f
  1: **for** i = 0 **to** N-1 **do**
  2:    ic[i] ← 0
  3:    f[i] ← 0
  4: **end for**
  5: nc ← 0
  6: time ← 0
  7: **for** i = 0 **to** N-1 **do**
  8:    **if** ic[i] = 0 **then**
  9:        nc ← nc + 1
10:        s ← DFS(ic,i,nc,f)
11:        sizes[nc] ← s
12:    **end if**
13: **end for**

Notice that DFS() visits each node and each of its links exactly once, so that it has time complexity $O(N+K)$. This is the same time complexity of an algorithm to find components based on BFS(). However, Algorithm 18 is more compact than Algorithm 14, thanks to the use of recursion.

We provide an implementation of Algorithm 18 and 19 in the program components, available for download at www.complex-networks.net, which takes as input a graph and reports on output the size (and composition) of the connected components of the graph. The program largest_component, instead, finds the connected components of a graph and returns as output only on the subgraph corresponding to the largest component.

A.8.2 Directed Graphs

As discussed in Section 1.3, directed graphs can have two different types of graph components, namely weakly connected and strongly connected components. Let us consider, for instance, the directed graph with $N = 8$ and $K = 12$ arcs shown in Figure A.24. The graph is not strongly connected, since there are pairs of nodes which are not mutually reachable.

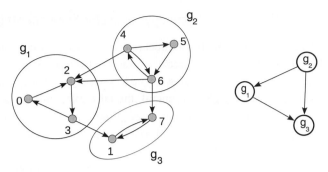

Fig. A.24 Components in a directed graph. The graph shown on the left-hand side is weakly connected, i.e. consists of a single weakly connected component, and is made of three strongly connected components g_1, g_2 and g_3. On the right-hand side we report the directed acyclic graph G^{SCC} of the strongly connected components.

For instance, it is possible to go from node 4 to node 3, but not vice versa. The graph consists of three strongly connected components g_1, g_2 and g_3, containing respectively nodes $\{0, 2, 3\}$, $\{4, 5, 6\}$ and $\{1, 7\}$. Instead the graph is weakly connected. In fact, all its eight nodes belong to a single weakly connected component.

In the remainder of this appendix, we will first focus on how to find the weakly connected components of a directed graph, and then we will discuss an algorithm for the strongly connected components.

Weakly Connected Components

The weakly connected components of a directed graph G are the components of the underlying undirected graph G^u (see Definition 1.12). Notice that, if A is the adjacency matrix of G, then the (i, j) element of the symmetric matrix $A + A^\top$ is different from zero if in the original graph G there is at least an arc from i to j, or an arc from j to i. Therefore, the underlying undirected graph G^u corresponds to a symmetric adjacency matrix, A^S, obtained by placing ones in all the non-zero entries of $A + A^\top$. The sparse **ij** representation of A^S can be obtained by the sparse **ij** representation of A by simply considering $A^S = (\mathbf{i}^S, \mathbf{j}^S, \mathbf{1})$, where

$$\mathbf{i}^s = \begin{pmatrix} \mathbf{i} \\ \mathbf{j} \end{pmatrix}, \quad \mathbf{j}^s = \begin{pmatrix} \mathbf{j} \\ \mathbf{i} \end{pmatrix}.$$

However, if in the original graph there are symmetric links, as between nodes 4 and node 6 in the graph in Figure A.24, then the resulting representation is redundant, since some arcs (e.g. both arcs $(4, 6)$ and $(6, 4)$ in the previous example) will appear twice. If one uses the matrices A and $A^\top$ in the **ij** sparse format, then a non-redundant representation of $\mathbf{i}^S$ and $\mathbf{j}^S$ can be obtained by the sparse representation of $A + A^\top$. In a lower-level implementation, the redundancy may be eliminated by using a variant of Algorithm 12 in which a new link is introduced only if it is not already present, i.e. lines 14–15 are executed only if $j[k]$ is not already contained in $\mathbf{j}_{\mathrm{CRS}}$. Once the sparse representation of A^S is obtained, we can use Algorithm 19 to find the weakly connected components of the graph.

Strongly Connected Components (SCC)

In order to check whether a directed graph is strongly connected, it is not enough to verify that all the nodes can be reached starting from a given node i (as it was in the case of undirected graphs). Instead, we need to verify that all nodes can be reached from each node i, with $i = 1, 2, \ldots, N$. In general, the DFS exploration introduced in Appendix A.8.1 to find the components in an undirected graph, can also be used for a directed graph. Invoking it from node i, the function will find the set of nodes reachable from i (the out-component of node i). Since we have to perform a DFS exploration starting from each node i of the graph, this is a very expensive procedure for a large graph.

A better strategy is based on the observation that the strongly connected component associated with node i is given by the intersection of the out-component and the in-component of node i. The out-component can be obtained by running Algorithm 18 starting from node i, while the in-component can be obtained by running the same algorithm starting from node i on a graph where the directions of all the arcs have been inverted. Notice that the graph with all directions inverted is a graph whose adjacency matrix is the transpose of the original matrix. Suppose, for example, that we want to find the SCC to which node 0 belongs in the graph in Figure A.24. We first perform DFS starting from node 0, marking all the nodes belonging to the out-component of node 0, namely nodes 1, 2, 3 and 7. Then, we start a slightly modified version of DFS from node 0 again, using the adjacency matrix $A^\top$, and visiting only the nodes which have been marked in the previous DFS run, i.e. node 2 and node 3. These nodes, together with node 0, are all the nodes of the strongly connected component g_1. To find another strongly connected component, the same procedure has to be repeated starting from a node not belonging to g_1, for instance from node 1.

The procedure discussed above can be expensive, since the strongly connected components are usually much smaller in size than the out-components of the starting nodes. Better algorithms can be obtained by noting that for any directed graph G, the graph G^{SCC} of the strongly connected components of G is a directed acyclic graph. This graph, as illustrated in Figure A.24, is obtained by collapsing each strongly connected component g_i of G into a single node, and placing a directed arc from g_i to g_j if there exists at least one node in g_i pointing to a node in g_j. Notice that the graph on the right-hand side of Figure A.24 does not have cycles. In fact, if a cycle existed in G^{SCC}, then all the nodes belonging to such a cycle would not be distinct strongly connected components of G. Suppose, for instance, that G^{SCC} in Figure A.24 was not acyclic, and had also an arc pointing from g_3 to node g_2. In this case, any node in any component of G would be strongly connected to any other node, so that G^{SCC} would consist of a single node. It is possible to prove that any directed acyclic graph is a rooted graph, and by performing a DFS on such a graph the highest finishing time is always assigned to one of the nodes of the connected component corresponding to the root of the associated G^{SCC}. Using these results, in 1972 Robert Tarjan found an optimal algorithm to extract strongly connected components in a directed graph [295]. Tarjan's algorithm performs a single DFS scan on the graph, and has time complexity $O(N + K)$. We describe here a simplified version of this algorithm, introduced by

---

Algorithm 20 `strong_components()`

Input: G
Output: ic,nc,sizes,f
 1: (ic,nc,sizes,f) ← `components`(G)
 2: nc1 ← nc
 3: ic1 ← ic
 4: **for** i = 0 **to** N-1 **do**
 5:    ic[i] ← 0
 6:    f1[i] ← f[i] {Copy all finishing times of the first step into f1}
 7:    f[i] ← 0
 8: **end for**
 9: nc ← 0
10: G ← `transpose`(G)
11: **for** i = N-1 t**down to** 0 **do**
12:    **if** ic[f1[i]] = 0 **then**
13:       nc ← nc + 1
14:       s ← DFS(ic,f1[i],nc,f)
15:       sizes[nc] ← s
16:    **end if**
17: **end for**

Sambasiva Rao Kosaraju and Micha Sharir, which has the same time complexity but performs two DFS scans. The pseudocode is presented in Algorithm 20, and works as follows:

1 We first run `components` on the adjacency matrix A, thus obtaining the vector `f[]` which contains the labels of the visited nodes in ascending order of finishing times.
2 Then, we compute matrix $A^\top$.
3 Finally, we scan the nodes in descending order of finishing time (lines 11–17), and call DFS for each unvisited node.

Notice that the loop at lines 11–17 in Algorithm 20 is equivalent to the main cycle of `components()`, except for the fact that the nodes are considered in descending order of finishing time. The proof that, by using Algorithm 20, one actually obtains all the SCCs of a graph is beyond the scope of this book, and can be found in the specialistic literature [86].

Algorithm 20 is implemented by the program `strong_conn`, available for download at `www.complex-networks.net`. The program takes as input a graph and returns the size (and, optionally, the composition) of all the strongly connected components of the graph. The program `node_components`, instead, takes as input a graph and a node i, and returns the in-component, the out-component, the weakly connected component and strongly connected component associated with i.

A.9 Random Sampling

Many algorithms make use of random numbers. This means that we need to be able to choose an element x from a set according to a certain probability density $p(x)$, such that if we repeat the selection a large number of times, then the frequency with which a particular element $\tilde{x}$ appears is proportional to the probability $p(\tilde{x})$. In this appendix, we focus on some general techniques to sample from continuous and discrete random variables. Such techniques will be used in the algorithms of graph sampling described in the next sections of the appendix. More information on random sampling can be found in books on Monte Carlo methods, such as for instance in Ref. [172] or Ref. [182].

A.9.1 Random Number Generators

All the random sampling methods that we will discuss in this appendix assume that the computer is able to generate a uniformly distributed random real number between 0 and 1. In general, a computer cannot generate truly random numbers, for several reasons. First of all, a digital computer is not able to represent a real number, which has an infinite number of significant digits, but can instead provide only an approximation based on a finite number of significant digits, e.g. by means of a floating-point notation. Second, any algorithm used by a computer works with finite objects and is by definition deterministic, i.e. *not random*. As a consequence, any algorithm can produce only a finite, predictable number of outputs, while truly random sequences are by definition infinite and non-reproducible.

For most practical purposes, it is not necessary to have *truly random numbers*, but at least a sequence of *pseudorandom numbers*. Pseudorandom numbers are generated by a deterministic procedure, but exhibit some key features of random numbers, and in particular:

- The values in the sequence are uniformly distributed in $[0, 1]$.
- The elements of the sequence are uncorrelated.
- It is impossible (or at least very hard) to guess the value of the next element of the sequence even knowing all the elements of the sequence generated so far.

Aside from these fundamental properties, a procedure to generate pseudorandom numbers should be computationally efficient. It is relatively easy to verify that a sequence of pseudorandom numbers $\{\xi_1, \xi_2, \ldots, \xi_R\}$ is uniformly distributed in $[0, 1]$. We just generate a high number R of elements and plot the frequency histogram of the values. As R increases, the histogram should quickly converge to a uniform distribution. In Figure A.25 we display the histogram of the values obtained by using a standard pseudorandom number generator for $R = 10^3, 10^4, 10^5$. Checking for the absence of correlations in a sequence of pseudorandom numbers is slightly more difficult, since we should in principle check for correlations of any order, but is still straightforward. In practice, the absence of pairwise correlation means that (ξ_n, ξ_{n+1}) is uniformly distributed in $[0, 1]^2$, the absence of three-terms correlations means that $(\xi_n, \xi_{n+1}, \xi_{n+2})$ is uniformly distributed in the unit cube, and

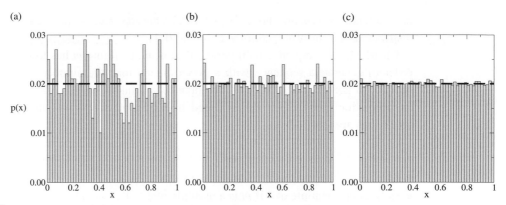

Fig. A.25 The values produced by a good random number generator should be uniformly distributed in [0, 1]. We show here the frequency distribution corresponding to a standard linear congruential generator, for a number of samples equal to (a) $R = 1000$, (b) $R = 10,000$, (c) $R = 100,000$. The histogram is constructed by dividing the interval [0, 1] into 50 bins.

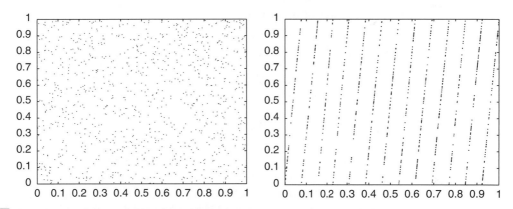

Fig. A.26 Examples of good (left) and bad (right) pseudorandom number generators. Each point in the plot represents a couple (ξ_n, ξ_{n+1}) of successive random numbers in the sequence.

so on. In Figure A.26 we show the distribution of (ξ_n, ξ_{n+1}) for good and bad random number generators.

Verifying that the sequence of numbers generated by a pseudorandom generator is not predictable is instead relatively difficult, and the majority of classical random number generators are known to produce predictable sequences, especially for specific values of their parameters.

A family of classical random number generators which is still in use in many applications is the so-called *Linear Congruential Generators* (LCG), which are defined by the recurrent relation

$$x_{n+1} = (ax_n + c) \bmod m, \quad n > 0, \tag{A.21}$$

with $x_0, a, c, m \in \mathbb{N}$. This equation generates a sequence of integers $x_1, x_2, \ldots$ between 0 and $m-1$. Dividing x_n by m, one obtains a rational number $\xi_n = x_n/m \in [0, 1-1/m]$. If the parameters x_0, a, c, m are chosen appropriately, the computed sequence ξ_n is approximately distributed as a uniform random variable in $[0, 1]$.

Clearly ξ_n is a sequence with period at most m, therefore m has to be large enough in order to generate a sequence with a period longer than the number of random numbers needed. In general, the quality of the obtained sequence depends on the choice of a, c, m, and in particular an LCG produces a sequence of period exactly m if and only if these three conditions are satisfied:

- m and c are *co-primes*, i.e. m and c do not have any common factor;
- $(a - 1)$ is divisible by all prime factors of m;
- $(a - 1)$ is a multiple of 4 if m is a multiple of 4.

Two possible good choices of such parameters are $m = 2^{32}, a = 1,664,525, c = 1,013,904,223$ or $m = 2^{31} - 1, a = 7^5, c = 0$, the latter being the set of parameters used to produce the histogram in Figure A.25 [182, 262]. It is important to note that, although being very simple to implement, linear congruential generators are not the best way to produce sequences of pseudorandom numbers. The state of the art today is represented by the Mersenne Twister [217], which is the pseudorandom number generator of choice of most of the programming languages and libraries for scientific computing. The procedure used by the Mersenne Twister is relatively complicated, and goes beyond the scoped of the present book. The interested reader can find more details in Ref. [217].

A.9.2 Inverse Function Method

Let $x \in \mathbb{R}$ be a random variable with *probability density function* $p_x(x)$, i.e. such that $p_x(x) \geq 0 \; \forall x \in \mathbb{R}$ and $\int_{\mathbb{R}} p_x(x) \, dx = 1$, and assume that we want to sample elements distributed according to $p_x(x)$. One of the simplest ways to sample from $p_x(x)$ is the so-called *inverse function method*. The method is based on the definition of *cumulative distribution function*:

$$P_x(\bar{x}) = Pr(x \leq \bar{x}) = \int_{-\infty}^{\bar{x}} p_x(t) \, dt$$

associated with $p_x(x)$, which corresponds to the probability $P(x \leq \bar{x})$ that a number sampled from $p_x(x)$ is smaller than or equal to $\bar{x}$. Notice that $P_x(x)$ takes values in $[0, 1]$. Let us consider the random variable

$$y = P_x(x)$$

and assume it has a uniform distribution in $[0, 1]$. Now, if we sample at random a value u in $[0, 1]$, the value:

$$\xi = P_x^{-1}(u) \tag{A.22}$$

is distributed according to $p_x(x)$. In fact, we can write:

$$Pr(P_x^{-1}(u) \leq x) = Pr(u \leq P_x(x)) = P_x(x)$$

where the last equality holds because, by definition, $Pr(u \leq k) = k$, since u is uniform in $[0, 1]$. This means that in order to obtain a number ξ distributed as $p_x(x)$ we can sample u uniformly in $[0, 1]$ and then obtain ξ as $\xi = P_x^{-1}(u)$.

Sampling from the Exponential Distribution

Let us assume that we want to sample form the exponential distribution $p_x(x) = \lambda e^{-\lambda x}$, $x \geq 0$ using the inverse function method. In this case we get

$$P_x(x) = \lambda \int_0^x e^{-\lambda t} \, dt = 1 - e^{-\lambda x}$$

whose inverse is

$$P_x^{-1}(u) = -\frac{1}{\lambda} \log (1 - u).$$

Hence, if we sample u uniformly in $[0, 1]$, we obtain a sample ξ distributed as $p_x(x) = \lambda e^{-\lambda x}$ by computing

$$\xi = -\frac{1}{\lambda} \log (1 - u).$$

Notice that in principle we might also compute

$$\xi = -\frac{1}{\lambda} \log(u)$$

since both u and $1 - u$ are uniformly distributed in $[0, 1]$.

Sampling from a Continuous Power-Law Distribution

If $p_x(x)$ is a power-law distribution $p_x(x) = (\gamma - 1)x^{-\gamma}$, $x \geq 1$, $\gamma > 1$, then we have:

$$P_x(x) = (\gamma - 1) \int_1^x t^{-\gamma} \, dt = 1 - x^{1-\gamma}$$

whose inverse function is:

$$P_x^{-1}(u) = (1 - u)^{\frac{1}{1-\gamma}}.$$

Hence, in order to sample ξ from the power-law $p_x(x) = (\gamma - 1)x^{-\gamma}$ we sample u uniformly in $[0, 1]$, and then we compute

$$\xi = (1 - u)^{\frac{1}{1-\gamma}}.$$

Notice that the cumulative distribution function $P_x(x)$ is a non-decreasing function of x, hence its inverse $P_x^{-1}(y)$ always exists. The only problem is that $p_x(x)$ might not have a primitive function. In that case, using the inverse function method is more complicated, since it is necessary to solve a non-linear integral equation to obtain each sample. A typical example is the Standard Gaussian distribution, defined as:

$$p_x(x) = \frac{1}{\sqrt{2\pi}} e^{-\frac{x^2}{2}}$$

whose primitive function can only be expressed by the integral

$$P_x(x) = \int_{-\infty}^{x} \frac{1}{\sqrt{2\pi}} e^{-\frac{t^2}{2}} \, dt$$

whose inverse cannot be expressed in closed form. There are indeed algorithms to sample form a Gaussian distribution, such as the Box–Muller algorithm and Marsaglia's algorithm, but they are specifically conceived for Gaussian distributions and cannot be used to sample from other continuous probability distributions. Instead, the acceptance–rejection method is a suitable alternative whenever it is difficult or impossible to invert $P_x(x)$.

A.9.3 Acceptance–Rejection Method

Sometimes, it may be expensive to compute the inverse of the cumulative distribution function $P_x(x)$ as required by the inversion method, e.g. because a non-linear equation has to be solved. Some other times, the primitive of $p_x(x)$ is an integral function, for which an inverse cannot be computed analytically. In these cases it may be convenient to use the so-called *acceptance–rejection technique*.

The idea is the following. We start by noting that sampling a random variable with distribution $p_x(x)$ has a clear geometric interpretation. Let us consider the region Ω_p of the plane between the x-axis and the function $p_x(x)$, as shown in Figure A.27. The region Ω_p has area equal to 1, since $\int_{\mathbb{R}} p_x(x) \, dx = 1$. Now let us sample a point uniformly in Ω_p. This means that the probability that a point lies in the region $\delta\Omega \subseteq \Omega_p$ is equal to the area of the region $\delta\Omega$ divided by the area of Ω_p (which in this case is equal to 1). If $(\bar{x}, \bar{y})$ are the coordinates of the sampled point, then the abscissa $\bar{x}$ is distributed with probability density $p_x(x)$. This is because the probability that the abscissa of a point lies in $[\bar{x} - \delta x/2, \bar{x} + \delta x/2]$ is equal to $\delta\Omega = p_x(\bar{x}) \, \delta x$, as shown in Figure A.27. Furthermore, the

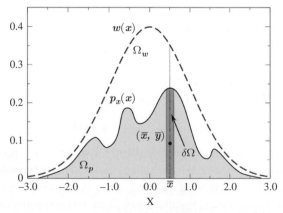

Acceptance–rejection sampling from $p_x(x)$ (solid line) is implemented by sampling a candidate $\bar{x}$ from the function $w(x)$ (dashed line), and accepting it with probability equal to the ratio $p(\bar{x})/w(\bar{x})$. The regions below the curves $w(x)$ and $p_x(x)$ are denoted as Ω_w and Ω_p, and their area are respectively equal to A and 1.

geometric interpretation of the distribution function $P_x(x)$ is the area of the set of points of Ω_p whose first coordinate is less than or equal to x.

Now, if we want to sample from $p_x(x)$, we look for a function $w(x) \geq p_x(x) \forall x \in \mathbb{R}$ whose primitive $W(x)$ is easily invertible. Consider the example shown in Figure A.27, where we indicated $p_x(x)$ with a solid line, $w(x)$ with a dashed line, and denoted by Ω_w and $\Omega_p \subset \Omega_w$, respectively, the region between the x-axis and the function $w(x)$, and the region between the x-axis and the function $p_x(x)$. We then sample uniformly a point $(\bar{x}, \bar{y})$ in Ω_w, and accept the sample if the point is inside Ω_p, while we reject it if it does not belong to Ω_p and repeat the sampling procedure until we accept a sample. By doing so, we are effectively sampling points uniformly at random inside Ω_p, and therefore the first coordinate of the point $(\bar{x}, \bar{y})$ sampled in this way is distributed as $p_x(x)$.

A practical way to implement this procedure is the following. Let $A = \int_{-\infty}^{\infty} w(x)\,dx$ and denote with $\bar{x}$ a random variable sampled from $w(x)/A$, e.g. using the inverse function method. Let $\bar{y}$ be sampled uniformly in $[0, w(\bar{x})]$. The sampled value $\bar{x}$ is accepted if $\bar{y}$ is less than $p_x(\bar{x})$, otherwise it is rejected. The procedure is shown in Figure A.27 and the pseudocode of the procedure is presented in Algorithm 21, where ξ_1 and ξ_2 are samples from two independent random variables, uniformly distributed in $[0, 1]$.

Algorithm 21 `acceptance_rejection_sampling()`

1: $\xi_1 \leftarrow \text{RAND}(0, 1)$
2: $\bar{x} = W^{-1}(A\xi_1)$
3: $\xi_2 \leftarrow \text{RAND}(0, 1)$
4: $\bar{y} \leftarrow w(\bar{x})\xi_2$
5: **if** $\bar{y} < p_x(\bar{x})$ **then**
6:      **return** $\bar{x}$ { accept the sample $\bar{x}$ and return it}
7: **else**
8:      **goto** 1 { reject the sample and repeat }
9: **end if**

The efficiency of the acceptance–rejection procedure depends on how easy it is to invert the function $W(x)$, and how frequently we accept the sample. As seen above, the fraction of accepted samples depends on the areas below the two curves $p_x(x)$ and $w(x)$ and is equal to $1/A$. Therefore, an efficient sampling would require that the function $w(x)$ has an easily invertible primitive $W(x)$, and that the area A under the curve $w(x)$ is as close as possible to 1.

A.9.4 Sampling from Discrete Distributions

Quite often in network science one has to sample from a set of integers, distributed according to a certain probability mass function p_k. For instance, if we want to create a graph with N nodes and a degree distribution which follows a certain functional form, we should first sample the sequence of integers $\{k_0, k_1, \ldots, k_{N-1}\}$ that will represent the degrees of the N nodes. Here we review a few methods to sample from discrete probability distributions.

Subdivision of the unit interval for discrete sampling from a given probability distribution $\{p_k\}$.

Simple Discrete Sampling

Let us suppose that $k \in \{1, \ldots, M\}$ is an integer random number, with probabilities $\{p_k\}$, and we want to sample an integer κ according to $\{p_k\}$. We can proceed as follows. We divide the interval $[0, 1]$ in M intervals so that the ith interval has length equal to p_i, as shown in Figure A.28. Then, we sample a random number ξ uniformly in $[0, 1]$, and detect the sub-interval k to which ξ belongs, and return k as a result of the sampling. This procedure is shown in Algorithm 22.

Algorithm 22 `discrete_sampling()`

1: Compute $P_i = \sum_{m=1}^{i} p_m$, $i = 1, \ldots, M$, $\quad P_0 = 0$;
2: find the integer k such that $P_{k-1} \leq \xi < P_k$.

The efficiency of this sampling scheme depends on how easy it is to identify the sub-interval to which the sample ξ belongs. In some specific cases, such interval can be determined analytically. A typical example is that of geometrically distributed random variables.

Geometrically distributed variables

We show here how to sample from a geometrically distributed random variable using discrete sampling. Let us consider $\tau \in (0, 1)$, and assume that the probability mass function p_k reads

$$p_k = (1 - \tau)\tau^{k-1}, \quad k = 1, \ldots, \infty$$

so that we have:

$$P_k = \sum_{j=1}^{k} p_j = 1 - \tau^k.$$

In this case the integer k that satisfies the condition $P_{k-1} \leq \xi < P_k$ can be determined analytically as

$$k = \lfloor \frac{\ln(1 - \xi)}{\ln \tau} \rfloor + 1, \tag{A.23}$$

where $\lfloor x \rfloor$ denotes the integer part of x. Therefore line 2 of Algorithm 22 reduces to Eq. (A.23). Notice that the existence of an analytical expression for k also allows us to manage the case of $M = \infty$.

Arbitrary probability mass functions

For an arbitrary set of probabilities $\{p_k\}$, the check at line 2 of Algorithm 22 can be easily performed by using binary search (see Algorithm 2 in Appendix A.3.1). In practice, we compute the values P_k of the cumulative probability distribution associated with $\{p_k\}$, and we store them in the array `cumul[]`, whose M components satisfy `cumul[k]` = P_k. When we have to check to which value of k a sample variable ξ in $[1, M]$ corresponds, we simply use Algorithm 2 on the array `cumul`. By doing so, the check at line 2 of Algorithm 22 can be performed in $O(\log_2 M)$ operations (see Appendix A.3.1 for details). This means that, in this case, the time complexity of Algorithm 22 is dominated by the computation of the cumulative distribution P_k, which has time complexity $O(M)$.

The complexity of line 1 in Algorithm 22 is $O(M)$, since we have to compute M values of P_i in order to find the interval k. In general, one is interested in sampling N values of k from a given $\{p_k\}$, hence line 2 of the algorithm has to be repeated N times. The overall complexity of this procedure is therefore $O(M) + O(N \log_2 M)$, and if the number M is very large, i.e. if $M \gtrsim N \log_2 M$, then most of the time will be spent in the computation of the $\{P_k\}$. In such a case a more efficient acceptance–rejection technique can be used, assuming that we can efficiently compute an estimate $\tilde{p}_{\max}$ of the largest p_k ($\tilde{p}_{\max} \geq p_k$, $k = 1, \ldots, M$). The technique consists in sampling an integer k chosen uniformly in $[1, M]$, and accepting the sample with probability $p_k/\tilde{p}_{\max}$. Geometrically, this corresponds to sampling a point uniformly in the interval $[0, M\tilde{p}_{\max}]$. If the point falls into one of the grey intervals (see Figure A.29), the corresponding value of k is sampled. The procedure is described in Algorithm 23.

Algorithm 23 `discrete_sampling_acceptance_rejection()`

1: select an integer random number uniformly in $[1, \ldots, M]$, i.e.:
2:     $\xi_1 \leftarrow \text{RAND}(0, 1)$
3:     $k \leftarrow \lfloor M\xi_1 \rfloor + 1$
4: $\xi_2 \leftarrow \text{RAND}(0, 1)$
5: **if** $p_k\xi_2 < \bar{p}$ **then**
6:     accept the sample
7: **else**
8:     reject it and `go to` 1
9: **end if**

Clearly, the procedure can be generalised to the case in which the estimate $\bar{p}$ depends on k, i.e. when $\bar{p}_k \geq p_k$, $k = 1, \ldots, M$. In such a case, which is the discrete version of

Discrete sampling by acceptance–rejection technique. The interval $[0, M\bar{p}]$ is divided into M equal intervals. The acceptance region is the union of the grey intervals.

the acceptance–rejection sampling described in Algorithm 21 for the continuous case, one would sample from $\{\bar{p}_k/\bar{P}\}$, where $\bar{P} = \sum_k \bar{p}_k$, and then accept the sample if $\xi p_k \leq \bar{p}_k$.

A.9.5 Sampling from a Discrete Power Law

As an example of what was discussed in Appendix A.9.4, we consider here in detail the problem of sampling from a power law distribution

$$p_k = \frac{k^{-\gamma}}{\zeta(\gamma)}, \quad k = 1, \ldots.$$

Since Algorithm 22 requires the evaluation of $\sum_{m=1}^{i} = k^{-\gamma}$, for which there is no simple expression, we will make use of the acceptance–rejection method. With reference to Figure A.30, we want to sample an integer k with a probability proportional to the area, $k^{-\gamma}$, of the kth rectangle. Let us consider the region below the solid black line, which is composed by two parts: a square of unit area, for $x < 1$, and a region of area $A = 1/(\gamma - 1)$ below the curve $y = x^{-\gamma}$, for $x > 1$. In order to sample from the grey region, we use the acceptance–rejection method. We sample a point (x, y) uniformly in the region below the black line. If the sampled point is inside the grey region, we accept the sample, otherwise we reject the sample and repeat the process. In practice, we first check whether the point is in the first region, the square of unit area. This happens with a probability equal to $1/(1 + A) = (\gamma - 1)/\gamma$. Otherwise, we sample a point in the second region: x is sampled by the inverse mapping, while y is sampled uniformly in $[0, x^{-\gamma}]$. For example, the sample point (x_1, y_1) shown in Figure A.30 is accepted, and we set $k = 3$, since the point lies in the second rectangle of the second region. Conversely the point (x_2, y_2) is rejected. This procedure is described in Algorithm 24. The program `power_law`, available for download at

---

Algorithm 24 `power_law_sample()`

Input: γ
Output: a sample from the discrete power-law $\frac{k^{-\gamma}}{\zeta(\gamma)}$
1:   $\xi_0 \leftarrow \text{RAND}(0, 1)$
2: **if** $\xi_0 < (\gamma - 1)/\gamma$ **then**
3:     $k = 1$
4: **else**
5:     $\xi_1 \leftarrow \text{RAND}(0, 1)$
6:     $\xi_2 \leftarrow \text{RAND}(0, 1)$
7:     $x \leftarrow (1 - \xi_1)^{1/(1-\gamma)}$
8:     $k_{\text{trial}} \leftarrow \lfloor x \rfloor + 1$
9:     $y \leftarrow \xi_2 x^{-\gamma}$
10:    **if** $y < k_{\text{trial}}^{-\gamma}$ **then**
11:      **return** k_{trial}
12:    **else**
13:      go to 1
14:    **end if**
15: **end if**

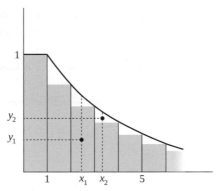

Fig. A.30 Sampling from a power law by acceptance–rejection technique. A sample point (x, y) is accepted if it falls inside the grey region.

www.complex-networks.net, can be used to sample from a discrete power law distribution with a given exponent, minimum degree and maximum degree.

A.9.6 Exclusive Sampling

Sometimes it is useful to extract m numbers, at random and without replacement, from a given sequence x[] of length M. A simple and efficient method to perform the sampling is illustrated in Algorithm 25. The idea behind this implementation of exclusive sampling is quite simple: at each step an element of x[] is sampled uniformly at random among the K remaining ones. Then, that element is swapped with the last element of the sequence x[] and the effective size of the sequence x[] is reduced by one. By doing so, the algorithm guarantees that the first K elements of x[] have not been sampled yet.

Algorithm 25 exclusive_sampling()

Input: x, m
Output: v
 1: $K \leftarrow M$
 2: **for all** i in 0 **to** m-1 **do**
 3:　　$\xi \leftarrow \text{RAND}(0, 1)$
 4:　　$j \leftarrow \lfloor K\xi \rfloor$
 5:　　v[i] $\leftarrow$ x[j]
 6:　　tmp $\leftarrow$ x[j]
 7:　　x[j] $\leftarrow$ x[K]
 8:　　x[K] $\leftarrow$ tmp
 9:　　$K \leftarrow K - 1$
10: **end for**

When the algorithm finishes, the vector v[] will contain m distinct integers randomly sampled from the sequence x[]. Of course, if $m = M$, the vector v[] will contain a random permutation of the input sequence. It is easy to realise that Algorithm 25 has time complexity $O(m)$. The main drawback of Algorithm 25 is that the input sequence `x[]` is modified by the m swaps, which means that we cannot rely on the actual ordering of the elements of the sequence `x[]` after the algorithm has been applied on it. A possible solution is to copy `x[]` into a temporary vector, even if this approach is not very efficient when the number of required samples m is much smaller than M. Another possibility is to use an algorithm based on rejection, such as that presented in Algorithm 26.

Algorithm 26 `exclusive_sampling_rejection()`

1: **for all** $i = 1$ **to** m **do**
2:     $\xi \leftarrow \text{RAND}(0, 1)$
3:     $j = \lfloor M\xi \rfloor$
4:     **if** $j \notin [\text{seq}_1, \ldots, \text{seq}_{i-1}]$ **then**
5:         $\text{seq}_i = j$
6:     **else**
7:         `go to 2:`
8:     **end if**
9: **end for**

Algorithm 26 is very effective in terms of memory, and its efficiency in terms of time depends on how line 4 is implemented. This line checks whether the new integer value j has already been sampled. If the check is performed sequentially, the check itself costs $O(i)$ operations, and the whole algorithm has time complexity $O(m^2)$ (for $m \ll M$).

A more efficient way to check if an integer has been already sampled is by means of a *binary search tree*, as described in Appendix A.3.3, which contains the sampled elements. In practice, when we sample a new element ξ we first check if the element is already in the tree, otherwise we insert ξ in the tree and return it. Since searching and inserting elements into a binary tree with H nodes has time complexity $O(\log H)$, this implementation of the check allows us to sample a new element in $O(\log M)$.

A.10 Erdős and Rényi Random Graph Models

We focus here on methods to sample Erdős and Rényi (ER) random graphs. We present two algorithms to sample ER random graphs according, respectively, to model A and model B introduced in Chapter 3. In the first case we will obtain graphs with N nodes and K links, while in the second case we will produce graphs with N nodes and a probability p that each pair of nodes is connected by an edge.

A.10.1 Sampling a Graph in $G_{N,K}^{ER}$ (ER Model A)

Several techniques can be used to obtain Erdős and Rényi random graphs with N nodes and K links. A simple method consists in sampling K random pairs (i, j) (i.e. i and j are uniformly sampled in $[0, \ldots, N-1]$). Each time a pair (i, j) is sampled, in order to avoid loops and multiple edges, we check if $i = j$ and if the edge (i, j) already exists. In such cases, the pair is rejected and a new pair of nodes is sampled. Otherwise, the edge (i, j) is added to the graph, and the edge counter is increased by one. The process is repeated until K links have been successfully sampled.

This method is very simple to describe and to implement, but it has the drawback that each time a new pair of nodes (i, j) is sampled, one has to check whether the edge (i, j) already belongs to the graph or not. If sampled edges are stored in an unordered list, each check costs on average $O(K)$. However, if we uniquely associate each of the $M = N(N-1)/2$ possible edges with an integer in the range $[1, N(N-1)/2]$, the sampling of K links becomes equivalent to the exclusive sampling of K integers out of a set of $N(N-1)/2$ integers. As explained in Appendix A.9.6, integer exclusive sampling with rejection can be efficiently implemented by using a binary search tree, which allows us to check for duplicate samples in $O(\log K)$ instead of $O(K)$. In Table A.2 we show a simple method to uniquely assign an integer in $[1, N(N-1)/2]$ to each pair (i, j), where $i, j \in [0, N-1]$. Without loss of generality, we always assume that $i > j$. The label k associated with each entry (i, j) of the table can be obtained from the values of i and j as follows:

Table A.2 Numbering all possible unordered pairs of nodes in a graph.

$i \backslash j$	0	1	2	3	4	$\ldots$
1	1					
2	2	3				
3	4	5	6			
4	7	8	9	10		
5	11	12	13	14	15	$\ldots$
$\vdots$	$\ldots$	$\ldots$	$\ldots$	$\ldots$	$\ldots$	

$$(i,j) \longrightarrow k = \frac{i(i-1)}{2} + j + 1. \tag{A.24}$$

In Algorithm 27 we present the pseudocode to sample Erdős and Rényi random graph from model A using integer sampling with rejection. Remember that the binary search tree `t` contains only the integer representation of each sampled edge (i,j), as obtained by formula A.24. Once all the K edges have been sampled, the algorithm dumps out these edges calling the function `dump_edges()` (line 17). This function transforms each entry k of the tree back into the corresponding pair (i,j), by inverting Eq. A.24 as follows. We first notice that, for a given value of i, $i > j$, the number k satisfies the bounds:

$$k_{\min}^i \equiv \frac{i(i-1)}{2} + 1 \leq k \leq \frac{i(i+1)}{2} = k_{\max}^i$$

Algorithm 27 `ER_A()`

Input: N, K
Output: G

1: t ← `empty_tree()`
2: n ← 0
3: **while** n < K **do**
4:     i ← $\lfloor \text{RAND}(0,1)N \rfloor$
5:     j ← $\lfloor \text{RAND}(0,1)N \rfloor$
6:     **if** i = j **then**
7:        **goto** 4
8:     **end if**
9:     $\bar{i}$ ← $max(i,j)$
10:    $\bar{j}$ ← $min(i,j)$
11:    k ← $(\bar{i})(\bar{i}-1)/2 + \bar{j} + 1$
12:    **if** k $\notin$ t **then**
13:       insert(k,t)
14:       n ← n + 1
15:    **end if**
16: **end while**
17: `dump_edges()`

---

therefore the quantity $2k$ satisfies the bounds

$$i(i-1) + 2 \leq 2k \leq i(i+1).$$

Let x denote the non-negative solution of the equation:

$$\frac{x(x+1)}{2} = k$$

i.e.

$$x = \frac{-1 + \sqrt{1 + 8k}}{2}. \tag{A.25}$$

Since:

$$\frac{(i-1)i}{2} < \frac{(i-1)i}{2} + 1 = k_{min}^i \le \frac{x(x+1)}{2} = k \le \frac{i(i+1)}{2} = k_{max}^i$$

then it follows that $(i-1) < x \le i$. Therefore, because of the monotonicity of $x(x+1)$ (for positive x) the value of i corresponding to a given label k is always obtained as:

$$i = \lceil x \rceil \qquad (A.26)$$

where $\lceil x \rceil$ is the smallest integer greater than or equal to x. Once we have computed i, we obtain j using Eq. A.24:

$$j = k - 1 - \frac{i(i-1)}{2}. \qquad (A.27)$$

Algorithm 27, which uses sampling with rejection, has time complexity $O(K \log K)$, because for each new edge one has to perform a check on the binary tree which costs $O(\log K)$ operations, while the memory storage requirement is $O(K)$. Algorithm 27 is implemented by the program er, available for download at www.complex-networks. net.

The integer rejection procedure used in Algorithm 27 is very efficient to sample sparse graphs, i.e. when K is $O(N)$, because rejections are quite rare. However, it can be too slow for dense graphs, because the rejection rate may become high. In such cases, a more efficient sampling algorithm can be obtained by using integer exclusive sampling without rejection, as described in Algorithm 25 in Section A.9.6. Using the above discussed one-to-one correspondence between the set of possible edges of a graph of order N and the first $M = N(N-1)/2$ natural numbers, a random dense graph can be easily sampled as follows:

1 Sample K integers $k_1, k_2, \ldots, k_K$ from $[1, \ldots, M]$ without repetition (as illustrated in Algorithm 25);
2 For each k_ℓ, $\ell = 1, \ldots K$, use Eqs. (A.25)–(A.27) to compute the indices of the non-zero entries of the adjacency matrix (i_ℓ, j_ℓ).

This method has time complexity $O(K)$ if we neglect the assignment of the initial integer vector of size $N(N-1)/2$, and is therefore more efficient than Algorithm 27. However, it needs a memory occupancy $O(N^2)$, since it uses a vector of size $N(N-1)/2$, which is much higher than that required by the other algorithm if the graph is sparse.

A.10.2 Sampling a Graph in $G_{N,p}^{ER}$ (ER Model B)

A method to sample Erdős and Rényi random graphs with N nodes and a probability p, is presented in Algorithm 28. The algorithm consists of a single loop over all the possible edges of the graph (lines 2–9). For each integer $k \in [1, M]$, we sample a value $\xi \in [0, 1]$. If $\xi \le p$, we compute the values of $\bar{i}$ and $\bar{j}$ corresponding to k using Eqs. (A.25)–(A.27), and we add the edge $(\bar{i}, \bar{j})$ to the graph. At the end of the loop, the vectors $\mathbf{i}$ and $\mathbf{j}$ contain the sparse representation of the adjacency matrix. Algorithm 28 is implemented in the program ER_B, available for download at www.complex-networks.net. The algorithm has time complexity $O(N^2)$. Consequently, this approach becomes inefficient for sparse graphs (i.e. when $p \ll 1$), because of the large fraction of rejections.

Algorithm 28 ER_B()

Input: N, p

Output: $K, \mathbf{i}, \mathbf{j}$

1: $K \leftarrow 0$
2: **for all** $k = 1$ **to** $N * (N-1)/2$ **do**
3:     $\xi \leftarrow \text{RAND}(0, 1)$
4:     **if** $\xi \leq p$ **then**
5:         compute $(\bar{i}, \bar{j})$ from k using Eqs. (A.25,A.26,A.27)
6:         $\mathbf{i}[K] \leftarrow \bar{i}$, $\mathbf{j}[K] \leftarrow \bar{j}$
7:         $K \leftarrow K + 1$
8:     **end if**
9: **end for**

An alternative method for sampling a graph in $G_{N,p}^{ER}$ consists in sampling an integer K^* from the binomial probability distribution:

$$p_K = \binom{M}{K} p^K (1-p)^{M-K}$$

where $K = pN$ and $M = N(N-1)/2$. Then, we sample a graph in the ensemble G_{N,K^*}^{ER}, using one of the two algorithms discussed in Section A.10.1. The interested reader may find a review of existing techniques to sample from a binomial distribution in [171]. For large values of N, the number K^* could be sampled, with a good approximation, from a Poisson distribution, with a time complexity $O(N)$ (see for instance [118]).

A.11 The Watts–Strogatz Small-World Model

We present here an algorithm to construct small-world graphs according to the Watts–Strogatz model. We first introduce a procedure to create an (N, m) circle graph, which is a graph with N nodes and $K = mN$ links as in Definition 4.5. Let us place the nodes of the graph on a circle and progressively label them with a number $i = 0, 1, 2, 3, \ldots, N - 1$, starting from an initial node and proceeding clockwise. This is shown in Figure A.31(a) for a graph with $N = 12$ nodes. Due to this ordering, the value of the label ℓ of each of the m nodes following a certain node i on the circle satisfies the relation:

$$\ell = (i + j) \mod (N), \quad j = 1, \ldots, m. \tag{A.28}$$

We observe that the $m = 3$ nodes following (preceding) node 4 in Figure A.31(b) have labels 5, 6 and 7 (1, 2 and 3), respectively. Now, to construct a circle graph, we perform a loop over all the nodes and, for each node i, we create only the m edges from i to the m neighbouring nodes *following* i on the circle. This procedure is illustrated in Figure A.31(c). We start from node 0 and add the three edges to nodes 1, 2 and 3, then we consider node 1 and add the three edges to nodes 2, 3 and 4, and so on. We notice that, when we arrive at node 3, and we add the three edges pointing to nodes 4, 5 and 6, all the $2m = 6$ edges of node 3 have been created. In fact, we have already created the edges pointing to node 3 from its m predecessors, namely $(0, 3)$, $(1, 3)$ and $(2, 3)$, in the

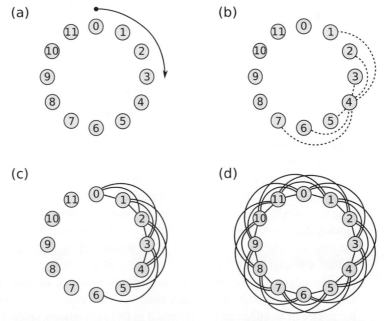

Fig. A.31 An example of the procedure to create a circle graph with $N = 12$ nodes and $m = 3$, using Algorithm 29. Nodes are put in a circle and labelled from 0, proceeding clockwise. At the end of the procedure, each node is connected to its $2m = 6$ closest neighbours.

previous steps, and we have also added in this step the m edges towards the successors of node 3. If we iterate this procedure over the N nodes we obtain the circle graph with $K = mN$ edges shown in Figure A.31(d). The function `create_circle()`, whose pseudocode is illustrated in Algorithm 29, uses this method and Eq. (A.28) to construct a circle graph with N nodes and $K = mN$ edges. Algorithm 29 loops over the nodes of the graph (line 2). For each node i, the algorithm uses Eq. (A.28) to compute the labels of the m nodes following i, and creates all the m edges from i to its m closest successors on the circle (lines 3–9).

Algorithm 29 `create_circle()`

Input: N, m
Output: R(N,m)
 1: $K \leftarrow 0$
 2: **for** i = 0 **to** N - 1 **do**
 3:     **for** j = 1 **to** m **do**
 4:         $\ell \leftarrow$ (i+j) mod(N)
 5:         $\mathbf{i}[K] \leftarrow i$; $\mathbf{j}[K] \leftarrow \ell$
 6:         $K \leftarrow K + 1$
 7:         $\mathbf{i}[K] \leftarrow \ell$; $\mathbf{j}[K] \leftarrow i$
 8:         $K \leftarrow K + 1$
 9:     **end for**
10: **end for**

We are now ready to describe an algorithm for the sampling of (N, m, p) Watts–Strogatz (WS) small-world graphs. The pseudocode presented in Algorithm 30 starts by calling the procedure `create_circle(N,m)` to initially create a (N, m) circle graph. Then, we perform a loop over the nodes (line 2) and, for each node i, we consider in turn each of its m successors. Let ℓ be one of them. With probability p decide to rewire the edge (i, ℓ) (line 5), and we select a node ℓ' at random such that $\ell' \neq i$ and $\ell' \neq \ell$, and we replace the edge (i, ℓ) with the edge (i, ℓ') only if (i, ℓ') does not already exist (line 12). Notice that, in this way, each edge is considered exactly once.

The time complexity of Algorithm 30 is $O(K + (C + R) \times pK)$, where C is the time complexity of the algorithm used to check the existence of an edge (line 9), and R is the time complexity of replacing an edge. By storing all the edges in a binary tree, as we have already done in Algorithm 27, we have $C = O(\log K)$ and $R = O(\log K)$, and consequently the time complexity of Algorithm 30 is dominated by $O(K \log K)$. Another possibility is to take advantage of the particular structure of the circle graph to speed up the check for existing edges and the rewire of an edge. In practice, we can store the circle graph in CRS format, maintaining for each node only the edges to its m successors. By doing so, the rewire of an edge can be performed in $O(1)$ operations, since it consists in updating the label of the endpoint of the edge in the vector $\mathbf{j}$ from ℓ to ℓ'. Similarly, it is possible to check whether the edge (i, ℓ') already exists by checking that none of the m labels corresponding

---

Algorithm 30 ws()

Input: N, m, p
Output: WS(G, p)

1: create_circle(N,m)
2: **for** $i = 0$ **to** N-1 **do**
3:   **for** $j = 1$ **to** m **do**
4:     $\ell \leftarrow (i+j) \mod (N)$
5:     $\xi \leftarrow$ RAND(0,1)
6:     **if** $\xi < p$ **then**
7:       $\xi_1 \leftarrow$ RAND(0,1)
8:       $\ell' \leftarrow \lfloor N\xi_1 \rfloor$
9:       **if** (i, ℓ') exists **or** $\ell' = i$ **or** $\ell' = \ell$ **then**
10:        **goto** 7
11:      **else**
12:        replace (i, ℓ) with (i, ℓ')
13:      **end if**
14:    **end if**
15: **end for**
16: **end for**

---

to neighbours of the current node i in **j** is equal to ℓ', which can be done in $O(m)$. The time complexity of the resulting algorithm is $O(Nm^2)$. In the program ws available for download at www.complex-networks.net we implemented a version of Algorithm 30 which uses the CRS representation of sparse matrices.

A.12 The Configuration Model

Here we present some algorithms to sample random graphs with a prescribed degree sequence from the configuration model introduced in Section 5.3. Let us start from an assigned degree sequence $\{k_1, k_2, \ldots, k_N\}$, corresponding to graphs with N nodes and $K = 1/2 \sum_{i=1}^{N} k_i$ edges. The first thing to sample graphs from this degree sequence is to assign to each node i, $i = 0, \ldots, N - 1$, a number of stubs (i.e. half-edges) equal to its degree k_i. This is shown, for instance, in the left part of Figure A.32 for a graph with four nodes and a degree sequence $\mathcal{K} = \{3, 3, 1, 1\}$. Then, to construct a graph we have to successively match pairs of stubs together, until there are no more unmatched stubs left. There are however different ways in which stub matching can be implemented.

The simplest algorithm is obtained by allowing the presence of loops and multiple edges. In this case, K pairs of stubs are connected, with each stub randomly sampled by the exclusive sampling discussed in Appendix A.9.6. The procedure is illustrated by Algorithm 31.

Algorithm 31 `conf_model_multigraph()`

Input: Degree sequence $\{k_1, k_2, \ldots, k_N\}$
Output: configuration model multigraph with degree sequence $\{k_1, k_2, \ldots, k_N\}$

1: S $\leftarrow$ 0
2: **for** $i = 1$ **to** N **do**
3:    **for** $j = 1$ **to** k_i **do**
4:       stubs[S] $\leftarrow$ i
5:       S $\leftarrow$ S+1
6:    **end for**
7: **end for**
8: **while** S > 0 **do**
9:    $\ell_1 \leftarrow \lfloor$ S $\times$ RAND$(0, 1)\rfloor$
10:   S $\leftarrow$ S-1
11:   $\ell_2 \leftarrow \lfloor$ S $\times$ RAND$(0, 1)\rfloor$
12:   S $\leftarrow$ S-1
13:   **if** $\ell_1 = \ell_2$ **then**
14:      {reject the sample: it's the same stub!!!}
15:      S $\leftarrow$ S+2
16:      **goto** line 8
17:   **end if**
18:   $i_1 \leftarrow$ stubs[ℓ_1]
19:   $i_2 \leftarrow$ stubs[ℓ_2]
20:   stubs[ℓ_1] $\leftarrow$ stubs[$S + 1$]
21:   stubs[ℓ_2] $\leftarrow$ stubs[S]
22:   create a link between node i_1 and node i_2
23: **end while**

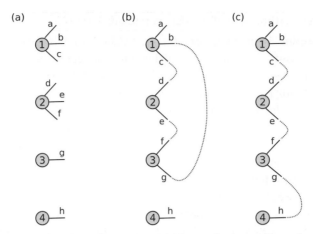

Fig. A.32 Stub matching for graph sampling in the configuration model. (a) A set of nodes and their stubs corresponding to the degree sequence $\mathcal{K} = \{3, 3, 1, 1\}$. In this case, there is no way to match the stubs in order to obtain a graph without loops and multiple edges. In the case of the degree sequence $\mathcal{K}' = \{3, 2, 2, 1\}$ it is instead possible to find a good matching (b), but also a bad matching (c) generating a loop.

The algorithm makes use of the array stubs[], of length equal to $2K$. The label of each node i is present in the array stubs[] a number of times equal to the number of stubs attached to i, i.e. k_i times. The exclusive sampling procedure on the array stubs[] has time complexity $O(K)$, since each stub is processed exactly once, and is thus very efficient. Unfortunately, it cannot guarantee that the resulting matching does not contain multiple edges or self-loops. This is because it is possible to sample two stubs ℓ_1 and ℓ_2 connected to the same node $i_1 = i_2$, thus producing a self-loop, or even to sample two stubs ℓ_1 and ℓ_2 which correspond to two nodes that are already connected, resulting in the creation of a multiple edge.

We are normally interested in generating simple graphs, so that matches producing loops or multiple edges should be discarded when they occur. This can be done as follows. Suppose that you have selected two candidate nodes, i_1 and i_2, as in the algorithm illustrated above. If the two nodes are distinct, and if they are not already connected, then a link will be created between them. Otherwise, the pair i_1, i_2 is rejected, and a new pair of candidate nodes is sampled, until a new link is formed. This procedure is implemented in Algorithm 32. However, this version of the stub matching algorithm requires the set of edges that have been created to be stored, in a similar way to what we did in Appendix A.10.1 to sample graphs from the ER model A. It is easy to show that this algorithm has time complexity $O(K \log K)$. Although Algorithm 32 avoids the formation of self-loops and multiple edges, it has another major drawback: it can enter an infinite loop. Consider, for instance, the case in which a node, say node 1, has more than $N - 1$ stubs. It is evident that in this case a loop or a multiple edge is unavoidable, since there are not enough nodes to be connected to the $k_1 > N - 1$ stubs of node 0. However, having $k_i \leq N - 1 \; \forall i$ is not a sufficient condition to guarantee that there exists at least one stub matching that corresponds to a simple graph. Consider again the degree sequence shown in Figure A.32(a). In this case, even if the total number of stubs is even, and the nodes with the largest number of stubs

---

Algorithm 32 conf_model_pair_reject()

Input: Degree sequence $\{k_1, k_2, \ldots, k_N\}$
Output: configuration model graph with degree sequence $\{k_1, k_2, \ldots, k_N\}$

1: S $\leftarrow$ 0
2: **for** $i = 1$ **to** N **do**
3:     **for** $j = 1$ **to** k_i **do**
4:         stubs[S] $\leftarrow$ i
5:         S $\leftarrow$ S+1
6:     **end for**
7: **end for**
8: **while** S > 0 **do**
9:     $\ell_1 \leftarrow \lfloor$S $\times$ RAND$(0, 1)\rfloor$
10:     S $\leftarrow$ S-1
11:     $\ell_2 \leftarrow \lfloor$S $\times$ RAND$(0, 1)\rfloor$
12:     S $\leftarrow$ S-1
13:     **if** $\ell_1 = \ell_2$ **then**
14:         {reject the sample - It's the same stub!}
15:         S $\leftarrow$ S+2
16:         **goto** line 8
17:     **end if**
18:     **if** $i_1 \neq i_2$ **and** (i_1, i_2) is not a link **then**
19:         create the link (i_1, i_2)
20:         stubs[ℓ_1] $\leftarrow$ stubs[S+1]
21:         stubs[ℓ_2] $\leftarrow$ stubs[S]
22:     **else**
23:         {reject the pair to avoid multiple edges or loops}
24:         S $\leftarrow$ S+2
25:         **goto** line 8
26:     **end if**
27: **end while**

---

have $N - 1 = 3$ stubs, we can easily check that there are no stub matchings producing graphs with no loops or multiple links.

There is a general way to test whether a given sequence of non-negative integers, corresponding to the node stubs, can indeed be considered as the degree sequence of a simple graph. Without loss of generality, let us label the nodes in non-increasing order of number of stubs: $k_1 \geq k_2 \geq \cdots \geq k_N$. As observed above, a necessary condition for the existence of at least one "good stub matching", i.e. a stub matching that gives rise to a simple graph, is $k_1 \leq N - 1$. If this condition is satisfied we can connect the stubs of the first node to k_1 distinct nodes. Once we have done this, we should check whether the same condition is satisfied for the remaining nodes, so that we can repeat the matching procedure until all the stubs are saturated. The nodes which are more likely to produce loops or multiple edges are those with the largest number of links. Therefore, a good way to match the stubs of the first node is to connect them with the k_1 remaining nodes with highest number of stubs, and to repeat this procedure recursively. It can be proved that if a good matching is not found by this procedure, then the given stub sequence does not allow any good matching. All this can be formalised mathematically, introducing the concept of a *successively compressible* degree sequence [121].

Definition A.18   *A sequence of integers $k_1 \geq k_2 \geq \cdots \geq k_N$ is said to be* compressible *if $k_1 \leq N - 1$, $k_{k_1+1} > 0$, and $k_N \geq 0$.*

Definition A.19   *A compressible set of integers is said to be compressed if the following operations are performed on the set:*

1 k_1 is removed from the set;
2 $k_2, k_3, \ldots, k_{k_1+1}$ are reduced by 1;
3 The set of integers resulting from the application of 1) and 2) are relabelled as $k_1, k_2, \ldots, k_{N-1}$ so that $k_1 \geq k_2 \geq \cdots \geq k_{N-1}$.

Definition A.20   *A set of integers is said to be* successively compressible *if it is compressible and if every set of integers resulting from successive compressions is compressible.*

Summing up, a necessary and sufficient condition for the existence of at least one simple graph corresponding to a given degree sequence $\mathcal{K}$ is the *successive compressibility* of $\mathcal{K}$. Going back to the degree sequence in Figure A.32(a), $\mathcal{K} = \{3, 3, 1, 1\}$ is compressible but not successively compressible, and therefore is not a good degree sequence. In fact, after matching the following pairs of stubs (a, d), (b, g), (c, h), we remain with three nodes and the sequence $\mathcal{K} = \{2, 0, 0\}$, which is not compressible. Conversely, if a degree sequence is successively compressible, then there exists a stub matching without loops or multiple edges. For instance, the sequence $\mathcal{K}' = \{3, 2, 2, 1\}$ shown in Figure A.32(b) is successively compressible. Imagine we want to construct a simple graph with this degree sequence, and we select the stubs in the following order d, c, f, e, g, b, a, h. The first pair of stubs, (d, c), forms a new edge joining nodes 1 and 2, and is accepted. The second pair of stubs, (f, e), forms a new edge between nodes 2 and 3, and is accepted. The third pair, (b, g), forms a new edge between nodes 1 and 3, and is accepted. At this point, the remaining two stubs, a and h, can be joined to form a new edge, completing the graph. However, the successive compressibility of a sequence does not guarantee that Algorithm 32 does not enter an infinite loop. Consider for instance the situation shown in Figure A.32(c), where we assume that the order of selected stubs is d, c, f, e, g, h, b, a. The first two matchings are the same as before. The third one creates the edge between nodes 3 and 4, which is a new edge and is accepted. However, the two stubs left at the end, a and b, can only form a self-loop on node 1. Since this choice is rejected in line 22 of Algorithm 32, because $i_1 = i_2$, the algorithm would keep proposing and rejecting the same matching over and over again! If this situation appears, then one possibility is to reject the whole construction, and start again from the degree sequence, provided it is successively compressible.

This leads to a third possible version of the matching procedure, which is implemented in Algorithm 33. The algorithm iteratively samples candidate pairs of stubs. If the pair corresponds to a loop or a multiple edge, it is rejected and the rejection counter `countreject` is increased by one. Otherwise the pair is accepted, and the rejection

---

Algorithm 33 conf_model_pair_reject_check()

Input: Degree sequence $\{k_1, k_2, \ldots, k_N\}$

Output: configuration model graph with degree sequence $\{k_1, k_2, \ldots, k_N\}$

```
 1: num_attempts ← 0, countreject ← 0
 2: if num_attempts > max_attempts then
 3:     print an error message
 4:     exit
 5: end if
 6: S ← 0
 7: for i = 1 to N do
 8:     for j = 1 to kᵢ do
 9:         stubs[S] ← i
10:         S ← S+1
11:     end for
12: end for
13: while S > 0 do
14:     ℓ₁ ← ⌊S × RAND(0,1)⌋
15:     S ← S-1
16:     ℓ₂ ← ⌊S × RAND(0,1)⌋
17:     S ← S-1
18:     if ℓ₁ = ℓ₂ then
19:         {reject the sample - It's the same stub!}
20:         S ← S+2
21:         goto line 13
22:     end if
23:     if i₁ ≠ i₂ and (i₁, i₂) is not a link then
24:         create the link (i₁, i₂)
25:         stubs[ℓ₁] ← stubs[S+1]
26:         stubs[ℓ₂] ← stubs[S]
27:         countreject ← 0
28:     else
29:         {reject the pair to avoid multiple edges or loops}
30:         S ← S+2
31:         countreject ← countreject + 1
32:         goto line 13
33:     end if
34:     if countreject > maxrejects then
35:         num_attempts ← num_attempts + 1
36:         goto line 1
37:     end if
38: end while
```

counter is reset to zero. If there are too many successive rejections, i.e. more than a preassigned number `maxreject`, it is very likely that the algorithm has ended up in a configuration which cannot lead to a simple graph, and it starts again from the initial degree sequence. In order to avoid infinite loops, the algorithm tries to construct a graph at most `max_attempts` times. Each attempt to construct a matching using Algorithm 33 has time complexity $O(K \log K)$, since each stub is considered exactly once, and for each pair of stubs we can check if the corresponding link exists in $O(\log K)$, by storing the edges in a binary tree. Several improvements in the efficiency of the algorithm are possible and are described in Ref. [121].

It is interesting to notice that, in the case of power-law degree distributions, the *structural cut-off* $k_{\max}^{\text{struct}}$ discussed in Section 5.5 determines whether a degree sequence can generate a graph without loops and multiple edges. In particular, a degree sequence with a maximum degree $k_{\max}$ larger than the structural cut-off reported in Eq. (5.48) will most probably not be successively compressible. In practice, a good practice in order to successfully generate graphs with power-law degree distributions using the configuration model is to reject sequences having a maximum degree larger than the structural cut-off $k_{\max}^{\text{struct}} \simeq \sqrt{\langle k \rangle N}$.

Algorithm 34 `check_successively_compressible()`

Input: `degrees[]`
Output: {TRUE | FALSE}

  1: **for** i **in** 0 **to** N-1 **do**
  2:    `sort_sequence_decreasing(degrees)`
  3:    **if** `degrees[0]` $\geq$ N-1-i **then**
  4:      **return** FALSE
  5:    **end if**
  6:    k $\leftarrow$ `degrees[0]`
  7:    `degrees[0]` $\leftarrow 0$
  8:    **for** j **in** 1 **to** k **do**
  9:      `degrees[j]` $\leftarrow$ `degrees[j]` - 1
10:    **end for**
11: **end for**
12: **return** TRUE

As a final remark, we note that verifying whether a degree sequence is successively compressible is in general computationally expensive. A procedure to check if a sequence is successively compressible is shown in Algorithm 34. The time complexity of this algorithm is $O(N \times S + K)$, where S is the time complexity of function `sort_sequence_decreasing()` to sort the array `degrees[]`, while $O(K)$ accounts for the execution of line 9 for each stub. Since sorting an array of length N has time complexity $O(N \log N)$ (see Appendix A.3.1), the overall time complexity of Algorithm 34 is $O(N^2 \log N)$. This is less efficient than a single iteration of Algorithm 32, which has time complexity $O(K \log K)$ for sparse graphs. In practice, it is usually more convenient to actually try to construct a graph by using Algorithm 33, without checking if the degree sequence is successively compressible. If Algorithm 33 fails to find a good matching within `max_attempts` iterations, one might decide to invest some time to check whether the degree sequence is indeed successively compressible or not.

You will find an implementation of Algorithm 33 in the program `conf_model_deg`, which takes as input a degree sequence and attempts to produce on output a simple graph sampled from the configuration model, within a certain number of attempts specified by the user. A second program, `conf_model_deg_nocheck`, implements Algorithm 31, and will in general produce graphs with loops and multiple edges, while the utility `check_compressible` checks whether a degree sequence given as input is successively compressible. All these programs are available for download at `www. complex-networks.net`.

A.13 Growing Unweighted Graphs

In Chapter 6 we introduced several different models to grow unweighted graphs. The common denominator of all such models is the fact that both the number of nodes and the number of edges change (usually grow) over time. This has to be taken into account when choosing how to represent a graph. Moreover, the probability $\Pi_{i \to j}$ for an existing node j to receive a new connection from a newly arrived node i at time t is not constant but changes over time. For instance, in all models of degree-based preferential attachment, this probability depends on the value of the degree of node j when i arrives. This means that, in principle, we should recompute the attachment probability vector $\Pi_{i \to j}$ after each node has been added, i.e. we should scan the list of edges and update the degree of all the nodes in the graph, which would require in general a number of operations of order $O(N^2)$ or more. In this appendix we will show that growing graph models can usually be implemented very efficiently, usually in time $O(K)$, where K is the final number of edges in the graph. This is due to the fact that each step of a growth model normally modifies the degree of a relatively small number of nodes, so that, at each time, it is necessary to update only a few entries of the vector of attachment probabilities $\Pi_{i \to j}$.

A.13.1 The Barabási–Albert (BA) Model

As explained in Section 6.2, the Barabási and Albert (BA) model generates graphs with a power-law degree distribution $p_k \sim k^{-\gamma}$, with $\gamma = 3$, by a suitable growth mechanism. The model starts with a complete graph of n_0 nodes, and adds a new node n at each time t, connecting it to $m \leq n_0$ already existing nodes. The probability $\Pi_{n \to i}$ for the new node n to be connected to an existing node i at time t is a linear function of the degree $k_{i,t-1}$, namely:

$$\Pi_{n \to i} = \frac{k_{i,t-1}}{2l_{t-1}}$$

where $k_{i,t-1}$ and l_{t-1} denote, respectively, the degree of node i and the total number of links in the network at time $t - 1$.

Notice that a naïve implementation of the BA model, which recomputes the whole vector $\Pi_{i \to j}$ at each time step t, and then performs a discrete sampling with rejection using Algorithm 23 (see Section A.9.4), will require a number of operations equal to:

$$\sum_{t=1}^{N} l_t = \frac{Nn_0(n_0 - 1)}{2} + \frac{mN(N + 1)}{2} \tag{A.29}$$

where we used Eq. (6.2). Notice that this number of operations is $O(N^2 m)$. However, there exists a more clever way of sampling a BA graph, which takes advantage of the ij-form representation of a sparse adjacency matrix and does not require the computation of the probabilities $\Pi_{i \to j}$. We note that given the vectors $\mathbf{i}$ and $\mathbf{j}$, each pair $(\mathbf{i}[k], \mathbf{j}[k])$, with $k = 1, 2, \ldots, K$ represents an edge of the graph. Since the graph is undirected, we can store each edge exactly once in $\mathbf{i}$ and $\mathbf{j}$, so that the total number of times that a given label ℓ appears either in **i** or in **j** is equal to the number of edges of node ℓ, namely to $k_{\ell,t-1}$. Now let us consider the vector V obtained as the concatenation of **i** and **j**, i.e. the vector whose first l_{t-1} components correspond to vector **i** and the following l_{t-1} components correspond to vector **j**. Since each node appears in vector V a number of times equal to its degree, by sampling an element v uniformly at random from V, the probability that v is equal to ℓ is exactly $k_{\ell,t-1}/(2l_{t-1})$. This observation is the core of the procedure ba(), whose pseudocode is presented in Algorithm 35.

Algorithm 35 ba()

Input: N, n_0, m
Output: **i**, **j**
1: $l \leftarrow 0$
2: **for** $i = 0$ **to** $n_0 - 1$ **do**
3:     **for** $j = i + 1$ **to** $n_0 - 1$ **do**
4:         **i**$[l] \leftarrow i$
5:         **j**$[l] \leftarrow j$
6:         $l \leftarrow l + 1$
7:     **end for**
8: **end for**
9: $n \leftarrow n_0$
10: $t \leftarrow 1$
11: **while** $t < N - n_0$ **do**
12:     **for** $j = 0$ **to** $m - 1$ **do**
13:         **i**$[l + j] \leftarrow t + n_0$
14:         $v \leftarrow$ sample_neighbour($\mathbf{i}, \mathbf{j}, l$)
15:         **while** check_neighbour($\mathbf{i}, \mathbf{j}, v, l, j$) = TRUE **do**
16:             $v \leftarrow$ sample_neighbour($\mathbf{i}, \mathbf{j}, l$)
17:         **end while**
18:         **j**$[l + j] \leftarrow v$
19:     **end for**
20:     $n \leftarrow n + 1, \ l \leftarrow l + m$
21: **end while**
22: **return** **i**, **j**

The algorithm starts by constructing a complete graph with n_0 nodes (lines 2–8). Then, nodes are added one by one, so that at time t we add the node labelled as $n = t + n_0$ and its new m stubs to the existing network with $n_{t-1} = n_0 + t - 1$ nodes and $l_{t-1} = n_0(n_0 - 1)/2 + m(t - 1)$ links. As a new node n arrives, we consider each of its m stubs (lines 12–19). For each stub j, we first append the label n at the end of array **i**, and then we select a candidate neighbour v by means of the sample_neighbour() function, whose pseudocode is presented in Algorithm 36. The sample_neighbour() procedure returns the label of one of the endpoints of all the edges added to the network up to time $t - 1$, selected at random. This is obtained by choosing with uniform probability one of the entries of vector V, concatenation of **i** and **j**. As we discussed above, this way of sampling is equivalent to the selection of each existing node with a probability linearly proportional to its degree. Notice that `sample_neighbour()` has time complexity $O(1)$, i.e. the number of operations performed in a call to `sample_neighbour()` does not depend on the actual size of the graph.

Algorithm 36 `sample_neighbour()`

Input: i, j, l

1: $v \leftarrow \lfloor \text{RAND}(0, 1) \times 2l \rfloor$
2: **if** $v < l$ **then**
3:      **return i**$[v]$
4: **else**
5:      **return j**$[v - l]$
6: **end if**

In order to avoid multiple edges, it is necessary to check whether the node v returned by `sample_neighbour()` has not yet been selected as neighbour of n when we attached one of the previous stubs of the node n. This check is performed efficiently by the procedure `check_neighbour()`, whose pseudocode is shown in Algorithm 37. In fact, we have only to verify that v does not appear in the last j positions of **j**, and consequently the function `check_neighbour()` has time complexity $O(m)$. If `check_neighbour()` returns FALSE, i.e. if node v is not already connected to n, then the label v is appended at the end of **j**, and the next stub is taken into account. Conversely, if `check_neighbour()` returns TRUE then the link (n, v) already exists, and a new candidate neighbour of n is repeatedly sampled until we find a node that is not already connected to n (line 16).

Algorithm 37 `check_neighbour()`

Input: i, j, v, l, j

Output: {TRUE | FALSE}

1: **for** k=0 to j-1 **do**
2:      **if** $v = $ **j**$[l + k]$ **then**
3:          **return** TRUE
4:      **end if**
5: **end for**
6: **return** FALSE

Figure A.33 explains how Algorithm 35 works in the case $n_0 = 4$, and $m = 3$. At time $t = 0$ (panel a) the graph is the complete graph with nodes $[0, 1, 2, 3]$, and the two arrays of integers denoting the edges of the graph are, respectively, **i** $= [0, 0, 0, 1, 1, 2]$, and **j** $= [1, 2, 3, 2, 3, 3,]$. The initial number of links is $l_0 = 6$. A new node arrives, it is assigned label $n = n_0 = 4$, and we start a loop over its $m = 3$ link stubs, indicated in grey. In the first iteration of the loop, the label of the new node is appended at the end of **i** (panel (b)) and a candidate neighbour v is selected by calling `select_neighbour()`. In this case we

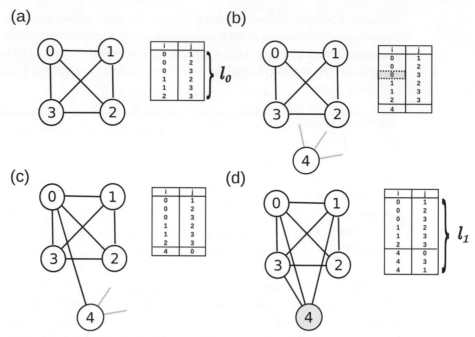

Fig. A.33 The first step of the BA model for $n_0 = 4$ and $m = 3$. If we store the adjacency matrix of the graph in ij-form, it is possible to sample graphs with N nodes from the BA model in $O(Nm^2)$, where m is the number of edge stubs arriving with each new node. This is due to the fact that the linear preferential attachment mechanism can be implemented through uniform sampling from the concatenation of the vectors i and j.

imagine that `select_neighbour()` has sampled the integer 2 and has therefore returned the candidate neighbour $v = 0$, which corresponds to the value of the label found in $i[2]$ (shaded in grey in panel b). By calling `check_neighbour()`, we discover that node 0 has not yet been selected as a neighbour of node 4, so that the sample can be accepted, the link $(4, 0)$ is added to the network by appending 0 to the end of j (panel c), and a new stub is taken into account. In panel (d) of Figure A.33 we show the state of the algorithm at the end of the first step, i.e. when node 4 has been finally connected to node 0, 3 and 1, respectively.

Notice that Algorithm 35 is equivalent to an iterated random sampling with rejection, in which the rejection test is performed, for each new node, only on a small portion of the array j. The time complexity of Algorithm 35 is $O(Nm^2)$. In fact, we need first to construct a complete graph of order n_0 (which has time complexity $O(n_0^2)$) and then, for each new node n, we call $O(m)$ times the functions `select_neighbour` and `check_neighbour`, whose time complexity are, respectively, $O(1)$ and $O(m)$. Consequently, the total number of operations performed to produce a network with N nodes grows with N and m as $O(n_0^2 + (N-n_0)m^2)$, which is dominated by $O(Nm^2)$. We implemented Algorithm 35 in the program `ba`, available for download from `www.complex-networks.net`.

A.13.2 Generalised Preferential Attachment Models

We discuss now a method for the efficient sampling of graphs from any growth model whose attachment probability is in the form:

$$\Pi_{n \to i} = \frac{f_i}{\sum_\ell f_\ell}$$

with f_i being a non-negative function of a structural property of node i. For instance, we have $f_i = k_i$ in the case of the BA model. Another example is the DMS model (see Definition 6.7 in Section 6.4), where the probability for an existing node i to receive an edge at time $t = 1, 2, 3, \ldots$ from the newly arriving node $n = n_0 + t$, reads:

$$\Pi_{n \to i} = \frac{k_{i,t-1} + a}{\sum_{l=1}^{n-1} \left(k_{l,t-1} + a \right)} \tag{A.30}$$

with $-m \le a$, so that in this case we have $f(k_i) = k_i + a$. As shown in Section 6.4, the DMS model produces graphs with power-law degree distributions $k^{-\gamma}$ where $\gamma = 3 + a/m$.

As in the case of linear preferential attachment, recomputing the attachment probabilities $\Pi_{n \to i} \forall i$ at each step would result in an inefficient algorithm with time complexity $O(N^2 m)$. A more efficient procedure to sample graphs from the DMS model is based on a discrete version of inverse sampling (see Section A.9.2). We start by noting that the probability distribution $\Pi_{n \to i}$ can be represented by the segment AB shown in Figure A.34, whose total length is $\sum_{i=1}^{n-1} \left(k_{i,t-1} + a \right)$. The segment AB consists of $n - 1$ sub-segments, one for each of the nodes of the graph at time t, whose lengths are respectively equal to $(k_{1,t-1} + a)$, $(k_{1,t-1} + a)$, ..., $(k_{n-1,t-1} + a)$. Now, if we sample a point in AB uniformly at random, the probability that the point will fall within the sub-segment associated with node i is proportional to $(k_{i,t-1} + a)$. Hence, it is possible to sample each node of the graph with probability given by Eq. (A.30) by sampling a point p uniformly at random in AB, looking for the segments within which $n - 1$ the point p falls, and selecting the node corresponding to that segment.

This sampling strategy can be implemented by using two arrays, namely an array `labels[]` of node labels, and an array `sum_widths[]` of real numbers whose ith entry `sum_widths[i-1]` contains the sum of the lengths of the first i sub-segments. We recall here that the usual convention of the C programming language requires that the first component of an array of length N has index 0 and the last one has index $N - 1$. Let us consider the network with $N = 6$ nodes in Figure A.35(a). The degree sequence of the network is $\{k_1 = 3, k_2 = 2, k_3 = 4, k_4 = 1, k_5 = 1, k_6 = 1\}$, where for convenience we have labelled the nodes from 1 to 6. Let us assume that this network is the result of the first few steps of the DMS model, where we started from a seed network of $n_0 = 3$ nodes, and we set $m = 1$ and $a = -0.5$. We show in the figure the segment with six sub-segments representing the probabilities $\Pi_{n \to i}$, and the corresponding vectors `labels[]` and `sum_widths[]`.

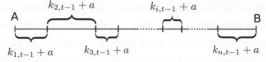

Fig. A.34 The distribution of attachment probabilities $\Pi_{n \to i}(t)$ in the DMS model can be represented by a segment AB of length $\sum_{i=1}^{n} \left(k_{i,t-1} + a \right)$, where each sub-segment of length $k_{i,t-1} + a$ is associated with node i. If we sample a point p uniformly at random in AB, the probability that p falls within the segment associated with node i is equal to $\Pi_{n \to i}$.

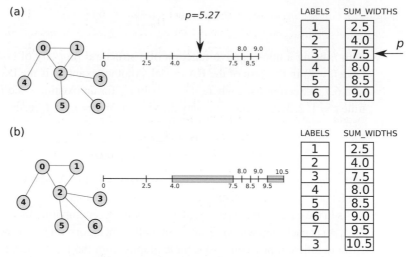

Fig. A.35 The representation of a cumulative distribution through the two arrays `labels[]` and `sum_widths[]` allows us to sample efficiently from the corresponding probability mass function. By doing so, the addition of each edge in the DMS model has time complexity $O(\log K)$.

For instance, node 1 is associated with a segment of length $k_{1,t-1} + a = 2.5$, as evident by the fact that the first element of the vector `labels[]` is 1, and the corresponding element of the vector `sum_widths[]` has value 2.5. Similarly, node 3 is associated with a segment of length $k_{3,t-1} + a = 3.5$. In fact, the label 3 appears in the third position of the vector `labels[]`, and the corresponding value of `sum_widths[]` is 7.5. Remember that, by construction, the width corresponding to the label at position i is obtained as `sum_widths[i]` - `sum_widths[i-1]`, so that in this case the width of the segment associated with node 3 is equal to `sum_widths[2]` - `sum_widths[1]` = 7.5 - 4.0 = 3.5. It is worth noticing that the pair of vectors `labels[]` and `sum_widths[]` are a compact representation for the cumulative degree distribution function of the network. Let us now assume that the new node with label 7 arrives at time t and has to attach $m = 1$ edge stubs. We sample a number p uniformly at random in [0, `sum_widths[5]`], since the total length of the segment is equal to the value of `sum_widths[5]`. In the figure we imagine that $p = 5.27$, so that p falls within the segment associated with node 3. Hence, we connect 7 with node 3 and obtain the graph in Figure A.35(b). As a result, we now have a new node in the graph, with degree $k_7 = 1$, which should be associated with a new sub-segment of length $k_7 + a = 0.5$. This is easily done by adding a new element to the vector `labels` with value equal to 7, and a new element to the vector `sum_widths[]` with value `sum_widths[5]` + 0.5 = 9.5.

However, since node 3 has acquired a new edge, we should also extend the length of the segment associated with it from 3.5 to 4.5. One possibility is to update the length of the segment associated with node 3 in the third position of the vector `sum_widths[]`, by setting `sum_widths[2]` = 8.5. But in this case we should update also the values of `sum_widths[]` for all the elements in the vector that are placed after node 3. Making these changes will cost on average $O(n)$ at each step t, which will result again in an overall time complexity $O(N^2 m)$ to construct a network with N nodes. Notice that the central idea of the sampling scheme we have introduced above is that each node i should be assigned a portion of the sampling space (i.e. of the segment representing the distribution $\Pi_{n \to i}$) whose width is proportional to $k_{i,t-1} + a$. It is not required that the portion of the sampling space associated with i has indeed to be a single segment of length $k_{i,t-1} + a$, or instead many segments whose total length is equal to $k_{i,t-1} + a$. Consequently, it is possible to *extend* the segment associated with node 3 by adding a new segment of length 1.0 at the end of the vector sum_widths[], and assigning it to node 3. This is exactly what we did in Figure A.35(b), where we added a new element to the vector labels[], with value 3 and a corresponding element to the vector sum_widths[] with value 10.5. Notice that this operation has time complexity $O(1)$. In the end, node 3 is associated with two segments, highlighted in grey in Figure A.35, whose total length is equal to (sum_widths[2] - sum_widths[1]) + (sum_widths[7] - sum_widths[6]) $= 3.5 + 1.0 = 4.5 = k_{3,t} + a$.

The procedure to grow a graph according to the DMS model is in all similar to the procedure to grow a BA model graph that we showed in Algorithm 35. The main difference is in the use of the additional vectors labels[] and sum_widths[], and of an appropriate function sample_neighbour_dms(), whose pseudocode is presented in Algorithm 38.

Algorithm 38 sample_neighbour_dms()

Input: i, j, l, K

  1: $p \leftarrow$ RAND$(0, 1) \times$ sum_widths[K-1]
  2: $i \leftarrow$ find_segment_label(p, labels, sum_widths)
  3: **return** i

The time complexity of sample_neighbour_dms() depends only on the time complexity of the function find_segment_label(), which finds the label associated with the segment within which the sampled value p lies. This function can be implemented through a slightly modified binary search on the vector sum_widths[], which has time complexity $O(\log K)$. As a result, the sampling of a graph in the DMS model has time complexity $O(Nm \log K) = O(K \log K)$, which is far better than the time complexity $O(N^2 m) = O(KN)$ of the naïve implementation that recomputes $\Pi_{n \to i}$ at each step.

We have implemented the DMS model in the program dms, available for download at www.complex-networks.net. By using a very similar strategy, we also implemented in the program bb_fitness the model of preferential attachment with fitness proposed by Bianconi and Barabási, and discussed in Section 6.5.

A.14 Random Graphs with Degree–Degree Correlations

In this appendix we discuss an algorithm to sample random graphs with an assigned joint probability distribution $p_{kk'}$. The algorithm is based on the hidden variable model introduced in Section 7.5. It takes as input a given graph G and produces a graph G' with the same degree–degree probability distribution as G. The corresponding pseudocode is presented in Algorithm 39.

Algorithm 39 `hidden_variable()`

Input: G
Output: i, j

```
 1: pkk[][] ← degree_corr_distr(G)
 2: rho[] ← compute_rho(pkk[])
 3: fhh[][] ← compute_fhh(pkk[])
 4: h[] ← sample_node_variables(rho[])
 5: K ← 0
 6: for n1 in 0 to N-1 do
 7:    h1 ← h[n1]
 8:    for n2 in i+1 to N-1 do
 9:       h2 ← h[n2]
10:       v ← RAND(0,1)
11:       if v < fhh[h1][h2] then
12:          i[K] ← n1
13:          j[K] ← n2
14:          K ← K + 1
15:       end if
16:    end for
17: end for
18: return i, j
```

Making use of Eq. (7.7), the function `degree_corr_distr()` computes the degree–degree probability distribution $p_{kk'}$ of the graph given as input from $\{e_{kk'}\}$, the number of edges between pairs of nodes of degree k and k'. The result is stored in the matrix `pkk[][]`. The degree–degree distribution is used to compute the values of $\rho(h)$ and $f(h, h')$ in Eq. (7.42) and Eq. (7.44), which are respectively stored in the array `rho[]` and in the matrix `fhh[][]`. The hidden variables to be associated with each node are sampled in the function `sample_node_variables()`, using a discrete inverse sampling similar to that described in Appendix A.13.2, and stored in the vector `h[]`.

The core of the algorithm consists of two nested loops, and resembles the procedure discussed in Appendix A.10.2 to sample graphs from the ER random graph model B. The main difference is that the probability of creating an edge between two nodes n1 and n2 is not constant but depends on the hidden variables h1 and h2. Notice that the number of edges of the resulting network can be different from that of the original graph. The algorithm has time complexity $O(N^2)$, since it considers each of the $\binom{N}{2}$ possible pairs of nodes exactly once, and for each pair performs a computation that has $O(1)$ time complexity. Notice that the initialisation of the algorithm, which includes the computation of $p_{kk'}$, $\rho(h)$, and $f(h, h')$, has time complexity $O(N \log N + K)$, which is dominated by $O(N^2)$. The program `hv_net` implements Algorithm 39, and is available for download at www. complex-networks.net.

A.15 Johnson's Algorithm to Enumerate Cycles

The problem of enumerating all the cycles of a given graph G with N nodes is known to have exponential time complexity in the number of nodes. The simplest algorithm to find cycles of a given length ℓ consists in considering all the possible ordered sequences of ℓ distinct nodes $\{i_0, i_1, i_2, \ldots, i_{\ell-1}\}$, and checking whether each sequence forms a cycle, i.e. if pairs (i_k, i_{k+1}) are such that $a_{i_k i_{k+1}} = 1$ for $k = 0, 1, \ldots, \ell - 2$, and also $a_{i_{\ell-1} i_0} = 1$. In a graph with N nodes it is possible to construct $N!/(N-\ell)!$ sequences of nodes of length ℓ. Since we are interested in finding *all* the cycles of the graph up to length $\ell = N - 1$, we have then to perform a check whose time complexity is $O(N!)$. We could argue that it is possible to restrict the search for cycles to sequences of nodes that are connected. We can then visit the graph and check if any of the paths originating from each of the nodes is indeed a cycle. However, the total number of paths in a generic graph with N nodes grows exponentially with N, even if the precise scaling depends on the particular structure of the graph.

Algorithm 40 `johnson_cycles()`

Input: G

Output: enumerate all the elementary cycles in the graph G

```
1: st ← stack_new()
2: for s = 0 to N-1 do
3:     for j = s to N-1 do
4:         blocked[j] ← FALSE
5:         B[j] ← ∅
6:     end for
7:     circuit(s, s)
8: end for
```

One of the classical algorithms to enumerate all the cycles of a graph was proposed by Donald Johnson in 1975 [169]. The pseudocode is presented in Algorithm 40. The main strength of Johnson's algorithm is that it explores each cycle exactly once. Notice, in fact, that if we have found a cycle originating at node i, then we would in principle find the same cycle every time we consider paths starting from any other node contained in the same cycle. The algorithm proposed by Johnson avoids unnecessary computations by assigning an *order* to each node of the graph, i.e. a number which identifies each node uniquely. Then, the search for cycles starts from the node with the lowest order, and proceeds by constructing paths containing exclusively nodes whose order is larger than the order of the starting node. This procedure guarantees that each cycle is visited exactly once, namely when we consider the paths originating at the node of the cycle having the smallest order. A natural way of ordering nodes is to use integers as node labels, and consider the order of the node labelled as i as greater than the order of another node whose label is j if $i > j$.

Algorithm 41 `circuit()`

Input: G, s, v

Output: enumerate all the elementary cycles starting in s and containing v

```
 1: f ← FALSE
 2: if v < s then
 3:     return f
 4: end if
 5: stack_push(st, v)
 6: blocked[v] ← TRUE
 7: for all w in neigh[v] do
 8:     if w = s then
 9:         stack_dump(st)
10:         f ← TRUE
11:     else
12:         if w > s then
13:             if blocked[w] = FALSE then
14:                 if circuit(s,w) = TRUE then
15:                     f ← TRUE
16:                 end if
17:             end if
18:         end if
19:     end if
20:     if f = TRUE then
21:         unblock(v)
22:     else
23:         for all w in neigh[v] do
24:             if w > s then
25:                 add v to B[w]
26:             end if
27:         end for
28:     end if
29: end for
30: stack_pop(st)
31: return f
```

Algorithm 40 looks simple because the core of Johnson's algorithm is the recursive function `circuit(s, v)`, whose pseudocode is presented in Algorithm 41. The basic idea is that each of the N calls to the function `circuit(s, s)` constructs the set of elementary paths starting at s that contain all the nodes reachable from s whose order is larger than s. In order to avoid exploring a path more than once, a node v is *blocked* when it is added for the first time to an elementary path starting at s, and remains blocked as long as every other path from v to s intersects the current elementary path at a node different from s. The node v is then unblocked as soon as the elementary path passing through v is contained in at least one cycle starting at s.

In more detail, the function `circuit()` maintains a stack in the variable `st` (see Section A.3.2) which contains the path that is currently being explored. Notice that a neighbour of the current node v is added to the current path (and to the stack) only if its order is larger than the order of v (line 12). When a cycle is found, the algorithm prints the current cycle by printing the whole stack stored in the variable `st` (line 9).

The algorithm also keeps track of the regions of the subgraph which do not contain cycles by maintaining, for each node w in the subgraph induced by nodes with order larger or equal than s, a list of its blocked predecessors `B[w]`. Notice that the current node v is unblocked only when a cycle has been detected in one of the paths containing v (line 20). Otherwise, v is added to `B[w]`, to indicate that there are no cycles in that portion of the graph. The procedure `unblock(v)`, detailed in Algorithm 42, takes care of recursively unblocking a node and all its blocked predecessors.

Algorithm 42 `unblock()`

Input: v

1: `blocked[v]` ← FALSE
2: **for all** w **in** `B[v]` **do**
3:     remove w from `B[v]`
4:     **if** `blocked[w]` = TRUE **then**
5:         `unblock(w)`
6:     **end if**
7: **end for**

It is possible to prove that the algorithm proposed by Johnson is correct, i.e. it actually enumerates all the cycles of the graph given as input, visiting each cycle exactly once. However, the proof is not immediate and is beyond the scope of this book. The interested reader can find it in the original paper by Johnson [169]. The usage of node blocking and of lists of blocked predecessors for each node guarantees that the time complexity of Algorithm 40 is $O((N+K)(c+1))$, where c is the number of cycles of the graph. However, this does not mean that Johnson's algorithm requires a time linear in the number of nodes and edges of the graph, since in general the total number of cycles in the graph depends on both N and K, and as we said before it could be even exponential in the number of nodes. Nevertheless, Johnson's algorithm is one of the most efficient procedures to find cycles in graphs available on the market.

The program `johnson_cycles`, available for download at www.complex-networks. net, implements Johnson's algorithm. Actually, since the algorithm has in general an exponential running time, the program `johnson_cycles` also accepts an optional parameter specifying the maximum length of the cycles to look for. The program can return as output both the total number of cycles up to the set maximum length and the list of all those cycles.

A.16 Motifs Analysis

We describe here a procedure to find 3-node motifs in directed networks, namely to compute the frequency of each of the thirteen different types of 3-node weakly connected directed graphs in Figure 8.9(b) as subgraphs of a given directed graph G, and to estimate whether each of them is under- or over-represented with respect to a *null model*.

The algorithm we present is a simplified version of the algorithm used in Ref. [219], and is based on the observation that the same unlabelled 3-node graph indeed corresponds to six different labelled graphs, each obtained by one of the six permutations of the three labels of the nodes. As an example, we show in Figure A.36 the unlabelled 3-node graph F10 and the six labelled graphs associated with it. If we consider the 3×3 adjacency matrix of any of those labelled graphs, e.g. the matrix A_1, it is always possible to find an appropriate permutation matrix which transforms that matrix into one of the other five. For instance, matrix A_1 can be transformed into A_2 through the permutation matrix:

$$P_{jl} = \begin{pmatrix} 1 & 0 & 0 \\ 0 & 0 & 1 \\ 0 & 1 & 0 \end{pmatrix} \tag{A.31}$$

which permutes node j and node l. In fact we have:

$$P_{jl}^{\top} A_1 P_{jl} = \begin{pmatrix} 1 & 0 & 0 \\ 0 & 0 & 1 \\ 0 & 1 & 0 \end{pmatrix} \begin{pmatrix} 0 & 1 & 0 \\ 1 & 0 & 1 \\ 1 & 0 & 0 \end{pmatrix} \begin{pmatrix} 1 & 0 & 0 \\ 0 & 0 & 1 \\ 0 & 1 & 0 \end{pmatrix} = \begin{pmatrix} 0 & 0 & 1 \\ 1 & 0 & 0 \\ 1 & 1 & 0 \end{pmatrix} = A_2 \tag{A.32}$$

This means that the graph F10 can indeed be represented by any of the six adjacency matrices $A_1, A_2, A_3, A_4, A_5, A_6$. If we choose one *representative adjacency matrix* for each of the thirteen graphs F1, F2, ..., F13 in Figure 8.9, the problem of deciding whether

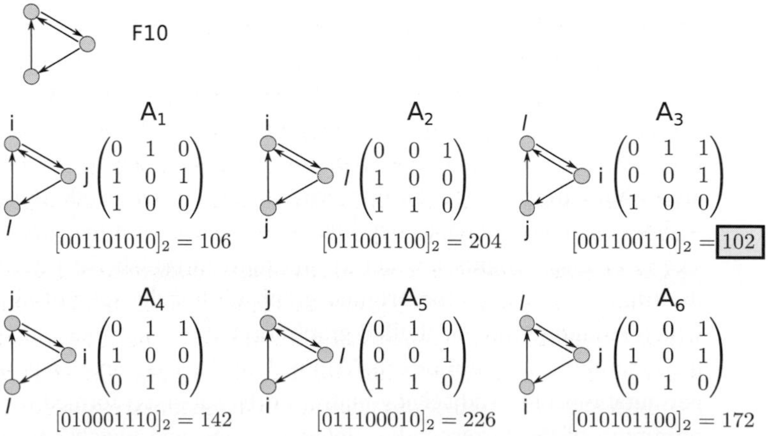

Fig. A.36 The unlabelled 3-node weakly connected directed graph F10, and the six corresponding labelled graphs. We present the adjacency matrix of each of the six labelled graphs, together with the associated integer number $I(A)$.

the directed subgraph H_{ijl} formed by a connected triple of nodes (i,j,l) corresponds to a specific unlabelled 3-node graph Fx, actually reduces to the problem of finding whether the adjacency matrix of H_{ijl} can be transformed, through appropriate permutations of node labels, in the representative adjacency matrix of that graph.

Even if the choice of a representative adjacency matrix for each of the 3-node weakly connected directed graphs is arbitrary, there is a particular choice that allows us to recognise a graph in a simple and efficient way. We notice that it is always possible to associate a unique integer number $I(A)$ with the adjacency matrix $A \in \mathbb{R}^{N \times N}$ of an unweighted graph G with N nodes, by defining:

$$I(A) = 2^0 a_{1,1} + 2^1 a_{1,2} + \cdots + 2^{N-1} a_{1,N} + 2^N a_{2,1} + 2^{N+1} a_{2,2} + \cdots + 2^{N^2-2} a_{N,N-1} + 2^{N^2-1} a_{N,N}.$$

This is equivalent to arranging the rows of the matrix one after the other and considering the resulting string of zeros and ones as a binary number, whose least significant bit is the element $a_{1,1}$, and most significant bit is $a_{N,N}$. Notice that $I(A)$ defines a bijection between the set of all the possible $N \times N$ unweighted adjacency matrices and the integers between 0 and $2^{N^2} - 1$. In Figure A.36 we show the binary numbers associated with each of the adjacency matrices of the six permutations of graph F10, and the corresponding representations in base ten. Notice that each binary number is obtained by reading the elements of matrix A in inverse ordering, going from the bottom right to the top left. As a convention, we choose as the representative adjacency matrix of a graph the adjacency matrix of the permutation of its labels associated with the smallest integer number. In the example shown in Figure A.36, the representative matrix for F10 is A_3, associated with the smallest integer number 102 (highlighted in grey). Notice that, by construction, each of the thirteen unlabelled 3-node graphs is associated with a distinct representative adjacency matrix, and to a distinct integer number. In this case, the integer numbers associated with the representative matrix of graphs F1, F2, ..., F13 in Figure 8.9 are 6, 12, 14, 36, 38, 46, 74, 78, 98, 102, 108, 110 and 238.

The procedure to find and count the different 3-node subgraphs of a given directed graph G is sketched in Algorithm 43. Starting from the original graph G, the algorithm constructs the underlying undirected graph G_u and performs a visit of all the connected triples of nodes in G_u. In order to guarantee that each connected triple is visited exactly once, the algorithm loops over the N nodes of the graph and, for each node i, considers all the triples (i, j, l) centred in i such that both j and l are neighbours of i in G_u, and i < j < l, i.e. the labels of the three nodes are in ascending order. Then, the algorithm calls the function get_motif_3() (pseudocode presented in Algorithm 44), which constructs the adjacency matrix associated with the subgraph formed by the triple (i,j,l) and recognises which of the thirteen (unlabelled) 3-node graphs is formed by the triple (i,j,l) in the original graph G. For the sake of simplicity, the pseudocode which recognises the unlabelled graph formed by the triple (i,j,l) was incorporated in the function motif_number() (Algorithm 45). This function performs all the six possible permutations of the adjacency matrix of the subgraph formed by (i,j,l), computes the minimum of the corresponding integer values, and matches it against the integer values of the representative adjacency matrix of each unlabelled graph F1, F2, ..., F13. We omitted the pseudocode of the function motif_represented_by(), which is responsible

---

Algorithm 43 `find_subgraphs_3()`

Input: G

Output: freq[] {vector of motifs counts}

  1: **for** n **in** 0 **to** 12 **do**
  2:    freq[n] = 0
  3: **end for**
  4: $G_u \leftarrow$ `undirected_graph`(G)
  5: **for** i **in** 0 **to** N-1 **do**
  6:    **for** j **in** `neighbours`(i, G_u) **do**
  7:      **if** j < i **then**
  8:        continue
  9:      **end if**
10:      **for** l **in** `neighbours`(i, G_u) **do**
11:        **if** l = j **or** l < i **or** l < j **then**
12:          continue
13:        **end if**
14:        I $\leftarrow$ `get_motif_3`(G, i, j, l)
15:        freq[I] $\leftarrow$ freq[I] + 1
16:      **end for**
17:    **end for**
18: **end for**
19: **return** freq

Algorithm 44 `get_motif_3()`

Input: G, i, j, l

Output: v {id of the motif defined by nodes i, j, l}

  1: nodes[0] $\leftarrow$ i
  2: nodes[1] $\leftarrow$ j
  3: nodes[2] $\leftarrow$ l
  4: **for** r **in** 0 **to** 2 **do**
  5:    **for** c **in** 0 **to** 2 **do**
  6:      **if** `is_neighbour`(nodes[r], nodes[c], G) **then**
  7:        m[r][c] $\leftarrow$ 1
  8:      **else**
  9:        m[r][c] $\leftarrow$ 0
10:      **end if**
11:    **end for**
12: **end for**
13: v $\leftarrow$ `motif_number`(m)
14: **return** v

Algorithm 45 `motif_number()`

Input: m

Output: motif_num {id of the motif corresponding to m}
  1: min $\leftarrow 2^9$
  2: **for** n **in** 0 **to** 5 **do**
  3:     m_perm $\leftarrow$ `permute(m, n)`
  4:     v $\leftarrow$ `integer_value(m_perm)`
  5:     **if** v < min **then**
  6:        min $\leftarrow$ v
  7:     **end if**
  8: **end for**
  9: motif_num $\leftarrow$ `motif_represented_by(min)`
10: **return** motifs_num

---

for mapping the integer numbers associated with the representative matrix of the subgraphs `F1, ..., F13`, since it is simple a sequence of `if/else` statements. Also, we did not include the pseudocode of the function `motifs_represented_by()`, which constructs the representative matrix associated with a given motif number by inverting the definition of $I(A)$.

The time complexity of algorithm `find_subgraphs_3()` is $O(n_\wedge)$, where $n_\wedge$ is the number of distinct connected triads of the graph. In fact, the algorithm processes each connected triad of the graph exactly once, and for each triad calls the function `get_motif_3()`. The latter function performs a number of operations which does not depend on the size of the graph, namely, the six permutations of the labels of the triad (which require a total of twelve 3×3 matrix multiplications), the computation of the corresponding integer values and of their minimum, and the matching of that minimum value with the values of the representative matrices of the thirteen 3-nodes unlabelled graphs. Since we have that $n_\wedge \leq \sum_i k_i(k_i - 1) = N\left(\langle k^2 \rangle - \langle k \rangle\right) = K\left(\langle k^2 \rangle / \langle k \rangle - 1\right)$, the time complexity of finding 3-node motifs can be rewritten as $O\left(K \langle k^2 \rangle / \langle k \rangle\right)$. Notice that the factor $\langle k^2 \rangle / \langle k \rangle$ cannot be discarded, since in general it is not constant and can be an increasing function of the order of the graph N. This happens, for instance, in the case of graphs with power-law degree distributions with exponent $2 < \gamma < 3$, where $\langle k^2 \rangle / \langle k \rangle$ diverges with N.

The program `f3m`, available for download at `www.complex-networks.net`, implements all the algorithms explained above. In particular, it takes as input a directed graph, and calls the function `find_subgraphs_3` to compute the frequency of each 3-node unlabelled graph as a subgraph of the graph given as input. Then, it constructs n random networks sampled from the configuration model defined by the degree sequence of the input graph, where n is an input provided by the user, calls `find_subgraphs_3()` on each of those synthetic graphs, and then computes the Z-score associated with each of the thirteen 3-node graphs, as defined in Definition 8.7.

A.17 Girvan–Newman Algorithm

The algorithm proposed by Girvan and Newman to find communities in networks is a divisive procedure based on the observation that, if a graph can be partitioned in a set of tightly connected communities with just a small percentage of edges connecting different communities, the edges which connect two communities will mediate a large number of shortest paths, i.e. they will be characterised by a large value of betweenness. Conversely, those edges whose end-points belong to the same community will in general be traversed by a relatively smaller fraction of shortest paths, and will be associated with smaller values of betweenness. Hence, the idea of successively removing edges with the largest betweenness, until the graph fragments into components.

The Girvan–Newman algorithm is based on the iteration of three steps, namely: *(1)* compute the betweenness of all the edges; *(2)* remove the edge of the graph with the largest betweenness; *(3)* perform a component analysis of the resulting graph. At each step, the algorithm prints on output the number of connected components, using the function `components()` defined in Algorithm 19, and the corresponding value of modularity. The Girvan–Newman algorithm terminates when all the edges have been removed and the graph consists of a number of components equal to the number of nodes N. The pseudocode of this algorithm is detailed in Algorithm 46.

Algorithm 46 `girvan_newman()`

Input: G(V, E)
Output: community dendrogram
  1: **for** k=0 **to** K-1 **do**
  2:    `brandes_edges(G)`
  3:    e ← `get_max_betweenness_edge(G)`
  4:    $G \leftarrow G \setminus \{e\}$
  5:    [ic,nc,sizes,f] ← `components(G)`
  6:    Q ← `compute_modularity(G, ic, nc)`
  7:    print nc, Q
  8: **end for**

Algorithm 46 consists of a single loop, which is repeated K times, so the time complexity is $O(K \times B)$, where B is the time complexity of the body of the loop. The core of the body algorithm consists in the computation of the edge betweenness. The algorithm proposed by Ulrik Brandes to compute node betweenness that we described in Appendix A.7 can be easily adapted to compute edge betweenness. We notice that, similarly to the case of node betweenness, the betweenness of an edge (i,j) can be computed recursively by appropriately summing up all the contributions to the dependency of the nodes i and j corresponding to shortest paths which traverse the edge (i,j). In practice, if the sparse adjacency matrix of the graph is stored in CRS form, we need to maintain a $2K$-dimensional vector `edge_bet[]` to store the betweenness of all the edges, and to modify only a few lines at the end of Algorithm 17 (precisely, from line 40 until the end of the program), as shown in Algorithm 47.

Algorithm 47 `brandes_edges()`

39: .....
40: **for** k = n - 1 **down to** 1 **do**
41:     w ← marked[k]
42:     **for all** i **in** predecessors(w) **do**
43:         v ← nj[i]/nj[w]*(1 + delta[w])
44:         delta[i] ← delta[i] + v
45:         n ← `get_edge_position(i,w)`
46:         `edge_bet[n]` ← `edge_bet[n]` + v
47:         n ← `get_edge_position(w,i)`
48:         `edge_bet[n]` ← `edge_bet[n]` + v
49:     **end for**
50:     cB[w] ← cB[w] + delta[w]
51: **end for**

It is easy to prove that the computation of the edge betweenness in an undirected graphs has time complexity $O(N(N+K \times P))$, where we have denoted as P the time complexity of `get_edge_position()`. This is because the time complexity of this algorithm is similar to the one of Algorithm 17, except that in the Brandes algorithm for the node betweenness the time complexity of the cycle between lines 40 and 47 is $O(NK)$, while in Algorithm 47 the time complexity of the cycle between lines 40 and 51 becomes $O(NKP)$. In particular, if we store the adjacency matrix of the graph in CRS format, and we sort the sub-vectors of **j** containing the neighbours of each node, `get_edge_position()` can be implemented through binary search on the list of neighbours of a node, which has time complexity $O(P) = O(\log N)$, even if in practice for sparse graphs the time required is often very close to the lower bound $\Omega(\log(K/N))$. After all the edge betweennesses have been found, the selection of the edge with maximal betweenness can be performed in $O(K)$. Therefore, we conclude that the first step of the Girvan–Newman algorithm has time complexity $O((NK \log N) + K) = O(NK \log N)$. Notice that in general there might be more than one edge with maximal betweenness. In those cases, ties are broken by selecting one of those edges uniformly at random.

The second step, corresponding to line 4 in Algorithm 46, requires the removal of the edge with maximum betweenness. The time complexity of this operation in general depends on the specific structure used to store the adjacency matrix. Since in our implementation we use the CRS format, it is convenient to adopt a $2K$-dimensional array `active`, where each component of the array corresponds to one of the edges of the original graph G, so that `active[k]` signals whether the edge is still active (1) or not (0). In this way, we simply need to initialise `active[i]` = 1 $\forall i = 0, 1, \ldots, K - 1$, and set `active[k]` = $\emptyset$ to indicate that the edge k has been removed from the graph. Consequently the operation of removal of the edge with maximum betweenness has $O(1)$ (constant) time complexity.

Algorithm 48 `compute_modularity()`

Input: G, part, nc
Output: Q {modularity of the node partition}

1: **for** i **in** 0 to N-1 **do**
2:    a[i] ← 0
3:    e[i] ← 0
4: **end for**
5: **for** i **in** 0 to N-1 **do**
6:    ci ← part[i]
7:    a[ci] ← a[ci] + degree(i)
8:    **for** j **in** neighbours(i) **do**
9:       cj ← part[j]
10:      **if** ci = cj **then**
11:         e[i] ← e[i] + 1
12:      **end if**
13:   **end for**
14: **end for**
15: Q ← 0
16: **for** n **in** 0 to nc **do**
17:    Q ← Q + e[n]/K - (a[n]/K)^2
18: **end for**
19: **return** Q

For the computation of the connected components in line 5 of Algorithm 46 we used a slightly modified implementation of the algorithms seen in Appendix A.8 in order to take into account only active edges, so that this step has time complexity $O(K)$. The modularity of a partition is computed by the function `compute_modularity()`, whose pseudocode is presented in Algorithm 48. The function `compute_modularity()` takes as input the graph, a partition into communities, which is encoded in an array `part[]`, and the number of communities, and computes the modularity of that partition using Eq. (9.17). In particular, `part[i]` is the label of the community to which node i belongs. For this purpose we can put `part[]` equal to the array `ic[]` returned by the function `components()`, since it has the property that `ic[i]` is the ID of the connected component to which node i belongs. It is easy to show that `compute_modularity()` has time complexity $O(K)$, since it visits each edge exactly twice. In conclusion, the time complexity of Algorithm 46 is K times the time complexity of the body of the loop, i.e. $O(K \times (NK \log N + 1 + K)) = O(K \times NK \log N) = O(K^2 N \log N)$.

One could object that indeed the number of edges in the graph decreases by one at each iteration. Hence, the effective time complexity of the body of the loop at the ith iteration is $O(N(K - i) \log N)$, and not $O(NK \log N)$. It is easy to show that this "detail" does not make any substantial difference for the computation of the time complexity. In fact, the expression for the scaling of the total number of elementary operations needed by Algorithm 46 reads:

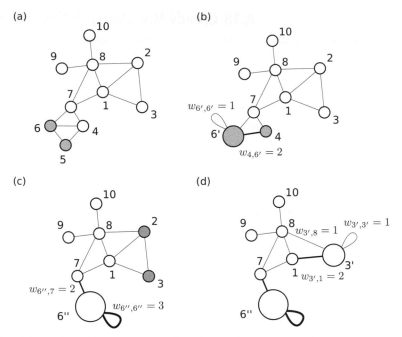

At each step of the greedy modularity optimisation algorithm, one chooses the two communities i and j whose merging would provide the highest increase (or the smallest decrease) of the modularity function, and replaces them with a new node representing all the nodes belonging to the two communities. The edges connecting this new node to the other communities are weighted according to the number of links connecting nodes in the new community to nodes in the other communities, while a weighted self-loop represents the edges lying entirely inside the newly formed community. The nodes shaded in grey in each panel are those selected to be merged at that step.

$N = 10$ nodes in Figure A.37(a), and assume that the merging of node 5 and node 6 (shaded in grey) will produce the largest increase of modularity. The algorithm merges the two nodes into the new node 6′, and creates the new links $(4, 6′)$ and $(7, 6′)$. The first edge has weight $w_{4,6'} = 2$ since both node 5 and node 6 were connected to node 4 before the merging, while the second edge has weight $w_{7,6'} = 1$. The newly created node has also a self-loop of weight $w_{6',6'} = 1$, accounting for the edge between node 5 and node 6 that now lies inside the new community 6′. The graph obtained after the first step has $N = 9$ nodes, and is shown in Figure A.37(b). Let us assume now that the maximum increase of modularity in this new graph would be obtained by merging node 4 with the newly formed node 6′. The two nodes are merged into a new node 6″, which is connected to node 7 by an edge with weight $w_{6'',7} = 2$, since both node 4 and node 6 in the original graph were connected to node 7, and has a self-loop of weight $w_{6'',6''} = 3$ accounting for the three edges among 4, 5 and 6. After the second step, we obtain the graph with $N = 8$ nodes in Figure A.37(c), where the merging of node 2 and node 3 would provide the maximum increase in modularity and produce the graph with $N = 7$ nodes in Figure A.37(d). The procedure continues until, after $N - 1$ steps, the graph consists of exactly one node. The result of the algorithm is a dendrogram, and the best partition in communities is the one corresponding to the highest value of modularity.

Algorithm 49 `greedy_modularity()`

Input: G(V,E)
Output: community dendrogram

  $Q \leftarrow 0$
  nc $\leftarrow$ N
  for i **in** 0 **to** N-1 **do**
    `a[i]` $\leftarrow k_i/2K$
    $Q \leftarrow Q +$`a[i]`^{2}
  end for
  for ℓ **in** 1 **to** N-1 **do**
    (i,j, ΔQ) $\leftarrow$ `find_best_merge()`
    $Q \leftarrow Q + \Delta Q$
    `merge_communities(i,j)`
    nc $\leftarrow$ nc - 1
    print nc, Q
  end for

It is easy to verify that the time complexity of Algorithm 49 is $O(N \times (O(F) + O(M)))$, where N is the number of nodes in the graph and $O(F)$ and $O(M)$ are respectively the time complexity of `find_best_merge()` and of `merge_communities()`. The naïve implementation of the algorithm originally proposed in Ref. [226] works directly on the adjacency matrix of the graph, so that the search for the best pair of communities to be merged has time complexity $O(K)$. Merging two communities i and j consists in the two following operations. First, we need to replace the row and column of the adjacency matrix associated with j with the sum the two rows and the two columns associated with i and j, respectively. Second, we have to remove from the matrix the row and the column associated with i. These operations can be performed in $O(N)$. Hence, the overall time complexity of the original algorithm proposed in Ref. [226] is $O(N(N + K))$.

A more efficient implementation of Algorithm 49 is described in Ref. [79]. In this version of the algorithm, the function `find_best_merge()` has time complexity $O(\log N)$, instead of $O(K)$, while the function `merge_communities()` has time complexity $O((|i| + |j|) \log N)$, instead of $O(N)$, where $|i|$ and $|j|$ are the number of neighbours of the two communities i and j. This is achieved by maintaining a sparse matrix $\Delta Q = \{\Delta Q_{ij}\}$ which contains, at each step, the potential variations of modularities resulting from the merge of each pair of communities. The matrix ΔQ is represented through an array `pq[]` of max-heap priority queues (see Appendix A.3.4), where the heap `pq[k]` contains the values of ΔQ_{kj} corresponding to the merge of k with each of the other communities present in the network at that step. Additionally, we also maintain a global max-heap priority queue H, which contains the maximum values of each of the priority queues `pq[k]` [79].

The use of priority queues allows the pair of communities yielding the largest increase of modularity to be selected efficiently at each step. In fact, the properties of priority queues discussed in Appendix A.3.4 guarantee that extracting the maximum value from H, which is the maximum increase of modularity achievable in the current graph by merging two communities, has time complexity $O(\log N)$. Hence the time complexity of the function find_best_merge() is $O(\log N)$.

The pseudocode of the function merge_communities() is presented in Algorithm 50. The algorithm merges community i into community j and updates the matrix ΔQ and the fraction of edges a[j] attached to all the nodes in the new community j. Notice that the time complexity of inserting and updating elements in a priority queue is dominated by $O(\log N)$, since the priority queue pq[k] of the generic node k can contain at most $N - 1$ elements. For each community k that is connected to either i or j (or to both), the function merge_communities() performs either two heap updates or two heap insertions (both having time complexity $O(\log N)$) and checks whether k is a neighbour of both i and j. If

---

Algorithm 50 merge_communities()

Input: G(V,E), i, j, a[], H, pq[]

Output: updated versions of a[], H, pq[]

  1: **for** k **in** neighs(i) **do**
  2:     **if** k **in** neighs(j) **then**
  3:         { j-i-k is a closed triangle }
  4:         $\Delta Q \leftarrow \Delta Q_{ik} + \Delta Q_{jk}$
  5:         update the key of node k in pq[j] to ΔQ
  6:         update the key of node j in pq[k] to ΔQ
  7:     **else**
  8:         { j-i-k is an open triad}
  9:         $\Delta Q \leftarrow \Delta Q_{ik} - 2a[j]a[k]$
10:         insert node k in pq[j] with key ΔQ
11:         insert node j in pq[k] with key ΔQ
12:         add k to the neighbourhood of j
13:         add j to the neighbourhood of k
14:     **end if**
15:     update the key of k in H
16: **end for**
17: **for** k **in** neighs(j) **do**
18:     **if** k **not in** neighs(i) **then**
19:         { i-j-k is an open triad}
20:         $\Delta Q \leftarrow \Delta_{jk} - 2a[i]a[k]$
21:         update the key of node k in pq[j] to ΔQ
22:         update the key of node j in pq[k] to ΔQ
23:     **end if**
24:     update the key of k in H
25: **end for**
26: update the key of j in H
27: remove i from H
28: a[j] $\leftarrow$ a[j]+a[i]
29: **return**

we use an additional binary search tree to maintain the list of neighbours of each node (see Section A.3.3), the latter check can be done on average in $O(\log N)$, and the insertion of a new neighbour in the adjacency list of j has time complexity $O(\log N)$, on average.

We notice that all the elementary operations performed by the inner loop of the function `merge_communities()` have time complexity $O(\log N)$, so that `merge_communities()` has a running time $O((|i| + |j|) \log N)$. As a result, the time complexity of the inner loop of Algorithm 49 is $O((|i| + |j|) \log N + \log N) = O((|i| + |j|) \log N)$, where i and j are the labels of the communities merged at each step. Hence, the time complexity of the improved implementation of Algorithm 49 is $O(S \log N)$, where S is the sum of the number of neighbours of the communities merged in each of the $N - 1$ steps. It is possible to show that $S \sim O(KD)$, where D is the depth of the dendrogram describing the community structure of the graph, i.e. the maximum distance from the root of the dendrogram to any of the leaves, which is equal to the maximum number of times any node of the network participates to a merge [79].

In the worst case, the depth of the dendrogram D might be $O(N)$, for instance if there exists a node which participates to all the $N - 1$ merges. This yields an overall time complexity $O(KN \log N)$, meaning that the improved algorithm would actually perform *slightly worse* that the naïve one, which has time complexity $O(N(N + K))$. However, the vast majority of real-world networks is characterised by a hierarchical community structure, for which $D \sim \log N$. In these graphs, the improved version of the algorithm has an effective time complexity $O(K \log^2 N)$, i.e. it is essentially linear in the number of edges of the graph, and can be used to find partitions in communities of graphs with several millions nodes and edges.

Our implementation of the algorithm proposed in Ref. [79] is included in the program `cnm`, available for download from at www.complex-networks.net, which takes as input a graph and provides as output the values of modularity at each step of the algorithm and the partition corresponding to the maximum value of modularity found. The program `cnm` also makes use of disjoint sets to store the partition in communities found at each step of the algorithm (see for instance Ref. [86] for details) and, more importantly, of randomised inserts in the binary search trees containing the neighbourhood of each node. By doing so, each of those binary search trees is, on average, balanced, meaning that the insert and lookup operations on the trees will require on average $O(\log N)$ time. Notice that at each step there might be more than one pair of communities whose merge produce the same increase of the modularity function. The algorithm implemented in the program `cnm` breaks those ties by selecting one of those pairs uniformly at random, so that subsequent invocations of `cnm` might in general provide slightly different partitions of the same graph.

A.19 Label Propagation

The label propagation algorithm discussed in Section 9.7 is a quite intuitive and efficient procedure to discover communities in network [266]. The main assumption is that if a graph is organised in communities and node i belongs to a community – let us call this community $C^{(i)}$ as we did in Section 9.6 – then, with high probability, the majority of the neighbours of i will also belong to $C^{(i)}$. Under this assumption, if one initially assigns different labels to the nodes, and then repeatedly updates the label of each node of the graph by setting it equal to the label of the majority of its neighbours, then all the nodes belonging to the same community should eventually have the same label.

The label propagation algorithm proceeds in *epochs*. During an epoch, the label of each node of the network is updated exactly once. There are two central aspects to be considered for the implementation of the label propagation. The first one is the *update strategy*, which determines when the labels of the nodes are updated during an epoch, while the second one is the *stopping criterion*, i.e. a condition which should signal the termination of the procedure.

Concerning update strategies, it is possible to choose between two approaches. The first one, called *synchronous update*, consists in computing the new labels to be assigned to all the nodes of the graph, and then updating all the labels at the end of each epoch. This strategy is relatively simple to implement, but quite inefficient in practice, since each node computes its new label based on the configuration of labels in the previous epoch. This fact usually produces periodic orbits, where neighbouring nodes flip between the same small set of labels in a periodic manner, so that quite often the algorithm remains trapped into configurations associated with small values of modularity. In the second approach, known as *asynchronous update*, one considers instead the nodes one by one, computes the new label of each node and updates it immediately, without waiting for the end of the epoch, thus effectively reducing the probability that the algorithm gets stuck in a cycle.

In Algorithm 51 we present the pseudocode of a possible implementation of label propagation with asynchronous update. The algorithm `label_propagation()` starts by assigning a different numeric label to each of the N nodes of the network. Then, the label of each node of the graph is updated and set equal to the label of the majority of the nodes in its neighbourhood. If two or more labels are present in the neighbourhood with the same number of occurrences, the node will adopt one of them with equal probability. The time complexity of the loop at lines 9–15 of is $O(K \log N)$. In fact, for each node i the computation of the most frequent label in its neighbourhood can be done in $O(k_i \log k_i)$, e.g. by using Algorithm 52. Notice that the time complexity of the algorithm `get_most_frequent_label()` is dominated by the time complexity of the function `sort_array()`, which is $O(k_i \log k_i)$ if we use a standard sorting algorithm like Quicksort (see Algorithm 3 in Appendix A.3.1). Consequently, the time complexity of the loop at lines 9–15 in Algorithm 51 is:

$$O\left(\sum_i k_i \log k_i\right) = O\left(\log(N) \sum_i k_i\right) = O(K \log N)$$

Algorithm 51 `label_propagation()`

Input: **j**, **r**, N
Output: labels {vector of node labels}
1: **for** i = 0 **to** N-1 **do**
2:      labels[i] ← i
3:      ids[i] ← i
4: **end for**
5: continue ← TRUE
6: **while** continue **is not** FALSE **do**
7:      `shuffle_array(ids)`
8:      continue ← FALSE
9:      **for** i = 0 **to** N -1 **do**
10:         neigh_label ← get_most_frequent_label(ids[i], **j**, **r**, N, labels)
11:         **if** neigh_label **not equal to** labels[ids[i]] **then**
12:             continue ← TRUE
13:             labels[ids[i]] ← neigh_label
14:         **end if**
15:     **end for**
16: **end while**

---

and the time complexity of `label_propagation()` is $O(E \times K \log N)$, where E is the number of epochs of the algorithm.

Indeed, the actual number of epochs E required by the algorithm to complete depends on many factors, the most important being the stopping criterion adopted. The most natural choice would be to let `label_propagation()` terminate only when none of the N nodes has changed its label during the last epoch, as done for instance in Algorithm 51. However, it is easy to realise that this criterion does not work well in practice. The reason is that, even with the use of asynchronous update, there is still the possibility that the algorithm remains trapped in a cycle, especially if the nodes are always visited in the same order in subsequent epochs. This problem can be somehow alleviated, but not avoided altogether, as we will see in a moment, by visiting the nodes in a different order at each epoch. This can be achieved by reshuffling the array of node ids at each epoch, e.g. using the procedure `shuffle_array()` presented in Algorithm 53.

However, even if the nodes are visited in a random order it is still quite improbable that the algorithm does not update any of the N labels during an epoch. Consider for instance the subgraph shown in Figure A.38, where node i has 8 neighbours. In the figure, we denoted the current label of i with grey, and we assumed that 4 of the neighbours of i have a white label while the remaining ones have a black label. In this configuration, the label of node i will oscillate periodically between white and black across epochs. In fact, since both the black and the white label are equally represented in the neighbourhood of i, the label propagation algorithm will break the tie by assigning to i one the two labels, chosen uniformly at random. However, whatever label we choose to associate with node i at a given epoch, the same tied configuration will most probably appear again in the

---

Algorithm 52 `get_most_frequent_label()`

Input: G, i, labels

Output: most frequent label in neighs(i)

  1: $k \leftarrow 0$

  2: **for** j **in** neighs[i] **do**

  3:    nlabel[k] $\leftarrow$ labels[j], $k \leftarrow k + 1$

  4: **end for**

  5: `sort_array(nlabel)`

  6: best_label $\leftarrow$ nlabel[0]

  7: best_freq $\leftarrow$ 1

  8: cur_label $\leftarrow$ nlabel[0]

  9: cur_freq $\leftarrow$ 1

10: **for** j **in** 1 **to** $k_i - 1$ **do**

11:    **if** nlabel[j] = cur_label **then**

12:        cur_freq $\leftarrow$ cur_freq + 1

13:    **else**

14:        **if** cur_freq > best_freq **then**

15:            best_label $\leftarrow$ cur_label

16:            best_freq $\leftarrow$ cur_freq

17:        **end if**

18:        cur_label $\leftarrow$ nlabel[j]

19:        cur_freq $\leftarrow$ 1

20:    **end if**

21: **end for**

22: **return** best_label

Algorithm 53 `shuffle_array()`

Input: $\mathbf{v}$, N

Output: an in-place random shuffling of the elements of $\mathbf{v}$

  1: **for** i = N-1 **down to** 0 **do**

  2:    j $\leftarrow$ RAND(0, i)

  3:    swap(v[i],v[j])

  4: **end for**

---

subsequent epoch. This is because the neighbours of i are tightly connected with other nodes having the same labels as theirs, thus it is very unlikely that any of the labels in the neighbourhood of i will be modified during an epoch. As a consequence, the label of node i will continuously flip between black and white. Unfortunately, configurations similar to that depicted in Figure A.38 are relatively common in real-world graphs, so that, by using such a stringent stopping criterion, Algorithm 51 might probably run for a very long time, if not forever.

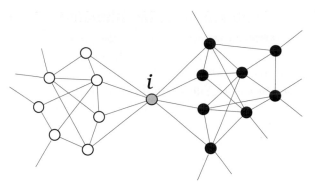

Fig. A.38 A typical configuration that introduces a periodic orbit in the label propagation algorithm. Notice that half of the neighbours of node i have a white label, while the other half have a black label. Since there is no clear majority of labels in the neighbourhood of i, the algorithm will set the label of node i either to white or to black, with equal probability. Moreover, the black and white communities are not connected through direct edges, so it is unlikely that any of the neighbours of i will change its label. As a result, a tie will occur again at the next epoch, and the label of i will be cyclically set to either black or white, with equal probability.

A common practical choice to solve this problem is to set a maximum number of epochs *a priori*, and to return the partition found either when no label update occurs during an entire epoch, or at the end of the last epoch, whichever occurs first. Although there is no known method to determine the maximum number of epochs, it has been observed that a number of epochs in the order of 10^2 is normally sufficient to find partitions with relatively high modularity in graphs with up to 10^6 nodes. If we choose *a priori* the number of epochs, and if this number does not depend on the size of the graph, then the time complexity of `label_propagation()` reduces to $O(K \log N)$, meaning that the algorithm is effectively linear in the number of edges of the graph and can be used to find communities in large networks.

An implementation of the label propagation algorithm can be found in the program `label_prop`, available for download from `www.complex-networks.net`. The program takes as input a graph and, optionally, a maximum number of epochs, and returns the partition found either when the algorithm reaches an equilibrium (i.e. a configuration in which no node changes its label during an entire epoch) or after the algorithm has run for the specified number of epochs, whichever occurs first. The program `label_prop` implements both the synchronous and the asynchronous update strategies. The reader will notice that synchronous update usually remains trapped in a cycle, and is thus of poor practical use.

A.20 Kruskal's Algorithm for Minimum Spanning Tree

A *minimum spanning tree* (MST) of a given weighted graph G is a connected subgraph of G consisting of all N nodes of the graph and of a set T of $N-1$ edges such that the sum of the weights of the edges in T is minimal. A classical algorithm able to find minimum spanning trees without enumerating all the spanning trees of a graph was proposed by Joseph Kruskal in 1956 [190]. The algorithm gets as input the weighted edge list of a graph and constructs a minimum spanning tree by iteratively considering all the edges in ascending order of weight, and placing an edge between two yet unconnected nodes only if this addition does not create a cycle. The algorithm relies on two simple observations. The first one is that, by adding a single edge to a tree, one always obtains a graph which is not a tree because it contains at least one cycle. The second observation is that if one connects two disjoint trees, i.e. two trees with no nodes in common, with a single edge, then such merging will result in another tree.

The pseudocode of Kruskal's algorithm is presented in Algorithm 54, and an example of how it works is shown in Figure A.39. The algorithm starts by considering N degenerate trees, each tree consisting of exactly one of the N nodes of the graph, and by setting to zero the edge counter `cnt` which will indicate how many edges have been already included in the spanning tree. Then, it uses the function `sort_edges()` to sort the list of edges of the graph in increasing order of their weight. In Figure A.39(a) we show a weighted graph with six nodes and its list of edges, while in panel (b) we presented the initial configuration of the algorithm with six degenerate trees and the list of edges of the graph, sorted in ascending

---

Algorithm 54 `kruskal()`

Input: `i, j, w`
Output: edges in the minimum spanning tree
  1: `K` ← `length(i)`
  2: `n` ← 0
  3: `cnt` ← 0
  4: `sort_edges(i, j, w)`
  5: **while** `cnt` < N-1 **do**
  6:    `(v1, v2, g)` ← `get_min_edge()`
  7:    `n` ← `n + 1`
  8:    `t1` ← `tree(v1)`
  9:    `t2` ← `tree(v2)`
10:    **if** `t1` ≠ `t2` **then**
11:       `cnt` ← `cnt + 1`
12:       `merge_trees(t1, t2)`
13:       print `v1, v2, g`
14:    **end if**
15: **end while**

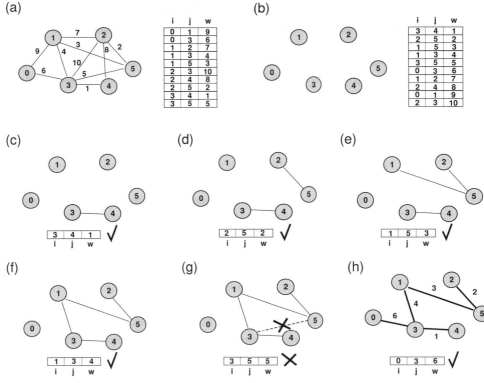

Fig. A.39 Kruskal's algorithm applied to the weighted graph with $N = 6$ nodes and $K = 10$ edges shown in panel (a). The algorithm starts by putting each node of the graph in a (degenerate) tree (b). Then, the edges are considered in increasing order of their weight, and added to the minimum spanning tree (MST) only if the addition does not create a loop (c)–(h). For instance, the edge (3, 5, 5) is not part of the MST, since its addition would create a cycle of three nodes, namely, node 1, node 3 and node 5.

order of their weight. The three vectors i, j and w contain, respectively, the two endpoints of each edge and the corresponding weight, so that the triple (i[ℓ], j[ℓ], w[ℓ]) will be the edge of the graph with the ℓth smallest weight. At each step n, the algorithm considers the weight of the nth edge g=w[n], the corresponding endpoints v1=i[n] and v2=j[n], and the trees to which those endpoints belong, respectively called t1 and t2. We assume that the (so far unspecified) function tree() returns a label representing the tree to which a node belongs. Then, if t1 ≠ t2 (i.e. if v1 and v2 belong to two distinct trees), the weighted edge (v1,v2,g) is added to the minimum spanning tree and printed on output, the function merge_trees() merges the two trees t1 and t2 into a new single tree, and cnt is incremented. This is what happens in panels (c)–(f) of Figure A.39, where the four edges $(3, 4, 1)$, $(2, 5, 2)$, $(1, 5, 3)$ and $(1, 3, 4)$ are successively added to the minimum spanning tree. Conversely, if v1 and v2 belong to the same tree, i.e. if t1=t2, the edge (v1, v2, g) is discarded, since it would create a cycle in the tree containing v1 and v2. This happens for instance in Figure A.39(g), where the edge $(3, 5, 5)$ is discarded, since node 3 and node 5 belong to the same tree, and the addition of an edge between them

---

Algorithm 55 `bbv()`

Input: N, n_0, m, w_0, δ
Output: **i, j, w**

1: $l \leftarrow 0$
2: **for** $i = 0$ **to** $n_0 - 1$ **do**
3:    **for** $j = i + 1$ **to** $n_0 - 1$ **do**
4:       $\mathbf{i}[l] \leftarrow i$
5:       $\mathbf{j}[l] \leftarrow j$
6:       $\mathbf{w}[l] \leftarrow w_0$
7:       $l \leftarrow l + 1$
8:    **end for**
9: **end for**
10: $n \leftarrow n_0$
11: $t \leftarrow 1$
12: **while** $t < N - n_0$ **do**
13:    **for** $j = 0$ **to** $m - 1$ **do**
14:       $\mathbf{i}[l + j] \leftarrow t + n_0$
15:       $v \leftarrow$ `sample_neighbour_bbv`$(\mathbf{i}, \mathbf{j}, l)$
16:       **while** `check_neighbour`$(\mathbf{i}, \mathbf{j}, v, l, j) =$ TRUE **do**
17:          $v \leftarrow$ `sample_neighbour_bbv`$(\mathbf{i}, \mathbf{j}, l)$
18:       **end while**
19:       $\mathbf{j}[l + j] \leftarrow v$
20:       $\mathbf{w}[l + j] \leftarrow w_0$
21:       `update_weights`$(v, \ w_0, \ \delta)$
22:       `update_strength`$(v, \ w_0 + \delta)$
23:    **end for**
24:    `update_strength`$(t, \ m \times w_0)$
25:    $n \leftarrow n + 1, \ l \leftarrow l + m$
26: **end while**

---

it is possible to obtain an implementation of Algorithm 55 which runs in $O(Nm \times (\log N + \log m + 2m)) = O(Nm \log N)$.

We have implemented the BBV model in the program `bbv`, available for download at `www.complex-networks.net`. The program requires as input a number of nodes N, the size of the initial seed graph, the number of edges m created by each new node, and the values of the parameters w_0 and δ.

List of Programs

Program name	Description	Reference
pm	Power method to compute the leading eigenvector of a matrix and the associated eigenvector	Chapter 2, Appendix A.5.1
shortest	Compute the distances and shortest paths from a node to all the other nodes of an unweighted network, using the Breadth-First Search algorithm	Chapter 2, Appendix A.6
shortest_avg_max_hist	Compute the distance from one node to all the other nodes of an unweighted network, and print on output the average and the maximum distance from the origin node to all the other nodes of the graph, together with the distribution of distances	Chapter 2, Appendix A.6
betweenness	Compute node betweenness	Chapter 2, Appendix A.7
bet_dependency	Compute the dependency of all the nodes of the graph due to the shortest paths originating at a subset of nodes	Appendix A.7
dijkstra	Compute all the distances and shortest paths from a node to all the other nodes of a weighted network, using Dijkstra's algorithm	Appendix A.6
er	Sample random graphs from Erdős and Rényi model A	Chapter 3, Appendix A.10
er_B	Sample random graphs from Erdős and Rényi model B	Chapter 3, Appendix A.10
components	Find the connected components of an undirected graph, using the Depth-First Search algorithm	Chapter 3, Appendix A.8
strong_conn	Find the strongly connected components of a directed graphs	Chapter 3, Appendix A.8
node_components	Determine the IN-component, the OUT-component, the weakly connected component and the strongly connected component of a node	Chapter 3, Appendix A.8
clust	Compute the average node clustering coefficient	Chapter 4
ws	Watts–Strogatz small-world network model	Chapter 4, Appendix A.11

Table A.3 List of programs provided on the website www.complex-networks.net

`power_law`	Sample from a discrete power-law distribution	Chapter 5, Appendix A.9
`fitmle`	Fit a distribution with a power-law function using the maximum-likelihood estimator	Chapter 5
`conf_model_deg`	Sample a graph with a given degree sequence from the configuration model	Chapter 5, Appendix A.12
`conf_model_deg_nocheck`	Sample a multigraph with a given degree sequence from the configuration model	Chapter 5, Appendix A.12
`ba`	Sample a graph from the Barabási–Albert model	Chapter 6, Appendix A.13
`dms`	Sample a graph from the Dorogovtsev–Mendes–Samukin model	Chapter 6, Appendix A.13
`bb_fitness`	Sample a graph from the Bianconi–Barabási fitness model	Chapter 6, Appendix A.13
`knn`	Compute the average nearest neighbour degree function of a graph	Chapter 7
`hv_net`	Sample a network with assigned degree–degree correlations using the hidden-variable model	Chapter 7, Appendix A.14
`johnson_cycles`	Count the cycles of a graph, using Johnson's algorithm	Chapter 8, Appendix A.15
`f3m`	Perform 3-node motif analysis on a graph	Chapter 8, Appendix A.16
`gn`	Find the communities of a graph using the Girvan–Newman algorithm	Chapter 9, Appendix A.17
`modularity`	Compute the modularity of a given partition of a network	Chapter 9
`cnm`	Find the communities of a graph using greedy modularity optimisation	Chapter 9, Appendix A.18
`label_prop`	Find the communities of a graph using the label propagation algorithm	Chapter 9, Appendix A.19
`clust_w`	Compute the weighted average node clustering coefficient	Chapter 10
`kruskal`	Compute the maximum/minimum spanning tree of a graph using Kruskal's algorithm	Chapter 10, Appendix A.20
`bbv`	Sample a weighted random graph using the Barrat–Barthélemy–Vespignani model	Chapter 10, Appendix A.21

References

[1] T. Achacoso and W. Yamamoto. *Ay's Neuroanatomy of* C. elegans *for Computation.* CRC Press, 1991.

[2] C. Aicher, A. Z. Jacobs and A. Clauset. "Learning latent block structure in weighted networks". *J. Complex Netw.* **3** (2015), 221–248.

[3] R. Albert and A.-L. Barabási. "Statistical mechanics of complex networks". *Rev. Mod. Phys.* **74** (2002), 47.

[4] R. Albert and A.-L. Barabási. "Topology of evolving networks: local events and universality". *Phys. Rev. Lett.* **85** (2000), 5234–5237.

[5] R. Albert, H. Jeong and A.-L. Barabási. "Internet: diameter of the World-Wide Web". *Nature* **401** (1999), 130–131.

[6] E. Almaas, P. Krapivsky and S. Redner. "Statistics of weighted treelike networks". *Phys. Rev. E* **71** (2005), 036124.

[7] N. Alon, R. Yuster and U. Zwick. "Finding and counting given length cycles". *Algorithmica* **17** (1997), 209–223.

[8] U. Alon. *An Introduction to Systems Biology: Design Principles of Biological Circuits.* Chapman & Hall/CRC Mathematical and Computational Biology. Taylor & Francis, 2006.

[9] L. A. N. Amaral et al. "Classes of small-world networks". *P. Natl. Acad. Sci. USA* **97** (2000), 11149–11152.

[10] T. Antal and P. Krapivsky. "Weight-driven growing networks". *Phys. Rev. E* **71** (2005), 026103.

[11] A. Arenas, A. Fernández and S. Gómez. "Analysis of the structure of complex networks at different resolution levels". *New J. Phys.* **10** (2008), 053039.

[12] A. Arenas et al. "Size reduction of complex networks preserving modularity". *New J. Phys.* **9** (2007), 176.

[13] S. Assenza et al. "Emergence of structural patterns out of synchronization in networks with competitive interactions". *Sci. Rep.* **1** (2011), 99.

[14] M. Baiesi and M. Paczuski. "Complex networks of earthquakes and aftershocks". *Nonlinear Proc. Geoph.* **12** (2005), 1–11.

[15] P. Ball. "Prestige is factored into journal ratings". *Nature* **439** (2006), 770–771.

[16] Y. Bar-Yam. *Dynamics of Complex Systems.* Westview Press, 2003.

[17] A.-L. Barabási. "Network science: luck or reason". *Nature* **489** (2012), 507–508.

[18] A. Barabási et al. "Evolution of the social network of scientific collaborations". *Physica A* **311** (2002), 590–614.

[19] A.-L. Barabási and R. Albert. "Emergence of scaling in random networks". *Science* **286** (1999), 509–512.

[20] A.-L. Barabási, R. Albert and H. Jeong. "Mean-field theory for scale-free random networks". *Physica A* **272** (1999), 173–187.

[21] A. Barrat and M. Weigt. "On the properties of small-world network models". *Eur. Phys. J. B* **13** (2000), 547–560.

[22] A. Barrat et al. "The architecture of complex weighted networks". *P. Natl. Acad. Sci. USA* **101** (2004), 3747–3752.

[23] A. Barrat, M. Barthélemy and A. Vespignani. "Modeling the evolution of weighted networks". *Phys. Rev. E* **70** (2004), 066149.

[24] A. Barrat, M. Barthélemy and A. Vespignani. "Weighted Evolving Networks: Coupling Topology and Weight Dynamics". *Phys. Rev. Lett.* **92** (2004), 228701.

[25] A. Barrat and R. Pastor-Satorras. "Rate equation approach for correlations in growing network models". *Phys. Rev. E* **71** (2005), 036127.

[26] M. Barthélemy. "Spatial networks". *Phys. Rep.* **499** (2011), 1–101.

[27] M. Barthélemy and L. A. N. Amaral. "Small-world networks: evidence for a crossover picture". *Phys. Rev. Lett.* **82** (1999), 3180–3183.

[28] M. Barthélemy et al. "Characterization and modeling of weighted networks". *Physica A* **346** (2005), 34–43.

[29] D. S. Bassett et al. "Adaptive reconfiguration of fractal small-world human brain functional networks". *P. Natl. Acad. Sci. USA* **103** (2006), 19518–19523.

[30] E. T. Bell. "Exponential numbers". *Am. Math. Mon.* **41** (1934), 411–419.

[31] E. A. Bender and E. R. Canfield. "The asymptotic number of labeled graphs with given degree sequences". *J. Comb. Theory* **A 24** (1978), 296.

[32] N. Berger et al. "Competition-induced preferential attachment". English. *Automata, Languages and Programming*. Ed. by J. Díaz et al. Vol. 3142. Lecture Notes in Computer Science. Springer Berlin Heidelberg, 2004, pp. 208–221.

[33] G Bianconi. "Emergence of weight-topology correlations in complex scale-free networks". *EPL-Europhys. Lett.* **71** (2005), 1029–1035.

[34] G. Bianconi and A.-L. Barabási. "Competition and multiscaling in evolving networks". *EPL-Europhys. Lett.* **54** (2001), 436.

[35] G. Bianconi and A.-L. Barabási. "Bose-Einstein condensation in complex networks". *Phys. Rev. Lett.* **86** (2001), 5632–5635.

[36] G. Bianconi, G. Caldarelli and A. Capocci. "Loops structure of the Internet at the autonomous system level". *Phys. Rev. E* **71** (2005), 066116.

[37] G. Bianconi and M. Marsili. "Loops of any size and Hamilton cycles in random scale-free networks". *J, Stat. Mech.-Theory E.* **2005** (2005), P06005.

[38] J. J. Binney et al. *The Theory of Critical Phenomena: An Introduction to the Renormalization Group*. New York, NY: Oxford University Press, Inc., 1992.

[39] P. Blanchard and D. Volchenkov. *Mathematical Analysis of Urban Spatial Networks*. Springer complexity. Springer, 2009.

[40] V. D. Blondel et al. "Fast unfolding of communities in large networks". *J, Stat. Mech.-Theory E.* **2008** (2008), P10008.

[41] S. Boccaletti, D.-U. Hwang and V. Latora. "Growing hierarchical scale-free networks by means of non-hierarchical processes". *Int. J. Bifurcat. Chaos* **17** (2007), 2447–2452.

[42] S. Boccaletti, V. Latora and Y. Moreno. *Handbook on Biological Networks*. World Scientific Lecture Notes in Complex Systems. World Scientific Publishing Company, Incorporated, 2009.

[43] S. Boccaletti et al. "Complex networks: structure and dynamics". *Phys. Rep.* **424** (2006), 175–308.

[44] N. Boccara. *Modeling Complex Systems*. New York: Springer-Verlag, 2004.

[45] M. Boguñá, R. Pastor-Satorras and A. Vespignani. "Cut-offs and finite size effects in scale-free networks". *Eur. Phys. J. B* **38** (2004), 205–209.

[46] M. Boguñá and R. Pastor-Satorras. "Class of correlated random networks with hidden variables". *Phys. Rev. E* **68** (2003), 036112.

[47] B. Bollobás. *Modern Graph Theory*. Corrected. Springer, 1998.

[48] B. Bollobás. *Random Graphs*. Cambridge studies in advanced mathematics. Academic Press, 1985.

[49] B. Bollobás. *Random Graphs*. Cambridge University Press, 2001.

[50] B. Bollobás and O. Riordan. "The diameter of a scale-free random graph". *Combinatorica* **24** (2004), 5–34.

[51] M. Bóna. *A Walk Through Combinatorics: An Introduction to Enumeration and Graph Theory*. New Jersey: World Scientific Pub. cop., 2006.

[52] P. Bonacich. "Factoring and weighting approaches to status scores and clique identification". *J. Math. Sociol.* **2** (1972), 113–120.

[53] P. Bonacich and P. Lloyd. "Eigenvector-like measures of centrality for asymmetric relations". *Soc. Networks* **23** (2001), 191–201.

[54] G. Bonanno, F. Lillo and R. Mantegna. "High-frequency cross-correlation in a set of stocks". *Quant. Financ.* **1** (2001), 96–104.

[55] J.-A. Bondy and U. S. R. Murty. *Graph Theory*. Graduate texts in mathematics. OHX. New York, London: Springer, 2007.

[56] U. Brandes et al. "Maximizing Modularity is hard" (2007).

[57] U. Brandes. "A Faster Algorithm for Betweenness Centrality". *J. Math. Sociol.* **25** (2001), 163–177.

[58] U. Brandes and T. Erlebach. *Network Analysis: Methodological Foundations*. Vol. 3418. World Scientific Lecture Notes in Complex Systems. Lecture Notes in Computer Science Tutorial, Springer-Verlag, 2005.

[59] A. Broder et al. "Graph structure in the web". *Comput. Netw.* **33** (2000), 309–320.

[60] J. Buhl et al. "Efficiency and robustness in ant networks of galleries". *Eur. Phys. J. B* **42** (2004), 123–129.

[61] E. Bullmore and O. Sporns. "Complex brain networks: graph theoretical analysis of structural and functional systems". *Nat. Rev. Neurosci.* **10** (2009), 186–198.

[62] E. Bullmore and O. Sporns. "The economy of brain network organization". *Nat. Rev. Neurosci.* (2012), 336–349.

[63] R. Burt. *Structural Holes*. Harvard University Press, 1995.

[64] G. Caldarelli. *Scale-Free Networks: Complex Webs in Nature and Technology*. Oxford Finance Series. Oxford: Oxford University Press, 2007.

[65] G. Caldarelli et al. "Scale-free networks from varying vertex intrinsic fitness". *Phys. Rev. Lett.* **89** (2002), 258702.

[66] D. S. Callaway et al. "Are randomly grown graphs really random?" *Phys. Rev. E* **64** (2001), 041902.

[67] R. F. i. Cancho and R. V. Solé. "Optimization in complex networks". *Lect. Notes Phys.* (2003), 114–126.

[68] A. Cardillo, S. Scellato and V. Latora. "A topological analysis of scientific coauthorship networks". *Physica A* **372** (2006), 333–339.

[69] A. Cardillo et al. "Structural properties of planar graphs of urban street patterns". *Phys. Rev. E* **73** (2006), 066107.

[70] C. Caretta Cartozo and P. De Los Rios. "Extended navigability of small world networks: exact results and new insights". *Phys. Rev. Lett.* **102** (2009), 238703.

[71] S. Carmi et al. "Asymptotic behavior of the Kleinberg model". *Phys. Rev. Lett.* **102** (2009), 238702.

[72] M. Catanzaro, M. Boguñá and R. Pastor-Satorras. "Generation of uncorrelated random scale-free networks". *Phys. Rev. E* **71** (2005), 027103.

[73] M. Chavez et al. "Functional modularity of background activities in normal and epileptic brain networks". *Phys. Rev. Lett.* **104** (2010), 118701.

[74] P. Chen et al. "Finding scientific gems with Google's PageRank algorithm". *J. Informetr.* **1** (2007), 8–15.

[75] T. Chow. *Mathematical Methods for Physicists: A Concise Introduction.* Cambridge University Press, 2000.

[76] F. Chung and L. Lu. "The average distances in random graphs with given expected degrees". *P. Natl. Acad. Sci. USA* **99** (2002), 15879.

[77] F. Chung and L. Lu. "The diameter of sparse random graphs". *Adv. Appl. Math* **26** (2001), 257–279.

[78] V. Ciotti et al. "Homophily and missing links in citation networks". *Eur. Phys. J. Data Sci.* **5** (2016).

[79] A. Clauset, M. E. J. Newman and C. Moore. "Finding community structure in very large networks". *Phys. Rev. E* **70** (2004), 066111.

[80] A. Clauset, C. R. Shalizi and M. E. J. Newman. "Power-law distributions in empirical data". *SIAM Rev.* **51**, (2007), 661–703.

[81] J. R. Clough et al. "Transitive reduction of citation networks". *J. Complex Netw.* **3** (2015), 189–203.

[82] R. Cohen and S. Havlin. "Scale-free networks are ultrasmall". *Phys. Rev. Lett.* **90** (2003).

[83] V. Colizza et al. "Detecting rich-club ordering in complex networks". *Nat. Phys.* **2** (2006), 110–115.

[84] V. Colizza, R. Pastor-Satorras and A. Vespignani. "Reaction-diffusion processes and metapopulation models in heterogeneous networks". *Nat. Phys.* **3** (2007), 276–282.

[85] A. Condon and R. M. Karp. "Algorithms for graph partitioning on the planted partition model". *Random Struct. Algor.* **18** (2001), 116–140.

[86] T. H. Cormen et al. *Introduction to Algorithms.* MIT Press, 2001.

[87] B. Cronin. *The Citation Process. The Role and Significance of Citations in Scientific Communication.* London: Taylor Graham, 1984.

[88] P. Crucitti, V. Latora and S. Porta. "Centrality in networks of urban streets". *Chaos* **16** (2006), 015113.

[89] P. Crucitti, V. Latora and S. Porta. "Centrality measures in spatial networks of urban streets". *Phys. Rev. E* **73** (2006), 036125.

[90] L. Danon et al. "Comparing community structure identification". *J. Stat. Mech. Theory E.* **2005** (2005), P09008.

[91] E. David and K. Jon. *Networks, Crowds, and Markets: Reasoning About a Highly Connected World.* New York, NY: Cambridge University Press, 2010.

[92] F. De Vico Fallani et al. "Graph analysis of functional brain networks: practical issues in translational neuroscience". *Phylos. T. R. Soc. B* **369** (2014).

[93] M. T. Dickerson et al. "Fast greedy triangulation algorithms". *Comp. Geom.-Theor. Appl.* **8** (1997), 67–86.

[94] E. W. Dijkstra. "A note on two problems in connexion with graphs". *Num. Math.* **1** (1959), 269–271.

[95] P. S. Dodds. "An experimental study of search in global social networks". *Science* **301** (2003), 827–829.

[96] S. N. Dorogovtsev and J. F. F. Mendes "Minimal models of weighted scale-free networks" arXiv:cond-mat/0408343.

[97] S. N. Dorogovtsev, J. F. F. Mendes and A. N. Samukhin. "Structure of growing networks with preferential linking". *Phys. Rev. Lett.* **85** (2000), 4633–4636.

[98] R. M. D'Souza et al. "Emergence of tempered preferential attachment from optimization". *P. Natl. Acad. Sci. USA* **104** (2007), 6112–6117.

[99] R. Durrett. *Random Graph Dynamics.* Cambridge Series in Statistical and Probabilistic Mathematics. Cambridge University Press, 2010.

[100] P. Erdős and A. Rényi. "On random graphs I". *Publ. Math.-Debrecen* **6** (1959), 290.

[101] P. Erdős and A. Rényi. "On the evolution of random graphs". *Publ. Math. Inst. Hungary. Acad. Sci.* **5** (1960), 17–61.

[102] E. Estrada. *The Structure of Complex Networks: Theory and Applications.* New York, NY: Oxford University Press, Inc., 2011.

[103] E. Estrada and N. Hatano. "Communicability in complex networks". *Phys. Rev. E* **77** (2008), 036111.

[104] E. Estrada and J. A. Rodríguez-Velázquez. "Subgraph centrality in complex networks". *Phys. Rev. E* **71** (2005), 056103.

[105] L. Euler. "Solutio problematis ad geometriam situs pertinentis". *Comment. Acad. Sci. U. Petrop.* **8** (1736), 128–140.

[106] J. A. Evans. "Future science". *Science* **342** (2013), 44–45.

[107] T. S. Evans and J. P. Saramäki. "Scale-free networks from self-organization". *Phys. Rev. E* **72** (2005), 026138.

[108] M. G. Everett and S. P. Borgatti. "The centrality of groups and classes". *J. Math. Sociol.* **23** (1999), 181–201.

[109] A. Fabrikant, E. Koutsoupias and C. H. Papadimitriou. "Heuristically optimized trade-offs: a new paradigm for power laws in the Internet". *Proceedings of the 29th International Colloquium on Automata, Languages and Programming.* ICALP '02. London, UK: Springer-Verlag, 2002, pp. 110–122.

[110] G. Fagiolo, J. Reyes and S. Schiavo. "World-trade web: topological properties, dynamics, and evolution". *Phys. Rev. E* **79** (2009), 036115.

[111] K. Falconer. *Fractal Geometry: Mathematical Foundations and Applications.* 2nd Ed. Wiley, 2003.

[112] F. D. V. Fallani et al. "Defecting or not defecting: how to 'read' human behavior during cooperative games by EEG measurements". *PLoS ONE* **5(12)**: (2011), 5:e14187(2010).

[113] F. D. V. Fallani and F. Babiloni. "The graph theoretical approach in brain functional networks: theory and applications". *Synthesis Lect. Biomed. Eng.* **5** (2010), 1–92.

[114] S. L. Feld. "The focused organization of social ties". *Am. J. Sociol.* **86** (1981), 1015–1035.

[115] S. L. Feld. "Why your friends have more friends than you do". *Am. J. Sociol.* **96** (1991), 1464–1477.

[116] M. Fiedler. "Algebraic connectivity of graphs". *Czech. Math. J.* **23** (1973), 298–305.

[117] R. A. Fisher. "The use of multiple measurements in taxonomic problems". *Ann. Eugenic.* **7** (1936), 179–188.

[118] G. S. Fishman. "Sampling from the Poisson distribution on a computer". *Computing* **17** (1976), 147–156.

[119] S. Fortunato. "Community detection in graphs". *Phys. Rep.* **486**, (2009), 75–174.

[120] S. Fortunato and M. Barthélemy. "Resolution limit in community detection". *P. Natl. Acad. Sci. USA* **104** (2007), 36–41.

[121] H. Frank and W. Chou. "Connectivity considerations in the design of survivable networks". *IEEE T. Circuits Syst.* **17** (1970), 486–490.

[122] L. Freeman. "A set of measures of centrality based on betweenness". *Sociometry* (1977).

[123] L. Freeman. "Centrality in social networks: conceptual clarification". *Soc. Networks* **1** (1979), 215–239.

[124] L. C. Freeman, S. P. Borgatti and D. R. White. "Centrality in valued graphs: A measure of betweenness based on network flow". *Soc. Networks* **13** (1991), 141– 154.

[125] G. Frobenius. "Über Matrizen aus nicht negativen Elementen". *S.-B. Deutsch. Akad. Wiss. Berlin. Math-Nat. Kl.,* (1912), 456–477.

[126] F. Gantmacher. *The Theory of Matrices.* Vol. 2. New York: Chelsea Publishing Company, 1959.

[127] E. Garfield. *Citation Indexing: Its Theory and Application in Science, Technology, and Humanities.* Information sciences series. Isi Press, 1979.

[128] D. Garlaschelli and M. Loffredo. "Patterns of link reciprocity in directed networks". *Phys. Rev. Lett.* **93** (2004), 268701.

[129] D. Garlaschelli et al. "The scale-free topology of market investments". *Physica A* **350** (2005), 491–499.

[130] C.-M. Ghim et al. "Packet transport along the shortest pathways in scale-free networks". *Eur. Phys. J. B* **38** (2004), 193–199.

[131] M. Girvan and M. E. J. Newman. "Community structure in social and biological networks". *P. Natl. Acad. Sci. USA* **99** (2002), 7821–7826.

[132] K.-I. Goh, B. Kahng and D. Kim. "Packet transport and load distribution in scale-free network models". *Physica A* **318** (2003), 72–79.

[133] K.-I. Goh, B. Kahng and D. Kim. "Universal behavior of load distribution in scale-free networks". *Phys. Rev. Lett.* **87** (2001), 278701.

[134] K.-I. Goh et al. "Classification of scale-free networks". *P. Natl. Acad. Sci. USA* **99** (2002), 12583–12588.

[135] K.-I. Goh et al. "Load distribution in weighted complex networks". *Phys. Rev. E* **72** (2005), 017102.

[136] S. R. Goldberg, H. Anthony and T. S. Evans "Modelling citation networks". *Scientometrics* **105** (2015), 1577–1604.

[137] G. Golub and C. Van Loan. *Matrix Computations.* Johns Hopkins Studies in the Mathematical Sciences. Johns Hopkins University Press, 2013.

[138] J. Gómez-Gardeñes and Y. Moreno. "From scale-free to Erdos-Rényi networks". *Phys. Rev. E* **73** (2006), 056124.

[139] J. Gómez-Gardeñes and Y. Moreno. "Local versus global knowledge in the Barabasi-Albert scale-free network model". *Phys. Rev. E* **69** (2004), 037103.

[140] B. H. Good, Y.-A. de Montjoye and A. Clauset. "Performance of modularity maximization in practical contexts". *Phys. Rev. E* **81** (2010), 046106.

[141] S. Goss et al. "Self-organized shortcuts in the Argentine ant". *Naturwissenschaften* **76** (1989), 579–581.

[142] M. S. Granovetter. "The strength of weak ties". *Am. J. Sociol.* **78** (1973), 1360.

[143] C. M. Grinstead and J. L. Snell. *Introduction to Probability.* Providence, RI: American Mathematical Society, 1997.

[144] J. Gross and J. Yellen. *Graph Theory and Its Applications, Second Edition.* Textbooks in Mathematics. Taylor & Francis, 2005.

[145] J. Guare. *Six Degrees of Separation: A Play.* Vintage Series. Vintage Books, 1990.

[146] R. Guimerá and L. A. N. Amaral. "Cartography of complex networks: modules and universal roles". *J, Stat. Mech.-Theory E.* **2005** (2005), P02001.

[147] R. Guimerá and L. A. N. Amaral. "Functional cartography of complex metabolic networks". *Nature* **433** (2005), 895–900.

[148] B. Gutenberg and C. Richter. "Magnitude and energy of earthquakes". *Nature* **176** (1955), 795.

[149] R. Gutiérrez et al. "Emerging meso- and macroscales from synchronization of adaptive networks". *Phys. Rev. Lett.* **107** (2011), 234103.

[150] F. Harary. *Graph Theory.* Addison-Wesley series in mathematics. Perseus Books, 1994.

[151] D. Hicks et al. "Bibliometrics: the Leiden manifesto for research metrics". *Nature* **520** (2015), 429–431.

[152] C. Hierholzer. "Über die Möglichkeit, einen Linienzug ohne Wiederholung und ohne Unterbrechung zu umfahren". *Math. Ann.* **6** (1873), 30–32.

[153] B. Hillier and J. Hanson. *The Social Logic of Space.* Cambridge University Press, 1984.

[154] J. E. Hirsch. "An index to quantify an individual's scientific research output". *P. Natl. Acad. Sci. USA* **102** (2005), 16569–16572.

[155] J. E. Hirsch. "Does the h index have predictive power?" *P. Natl. Acad. Sci. USA* **104** (2007), 19193–19198.

[156] P. Holme and B. J. Kim. "Growing scale-free networks with tunable clustering". *Phys. Rev. E* **65** (2002), 026107.

[157] J. Hopcroft and R. Tarjan. "Efficient planarity testing". *J. ACM* **21** (1974), 549–568.

[158] *http://geant3.archive.geant.net*.

[159] *http://ghr.nlm.nih.gov*.

[160] *http://www.caida.org*.

[161] B. Hu et al. "A weighted network model for interpersonal relationship evolution". *Physica A* **353** (2005), 576–594.

[162] E. Isaacson and H. Keller. *Analysis of Numerical Methods*. Dover Books on Mathematics Series. Dover Publications, 1994.

[163] M. Jackson. *Social and Economic Networks*. Princeton University Press, 2010.

[164] A. Jacobs. *Great Streets*. MIT Press, 1993.

[165] H. Jeong et al. "Lethality and centrality in protein networks". *Nature* **411** (2001), 41–42.

[166] H. Jeong et al. "The large-scale organization of metabolic networks". *Nature* **407** (2000), 651–654.

[167] B. Jiang and C. Claramunt. "Topological analysis of urban street networks". *Environ. Plann. B* **31** (2004), 151–162.

[168] E. Jin, M. Girvan and M. Newman. "Structure of growing social networks". *Phys. Rev. E* **64** (2001), 046132.

[169] D. B. Johnson. "Finding all the elementary circuits of a directed graph". *SIAM J. Comput.* **4** (1975), 77–84.

[170] S. C. Johnson. "Hierarchical clustering schemes". *Psychometrika* **32** (1967), 241–254.

[171] V. Kachitvichyanukul and B. W. Schmeiser. "Binomial random variate generation". *Commun. ACM* **31** (1988), 216–222.

[172] M. Kalos and P. Whitlock. *Monte Carlo Methods*. Wiley, 1986.

[173] T. Kamada and S. Kawai. "An algorithm for drawing general undirected graphs". *Inform. Process. Lett.* **31** (1989), 7–15.

[174] L. Katz. "A new status index derived from sociometric analysis". English. *Psychometrika* **18** (1953), 39–43.

[175] M. G. Kendall. "A new measure of rank correlation". *Biometrika* **30** (1938), 81–93.

[176] M. G. Kendall. *Rank Correlation Methods*. London, Griffin, 1970.

[177] J. M. Kleinberg. "The convergence of social and technological networks". *Commun. ACM* **51** (2008), 66.

[178] J. M. Kleinberg. "The small-world phenomenon". *Proceedings of the thirty-second annual ACM symposium on Theory of computing – STOC '00*. ACM Press, 2000.

[179] J. M. Kleinberg. "Authoritative sources in a hyperlinked environment". *J. ACM* **46** (1999), 604–632.

[180] J. M. Kleinberg. "Navigation in a small world". *Nature* **406** (2000), 845–845.

[181] D. E. Knuth. *The Art of Computer Programming, Volume I: Fundamental Algorithms, 3rd Edition*. Addison-Wesley, 1997.

[182] D. E. Knuth. *The Art of Computer Programming, Volume II: Seminumerical Algorithms, 3rd Edition*. Addison-Wesley, 1997.

[183] D. E. Knuth. *The Art of Computer Programming, Volume III: Sorting and Searching, 2nd Edition*. Addison-Wesley, 1973.

[184] G. Kossinets and D. J. Watts. "Empirical analysis of an evolving social network". *Science* **311** (2006), 88–90.

[185] P. Krapivsky and S. Redner. "A statistical physics perspective on Web growth". *Comput. Netw.* **39** (2002), 261–276.

[186] P. Krapivsky and S. Redner. "Organization of growing random networks". *Phys. Rev. E* **63** (2001), 066123.

[187] P. Krapivsky, S. Redner and F. Leyvraz. "Connectivity of growing random networks". *Phys. Rev. Lett.* **85** (2000), 4629–4632.

[188] P. Krapivsky, G. Rodgers and S. Redner. "Degree distributions of growing networks". *Phys. Rev. Lett.* **86** (2001), 5401–5404.

[189] V. Krebs. "Mapping networks of terrorist cells". *Connections* **24** (2002), 43–52.

[190] J. B. Kruskal. "On the shortest spanning subtree of a graph and the traveling salesman problem". *P. Am. Math. Soc.* **7** (1956), 48–48.

[191] J. M. Kumpula et al. "Emergence of communities in weighted networks". *Phys. Rev. Lett.* **99** (2007), 228701.

[192] L. Lacasa, V. Nicosia and V. Latora. "Network structure of multivariate time series". *Sci. Rep.* **5** (2015), 15508.

[193] L. Lacasa et al. "From time series to complex networks: The visibility graph". *P. Natl. Acad. Sci. USA* **105** (2008), 4972–4975.

[194] L. Leydesdorff. *The Challenge of Scientometrics: The Development, Measurement, and Self-Organization of Scientific Communications*. Universal-Publishers, 2001.

[195] R. Lambiotte, J. C. Delvenne and M. Barahona. "Laplacian dynamics and multiscale modular structure in networks" (2008).

[196] A. Lancichinetti and S. Fortunato. "Consensus clustering in complex networks". *Sci. Rep.* **2** (2012).

[197] A. Lancichinetti, S. Fortunato and F. Radicchi. "Benchmark graphs for testing community detection algorithms". *Phys. Rev. E* **78** (2008), 046110.

[198] S. Lang. *Linear Algebra*. Springer Undergraduate Texts in Mathematics and Technology. Springer, 1987.

[199] V. Latora and M. Marchiori. "A measure of centrality based on network efficiency". *New J. Phys.* **9** (2007), 188.

[200] V. Latora and M. Marchiori. "Economic small-world behavior in weighted networks". *Eur. Phys. J. B* **32** (2003), 249–263.

[201] V. Latora, V. Nicosia and P. Panzarasa. "Social cohesion, structural holes, and a tale of two measures". English. *J. Stat. Phys.* **151** (2013), 745–764.

[202] V. Latora and M. Marchiori. "Efficient behavior of small-world networks". *Phys. Rev. Lett.* **87** (2001), 198701.

[203] V. Latora and M. Marchiori. "Vulnerability and protection of infrastructure networks". *Phys. Rev. E* **71** (2005), 015103(R).

[204] V. Latora et al. "Identifying seismicity patterns leading flank eruptions at Mt. Etna Volcano during 1981–1996". *Geophys. Res. Lett.* **26** (1999), 2105–2108.

[205] S. Lehmann, A. D. Jackson and B. E. Lautrup. "Measures for measures". *Nature* **444** (2006), 1003–1004.

[206] J. Leskovec et al. "Community structure in large networks: natural cluster sizes and the absence of large well-defined clusters". *Internet Math.* **6** (2009), 29–123.

[207] D. Liben-Nowell et al. "Geographic routing in social networks". *P. Natl. Acad. Sci. USA* **102** (2005), 11623–11628.

[208] F. Liljeros et al. "The web of human sexual contacts". *Nature* **411** (2001), 907–908.

[209] M.-E. Lynall et al. "Functional connectivity and brain networks in schizophrenia". *J. Neurosci.* **30** (2010), 9477–9487.

[210] I. A. S. M. Abramowitz. *Handbook of Mathematical Functions: With Formulas, Graphs, and Mathematical Tables.* Dover Publications; 1965.

[211] A. Ma and R. J. Mondragón. "Rich-cores in networks". *PLoS ONE* **10** (2015).

[212] A. Ma, R. J. Mondragón and V. Latora. "Anatomy of funded research in science". *P. Natl. Acad. Sci. USA* **112** (2015), 14760–14765.

[213] P. J Macdonald, E Almaas and A.-L Barabási. "Minimum spanning trees of weighted scale-free networks". *EPL-Europhys. Lett.* **72** (2005), 308–314.

[214] S. Mangan and U. Alon. "Structure and function of the feed-forward loop network motif". *P. Natl. Acad. Sci. USA* **100** (2003), 11980–11985.

[215] R. Mantegna. "Hierarchical structure in financial markets". *Eur. Phys. J. B* **11** (1999), 193–197.

[216] R. Mantegna and H. Stanley. *Introduction to Econophysics: Correlations and Complexity in Finance.* Cambridge University Press, 1999.

[217] M. Matsumoto and T. Nishimura. "Mersenne Twister: a 623-dimensionally equidistributed uniform pseudo-random number generator". *ACM T. Model Comput S.* **8** (1998), 3–30.

[218] C. W. Miller. "Superiority of the h-index over the impact factor for physics" (2007).

[219] R. Milo et al. "Network motifs: simple building blocks of complex networks". *Science* **298** (2002), 824–827.

[220] R. Milo et al. "Superfamilies of evolved and designed networks". *Science* **303** (2004), 1538–1542.

[221] M. Mitrović and B. Tadić. "Spectral and dynamical properties in classes of sparse networks with mesoscopic inhomogeneities". *Phys. Rev. E* **80** (2009), 026123.

[222] M. Molloy and B. Reed. "A critical point for random graphs with a given degree sequence". *Random Struct. Algor.* **6** (1995), 161–180.

[223] M. Molloy and B. Reed. "The size of the giant component of a random graph with a given degree sequence". *Comb. Probab. Comput* **7** (1998), 295–305.

[224] T. Nakagaki, H. Yamada and A. Tóth. "Intelligence: maze-solving by an amoeboid organism". *Nature* **407** (2000), 470–470.

[225] M. E. J. Newman. "Clustering and preferential attachment in growing networks". *Phys. Rev. E* **64** (2001), 025102.

[226] M. E. J. Newman. "Fast algorithm for detecting community structure in networks". *Phys. Rev. E* **69** (2004), 066133.

[227] M. E. J. Newman. "Random graphs with clustering". *Phys. Rev. Lett.* **103** (2009), 058701.

[228] M. E. J. Newman and D. J. Watts. "Scaling and percolation in the small-world network model". *Phys. Rev. E* **60** (1999), 7332–7342.

[229] M. E. J. Newman. "Analysis of weighted networks". *Phys. Rev. E* **70** (2004), 056131.

[230] M. E. J. Newman. "Assortative mixing in networks". *Phys. Rev. Lett.* **89**, (2002), 208701.

[231] M. E. J. Newman. "Handbook of graphs and networks". Wiley-VCH, 2003. Chap. Random graphs as models of networks, p. 35.

[232] M. E. J. Newman. "Mixing patterns in networks". *Phys. Rev. E* **67**, (2003), 026126.

[233] M. E. J. Newman. "Scientific collaboration networks. II. Shortest paths, weighted networks, and centrality". *Phys. Rev. E* **64** (2001), 016132.

[234] M. E. J. Newman. "Scientific collaboration networks. I. Network construction and fundamental results". *Phys. Rev. E* **64** (2001), 016131.

[235] M. E. J. Newman. "The structure and function of complex networks". *SIAM Rev.* **45**, (2003), 167–256.

[236] M. E. J. Newman, A.-L Barabási and D. J. Watts, eds. *The Structure and Dynamics of Networks*. Princeton studies in complexity. Princeton, Oxford: Princeton University Press, 2006.

[237] M. E. J. Newman and M. Girvan. "Finding and evaluating community structure in networks". *Phys. Rev. E* **69**, (2004), 026113.

[238] M. E. J. Newman, S. H Strogatz and D. J. Watts. "Random graphs with arbitrary degree distributions and their applications". *Phys. Rev. E* **64**, (2001), 026118.

[239] M. E. J. Newman. *Networks: An Introduction*. New York, NY: Oxford University Press, Inc., 2010.

[240] M. E. J. Newman. "A measure of betweenness centrality based on random walks". *Soc. Networks* **27** (2005), 39–54.

[241] M. E. J. Newman. "Communities, modules and large-scale structure in networks". *Nat. Phys.* **8** (2012), 25–31.

[242] M. E. J. Newman. "Power laws, Pareto distributions and Zipf's law". *Contemp. Phys.* **46** (2005), 323–351.

[243] V. Nicosia et al. "Phase transition in the economically modeled growth of a cellular nervous system". *P. Natl. Acad. Sci. USA* **110** (2013), 7880–7885.

[244] V. Nicosia et al. "Controlling centrality in complex networks". *Sci. Rep.* **2** (2011).

[245] E. L. N. L. Biggs and R. Wilson. *Graph Theory 1736–1936*. Oxford: Clarendon Press, 1976.

[246] J. Noh and H. Rieger. "Stability of shortest paths in complex networks with random edge weights". *Phys. Rev. E* **66** (2002), 066127.

[247] J.-P. Onnela et al. "Dynamics of market correlations: Taxonomy and portfolio analysis". *Phys. Rev. E* **68** (2003), 056110.

[248] J.-P. Onnela et al. "Intensity and coherence of motifs in weighted complex networks". *Phys. Rev. E* **71** (2005), 065103.

[249] T. Opsahl et al. "Prominence and control: the weighted rich-club effect". *Phys. Rev. Lett.* **101** (2008), 168702.

[296] A. Tero, R. Kobayashi and T. Nakagaki. "Physarum solver: a biologically inspired method of road-network navigation". *Physica A* **363** (2006), 115–119.

[297] P. Tieri et al. "Quantifying the relevance of different mediators in the human immune cell network". *Bioinformatics* **21** (2005), 1639–1643.

[298] J. Travers and S. Milgram. "An experimental study of the small world problem". *Sociometry* **32** (1969), 425–443.

[299] M. Tumminello et al. "A tool for filtering information in complex systems". *P. Natl. Acad. Sci. USA* **102** (2005), 10421–10426.

[300] A. M. Turing. "On computable numbers, with an application to the entscheidungsproblem". *P. Lond. Math, Soc.* **42** (1936), 230–265.

[301] R. S. Varga. *Matrix Iterative Analysis.* Englewood Cliffs, NJ: Prentice Hall Inc., 1962.

[302] S. Varier and M. Kaiser. "Neural development features: spatio-temporal development of the *Caenorhabditis elegans* neuronal network". *PLoS Comput. Biol.* **7** (2011). Ed. by K. J. Friston, e1001044.

[303] L. R. Varshney et al. "Structural properties of the *Caenorhabditis elegans* neuronal network". *PLoS Comput. Biol.* **7** (2011). Ed. by O. Sporns, e1001066.

[304] A Vazquez. "Disordered networks generated by recursive searches". *EPL-Europhys. Lett.* **54** (2001), 430–435.

[305] A. Vázquez. "Growing network with local rules: Preferential attachment, clustering hierarchy, and degree correlations". *Phys. Rev. E* **67** (2003), 056104.

[306] A. Vázquez, R. Pastor-Satorras and A. Vespignani. "Large-scale topological and dynamical properties of Internet". *Phys. Rev. E* **65,** (2002), 066130.

[307] A. Vázquez et al. "Modeling of protein interaction networks". *Complexus* **1** (2003), 38–44.

[308] S. Wasserman and K. Faust. *Social Network Analysis: Methods and Applications.* Vol. 8. Cambridge University Press, 1994.

[309] D. J. Watts. *Small Worlds: The Dynamics of Networks Between Order and Randomness.* Princeton, NJ: Princeton University Press, 1999, xv, 262 p.

[310] D. J. Watts. *Small Worlds: The Dynamics of Networks Between Order and Randomness.* Menasha, Wisc.: The Association, 2003.

[311] D. J. Watts and S. H. Strogatz. "Collective dynamics of 'small-world' networks". *Nature* **393** (1998), 440–442.

[312] S. Weber and M. Porto. "Generation of arbitrarily two-point-correlated random networks". *Phys. Rev. E* **76** (2007), 046111.

[313] D. West. *Introduction to Graph Theory.* Prentice Hall PTR, 2007.

[314] J. G. White et al. "The structure of the nervous system of the nematode *Caenorhabditis elegans*". *Phylos. T. R. Soc. B* **314** (1986), 1–340.

[315] H. Wielandt. "Unzerlegbare nicht negativen Matrizen". *Math. Z.* **52** (1950), 642–648.

[316] C. R. Woese, O. Kandler and M. L. Wheelis. "Towards a natural system of organisms: proposal for the domains Archaea, Bacteria, and Eucarya". *P. Natl. Acad. Sci. USA* **87** (1990), 4576–4579.

[317] R. Xulvi-Brunet and I. M. Sokolov. "Reshuffling scale-free networks: from random to assortative". *Phys. Rev. E* **70** (2004), 066102.

[318] S. Yook et al. "Weighted evolving networks". *Phys. Rev. Lett.* **86** (2001), 5835–5838.

[319] G. U. Yule. "A mathematical theory of evolution, based on the conclusions of Dr. J. C. Willis, F.R.S." *Phylos. T. R. Soc. B* (1924).

[320] W. W. Zachary. "An information flow model for conflict and fission in small groups". *J. Anthropol. Res.* **33** (1977), 452–473.

[321] D. Zheng et al. "Weighted scale-free networks with stochastic weight assignments". *Phys. Rev. E* **67** (2003), 040102.

[322] C. Zhou and J. Kurths. "Dynamical weights and enhanced synchronization in adaptive complex networks". *Phys. Rev. Lett.* **96** (2006), 164102.

[323] S. Zhou and R. J. Mondragón. "The rich-club phenomenon in the Internet topology". *IEEE Commun. Lett.* **8** (2004), 180–182.

[324] T. Zhou et al. "Bipartite network projection and personal recommendation". *Phys. Rev. E* **76** (2007), 046115.

[325] G. K. Zipf. *Human Behavior and the Principle of Least Effort*. Addison-Wesley, Reading, MA (USA), 1949.

Author Index

Index